DEPT. OF O[BSTETRICS &]
GYNAECOLO[GY]
'D' FLOOR CLARENDON WING
BELMONT GROVE
LEEDS LS2 9NS

GW01558189

Ovarian Cancer 2

Helene Harris

Ovarian Cancer 2
Biology, diagnosis and management

Edited by

F. Sharp
Professor of Obstetrics and Gynaecology
Northern General Hospital
Sheffield
UK

W.P. Mason
Consultant Gynaecologist
St Mary's Hospital
London
UK

W. Creasman
Professor of Surgery
Medical University of South Carolina
USA

CHAPMAN & HALL MEDICAL
London · Glasgow · New York · Tokyo · Melbourne · Madras

Chapman & Hall, 2–6 Boundary Row, London SE1 8HN

Chapman & Hall, 2–6 Boundary Row, London SE1 8HN, UK

Blackie Academic & Professional, Wester Cleddens Road, Bishopbriggs, Glasgow G64 2NZ, UK

Chapman & Hall Inc., 29 West 35th Street, New York, NY10001, USA

Chapman & Hall Japan, Thomson Publishing Japan, Hirakawacho Nemoto Building, 6F, 1–7–11 Hirakawa-cho, Chiyoda-ku, Tokyo 102, Japan

Chapman & Hall Australia, Thomas Nelson Australia, 102 Dodds Street, South Melbourne, Victoria 3205, Australia

Chapman & Hall India, R. Seshadri, 32 Second Main Road, CIT East, Madras 600 035, India

First edition 1992

© 1992 Chapman & Hall

Typeset in 10/12 Palatino by Photoprint, Torquay, Devon
Printed in Great Britain at the University Press, Cambridge

ISBN 0 412 45190 5

Apart from any fair dealing for the purposes of research or private study, or criticism or review, as permitted under the UK Copyright Designs and Patents Act, 1988, this publication may not be reproduced, stored, or transmitted, in any form or by any means, without the prior permission in writing of the publishers, or in the case of reprographic reproduction only in accordance with the terms of the licences issued by the Copyright Licensing Agency in the UK, or in accordance with the terms of licences issued by the appropriate Reproduction Rights Organization outside the UK. Enquiries concerning reproduction outside the terms stated here should be sent to the publishers at the UK address printed on this page.
   The publisher makes no representation, express or implied, with regard to the accuracy of the information contained in this book and cannot accept any legal responsibility or liability for any errors or omissions that may be made.

A catalogue record for this book is available from the British Library

Library of Congress Cataloging-in-Publication Data available

Contents

	Contributors	xv
	Editors' preface	xix
	HHMT introduction	xxi

Part One   GENETICS AND IMMUNOLOGY

1	**Familial ovarian cancer**	3
	B.A.J. Ponder, J. Peto and D.F. Easton	
1.1	The evidence for inherited predisposition	3
1.2	Implications of inherited predisposition	4
1.3	Future studies	7
	References	7
2	**Hereditary ovarian cancer**	9
	P. Watson and H.T. Lynch	
2.1	Introduction	9
2.2	Hereditary non-polyposis colorectal cancer	9
2.3	Hereditary site-specific ovarian cancer	11
2.4	Hereditary breast and ovarian cancer syndrome	11
2.5	Clinical implications of hereditary ovarian cancer	14
2.6	Conclusion	14
	Acknowledgement	15
	References	15
3	**Cytogenetic changes in human epithelial ovarian cancer**	17
	H.H. Gallion, D.E. Powell, L.W. Smith, C.C. Vaugh and E.A. Case	
	References	22
4	**Molecular genetics of ovarian cancer**	23
	W. Foulkes, G. Stamp and J. Trowsdale	
4.1	Introduction	23
4.2	Allele loss	25
4.3	Oncogene activation	30
4.4	Germline mutations and inherited predisposition	31
4.5	Conclusions	31

Contents

	Acknowledgements	32
	References	32
5	**Molecular pathogenesis of ovarian cancer**	**35**
	W.S. Lowry, S.E.H. Russell, G.I. Hickey and R.J. Atkinson	
5.1	Allele loss	35
5.2	Incessant ovulation	36
5.3	Ovarian pathophysiology	36
5.4	Epidemiological data	36
5.5	Conclusion	37
	References	38
6	**Molecular structure and clinical applications of a cancer-associated mucin**	**39**
	J. Taylor-Papadimitriou, D. Allen, M. Granowska, N. Peat, T. Duhig, A. Spicer, J. Burchell and S.J. Gendler	
6.1	Introduction	39
6.2	General features of mucins	40
6.3	The *MUC1* gene and its products	41
6.4	Possible functions of mucins	43
6.5	Variations in glycosylation of the same core protein in different tissues	43
6.6	Aberrant glycosylation of *MUC1* in normal and malignant epithelial cells	44
6.7	Use of the SM-3 antibody for imaging ovarian cancer	45
6.8	T cell epitopes in PEM	46
6.9	Animal models for testing antigens based on *MUC1* in tumour rejection	47
6.10	Conclusions	48
	References	48

Part Two ONCOGENES, GROWTH FACTORS AND CYTOKINES

7	**Expression of the epidermal growth factor receptor, HER-2/*neu* and p53 in ovarian cancer**	**53**
	A. Berchuck, J.R. Marks and R.C. Bast Jr	
	References	58
8a	**Factors regulating the growth of normal and malignant ovarian epithelium**	**61**
	R.C. Bast Jr, G.C. Rodriguez, S. Wu, C.M. Boyer and A. Berchuck	
	Acknowledgements	65
	References	65
8b	**Inhibition of breast and ovarian tumour cell growth by antibodies and immunotoxins reactive with distinct epitopes on the extracellular domain of HER-2/*neu* (c-*erb*B2)**	**67**
	R.C. Bast Jr, F.J. Xu, G.C. Rodriguez, R. Whitaker, M. Boente, A. Berchuck, S. McKenzie, L.L. Houston and C.M. Boyer	
	Acknowledgements	70
	References	70

9	**Cloning of genes encoding ovarian carcinoma-specific antigens**	73
	I.G. Campbell, W.D. Foulkes, T.A. Jones and J. Trowsdale	
9.1	Introduction	73
9.2	cDNA cloning strategy	74
9.3	Methods and results	74
9.4	Discussion	79
	References	84
10	**Cytokines and ovarian cancer**	87
	S.T.A. Malik, M.S. Naylor and F.R. Balkwill	
10.1	Introduction	87
10.2	Tumour necrosis factor and intraperitoneal tumours in animals	87
10.3	Cytokine production in human ovarian cancer	88
10.4	Therapeutic potential of IFN-γ	90
10.5	Conclusion	91
	References	91
11	**Characterization of cytokines produced by ovarian cancer cells**	93
	L.F. Carson, M.M. Moradi, B-Y. Li, M.C. Olson, D. Mohanraj, S.A. Elg and S. Ramakrishnan	
11.1	Introduction	93
11.2	Interaction between the tumour cells and the host	93
11.3	Expression of growth factors/growth factor receptors	94
11.4	Studies on macrophage colony stimulating factor (M-CSF)	94
11.5	Properties of M-CSF	96
11.6	Studies on other cytokines	96
11.7	Conclusions	97
	Acknowledgements	97
	References	97
12	**Role of interleukin-6 in ovarian cancer**	101
	J.S. Berek, J.M. Watson, and O. Martínez-Maza	
12.1	Introduction	101
12.2	Epithelial ovarian cancer cells produce IL-6	103
12.3	IL-6 production by ovarian cancer cells can be modulated by exposure to cytokines	107
12.4	IL-6 is produced by freshly isolated primary ovarian cancer cells	107
12.5	Elevated levels of IL-6 are seen in the ascitic fluid of women with ovarian cancer	107
12.6	Elevated serum levels of IL-6 are associated with ovarian cancer	108
12.7	Conclusions	109
	Acknowledgements	111
	References	111
13	**CSF-1 and its receptor in ovarian and other gynaecological neoplasms**	115
	B.M. Kacinski	
13.1	Introduction	115

13.2	Our own investigations	116
13.3	Ongoing research and future directions	122
	Acknowledgements	123
	References	123

14 Regulation of growth of human ovarian cancer cells — 127
G.B. Mills, S. Hashimoto, J. Hurteau, R. Schmandt, S. Campbell, C. May, M. Hill, P. Shaw, R. Buckman and D. Hogg

14.1	Introduction	127
14.2	Ovulation and ovarian cancer	128
14.3	Growth factors and their receptors in ovarian cancer	130
14.4	Growth inhibitors	140
14.5	TTK	141
14.6	Summary, conclusions and future directions	143
	References	143

Part Three DRUG RESISTANCE AND EXPERIMENTAL THERAPEUTICS

15 Signal transduction pathway regulation of cisplatin sensitivity — 149
S.B. Howell, R.D. Christen, P.A. Andrews, S.C. Mann and D. Hom

15.1	Introduction	149
15.2	Protein kinase A pathway	149
15.3	Epidermal growth factor receptor pathway	151
15.4	Conclusion	151
	References	152

16 Immunotoxin therapy in ovarian cancer — 153
M.A. Bookman

16.1	Introduction	153
16.2	Selection of immunotoxins	153
16.3	Regional versus systemic therapy	155
16.4	Clinical trials with intraperitoneal immunotoxins	156
16.5	New immunotoxin targets	159
	References	159

17 Effects of granulocyte macrophage colony stimulating factor in cyclophosphamide- and carboplatin-treated patients — 161
J.H. Edmonson, G. Colon-Otero, H.J. Long, T.R. Fitch, L.C. Hartmann, J.A. Jefferies and T.A. Braich

17.1	Introduction	161
17.2	Materials and methods	161
17.3	Results	162
17.4	Conclusion	165
	References	166

18	**Mitochondrial poisons and ovarian cancer**	**167**
	A. Manetta, D. Emma and G. Gamboa	
18.1	Introduction	167
18.2	Materials and methods	168
18.3	Results	169
18.4	Conclusion	171
	References	172
19	**TNFα mediated lysis in gynaecological malignancies**	**175**
	J.L. Collins and D.G. Mutch	
	References	184
20	**Chemo-immunotherapy with a combination of low-dose cisplatin and recombinant interleukin 2**	**189**
	M. Bernsen, H.F.J. Dullens, W. den Otter and A.P.M. Heintz	
20.1	Introduction	189
20.2	The effect of low-dose rIL-2 and low-dose cisplatin therapy in the syngeneic murine tumour model DBA/2-SL2	190
20.3	The effects of cisplatin on rIL-2 in the syngeneic murine tumour model DBA/2-P185	192
20.4	Combination therapy of cisplatin and rIL-2 in the syngeneic murine tumour model C3HeB/FeJ-MOT	193
20.5	Discussion	194
	References	194

Part Four PATHOLOGY, EARLY DETECTION AND PROGNOSIS

21	**Early ovarian cancer**	**199**
	R.E. Scully	
	References	204
22	**New techniques in the pathological assessment of ovarian cancer**	**207**
	M. Wells	
22.1	Immunohistochemistry	207
22.2	Flow cytometry	209
22.3	Morphometry	209
22.4	Nucleolar organizer regions	210
	References	210
23	**Flow cytometry in ovarian cancer**	**213**
	P.S. Braly	
23.1	Introduction	213
23.2	Review of technique	214
23.3	Ploidy studies in ovarian cancer	215
23.4	FCM in the evaluation of effusions	217
23.5	Treatment monitoring and modification	218
23.6	Summary	219
	References	220

Contents

24	**Ultrasound for early cancer screening**	225
	W.P. Collins, T.H. Bourne, K. Reynolds, V. Bhan, J. Hampson,	
	P. Royston, M.I. Whitehead and S. Campbell	
24.1	Introduction	225
24.2	Ultrasonography	225
24.3	Transabdominal screening	226
24.4	Transvaginal screening	229
24.5	Conclusion	235
	Acknowledgements	235
	References	235
25	**Role of colour Doppler in an ultrasound-based screening programme**	237
	S. Campbell, T.H. Bourne, K. Reynolds, J. Hampson, P. Royston,	
	M.I. Whitehead and W.P. Collins	
25.1	Introduction	237
25.2	Principles of colour Doppler	237
25.3	End-points	238
25.4	Diagnosis of overt ovarian masses	238
25.5	Normal ovarian function	239
25.6	Early familial cancer screening	239
25.7	General population screening	241
25.8	Positive predictive values	243
25.9	Other gynaecological cancers	244
25.10	Conclusions	245
25.11	Technological developments	245
	Acknowledgements	246
	References	246
26	**Transvaginal colour Doppler in the differentiation between benign and malignant ovarian masses**	249
	A. Kurjak and I. Zalud	
26.1	Introduction	249
26.2	Tumour neovascularization	249
26.3	Tissue characterization by pulsed and colour Doppler	250
26.4	Clinical methods for differentiation of ovarian masses	251
26.5	Experience in Zagreb	252
26.6	Conclusion	261
	References	263
27	**Role of CA-125 in screening for ovarian cancer**	265
	I. Jacobs, A. Prys Davies and D. Oram	
27.1	The CA-125 antigen	265
27.2	Elevation of serum CA-125 levels is related to disruption of tissue barriers	266
27.3	Serum CA-125 levels are elevated in some cases of preclinical asymptomatic ovarian cancer	267
27.4	The London Hospital Ovarian Cancer Screening Study	268

27.5	Estimating the lead time achieved by screening with CA-125	271
27.6	Summary	273
	References	274

28	**Advantages and disadvantages of randomized controlled trials of ovarian cancer screening**	**277**
	Ian V. Scott	
28.1	Introduction	277
28.2	Justification for screening	277
28.3	Why randomized trials?	279
28.4	What size?	279
28.5	Target population	279
28.6	Which tests?	281
28.7	Timing of trials	285
28.8	The purpose of a proposed randomized trial	285
28.9	Conclusions	285
	References	286

Part Five MANAGEMENT

29	**Intraperitoneal therapy in the management of ovarian cancer**	**291**
	M. Markman	
29.1	Introduction	291
29.2	Important lessons learned from a decade of research with intraperitoneal therapy of ovarian cancer	291
29.3	The future of intraperitoneal therapy in the management of ovarian cancer	294
	References	295

30	**Intraperitoneal cisplatin and interferon-α in persistent epithelial ovarian cancer: summary of phase I–II trials**	**297**
	J.S. Berek and M. Markman	
30.1	Introduction	297
30.2	Intraperitoneal phase I–II trials	298
30.3	Discussion	307
	Acknowledgements	311
	References	311

31	**Overview of randomized chemotherapy trials for advanced ovarian cancer**	**315**
	Advanced Ovarian Cancer Trialists Group (AOCTG)	
31.1	Introduction	315
31.2	Methods and data	316
31.3	Results	318
31.4	Discussion	319
31.5	Future studies	322
	Acknowledgements	323

	References	323
	Appendix A	328
	Appendix B	330

32 Meta-analysis of treatment for advanced ovarian cancer — 331
R.J. Osborne

32.1	Introduction	331
32.2	Methodology	333
32.3	Analysis	335
32.4	Discussion	342
	References	343

33 Impact of maximal cytoreductive surgery on survival in advanced ovarian cancer — 345
W.P. Soutter, R.W. Hunter and N.D.E. Alexander

33.1	A definition of maximal cytoreductive surgery	345
33.2	The rationale behind MCS	345
33.3	Studies without controls	345
33.4	A non-randomized study with controls	346
33.5	Evidence against survival benefit	346
33.6	A meta-analysis of the effect of MCS on median survival time	346
33.7	Relief of symptoms by MCS	348
33.8	Comment	348
	References	349

34 Management and outcome of stage III epithelial ovarian cancer — 351
N.F. Hacker, G.V. Wain and J.B. Trimbo

34.1	Introduction	351
34.2	Materials and methods	351
34.3	Results	353
34.4	Discussion	354
	References	355

35 The role of surgery in epithelial ovarian cancer — 357
D. Luesley, C. Finn and R. Varma

35.1	Introduction	357
35.2	The objectives of the primary laparotomy	358
35.3	Cytoreduction	361
35.4	West Midlands' studies	364
35.5	Super-radical surgery	365
35.6	Palliative surgery and conclusion	365
	References	366

36 Debulking surgery at second-look laparotomy — 369
W.J. Hoskins

	References	372

37	**Evaluation of debulking surgery at second-look laparotomy**	**375**
	W.T. Creasman	
37.1	Introduction	375
37.2	Material and methods	375
37.3	Results	376
37.4	Discussion	377
37.5	Conclusions	382
	References	383
38	**Second-look laparotomy in the routine management of ovarian cancer**	**385**
	A.J. Ferrier and A.D. De Petrillo	
38.1	Introduction	385
38.2	Indications for second-look laparotomy	385
38.3	Outcome of second-look laparotomies	386
38.4	Second-look laparotomy and survival	386
38.5	Pathological findings and outcome after second-look laparotomy	387
38.6	Second-look laparotomy as a diagnostic test	388
38.7	Second-look laparotomy as a therapeutic manoeuvre	391
38.8	Conclusion	392
	References	392

Recommendations and conclusions	**395**
Genetic aspects	395
Growth factors and cytokines	396
Immunology	397
Drug resistance	398
Experimental therapeutics	399
Pathology	400
Early detection	400
New therapies	401
Prognostic factors	401
Meta-analysis of treatment	402
Surgery	402

Selected arguments	**405**
Familial and hereditary ovarian cancer	405
Genetic changes	407
Growth factors and cytokines	415
Immunology	419
Drug resistance	420
Immunotoxins	422
Experimental therapeutics	426
Early detection	428
Meta-analysis of treatment	433
Surgery	436

Index	**443**

Contributors

Chapter 1

B.A.J. Ponder
CRC Human Cancer Genetics Research
 Group
Department of Pathology
University of Cambridge, UK

J. Peto and D.F. Easton
Institute of Cancer Research
Sutton
Surrey, UK

Chapter 2

P. Watson and H.T. Lynch
Department of Preventive Medicine and
 Public Health
Creighton University School of Medicine
Nebraska, USA

Chapter 3

H.H. Gallion and E.A. Case
Department of of Obstetrics and
 Gynecology

D.E. Powell, L.W. Smith and C.C. Vaugh
Department of Pathology
University of Kentucky Medical Centre,
 USA

Chapter 4

W. Foulkes, G. Stamp and J. Trowsdale
Imperial Cancer Research Fund
London, UK

Chapter 5

W.S. Lowry, S.E.H. Russell, G.I. Hickey
and R.J. Atkinson
Department of Oncology
The Queen's University of Belfast, Northern
 Ireland

Chapter 6

J. Taylor-Papadimitriou, D. Allen,
M. Granowska, N. Peat, T. Duhig,
A. Spicer, J. Burchell and S.J. Gendler
Imperial Cancer Research Fund
London, UK

Chapter 7

A. Berchuk, J.R. Marks and R.C. Bast Jr
Department of Obstetrics and Gynecology
Duke University Medical Center
North Carolina, USA

Chapter 8a

R.C. Bast Jr, G.C. Rodriguez, S. Wu,
C.M. Boyer and A. Berchuk
Departments of Medicine, and Obstetrics
 and Gynecology
Duke University Medical Center and the
 Duke Comprehensive Cancer Center
North Carolina, USA

Contributors

Chapter 8b

R.C. Bast Jr, F.J. Xu, G.C. Rodriguez,
R. Whitaker, M. Boente, A. Berchuk,
L.L. Houston and C.M. Boyer
Departments of Medicine, Microbiology/
  Immunology, and Obstetrics and
  Gynecology
Duke University Medical Center and the
  Duke Comprehensive Cancer Center
North Carolina, USA

Chapter 9

W.D. Foulkes, T.A. Jones and
J. Trowsdale
Imperial Cancer Research Fund
London, UK

Chapter 10

S.T.A. Malik, M.S. Naylor and
F.R. Balkwill
Imperial Cancer Research Fund
London, UK

Chapter 11

L.F. Carson, M.M. Moradi and S.A. Elg
Women's Cancer Center

B.-Y. Li, M.C. Olson, D. Mohanraj and
S. Ramakrishnan
Department of Pharmacology
University of Minnesota, USA

Chapter 12

J.S. Berek
Department of Obstetrics and Gynecology

J.M. Watson and O. Martínez-Maza
Department of Microbiology and
  Immunology
UCLA Medical School
Los Angeles
California, USA

Chapter 13

B. Kacinski
Yale University School of Medicine
Connecticut, USA

Chapter 14

G.B. Mills, S. Hashimoto, J. Hurteau,
R. Schmandt, S. Campbell, C. May,
M. Hill, P. Shaw, R. Buckman and
D. Hogg
Oncology Research
Toronto General Hospital
Ontario, Canada

Chapter 15

S.B. Howell, R.D. Christen,
P.A. Andrews, S.C. Mann and D. Hom
Department of Medicine
University of California
La Jolla
California, USA

Chapter 16

M.A. Bookman
Department of Medical Oncology
Fox Chase Cancer Center
Pennsylvania, USA

Chapter 17

J.H. Edmonson, H.J. Long, L.C. Hartmann
and J.A. Jefferies
Mayo Clinic
Rochester
Minnesota, USA

G. Colon-Otero
Mayo Clinic
Jacksonville
Florida, USA

T.R. Fitch and T.A. Braich
Mayo Clinic
Scottsdale
Arizona, USA

Chapter 18

A. Manetta, D. Emma and G. Gamboa
Department of Obstetrics and Gynecology
UCI Medical Center
California, USA

Chapter 19

J.L. Collins and D.G. Mutch
Division of Gynecologic Oncology
Washington University School of Medicine
Missouri, USA

Chapter 20

M. Bernsen and A.P.M. Heintz
Department of Obstetrics and Gynaecology

H.F.J. Dullens and W. den Otter
Department of Experimental Pathology
University of Utrecht, The Netherlands

Chapter 21

R.E. Scully
Department of Pathology
Harvard Medical School
Massachusetts General Hospital, USA

Chapter 22

M. Wells
Department of Pathology
University of Leeds, UK

Chapter 23

P.S. Braly
Department of Reproductive Medicine
University of California
San Diego
California, USA

Chapters 24 and 25

W.P. Collins, T.H. Bourne, K. Reynolds,
V. Bhan (Chapter 24 only), J. Hampson,
M.I. Whitehead and S. Campbell
Department of Obstetrics and Gynaecology
King's College School of Medicine and
  Dentistry
London, UK

P. Royston
Department of Medical Physics
Royal Postgraduate Medical School
London, UK

Chapter 26

A. Kurjak and I. Zalud
Ultrasonic Institute Medical School
University of Zagreb
Croatia

Chapter 27

I. Jacobs, A. Prys Davies and D. Oram
Department of Obstetrics and Gynaecology
The London Hospital, UK

Chapter 28

I.V. Scott
Derby City Hospital, UK

Chapter 29

M. Markman
Department of Medicine
Memorial Sloan-Kettering Cancer Center
New York, USA

Chapter 30

J.S. Berek (see Chapter 12)
and M. Markman (see Chapter 29)

Chapter 31

Advanced Ovarian Cancer Trialists Group
(AOCTG) (see p.315 for list of Members)

Contributors

Chapter 32

R.J. Osborne
Division of Gynecologic Oncology
University of Toronto
Ontario, Canada

Chapter 33

W.P. Soutter and R.W. Hunter
Institute of Obstetrics and Gynaecology

N.D.E. Alexander
Department of Medical Physics
Royal Postgraduate Medical School
London, UK

Chapter 34

N.F. Hacker, G.V. Wain and J.P. Trimbo
(Visiting Professor from the University of Leiden, The Netherlands)
Department of Gynaecologic Oncology
Royal Hospital for Women
Sydney, Australia

Chapter 35

D. Luesley, C. Finn and R. Varma
West Midlands Ovarian Cancer Group
Birmingham, UK

Chapter 36

W.J. Hoskins
Gynecology Service
Memorial Sloan-Kettering Cancer Center
New York, USA

Chapter 37

W.T. Creasman
Professor of Surgery
Medical University of South Carolina, USA

Chapter 38

A.J. Ferrier and A.D. De Petrillo
Department of Obstetrics and Gynecology
Ontario Cancer Institute
Princess Margaret Hospital
Ontario, Canada

Editors' preface

This book is the result of the Third Helene Harris Memorial Trust Biennial International Forum on Ovarian Cancer which took place in Charleston, South Carolina, USA on 16–20 April 1991. The Forum was regarded as a particularly important one in the series to date as it was the first to be held in North America.

Ovarian cancer remains a significant cause of morbidity and mortality in women, and continues to present major problems in detection and management. As on previous occasions, improved understanding of the biology of the disease seemed the key to future progress, and so the Forum sought to interface a distinguished group of international experts from the fields of genetics, molecular biology, biochemistry, pathology and clinical gynaecological oncology.

The major topics addressed include scientific and diagnostic principles with a major emphasis on molecular genetics and pathogenesis; oncogenes, growth factors and cytokines; drug resistance and experimental therapeutics; early detection, including the present state of the art in relation to ultrasound and tumour markers; and selected aspects of the management of ovarian cancer.

This book, which brings together in chapter form the main content of the deliberations, would not be complete without mention of the recommendations and conclusions from the Forum. The progress being made on the genetic front is rapidly moving and it is likely that it will be from this area that major steps forward in our understanding of the biology of this important cancer will be made. It seems likely that in the not too distant future the genetic basis of the disease will be revealed. Such information will logically produce important indicators for advancing our knowledge of the associated molecular processes. A number of strategies to overcome drug resistance during chemotherapy for ovarian cancer are being evaluated, based on principles of dose intensity, known mechanisms of drug resistance, and alternative biological approaches. A major hiatus in our knowledge concerns early tumourigenesis. Further work in this field is required. Morphologically identifiable preneoplastic changes in the ovary remain enigmatic and elusive, and it is likely that a range of ancillary techniques such as immunocytochemistry, flow cytometry and ploidy measurement, will feature if additional help is to be given in this area in the routine histopathology laboratory. Meta-analysis provides a technique for reviewing data from clinical trials by pooling to increase statistical power, resolve uncertainty, or reconcile differences in outcome between trials of similar design. Surgery still appears to have an important role in the management of patients with ovarian cancer. The surgical management of advanced ovarian cancer remains controversial, because there has never been a prospective randomized study

Editors' preface

to adequately define the role of cytoreduction. It has an undeniable role in the palliation of patients with recurrent disease, especially with associated bowel obstruction.

It is to be hoped that a number of the concepts presented in the following chapters will soon filter down into routine clinical practice. This book attempts to present a state-of-the-art résumé of progress in the different areas addressed and we are confident that it will influence research trends over the next few years. There is still much to do but we hope that all workers in the scientific and clinical fields of ovarian cancer will be encouraged by our statements.

We thank Beryl Stevens of the Royal College of Obstetricians and Gynaecologists for invaluable help in transcribing the tapes of our dialogue. We also thank the editorial staff at Chapman & Hall for the way they produced a handsome book from our rough manuscript.

F. Sharp, Sheffield
W. Mason, London
W. Creasman, Charleston

Helene Harris Memorial Trust Introduction

The Helene Harris Memorial Trust was formed in 1986 to promote communication, discussion and research into ovarian cancer on an international scale. The dialogue between scientists and clinicians has proved most rewarding and I am gratified that the HHMT has been responsible for assisting progress in a number of vital areas.

I would like to thank our editors, Professor Frank Sharp, Mr Peter Mason and Dr William Creasman for their dedicated work which has provided a volume which will prove indispensable to everyone involved in the latest treatment and research into ovarian cancer.

My thanks also to our many distinguished contributors who have voluntarily donated the presentations from which this book is compiled. A further debt of gratitude is owed to Mrs Shirley Claff and Ms Beryl Stevens for their fine efforts in administering, liaising and transcribing the work of our varied contributors from many countries.

I would also like to extend my appreciation to the Medical University of South Carolina and Dr William T. Creasman, who were our hosts for the Charleston Forum and provided the back up and the warm Southern hospitality which was vital to a successful meeting.

The work of the HHMT will continue. Much has been accomplished, but even more needs to be done until we can say that we can both prevent and cure ovarian cancer. I trust that all who read this book will be assisted in that task.

J.E. Harris

Part One
Genetics and Immunology

Chapter 1
Familial ovarian cancer

B.A.J. PONDER, J. PETO and D.F. EASTON

Case reports[1,2] and population-based epidemiological studies[3] indicate that inherited predisposition is a significant factor in a minority of epithelial ovarian cancer. We will review briefly the evidence for inherited predisposition and its extent, and consider the implications for research and for clinical practice.

1.1 THE EVIDENCE FOR INHERITED PREDISPOSITION

An increasing number of families are reported in which several cases of ovarian cancer have occurred in a pattern which suggests dominantly inherited predisposition. Our own experience suggests that about one patient in 100 with ovarian cancer will give a family history of two or more affected close relatives. Clinical case reports are, however, difficult to interpret: an unbiased assessment of the role of inherited factors requires a population-based study.

Several case-control studies, notably the cancer and steroid hormone (CASH) study[3], show about a threefold increased risk for ovarian cancer in close relatives of cases. However, these data do not prove a genetic cause; nor, supposing a genetic predisposition is responsible, do they allow one to distinguish between a slightly increased risk to the relatives of many cases, or a very high risk to the relatives of a small number. Either could produce the observed result in the population as a whole. The distinction can be made by asking what is the risk to a woman who has already two affected relatives, compared to women who have only one. If there is a small risk widely distributed, the two will not be greatly different; but if a high risk applies to a small number of families, these would be marked as families with two or more cases in close relatives, and the risk to a subsequent relative would be high.

Preliminary analysis of data from the OPCS study, based on the records of the Office of Population Censuses and Surveys in the UK, show that whereas the lifetime risk of death from ovarian cancer to women with one affected relative is about 1 in 40 (three times the risk in the general population), the risk to a woman who has two affected close relatives may be as much as 30–40%. The data are therefore consistent with predisposition in these families by a single autosomal dominant gene, conferring a lifetime risk in excess of 50% in those who inherit the gene.

We have previously estimated that this gene might account for up to 5% of cases of ovarian cancer occurring before the age of 50; or about 40 cases per year in England and Wales (population roughly 50 million). The risk of ovarian cancer to a woman who has inherited the gene is about 2% per year from the age of 30[4].

Conclusions about inherited predisposition from epidemiological data are inevitably tentative, and can only be proved by the demonstration of linkage between ovarian cancer and the inheritance of a known genetic marker in the families. Linkage has recently been reported to markers on chromosome 17 in three families with multiple cases of breast and ovarian cancer[5], lending strong support to the hypothesis of inherited predisposition.

There is also evidence of an increased risk of ovarian cancer in the relatives of breast cancer patients, and vice versa, but these familial risks (of the order 1.5 fold) are lower than the 'site-specific' risks. This suggests that some gene or genes predispose to both ovarian and breast cancer, whereas other genes are more site specific. This epidemiological evidence of heterogeneity is supported by observations in large families: some families have multiple cases of both cancers[6], whereas in other families there appears to be predisposition only to breast cancer or only to ovarian cancer.

Although the epidemiological data are consistent with a single major gene effect, it is quite possible (or even likely) that several predisposing genes will be involved (genetic heterogeneity); this situation can only be resolved using genetic linkage studies, by identifying genetic markers in two or more distinct transformed regions linked to the disease in different families (see below). It is important to note that the gene (or genes) causing the spectacular families are not necessarily the same as those giving rise to the elevated risks in the population studies. If different genes confer very different risks, then many of the impressive families could be the result of very rare genes conferring near complete penetrance, whereas the familial aggregation observed in population studies may be caused by commoner genes causing lower risks. This situation of multiple genes conferring different penetrances appears to hold for familial breast cancer[5], but as yet it is not clear whether it is also true of ovarian cancer.

1.2 IMPLICATIONS OF INHERITED PREDISPOSITION

1.2.1 SEARCHING FOR THE PREDISPOSING GENES

Provided that the inherited predisposition in any one family is largely due to the effects of a single gene, it is possible to search for the gene using genetic linkage. Once a linkage is found, it is possible to test other families (if they are large enough to yield a significant result) for linkage at the gene locus; and in families which show linkage, to identify the family members who have inherited the predisposition and those who have not. The next stage is to identify the predisposing gene itself, using the techniques of genetic mapping. When this has been done, a direct test for susceptible individuals may be possible by screening for mutations in the gene. Elucidation of the normal function of the gene and the way in which this is altered by mutation may improve the understanding of the pathogenesis of the disease, and may suggest strategies for prevention or treatment. These steps are straightforward in principle, and the predisposing genes for several inherited syndromes associated with cancer (e.g. retinoblastoma, Wilms' tumour, neurofibromatosis type I) have already been identified. Two problems may make the task in ovarian cancer a little harder.

The need for large families

Linkage analysis depends upon demonstrating a significant association between the inheritance of DNA markers and the inheritance of the disease. Families in which there

Implications of inherited predisposition

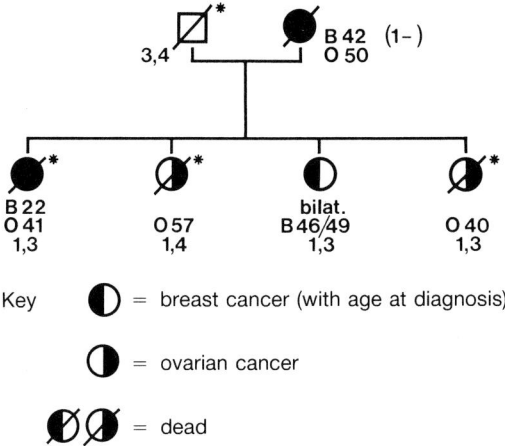

Figure 1.1 Family with breast and ovarian cancer. Genetic linkage analysis was possible in this family, even though all but one of the affected individuals had died, by using DNA extracted from pathology blocks from the individuals marked by an asterisk. The alleles scored at the growth hormone locus on chromosome 17q are shown (1,3,4). The partial genotype of the mother (1−) was inferred from the results in the rest of the family, as no pathology material was available in her case.

are several surviving members with ovarian cancer are rare. In order to obtain material from sufficient family members, it will almost certainly be necessary to collect archival pathology material for DNA extraction, and to 'reconstruct' the genotype of other family members who are unobtainable, using samples from their close relatives (Figure 1.1). Unless very large families can be found, results from several families must be combined to yield a significant result; which raises the second problem, genetic heterogeneity.

Genetic heterogeneity

If different genes predispose to ovarian cancer in different families, attempts to combine linkage results from several families will fail (or at least, will require sophisticated analysis to detect the heterogeneity). Unpublished results from the families with breast and ovarian cancer collected by Lynch (Feunteun and Lenoir, personal communication) already suggest that there will be a problem: of five families with similar clinical features, three showed evidence for linkage to markers on chromosome 17, and two did not. Several supposedly distinct clinical syndromes of ovarian cancer are described; for example, site-specific ovarian cancer, ovarian and breast cancer and ovarian cancer as part of the 'family cancer syndrome' with colorectal, breast and possibly other cancers. The results from the Lynch families suggest, however, that it will not be possible to select a genotypically homogeneous set of families using these clinical criteria. Moreover, the same 17q markers that show linkage in the families with breast and ovarian cancer are also linked to some early onset site-specific breast cancer families[5], suggesting that the same gene may be responsible for the different syndromes.

1.2.2 RECOGNITION AND MANAGEMENT OF INDIVIDUALS AT RISK

Recognition

In many inherited cancers, a characteristic phenotype indicates clearly which individuals have inherited the predisposition. For example, an individual with colon cancer who has multiple colonic polyps almost certainly has an inherited polyposis syndrome, and a search for the polyps is the basis for a screening test for individuals at risk in families. In ovarian cancer, no such phenotypic or genotypic marker of predisposition is as yet available. Field change in the coelomic epithelium is well recognized in the form of metaplasias (endosalpingiosis, endometriosis), multifocal 'borderline' tumours, and possibly the origin of some malignant tumours from multiple peritoneal sites without primary ovarian involvement (though the latter group is controversial). So far, however, there is no indication that any of these is associated with a family history of ovarian cancer, although the reports of primary extra-ovarian cancer after prophylactic oophorectomy (see below) may indicate that a field change of some type is present in predisposed cases. Even a recognizable preneoplastic change has not been described in ovarian epithelium. Progress in this area would clearly be of biological as well as practical interest, and study of peritoneal biopsies and prophylactic oophorectomy specimens from women at high risk may possibly provide some clues.

In the absence of pathological or genetic markers, family history is currently the only criterion of risk. Two or more affected close relatives, especially if they are young, or if one of them has breast cancer as well, strongly suggests inherited predisposition. Assuming dominant inheritance, the risks that each family member has inherited the gene can be estimated from their relationship to affected family members. The risk that such an individual will develop ovarian cancer is, of course, related to age: even in multiple-case families the first cancers rarely develop before the age of 30, and the limited data which are so far available suggest that risk is evenly spread after that age. Thus, a daughter or sister of an affected woman in a family has a 50% chance of inheriting the gene; since not all gene carriers develop cancer, her lifetime risk is about 40%, and this is spread at roughly 1% per year from age 30 onwards. In some families there will also be a risk of other cancers, such as breast and colorectal: but there is not sufficient information to quantify the risks for each cancer separately.

Management

The options for women at risk are: (i) do nothing, (ii) screening, (iii) prophylactic oophorectomy. Most women who seek advice will want to take some positive action, but one must remember that at present neither screening nor prophylactic oophorectomy is of proven benefit. Each carries some morbidity (screening, as a result of anxiety and the possibility of false positive results which lead to unnecessary investigations) and some women, especially those in the family who have not themselves sought advice, may prefer to do nothing.

Screening usually consists of clinical examination (directed also to possible associated cancers), abdominal or vaginal ultrasound, and possibly estimation of serological markers such as CA-125. Women who have an abnormal result may be evaluated further by a second screening test of higher specificity, such as colour Doppler blood flow measurement[7]. The recommended starting age is 25 years, and frequency one year. At present our policy is to offer screening to all women in families with two or more cases, who have either an affected first or second degree relative. We estimate that in the UK there are some 5500 women between the ages of 25 and 70 who have two or more relatives affected with ovarian cancer; in this group there will be an expected 36 cases of ovarian cancer per year. This group contains roughly one-third of all the women who are at high risk through inherited predisposition. The figures provide some indication both of the scale of a screening programme for families, and of the feasibility of evaluating screening by a randomized trial, if this were thought to be acceptable on ethical grounds.

Because the efficacy of screening is uncertain, prophylactic oophorectomy may be considered for women at high risk, once childbearing is complete. Even this, however, does not guarantee freedom from risk. Primary cancer arising from the coelomic epithelium is well documented, and a few cases have been reported in women who have undergone prophylactic oophorectomy because of their strong family history[8]. This raises the possibility that inherited predisposition may increase the malignant potential of the whole coelomic epithelium, in which case prophylactic oophorectomy might offer only limited protection. Follow-up of a carefully documented group of women who have had prophylactic oophorectomy because of their family history is required to answer this question. This could be achieved as part of the international collaboration proposed below.

1.3 FUTURE STUDIES

Familial ovarian cancer has been under-recognized; but even when they are specifically sought, large families are uncommon. Genetic linkage, studies of the pathology of prophylactically removed ovaries and biopsies of coelomic epithelium from women at risk, evaluation of screening and evaluation of prophylactic oophorectomy, are all likely to need material and international collaboration to achieve sufficient numbers. Familial ovarian cancer registers have started in several European countries, including the UK, and in the USA; and joint studies are being planned. Clinicians who look after families, and anyone who is interested in contributing to the work, are invited to contact the authors, who will put them in touch with the appropriate local or national group.

REFERENCES

1. Fraumeni, J.F. Jr, Grundy, E.W., Creagan, E.T. and Everson, R.B. (1976) Six families prone to ovarian cancer. *Cancer*, **36**, 364–9.
2. Franchesci, S., LaVecchia, C. and Managioni, C. (1982) Familial ovarian cancer: eight more families. *Gynaecol. Oncol.*, **13**, 31–6.
3. Schildkraut, J.M. and Thompson, W.D. (1988) Familial ovarian cancer: a population based case-control study. *Am. J. Epidemiol.*, **128**, 456–66.
4. Ponder, B.A.J., Easton, D.F. and Peto, J. (1989) Risk of ovarian cancer associated with a family history, in *Ovarian Cancer* (eds F. Sharp, W.P. Mason and R.E. Leake), Chapman & Hall, London, pp. 3–6.
5. Hall, J.M., Lee, M.K., Newman, B. *et al.* (1990) Linkage of early-onset familial breast cancer to chromosome 17 or 21. *Science*, **250**, 1684–9.
6. Lynch, H.T., Watson, P., Bewtra, C. *et al.* (1991) Hereditary ovarian cancer: heterogeneity in age at diagnosis. *Cancer*, **67**, 1460–6.
7. Bourne, T., Campbell, S., Steer, C. *et al.* (1989) Screening techniques for ovarian cancer. *Br. Med. J.*, **299**, 1367–70.
8. Tobacman, J.K., Tucker, M.A., Kase, R. *et al.* (1982) Intra-abdominal carcinomatosis after prophylactic oophorectomy in ovarian cancer-prone families. *Lancet*, **ii**, 795–7.

Chapter 2

Hereditary ovarian cancer

P. WATSON and H.T. LYNCH

2.1 INTRODUCTION

Case-control studies have demonstrated the importance of a family history of ovarian cancer[1] and a family history of other cancers[2] on the risk for ovarian cancer. Multiple genetic and non-genetic mechanisms may be involved in this association. One type of mechanism is the particular focus of this report: hereditary ovarian cancer (HOC). Conceptually, HOC includes any ovarian cancer occurring as a manifestation of an inherited disorder caused by a major gene. As a practical matter, HOC includes ovarian cancer occurring in a woman with a family history or a personal medical history indicative of an inherited disorder believed to cause ovarian cancer.

Hereditary ovarian cancer is heterogeneous as evidenced by the several distinctive inherited disorders which are believed to cause a substantial increase in risk of ovarian tumours. For example, women with the basal cell naevus syndrome, an autosomal dominant trait, are at high risk for benign ovarian fibromas and to a lesser extent for ovarian carcinomas[3,4], while those with the Peutz–Jegher's syndrome, also an autosomal dominant trait, are predisposed to sex cord tumours with annular tubules[5,6].

Our purpose in this report is to describe our recent studies of families with three types of inherited disorder, where the major or only known manifestation of the disorder is cancer, and where ovarian cancer is believed to be a manifestation of the disorder. These disorders are hereditary non-polyposis colorectal cancer (HNPCC)[7], hereditary breast and ovarian cancer syndrome (HBOC)[8], and hereditary site-specific ovarian cancer[9].

2.2 HEREDITARY NON-POLYPOSIS COLORECTAL CANCER

HNPCC[7] is hypothesized to be a highly penetrant, autosomal dominantly inherited disorder. The characteristic manifestation of the disorder is colorectal cancer, often occurring at an early age (mean age at diagnosis 45 years). Endometrial cancer is reported very commonly in HNPCC families, and is accepted as a manifestation of the disorder. HNPCC families have been described in which multiple cases of other types of cancer have been diagnosed in high risk family members, including cancer of the ovary[10]; however, the extent to which such observations might be due to chance, given the incidence of cancer in the general population, was uncertain. (See Figure 2.1 for typical HNPCC pedigrees with ovarian cancer.)

We recently examined the occurrence of some of these cancer types in high risk members of 23 large Lynch syndrome famil-

Hereditary ovarian cancer

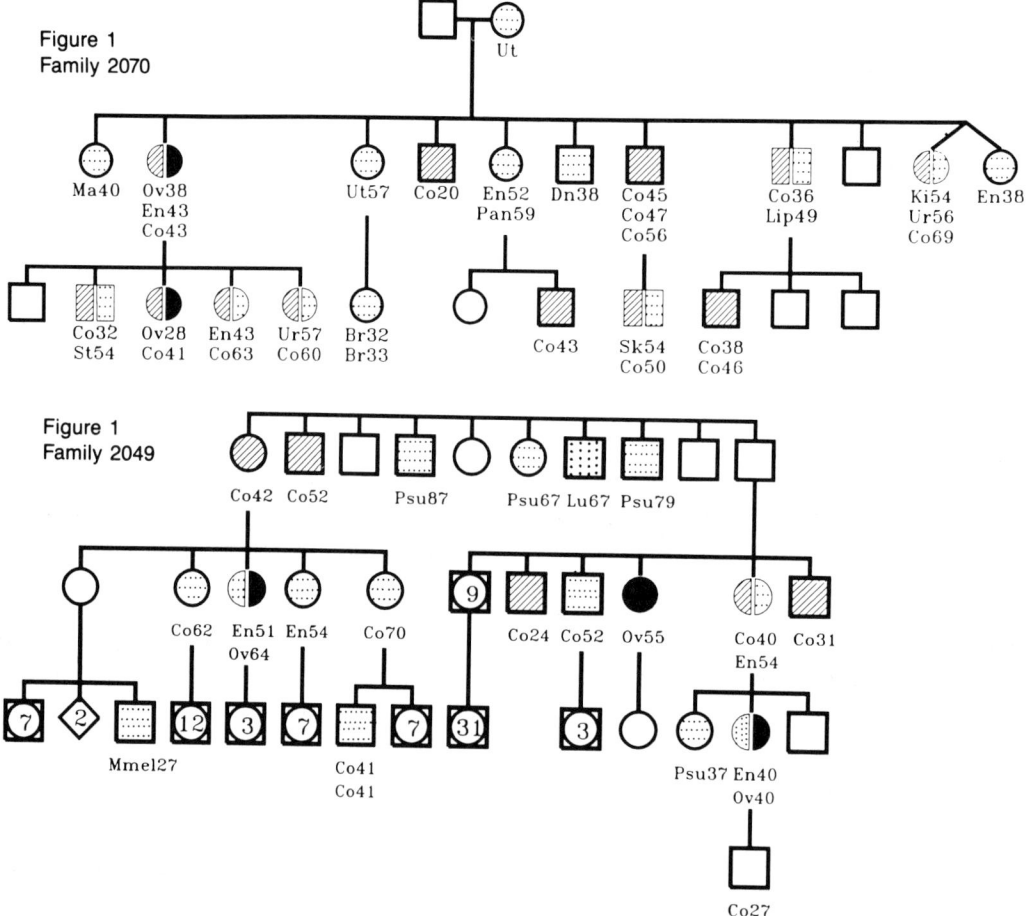

Figure 2.1 Pedigrees of hereditary non-polyposis colorectal cancer families with ovarian cancer. Males are indicated by squares; females by circles. Cancer-affected cases are shaded, with cancer site/type and age at diagnosis indicated. Cancer site/type abbreviations: Br, breast; Co, colon; Dn, duodenum; En, endometrium; Ki, kidney; Lu, lung; Ma, maxilla; Mmel, malignant melanoma; Ov, ovary; Pan, pancreas; Psu, unknown site; Sk, skin; St, stomach; Ur, ureter; Ut, uterus.

ies (P. Watson & H.T. Lynch, in preparation). Family members were classified as high vs. low risk according to their position in the pedigree relative to cases with colorectal or endometrial cancer. The high risk group included persons with these cancers, and the common ancestors and first degree relatives of these persons. We compared the observed frequency of cancer at a specific site with the expected number, based on general population incidence rates and adjusted for the family size and the age and sex of the family members[11,12].

Ovarian cancer occurred among high risk women more commonly than expected, based on the general population incidence rates ($P<0.05$); specifically, it occurred at 3.5 times the expected rate. The median age at diagnosis in the 13 affected women with histologically verified ovarian cancer was 40 years. All cases were adenocarcinomas, usually serous cystadenocarcinoma. Other cancer was diagnosed in other organs excessively frequently as well, including the stomach, small bowel, and upper renal tract (the renal pelvis and ureter) (P. Watson & H.T. Lynch, in preparation).

We also evaluated the question of whether these cancers tended to aggregate in a subset of the families. This was done by computing an index for each family indicating the frequency of cancer in that family relative to the other families, then computing the variances of these indices. If this variance was larger than expected, under the null hypothesis of homogeneous risk among families, this was an indication that cancers clustered in certain families. We used a simulation process to determine the distribution of variances given random distribution of cancers among families, and if the actual variance was higher than nearly all of these simulated variances, we concluded that the cancer did aggregate in certain families (P. Watson & H.T. Lynch, in preparation).

In the case of ovarian cancer, we did see significant familial aggregation. The 13 ovarian cancers occurred in 8 families (so that 15 families had no instances of the diagnosis, including some very large families). We tentatively conclude that some Lynch syndrome families may have a relative risk of ovarian cancer that is higher than 3.5; others may have a relative risk which is lower than 3.5. The clinical implication of this observation is that ovarian cancer screening (beginning at age 25 years) and prevention (prophylactic total abdominal hysterectomy and bilateral salpingo-oophorectomy TAH-BSO) should be recommended to high risk women in those Lynch syndrome families with a family history of ovarian cancer.

2.3 HEREDITARY SITE-SPECIFIC OVARIAN CANCER

Occasionally we encounter a family in which multiple family members are affected by ovarian cancer, without multiple cases of cancer of the breast or colon, and without evidence of another hereditary syndrome associated with ovarian cancer[9]. We refer to such families as hereditary site-specific ovarian cancer families, but this is a provisional classification. In our experience, such families are rare relative to Lynch syndrome or breast and ovary families. Any particular instance may be a coincidence of independent events; it may be a Lynch syndrome family or breast and ovary family in which a few or none of the usually predominant cancer types have been expressed (see Figure 2.2 for illustrative pedigrees).

We have reported results from a study of four site-specific ovarian families in which 18 cases of ovarian cancer had occurred[10]. The mean age at diagnosis was approximately 49 years, and like ovarian cancer in HNPCC and in HBOC (see below), the histological type, when known, was nearly always serous cystadenocarcinoma.

2.4 HEREDITARY BREAST AND OVARIAN CANCER SYNDROME

Many hereditary breast cancer families have multiple cases of ovarian cancer, including cases where women with breast cancer also develop ovarian cancer[8]. Any hereditary breast cancer-prone family which includes two or more ovarian cancer cases among the breast cancer-affected women or their first degree relatives is classified as an HBOC family. However, this cut-off is arbitrary; breast cancer families with a single case of ovarian cancer have not been shown to differ systematically from HBOC families in any other way. HBOC pedigrees are illustrated in Figure 2.3.

Analysis of 45 documented ovarian cancer cases in 12 HBOC families[10] indicated a mean age of diagnosis of 52 years, significantly lower than the population mean of 59 years. The age range was very wide (32–79

Hereditary ovarian cancer

Figure 2
Family 133

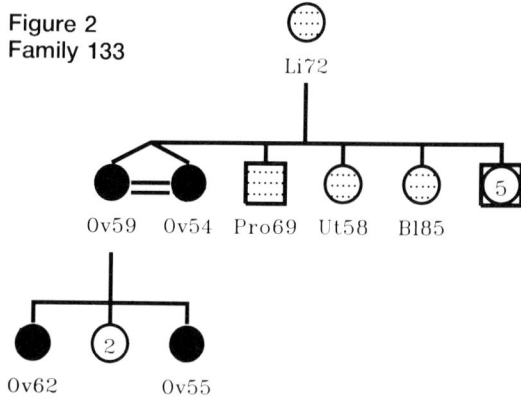

Figure 2
Family 1704

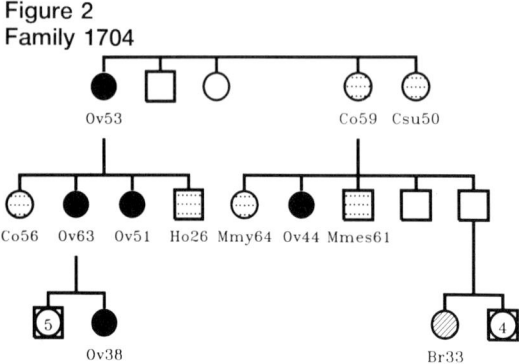

Figure 2.2 Pedigrees of site-specific ovarian cancer families. Males are indicated by squares; females by circles. Cancer-affected cases are shaded, with cancer site/type and age at diagnosis indicated. Cancer site/type abbreviations: Bl, bladder; Br, breast; Co, colon; Csu, unknown; Ho, Hodgkin's disease; Li, liver; Mmes, malignant mesothelioma; Mmy, malignant myeloma; Ov, ovary; Pro, prostate; Ut, uterus.

Table 2.1 Cumulative incidence of breast and ovarian cancer in Family 1816

Age (years)	Cumulative incidence (%)		
	Breast cancer	Ovarian cancer	Any cancer
20	0.0	0.0	0.0
30	2.6	0.0	2.6
40	23.1	2.9	26.7
50	45.4	39.2	64.5
60	64.1	48.4	84.5
70	76.4	62.8	93.2

years). The vast majority of diagnoses are serous cystadenocarcinoma.

We have recently observed genetic linkage between the development of breast and/or ovarian cancer in HBOC and a polymorphic marker (CMM86) mapped to the long arm of chromosome 17 (Narod *et al.*, in preparation). Genetic linkage studies aim to localize the gene for a disorder by searching for co-segregation of the disorder with specific alleles of mapped polymorphic systems. We tested the chromosome 17 marker in five families. The overall lod score was 3.03 under the assumption of heterogeneity. Three families had positive lod scores and two had large negative lod scores. Examination of the clinical records associated with the five families revealed no consistent differences between the three positive families and the two negative ones. This same marker has been linked to early onset breast cancer in a study by Mary-Claire King and colleagues at Berkeley[13].

Family 1816 (Figure 2.3) was among the linkage-positive families, and had the highest single-family lod score (2.7 at recombination fraction = 0.10). The implications of finding linkage in an HBOC family are many. Linkage will enable us, for the first time, to identify women who are likely to be gene carriers based on the presence of the marker allele. This identification process is presently not sufficiently accurate for clinical application. However, further linkage study in the region can be expected to identify more tightly linked markers.

In our initial study, we identified 39 women who were either affected with a syndrome cancer or 'marked' as gene carriers, and used these in a life table analysis to estimate the penetrance of the disorder (Narod *et al.*, in preparation). Table 2.1 shows the results of this analysis. Included in the

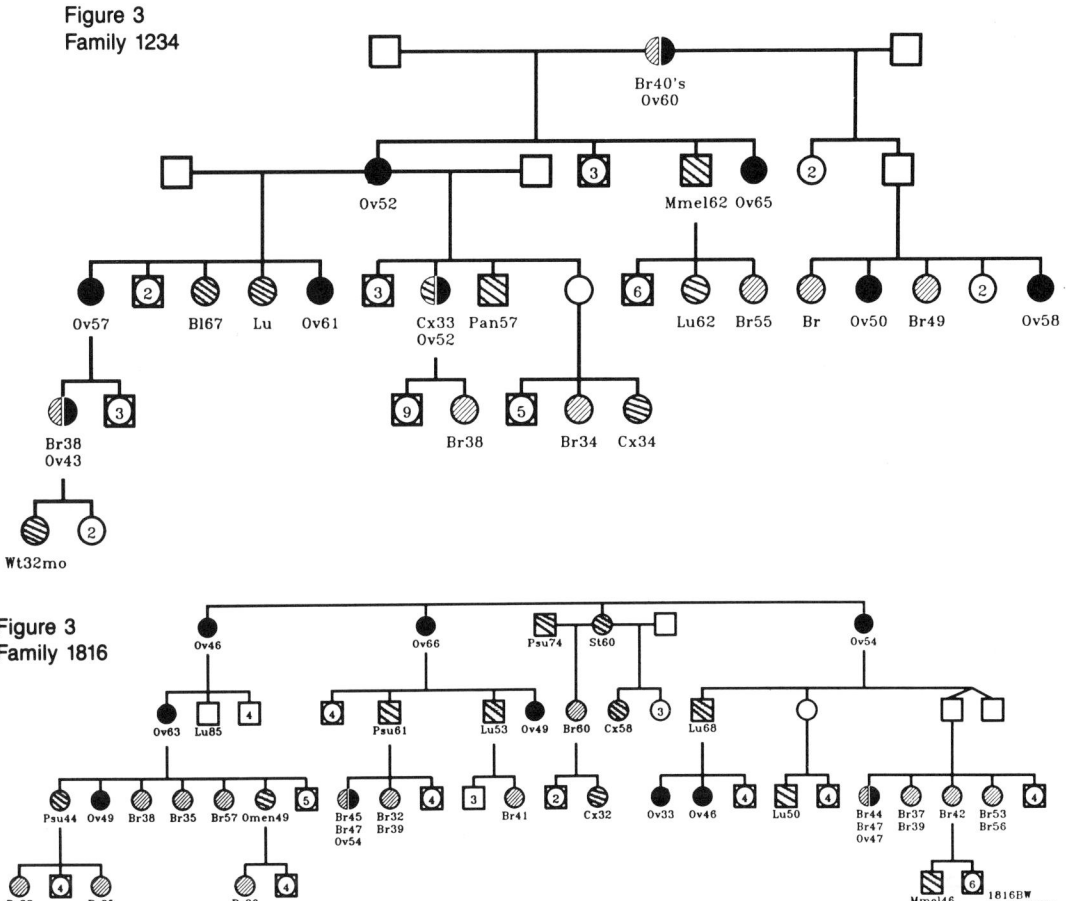

Figure 2.3 Pedigrees of hereditary breast and ovarian cancer families. Males are indicated by squares; females by circles. Cancer-affected cases are shaded, with cancer site/type and age at diagnosis indicated. Cancer site/type abbreviations: Bl, bladder; Br, breast; Cx, cervix uteri; Lu, lung; Mmel, malignant melanoma; Omen, omentum; Ov, ovary; Pan, pancreas; Pro, prostate; Psu, unknown primary site; St, stomach; Ut, uterus; Wt, Wilms' tumour.

category of 'any known cancer' are one case of adenocarcinoma of the abdomen, primary site unknown, and one case of cancer of the omentum which occurred in a woman who had previously undergone prophylactic oophorectomy. These figures indicate the high penetrance of cancer in this family and the early age of cancer onset.

There are 46 women in Family 1816 who are presently cancer-free, between the ages of 20 and 45, and who are either at 50% risk of carrying the gene (having a cancer-affected mother or sister) or whose father is at 50% risk. We are in the process of completing collection of specimens from all these women, with the aim of classifying them as likely gene carriers according to the marker allele. We hope that this group (and similar groups from other families with unambiguous linkage) will be useful in research on early cancer detection methods and on biomarkers of cancer susceptibility. We hope also, in the near future, to identify better markers in this region, so that we can provide these women with highly accurate risk information which can be used in planning surveillance and prevention.

2.5 CLINICAL IMPLICATIONS OF HEREDITARY OVARIAN CANCER

The most important step in the recognition of women at risk for HOC is the compilation of an accurate family history which embraces cancer of all anatomical sites. In our experience, a nurse who is taught to gather this information effectively can free up the physician's time for pedigree interpretation. In most circumstances, the modified nuclear pedigree focusing on all first degree relatives and selected second degree relatives (namely, both sets of grandparents, aunts, and uncles) may be sufficient. In certain circumstances, extension of the pedigree even further may be necessary. A crucial consideration for HOC diagnosis is that the physician be cognizant of the natural history of hereditary cancer, particularly its early age of onset, integral association in pedigrees with specific other tumours, excess of cases with multiple primary cancer, and association with specific phenotypic signs as in the cancer-associated genodermatoses (e.g. the multiple naevoid basal cell carcinoma syndrome and the Peutz–Jeghers syndrome).

When a hereditary cancer family is recognized, genetic counselling is vital. We consider genetic counselling to encompass the provision of accurate genetic risk information in concert with education about the natural history of hereditary cancer, with particular emphasis on all facets of its surveillance and management. We initiate the educational process in the teens. Establishing an empathetic patient/physician relationship is vital for compliance, which ultimately is based on two-way communication between patient and physician. Clearly, surveillance/management for the several differing hereditary ovarian carcinoma syndromes should be extended to those cancer types which predominate in the disorders of concern. Thus, in the case of HNPCC, focus is given to the colon, and due to the early onset and proximal predominance, colonoscopy is mandatory, with initiation at age 25 years. When these patients manifest initial colorectal cancer, we perform no less than a subtotal colectomy because of the excess of synchronous colorectal cancer and the lifetime problem of metachronous colorectal cancer. Women in HNPCC families with a family history of endometrial and/or ovarian cancer are candidates for prophylactic TAH-BSO at the time of surgery for colorectal cancer. With respect to HBOC, we initiate formal surveillance by age 25 years. This involves instruction in breast self-examination, mammography, and examination of the breasts by a physician twice a year. We perform mammography every other year until age 35 years and then annually thereafter. We are currently experimenting with a programme for surveillance of ovarian carcinoma. We initiate annual serum CA-125 assessment and pelvic probe ultrasound screening, beginning at age 25 years. We perform these examinations annually. We advise high risk women that our surveillance for ovarian cancer has many limitations and thereby prepare them for eventual consideration of prophylactic TAH-BSO as soon as they have completed their families. We always advise these women that even if their ovaries are histologically normal at the time of the TAH-BSO, we cannot give them absolute assurance that they will never manifest extra-pelvic mesothelial cystadenocarcinoma of the ovary. Indeed, this has occurred in several of our patients[14].

2.6 CONCLUSION

Ovarian cancers observed in HNPCC, hereditary site-specific ovarian cancer, and hereditary breast cancer have been found to differ from ovarian cancers in the general population by having an earlier age at diagnosis. Furthermore, we have found a statistically significant difference among the hereditary subgroups in age at diagnosis, with the earliest diagnoses occurring in HNPCC[10]. Cancers in all subgroups have been found to be similar to ovarian cancers seen in the general population in histological type. Study of medical records of affected cases has revealed no differences among these hereditary subgroups, nor between the hereditary cases and the general population in terms of clinical presentation or course after diagnosis. The high risk for ovarian cancer, and the early age of onset of the disease, has led to the recommendation for screening with pelvic ultrasound and serum markers in an attempt to diagnose early lesions, and for prophylactic TAH-BSO. Our ability to use these cancer control measures may increase with the advance of knowledge of the genetics of these disorders, especially when we gain the ability to identify gene carriers prior to cancer development.

ACKNOWLEDGEMENT

Support for this work was provided by grants from the National Cancer Institute, #5 RO1 CA48802 and #1 RO1 CA41371.

REFERENCES

1. Schildkraut, J.M., Risch, N. and Thompson, W.D. (1989) Evaluating genetic association among ovarian, breast, and endometrial cancer: evidence for a breast/ovarian cancer relationship. *Am. J. Hum. Genet.*, **45**, 521–9.
2. Schildkraut, J.M. and Thompson, W.D. (1988) Familial ovarian cancer: a population-based case control study. *Am. J. Epidemiol.*, **128**, 456–66.
3. McKusick, V.A. (1988) *Mendelian Inheritance in Man*, 8th edn, Johns Hopkins University Press, Baltimore.
4. Berlin, N.I., Van Scott, E.J., Clendenning, W.E. *et al.* (1966) Basal cell nevus syndrome. *Ann. Intern. Med.*, **64**, 403–21.
5. Giardiello, F.M., Welsh, S.B., Hamilton, S.R. *et al.* (1987) Increased risk of cancer in the Peutz–Jeghers syndrome. *N. Engl. J. Med.*, **316**, 1511–14.
6. Scully, R.E. (1970) Sex cord tumor with annular tubules: a distinctive ovarian tumor of the Peutz–Jeghers syndrome. *Cancer*, **25**, 1107–21.
7. Lynch, H.T., Kimberling, W.J., Albano, W.A. *et al.* (1985) Hereditary nonpolyposis colorectal cancer, Parts I and II. *Cancer*, **56**, 939–51.
8. Lynch, H.T., Harris, R.E., Guirgis, H.A. *et al.* (1978) Familial association of breast/ovarian carcinoma. *Cancer*, **41**, 1543–8.
9. Lynch, H.T., Albano, W.A., Black, L. *et al.* (1981) Familial excess of cancer of the ovary and other anatomic sites. *JAMA*, **245**, 261–4.
10. Lynch, H.T., Watson, P., Bewtra, C. *et al.* (1991) Hereditary ovarian cancer – heterogeneity in age at diagnosis. *Cancer*, **67**, 1460–6.
11. Monson, R.R. (1974) Analysis of relative survival and proportional mortality. *Comput. Biomed. Res.*, **7**, 325–32.
12. Young, J.L., Percy, C.L. and Asire, A.J. (1981) Surveillance, epidemiology, and end results: incidence and mortality data, 1973–77. *NCI Mono*, **57** (whole issue).
13. Hall, J.M., Lee, M.K., Newman, B. *et al.* (1990) Linkage of early-onset familial breast cancer to chromosome 17q21. *Science*, **250**, 1684–9.
14. Lynch, H.T., Bewtra, C. and Lynch, J.F. (1986) Familial ovarian carcinoma: clinical nuances. *Am. J. Med.*, **81**, 1073–6.

Chapter 3

Cytogenetic changes in human epithelial ovarian cancer

H.H. GALLION, D.E. POWELL, L.W. SMITH,
C.C. VAUGH and E.A. CASE

Cytogenetic analysis of tumour cells has revealed the presence of chromosomal abnormalities in the majority of human malignancies[1]. The identification of consistent translocations, duplications and deletions has provided useful information regarding the location of genetic alterations involved in human carcinogenesis. These cancer-related genes, oncogenes and tumour suppressor genes, are normal cellular genes with indispensable roles in controlling cell growth. Oncogenes normally promote cell growth and activation of one allele by mutation, translocation or amplification results in increased gene activity and uncontrolled cell proliferation. Tumour suppressor genes, in contrast, function to suppress cell proliferation and loss or inactivation of both alleles removes the normal constraint to cell growth. Available data clearly indicate that human carcinogenesis is a multistep process which results in the accumulation of a number of genetic alterations within a single cell, ultimately converting a normal cell to its malignant counterpart.

The Philadelphia chromosome was the first tumour-specific chromosome abnormality to be identified in a human cancer[2]. This characteristic chromosome aberration, which results in the translocation of the c-*abl* oncogene from its normal position on chromosome 9 to a new location on chromosome 22 is observed in the majority of patients with chronic myelogenous leukaemia. Consistent chromosome translocations have likewise been identified in several other hematological malignancies including Burkitt's lymphoma, chronic myelocytic leukaemia and acute myeloblastic leukaemia[3]. The oncogenes c-*myc*, c-*abl* and c-*mos* have been localized to specific breakpoints present in these malignancies indicating that acquired alterations of these oncogenes are associated with tumour induction or progression[4].

Although consistent aberrations are seen most frequently in hematological diseases, characteristic cytogenetic changes have also been identified in several childhood tumours including retinoblastoma and Wilms' tumours [5,6]. It was through the cytogenetic study of these rare heritable embryonal tumours that the essential role of tumour suppressor genes in human carcinomas was first established. The clue as to the location of the retinoblastoma tumour suppressor gene came from early cytogenetic studies which revealed visible constitutive chromosomal losses within the long arm of chromosome 13 in certain patients with the inherited form of the disease[5]. These losses, if present, were variable but always included one particular chromosome band. Subsequent cytogenetic analysis of the tumours themselves identified frequent losses in the same region of chromosome 13, both in patients with the inherited and the sporadic form of this disease[7–9]. Ultimately, molecular genetic studies utilizing restriction fragment polymorphism analysis confirmed 13q14 as the location of the critical tumour suppressor gene responsible for the development of retinoblastoma[10].

Although carcinomas represent over 80% of human malignancies, there are relatively few cytogenetic data concerning solid tumours in adults. The presence of specific chromosomal abnormalities in ovarian cancer was first reported in 1979 by Trent and Salmon[11]. These authors noted frequent deletions of chromosomal material from the long arm of chromosome 6. Subsequent investigators studying cytogenetic changes in ovarian cancer have observed a number of different complex karyotypic alterations which may affect genes important in ovarian carcinogenesis. Wake, for example, reported frequent deletions of 6q and gains on 14q in a series of 19 cases of ovarian cancer[12]. These authors observed one or both of these specific markers in all cases, and suggested that a specific translocational rearrangement between chromosome 6 and 14 was a primary change specifically associated with serous carcinoma of the ovary. Wang-Peng, in a study of cytogenetic changes in cells from ascites and effusion specimens from 44 patients with ovarian cancer, reported that chromosomes 1, 2, 3, 4, 9, 10, 15, 19, 6, and 11 were the most frequently involved in structural abnormalities[13]. However, the specific translocation initially reported by Wake was not observed. Panneni and Atkin, in similar studies, also noted frequent involvement of chromosomes 1, 3, and 6 [14,15]. Finally, Tanaka, in a recent detailed analysis of 13 specimens from nine untreated patients with advanced ovarian cancer, identified frequent complex aberrations involving chromosomes, 1, 3, 6, 7, 10 and 12[16].

The current study is unique in that only specimens from the primary ovarian tumour of patients with untreated epithelial ovarian cancer were utilized. Therefore, secondary changes induced by therapy or abnormalities occurring late in the course of disease progression should be less prominent. To date, we have analysed cytogenetic changes in the primary ovarian tumours of 49 patients with epithelial ovarian cancer. Tumour specimens were prepared as both short-term suspension and long-term monolayer cultures. Short-term cultures were incubated overnight and were harvested immediately, whereas long-term cultures were incubated for 5–7 days prior to harvest. Slides were made of all cultures in which divisions were seen. Cells were stained according to the Giemsa trypsin banding technique, analysed microscopically and photographed for karyotyping.

The histological characteristics of the tumours studied are presented in Table 3.1. The most common cell type was serous cystadenocarcinoma followed by endometrioid carcinoma, and mucinous cystadenocarcinoma. Ten patients had tumours of low malignant potential. Seven tumours were grade 1, 14 were grade 2 and 18 were grade 3. Fourteen patients had FIGO stage I disease, 7 patients stage II disease and 28 patients had stage III or IV tumours.

Of the 49 tumours studied, 37 (76%) had sufficient cell growth for cytogenetic analysis.

Table 3.1 Histological characteristics of tumours studied ($n = 49$)

	No. of tumours
Cell type	
Serous cystadenocarcinoma	25
Endometrioid carcinoma	10
Mucinous cystadenocarcinoma	8
Undifferentiated carcinoma	5
Clear cell carcinoma	1
Grade	
Low malignant potential	10
1	7
2	14
3	18

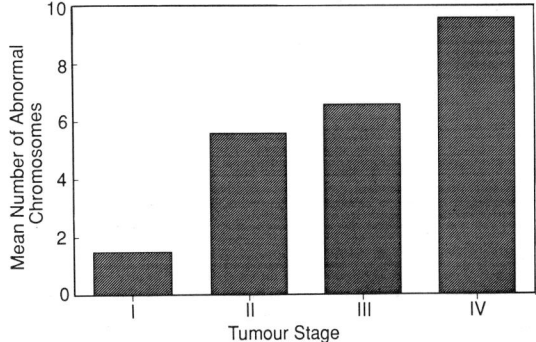

Figure 3.1 Chromosomal abnormalities related to histological stage in ovarian cancer ($n = 23$).

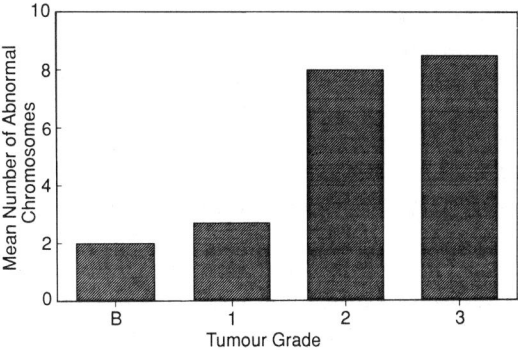

Figure 3.2 Chromosomal abnormalities related to grade in ovarian cancer ($n = 23$).

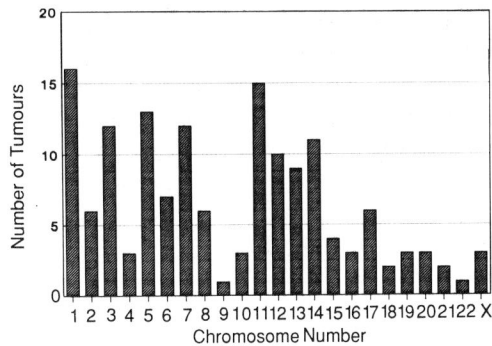

Figure 3.3 Frequency of structurally abnormal chromosomes in ovarian malignancies ($n = 23$).

Fourteen tumours had normal karyotypes. The presence of cytogenetically normal chromosomes, although occasionally present in poorly differentiated tumours, was observed most frequently in tumours of low malignant potential and grade 1 tumours. This finding may reflect either the overgrowth of benign stromal cells or the presence of genetic abnormalities too small to be detected cytogenetically. The remaining 23 tumours exhibited visible chromosomal aberrations and are the basis of this report. A mean of 7 different chromosomes per tumour had identifiable abnormalities (range 1–14). A direct relationship between the number of chromosomes with structural changes and advancing tumour grade and stage was observed and is illustrated in Figures 3.1 and 3.2.

The sites and frequencies of abnormal chromosomes are illustrated in Figure 3.3. The most commonly identified abnormalities were deletions, translocations and additions. Double minute chromosomes or homogeneous staining regions were rarely observed. Chromosomes 1 and 11 were the most commonly involved, being abnormal in 84% and 79% of tumours respectively. Abnormalities in chromosomes 5, 3 and 7 were also

Cytogenetic changes in human epithelial ovarian cancer

Table 3.2 Clinical and cytogenetic findings in patients with borderline epithelial ovarian cancer

Case no.	Cell type	Stage	Karyotype
1.	Serous	IIB	49 XX, +3C group
2.	Serous	IIB	47 XX, +8, inv 5q
3.	Serous	IIIB	48 XX, +5, +8
4.	Serous	IIIB	48 XX, +7

noted with some frequency. No single unique breakpoint was identified. Although structural aberrations of chromosome 14 were noted in 11 tumours, the characteristic 6–14 translocation reported by Wake was not observed. In several tumours, breakage at the 11p13 locus which is characteristic of Wilms' tumour was seen.

All 10 tumours of low malignant potential had adequate cells for analysis. Although the majority of these tumours exhibited normal karyotypes, chromosomal abnormalities were identified in four. With the exception of one tumour, the karyotypic changes were simple duplications of whole chromosomes as shown in Table 3.2. This finding is in direct contrast to the multiple complex changes seen in poorly differentiated tumours. Representative tumour karyotypes from a borderline and a poorly differentiated serous ovarian tumour are illustrated in Figures 3.4 and 3.5.

The frequent finding of abnormalities on chromosome 1 is consistent with observations in most other solid tumours including lung, breast, colon and cervix and does not appear to be specific for ovarian cancer [17,18]. Abnormalities of chromosome 11 were the second most commonly observed finding in our series. Although structural changes of chromosome 11 have been reported in previous studies of malignant ovarian tumours, they have not been observed as frequently as in the current series. This chromosome is of considerable interest since it is known to harbour the tumour suppressor gene for Wilms' tumour

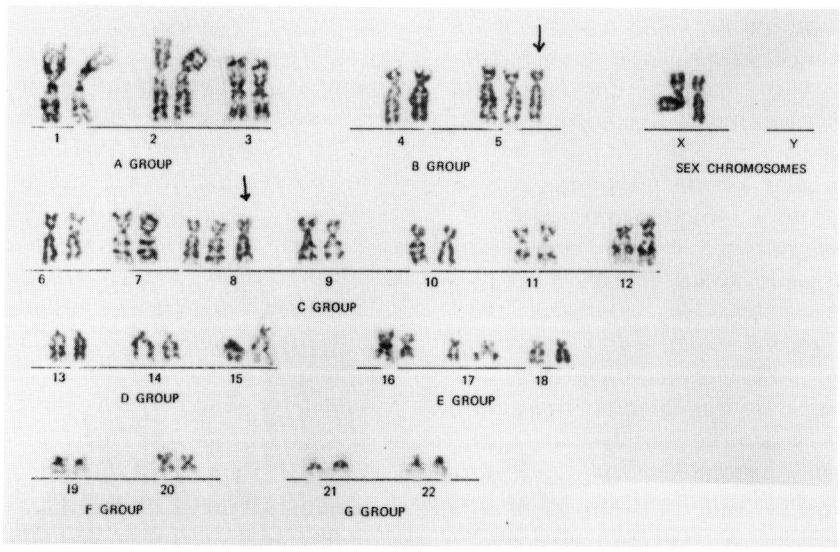

Figure 3.4 Trisomy of chromosomes 5 and 8 are present in this karyotype from a Stage IIIB borderline serous tumour.

Cytogenetic changes in human epithelial ovarian cancer

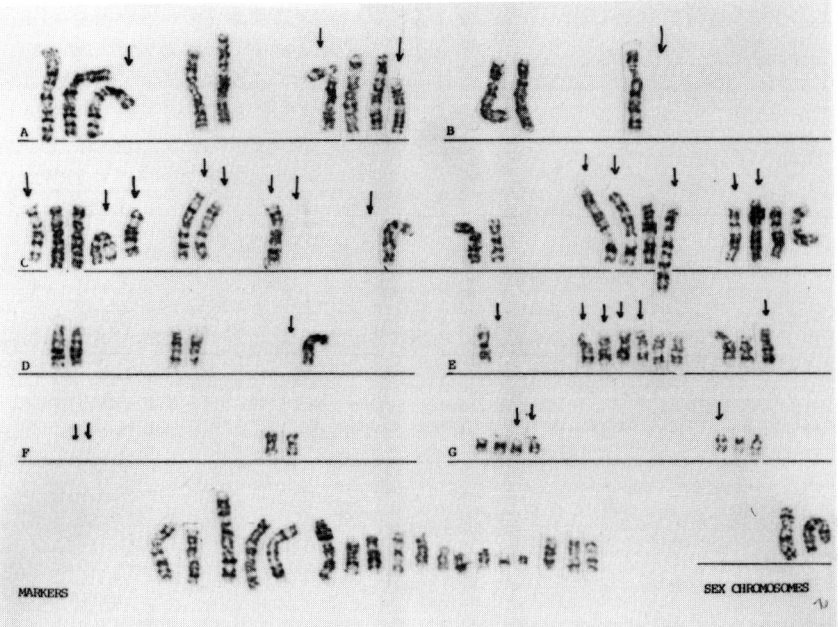

Figure 3.5 There are numerous complex changes in this karyotype from an advanced stage, poorly differentiated ovarian tumour. Note abnormalities in chromosomes 1 and 11 and unidentified marker chromosomes.

and has been implicated as the site of a putative tumour suppressor gene in bladder and breast cancer[19–21]. It is striking that no tumour of borderline malignant potential had abnormalities of chromosome 11. In contrast, abnormalities of chromosome 11 were present in 79% of the invasive tumours. This observation raises the possibility that genes located in chromosome 11 may be involved in the aetiology or biological aggressiveness of frankly invasive tumours.

Recently, several authors have reported the presence of non-random losses of genetic material from specific chromosomal segments in ovarian cancer too small to be seen on cytogenetic analysis. These molecular analytical studies may provide additional relevant information regarding the location of potential tumour suppressor genes. Lee and coworkers have reported the presence of frequent tumour-specific allelic losses on chromosome segments 6q, 11p, and 17p in human ovarian carcinomas[22]. Likewise, Ehlen and Dubeau have observed losses affecting loci on 3p, 6q, and 11p in a high percentage of ovarian tumours[23]. These data suggest that inactivation of genes located on these chromosomes may play a role in ovarian carcinogenesis. Of interest, these authors also noted that losses of chromosomes 3 and 11 were associated with high grade tumours[24]. This finding is consistent with our data and suggests that these chromosomes may control tumour functions resulting in a biologically aggressive phenotype.

Further cytogenetic and molecular genetic studies of ovarian carcinomas should not only provide insight into the specific genetic alterations involved in the pathogenesis of ovarian cancer but may also be useful in the identification of patients who are at high risk for developing this highly aggressive malignancy.

REFERENCES

1. DeVita, V.T., Jr, Hellman, S. and Rosenberg, S.A. (1985) *Cancer Principles and Practice in Oncology*, 2nd edn, J.B. Lippincott, Philadelphia, pp. 73–7.
2. Rowley, J.D. (1973) A new consistent chromosomal abnormality in chronic myelogenous leukemia identified by quinacrine fluorescence and Giemsa staining. *Nature*, **243**, 290.
3. Heim, S. and Mitelman, F. (1987) Chromosomal abnormalities in specific disorders, in *Cancer Cytogenetics*, Alan R. Liss, New York, pp. 41–141.
4. Rowley, J.D. (1983) Human oncogene locations and chromosome aberrations. *Nature*, **301**, 290.
5. Yunis, J.J. and Ramsay, N. (1978) Retinoblastoma and sub-band deletion of chromosome 13. *Am. J. Dis. Child.*, **132**, 161.
6. Riccardi, V.M., Sujansky, E., Smith, A.C. *et al.* (1978) Chromosomal imbalance in the Aniridia–Wilms' tumor association: 11p interstitial deletion. *Pediatrics*, **61**, 604–10.
7. Balaban, G., Gilbert, F., Nichols, W. *et al.* (1982) *Cancer Genet. Cytogenet.* **6**, 213.
8. Balaban-Malenbaum, G. *et al.* (1981) *Cancer Genet. Cytogenet.*, **3**, 243.
9. Benedict, W.F., Banerjee, A., Mark, C. *et al.* (1983) *Cancer Genet. Cytogenet.*, **10**, 311.
10. Cavence *et al. Nature*, **305**, 779.
11. Trent, J.M. and Salmon, S.E. (1979) Karyotypic analysis of human ovarian carcinoma cells cloned in agar. *Am. J. Hum. Genet.*, **31**, 112A.
12. Wake, N., Hreshchyshyn, M.M., Piver, S.M. *et al.* (1980) Specific cytogenetic changes in ovarian cancer involving chromosomes 6 and 14. *Cancer Res.*, **40**, 4512–8.
13. Whang-Peng, J., Knutsen, T., Douglass, E.C. *et al.* (1984) Cytogenetic studies in ovarian cancer. *Cancer Genet. Cytogenet.*, **11**, 91–106.
14. Panani, A. and Ferti-Passantonopoulou, A. (1985) Common marker chromosomes in ovarian cancer. *Cancer Genet. Cytogenet.*, **16**, 65–71.
15. Atkin, N.B. and Baker, M. (1987) Abnormal chromosomes including small metacentrics in 14 ovarian cancers. *Cancer Genet. Cytogenet.*, **26**, 355–61.
16. Tanaka, K., Boice, C.R. and Testa, J.R. (1989) Chromosome aberrations in nine patients with ovarian cancer. *Cancer Genet. Cytogenet.*, **43**, 1–14.
17. Kovacs, G. (1978) Abnormalities of chromosome No. 1 in human solid malignant tumours. *Int. J. Cancer.*, **21**, 688–94.
18. Rowley, J.D. (1977) Mapping of human chromosomal regions related to neoplasia: Evidence from chromosomes 1 and 17. *Proc. Natl Acad. Sci. USA*, **74**, 5729–33.
19. Weissman, E.E., Saxon, P.J., Pasquale, S.R. *et al.* (1987) Introduction of a normal human chromosome 11 into a Wilms' tumor cell line controls its tumorigenic expression. *Science*, **236**, 175–80.
20. Fearon, E.R., Feinberg, A.P., Hamilton, S.H. *et al.* (1985) Loss of genes on the short arm of chromosome 11 in bladder cancer. *Nature*, **318**, 377–80.
21. Ali, J.U., Lidereau, R., Theillet, C. *et al.* (1987) Reduction to homozygosity of genes on chromosome 11 in human breast neoplasia. *Science*, **221**, 185–8.
22. Lee, J.H., Kavanaugh, J.J., Wildrick, D.M. *et al.* (1990) Frequent loss of heterozygosity on chromosomes 6q, 11 and 17 in human ovarian carcinomas. *Cancer Res.*, **50**, 2724–8.
23. Ehlen, T. and Dubeau, L. (1991) Allelic losses and oncogene amplification in human ovarian carcinomas of different histopathologic grades. *Oncogene*, **5**, 219–23.
24. Ehlen, T., Robinson, W.R. and Zheng, J. (1990) Relationship between losses of heterozygosity on chromosomes 3, 6, 11 and grade of human ovarian carcinomas. Presented at the Fox Chase Cancer Center Conference, Marble Island, Vermont, October 1990.

Chapter 4
Molecular genetics of ovarian cancer

W. FOULKES, G. STAMP and J. TROWSDALE

4.1 INTRODUCTION

Several recent studies have begun to confirm ideas of multistep gene alterations in the development of adult epithelial malignancies. In particular, colon cancer, which has easily accessible, identifiable stages, has provided a framework for devising ways of investigating genes involved in ovarian malignancy[1]. This model proposes that colorectal tumours result from two types of genetic event: activation of oncogenes and inactivation of tumour suppressor genes. These mutations may be thought of as dominant, or recessive, respectively, although this situation is clearly more complex (see below). Tumours may need mutations in several different genes in order to become malignant. Less may be needed for benign lesions such as predisposing adenomas. It also appears that the order of the different mutations may not be as important as the overall accumulation. Mutations may be inherited and therefore present in the germline or may be acquired somatically, as the result of carcinogens, for example.

How well does ovarian malignancy fit this framework? First, mutations in dominant oncogenes such as *ras* have been found in ovarian tumours, as well as amplification of other oncogenes thought to provide a growth advantage, in a small number of studies so far[2–4]. Second, there is accumulating evidence for inactivation of tumour suppressor genes in ovarian cancer.

4.1.1 TUMOUR SUPPRESSOR GENES

The tumour suppressor genes have excited considerable interest. Since inactivation of genes by mutation is more frequent than activation, mutations in these genes may play a highly significant role in the development of a tumour. They may be picked up by observing loss of specific chromosomal regions in tumours. Such losses have been interpreted to signify loss of expression of products which normally regulate growth or differentiation and thus indirectly suppress tumour development. The statistical 'two-hit' hypothesis of carcinogenesis in retinoblastoma offered a verifiable model of how inactivation in tumours might occur[5]. Initially, the model did not state where the second 'hit' might occur. A priori the second gene change or 'hit' could occur on: (i) the same gene that suffered the first 'hit'; (ii) the corresponding allele on the homologous chromosome; or (iii) an entirely different gene[6]. The model was clarified to suggest that tumours can develop either from an inherited mutation in a gene on one chromosome 13, with subsequent loss of the other normal allele on the other chromosome, or a *de novo* somatic mutation occurs in the same gene in the retinoblasts of an otherwise normal individual and a tumour arises when loss of the normal allele takes place on the other chromosome[7]. Allele loss may occur by a number of different genetic mechanisms, including deletion and mitotic recombination. The presence of an inherited, germline mutation means that there is only one normal copy of the gene in all cells and therefore malignancy is likely to be a more frequent and earlier occurrence in these individuals. This is in fact the case for retinoblastoma. Later experimental work demonstrated that there was, in retinoblastoma, allele loss of markers on chromosome 13q14, adjacent to a region that had been shown to be constitutionally deleted by cytogenetic analysis[8]. Following this it has been shown that the gene retained in the tumour, or its product, is abnormal[9].

4.1.2 DOMINANCE AND RECESSIVITY

It is worth making two points here about dominance and recessivity in the context of tumorigenesis. First, these terms are usually used to describe effects at the cellular level. Thus, one can think of mutations in the *Rb* gene as being recessive to the normal gene product, since as stated above this must be lost or abnormal for the phenotype to be manifested. However, as far as is known, most individuals heterozygous for the *Rb* mutation and the normal allele develop retinoblastoma. The reason for this is probably that the cellular growth necessary for development of the retina means that loss of the normal copy of the *Rb* gene is statistically likely to occur in at least one cell. The second point is that the notion of recessivity of tumour suppressor genes is an oversimplication. It is difficult to envisage how, in the case of a sporadic tumour, cells incurring the first mutation would expand sufficiently to provide enough cells for the probability of loss of the second allele to take place. This problem may be overcome by proposing that the suppressor mutations provide a growth advantage even in the presence of the normal allele. There is evidence of this in the case of the p53 mutation, on chromosome 17p, which appears to behave in a dominant negative rather than a recessive mode[10]. There is also recent evidence of an inherited p53 mutation in the Li–Fraumeni syndrome[11].

4.1.3 FUNCTIONS OF CANCER GENES

The oncogenes and tumour suppressor genes of known function appear to be involved in regulating the growth, differentiation and division of cells[12]. Other genes may be important in subsequent progression of the tumour, and mutations in these may tend to occur at later stages in the tumour's development. It appears that malignancies at particular sites may share affected genes with other tumours; p53 alterations are found in lung, colon and breast cancer as well as other tumours such that they have been shown to be the commonest mutational event in human cancer[13]. In addition, there may be tissue-specific alterations, due to local cellular requirements. Which kinds of genes would be expected to be altered in ovarian cancer? Two speculations may be worth making: (i) it is not unreasonable to propose that hormonal pathways will be affected and (ii) some mutations may be common to both breast and ovarian cancer.

4.1.4 OVARIAN CANCER

With this framework, one can devise a strategy to search for genes which may be involved in the development of ovarian malignancies. We may expect to find examples of germline mutations, most likely in a subset of families with multiple cases of ovarian cancer. There will probably be several examples of allele loss associated with tumour suppressor genes, and we will find activated oncogenes. The benefits from knowledge of these genes may be considerable. If we can find genes with a germline involvement, we may then be in a better position to identify and to help those at risk. On the other hand, knowledge of genes with a somatic or germline association may help to develop therapies. We have started experimental work attempting to map genes involved in growth of ovarian tumours. In this paper we will review some of our work and the work of others, particularly investigating somatic allele loss in ovarian tumours, to search for tumour suppressor genes.

4.2 ALLELE LOSS

4.2.1 MATERIALS AND METHODS

In order to look for allele loss in ovarian tumours it is necessary to collect DNA from tumours as well as normal (usually blood) tissue samples from the same individual. The DNA is diagnosed with appropriate restriction enzymes, separated by electrophoresis and then blotted onto nitrocellulose membranes and probed with sequences from the chromosome areas of interest. The tumour tissue sample must be free of normal tissue for this procedure to be effective and should be from tumours of a defined histological type. In practice, unlike the situation for colorectal tumours, there is marked variation in histology in ovarian tumours, even for the common adenocarcinomas, which comprise about 80% of our samples. Each tumour is therefore examined histologically and its features are accurately recorded, along with details of the operative staging. This information may be useful at a later stage when it may be possible to uncover type- and grade-specific changes and when more may be known about early stages of the disease. Normal stromal tissue is carefully removed from the tumour wherever possible and the level of 'contamination' is recorded. Table 4.1 provides a list of the histopathological types and their frequencies from the first 30 samples that we analysed and records the range of percentages of tumour compared with stroma in each of these samples. This also emphasizes that not all samples are tumours and not all 'tumours' are cancer! These first 30 samples are an accurate representation of all the samples that we have collected so far. Examples of sectioned tumour material are shown in Figures 4.1 and 4.2. Our results are presented in the next section.

4.2.2 ALLELE LOSS ON 6q, 6p AND 17q

We have investigated 30 malignant tumour/normal DNA paired samples to date for allele loss. We have concentrated on chromosome 6, because of cytogenetic evidence for loss of part of the long arm in a significant proportion of ovarian cancers, and in one paper of a specific translocation[14]. We have also looked at 17q. Not all pairs of samples are informative for different probes; this depends on the presence of two different alleles in the sample, i.e. heterozygosity, detected with the particular probe/enzyme combination. An example of allele loss on a Southern blot is shown in Figure 4.3. Here we show an autoradiograph that has been probed with the oestrogen receptor probe, pOR3, which gives two alleles of 1.5 and 0.7 kb and five invariant bands. Loss of either of the two alleles is seen in the tumour sample of four specimens. Specimen 1 shows complete absence of the 1.5-kb allele, whereas in the other three specimens where allele loss is seen (8, 9 and 10), there is partial retention of the relevant band. This is probably due to the

Table 4.1 Epithelial ovarian tumours: subtype and stromal 'contamination'

Histological subtype	Number	Percent tumour in sample
Malignant tumours		
Solid undifferentiated adenocarcinoma*	7	50–90
Papillary undifferentiated adenocarcinoma	5	25–90†
Serous papillary adenocarcinoma	5	50–75
Mucinous adenocarcinoma	3	50–80
Endometrioid adenocarcinoma	2	50 & 80
Borderline tumours		
Serous	2	20 & 50
Mucinous	1	45
Miscellaneous specimens		
Mucinous cystadenoma	2	10 & 30
Leiomyoma	1	80
Endometriosis	1	–
Normal ovary	1	–

* 'Undifferentiated' does **not** imply an anaplastic, and therefore more malignant, phenotype.
† Where there is less than 50% tumour in the sample, true allele loss may be missed unless quantitative methods are used. See Figure 4.1.

stromal component. Alternatively, loss may not have occurred in all the tumour cells.

Our data are presented in Figure 4.4 (chromosome 6) and Figure 4.5 (chromosome 17). These figures demonstrate the allele loss that we have seen using one probe on each arm of chromosome 6 and one probe on 17q, lined up with the positions to which they map. On the right, we show the fractional loss, i.e. the proportion of samples showing loss (the top number) compared with the number of samples that are informative for that probe (the bottom number), i.e. the specimen is heterozygous for the marker used, and therefore loss of one allele can be detected (see Figure 4.3). Figure 4.4 shows that 6q is lost more frequently than 6p. The two tumours that show loss with the HLA probe also show loss with 6q and therefore it may be that the loss on 6p is due to the whole chromosome being lost and that 6q is 'driving' the loss. This needs to be investigated further. Figure 4.5 demonstrates a high frequence of loss using a marker (CMM86) that is thought to be closely linked with early-onset familial breast cancer[15]. The possible significance of this is discussed below.

4.2.3 COMPILATION OF ALLELE LOSS DATA

Epithelial ovarian cancer (EOC) has not been as extensively investigated as other adult epithelial malignancies and the first paper specifically concerned with allele loss in ovarian cancer was published in 1989 by Lee *et al.*[16]. In Table 4.2 we present a summary of findings from all evaluable published work, plus our own data. As for Figures 4.4 and 4.5 we present the data as proportional or fractional loss. Where there is a high degree of concordant loss of alleles on both arms it is difficult to know which event, if any, is primary. One loss may make the other more likely or they may be completely independent events. In the paper by Lee and

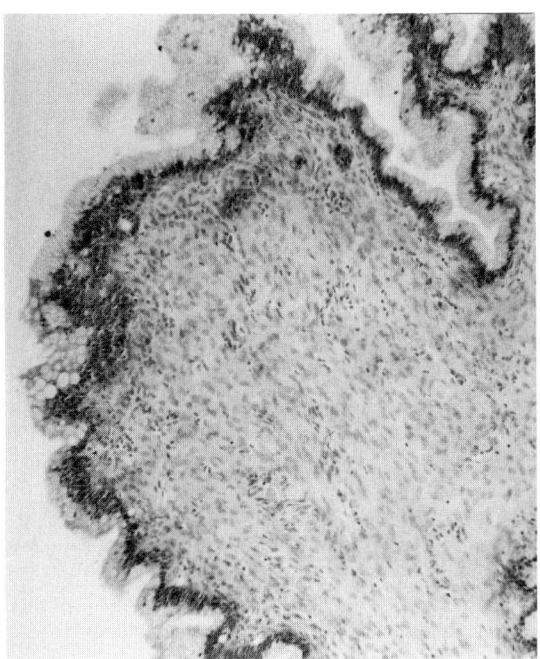

Figure 4.1 Medium power view of borderline mucinous tumour. Note the high proportion of stroma to the surface mucinous epithelium, which is the neoplastic component of such tumours.

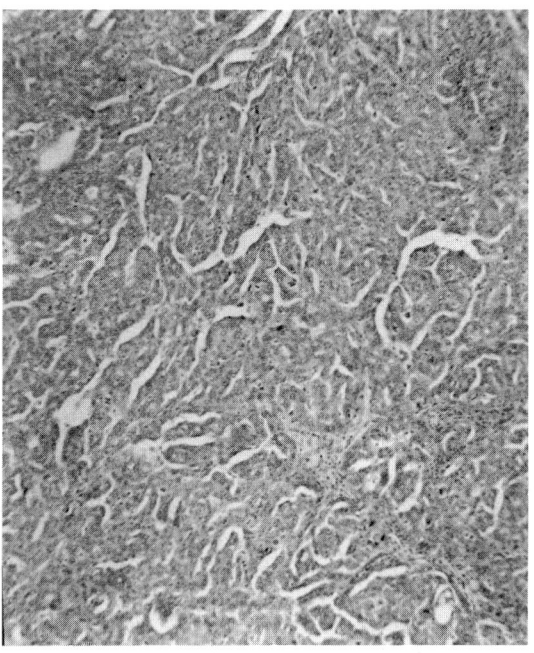

Figure 4.2 Low power view of solid undifferentiated adenocarcinoma. This specimen shows much less stroma than Figure 4.1 and is comprised of cells that are attempting, in some areas, to form a papillary architecture. The cells are not following any specific differentiation pathway.

colleagues, 50% loss at the c-Ha-*ras*1 locus on 11p14-15 in 10 informative cases of EOC was seen. This is interesting for several reasons. Firstly, the *ras* family of genes are thought to be transforming in a dominant fashion, as discussed above, whereas allele loss at this site would at least suggest a dominant negative effect. Secondly, this region is adjacent to a gene at 11p13, *WT1*, that fits many of the criteria for a tumour suppressor gene[17] which is expressed in the genital ridge, fetal gonad and mesothelium and therefore might be expected to be involved with ovarian development[18]. The kidney and ovary have common embryological origins and they may share tumour suppressor genes. No studies have yet been published on *WT1* in EOC.

Ehlen and Dubeau confirmed these results and also added new data on chromosomes 3p, 5p, 6 and 21q[19]. In a small series they showed slightly less allele loss at 11p15 than Lee *et al.* (5/15). On 3p, samples that were informative for both probes used showed preferential loss distal to 3p21 (3/4). The commonest molecular abnormality in renal cell cancer is allele loss of markers on 3p[20], but it would appear that the breakpoints that have been seen cytogenetically in renal cell cancer are too proximal to include the putative ovarian tumour suppressor gene[21], but the numbers are so far very small. On chromosome 6 they used three probes, one of which definitely maps to 6q, whereas the other two were designated 6p21-qter. The 6q probe, hER, was lost in only 2 out of 7 samples, but 7 out of 12 specimens (including

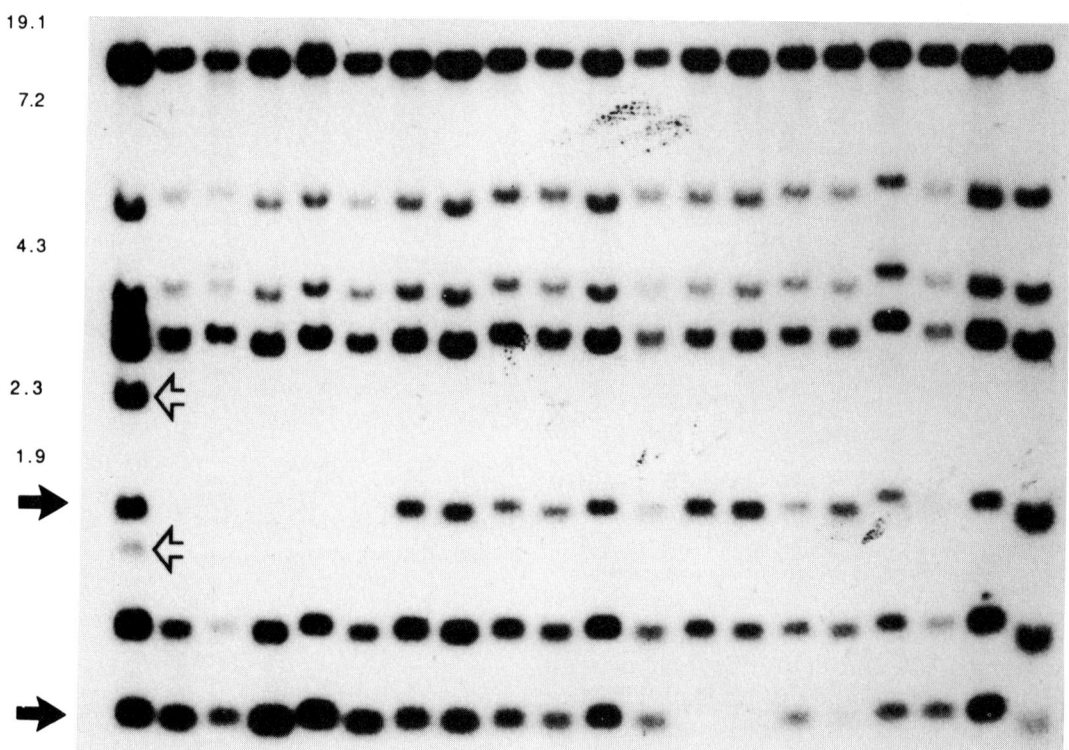

Figure 4.3 Autoradiograph of Southern blot of DNA from normal–tumour (N,T) pairs of samples. The filter was probed with ^{32}P-labelled pOR3, mapping to 6q24–27 and exposed for 110 hours. Molecular weight markers (in kilobases) are shown along the left hand side of the figure. The polymorphic bands are denoted by filled-in arrows. Allele loss is seen in samples 1,8,9 and 10. Sample 1N is a partial digest, denoted by open arrows. Sample 10T shows a small signal from the lowest band: this probably due to stroma (see text and Table 4.1).

one from the above group) showed loss with the undesignated probes. Until more is known about these two probes' location, one cannot fully interpret the results. We have been able to show, however, that the loss is mainly on 6q rather than 6p, where, for example, the major histocompatibility complex genes lie (see Figure 4.4). Indeed, there were no tumours where 6p was lost without 6q also being lost. Figure 4.4 also shows that there was no allele loss seen in the benign or borderline tumours. In addition to the above chromosomes, Ehlen and Dubeau studied 5p and 21q and did not see any loss on either chromosome.

Lee and colleagues, continuing their work

SAMPLE Totals	Malignant 23	Benign 6	Borderline 4	Normal 1
	Allele loss / Informative	Allele loss / Informative	Allele loss / Informative	Allele loss / Informative
MHC; HLA loci	2/14	0/3	0/3	0/1
pOR3	8/14	0/3	0/3	0/1

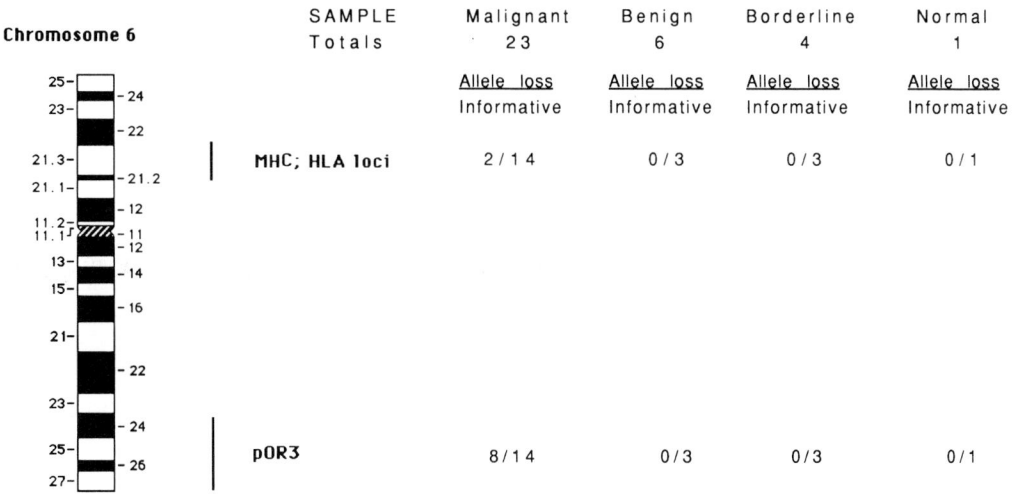

Figure 4.4 A representation of allele loss seen on chromosome 6 from our own work. The bars show the position of the probes, and the proportion of loss seen is shown on the right.

SAMPLE Totals	Malignant 30	Benign 6	Borderline 4	Normal 1
	Allele loss / Informative	Allele loss / Informative	Allele loss / Informative	Allele loss / Informative
CMM86	17/29	0/3	0/3	0/1

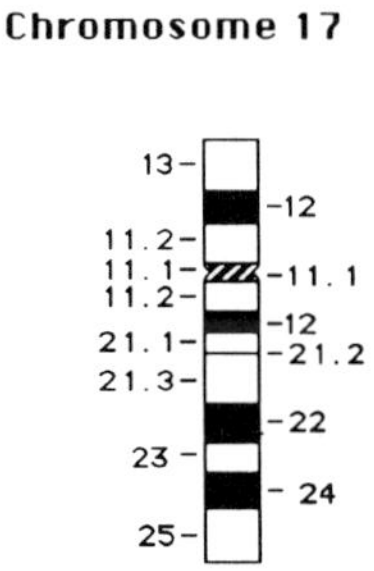

Figure 4.5 A representation of allele loss seen on chromosome 17 from our own work. The small bar shows the position of the probe used.

on ovarian cancer, demonstrated similar loss on 11p as in their previous paper, but also studied 1q, 2q, 5q, 6p, 6q, 7q, 11q, 12p, 17p, 17q, 19q and 22q[22]. They found no significant loss on 1q, 2q, 5q, 6p, 7q, 11q, 12p, 19q or 22q. They did however note losses above 60% on 6q and 17p. They also found that although the loss at D17S26 on 17q was only 31%, when tumours informative for both probes were analysed, 6 out of 14 (43%) had loss on both arms of this chromosome. Their most interesting finding was that, like the data on 3p discussed above, on 6q, allele loss using the probes c-*myb* (6q22-23) and pOR3 (6q24-27) was preferentially seen distally. This is potentially interesting as pOR3 is a cDNA probe within the oestrogen receptor (ER). Although tumour ER levels are thought to be important in breast cancer, no rearrangement or amplification of the ER gene was seen by Southern blot analysis, nor was there significant allele loss at this locus in a

Table 4.2 Epithelial ovarian cancer: compilation of allele loss data*

Chromosome	Arm p	Arm q	Concordant loss of p+q arms
Allelle loss >35% on one or other arm of the relevant chromosome			
3	4/7	–	–
6	2/23	21/35	2/22
11	14/30	0/6	0/3
17	15/29	40/69†	10/18

Allele loss <25%
Chromosomes 1q, 2q, 5p, 5q, 7q, 12p, 13q, 19q, 21q, 22q in addition to the data shown above‡

* Data have been compiled from our own unpublished results plus those from published work. In each study, if more than one probe for each arm has been used, we have taken the highest value for allele loss. Refer to the text for the source of the data.
† The allele loss seen here is lower than that seen in the two most recent publications. For details, see text.
‡ For references, see text.

series of breast cancers[23]. This notwithstanding, there is no doubt that ER remains a candidate gene in ovarian cancer.

In view of the involvement of the retinoblastoma gene, *Rb*, in a number of adult cancers, Sansano and colleagues studied EOC but failed to find any involvement of this gene[24]. However, Li *et al.* found 30% allele loss at the *Rb* locus in 20 tumours. They also noted a diminished amount of *Rb* protein in 3 out of 10 fresh tumour samples[25]. These conflicting results require clarification.

Two groups from Belfast and Edinburgh have recently provided very strong evidence for the presence of a tumour suppressor gene distally on chromosome 17q[26,27]. They both used the highly polymorphic probe pTHH59, which has now been mapped to 17q25.1. Taken together there was 69% allele loss in 26 EOC samples. Our own data using CMM86 mapping to 17q22 and that of Lee *et al.* using D17S26 at 17q25.2 show less frequent allele loss in a way that suggests that the putative gene may well be between CMM86 and pTHH59. This explains why, in Table 4.2, the proportion of EOC with loss is lower than in these two recent papers. The very recent data that show linkage to 17q in early onset familial breast cancer[15] make this a fascinating area to be working in and given the fact that breast and ovarian cancer often coexist in breast cancer-prone families, may mean that a gene for familial ovarian cancer is situated here.

All these results taken together underline the notion that a large number of chromosomes may harbour tumour suppressor genes. From Table 4.2 it seems that 6q and 17q are the most frequently lost chromosome arms but with more specific probes and larger numbers of samples, 3p, 11p and 17p may also prove to be lost in a higher percentage of EOC. Whilst some tumour suppressor genes may be unique to a particular cancer, many are involved in numerous tumours and may therefore play a central role in initiation or maintenance of the malignant phenotype. We would like to map the regions of maximum loss in EOC more fully and isolate the genes that are lost or mutated.

4.3 ONCOGENE ACTIVATION

Oncogenes can become carcinogenic by the overexpression of a normal product without amplification (e.g. by loss of a controller element or increased RNA stability, etc.); secondly by amplification of the gene itself; and finally by mutation of the gene, such that the product has a greater transforming capability.

A number of oncogenes have been studied in EOC, but the results have generally not been as dramatic as those seen when looking for allele loss. However, a number of observations appear to have particular relevance. Gullick[28] noted that c-*erb*B2, Ki-*ras* and c-*myc* appeared to be involved in significant numbers of EOCs. Since this review, his own group and others have confirmed these findings[2,29,30].

Others have attempted to find out the mechanism by which these oncogenes and their products act. Smith and colleagues, noting the increased expression of *fos*, *myc*, Ha-*ras* and Ki-*ras* analysed 14 serous adenocarcinomas to see if the increased expression was due to amplification or rearrangement. Using restriction endonucleases and DNA-specific probes they found no evidence of either genetic rearrangement or amplification[3]. Of course, with these techniques one is not looking at the physical structure of the gene itself, unless the restriction enzymes happen to cut within an altered part of the gene, which is a highly unlikely event. Therefore it is important to look for mutations within the gene itself. Enomoto and colleagues[31] investigated EOC by polymerase chain reaction amplification of specific Ki-*ras* gene sequences. They found mutations in codon 12 of this gene in only 2 out of the 13 EOCs studied. In addition 7 EOCs were assessed for transforming activity by NIH 3T3 transfection, but none were positive, and therefore it seems from this study that Ki-*ras* mutations are not generally the cause of the overexpression seen. The significance of these findings is uncertain at the moment.

4.4 GERMLINE MUTATIONS AND INHERITED PREDISPOSITION

The studies detailed above describe how a start has been made in searching for genes involved in ovarian cancer. When more data have been obtained, and the genes uncovered by allele loss identified, it will be desirable to search for evidence of inherited mutations. However, not all tumour suppressor genes would be expected to have inherited counterparts, because their effects may be deleterious to the development of the whole organism.

Between 1 and 10% of all EOC is inherited[32,33]. Data so far collected would suggest that a single dominant gene is responsible for the EOC in these families, but most familial ovarian cancer occurs in families where other tumours also occur in members and different genes may be involved in different families, e.g. early and late onset familial breast cancer do not appear to be linked to the same chromosomal region[15]. Another problem is that in some families where a susceptibility gene has been inherited there may be cases of ovarian cancer that are not actually caused by inheritance of this gene, but might reasonably be thought to be so caused. This may obscure linkage and confuse analysis. However, despite these caveats, family data will be very useful in isolating regions where tumour suppressor genes may lie. The way to proceed would be to first identify families with at least two affected members and attempt to carry out linkage studies with candidate probes, rather than looking randomly through the genome.

Partly due to the aggressive nature of EOC, large informative families are hard to come by and national and possibly international collaboration may be required to accumulate sufficient data. In the UK, the Cancer Family Study Group is coordinating the collecting of such families and members have also been collaborating with North American colleagues[34]. Hopefully these collaborative efforts will lead to the sort of advances that have been seen recently for other family cancers[11,15]. Obviously the hope is that demonstrating linkage will lead to the gene(s) involved and this in turn will result in improved screening of those at particularly high risk, earlier diagnosis and better treatment.

4.5 CONCLUSIONS

The genetic alterations that lead to ovarian cancer are still relatively unknown compared with other adult tumours. The reasons for this have been outlined. We have reviewed the published work on allele loss and have added some data from our laboratory. So far it appears that chromosomes 3p, 6q, 11p and 17 show the greatest degree of allele loss, but much of the work remains preliminary and more probes need to be used to try to localize the regions of maximum loss accurately. Oncogenes are also involved in the pathogenesis of EOC but it is not yet clear how important a role they play. There is strong evidence for the presence of inherited genes that predispose to ovarian cancer and, as in colorectal cancer, the collection of such family data may be crucial in isolating genes that are involved in both inherited and non-inherited ovarian cancer. Whatever the outcomes of this work are, there is no doubt that an increase in understanding of the basic biology of this disease will ultimately benefit future patients.

ACKNOWLEDGEMENTS

Without the help of many hospitals in and around London, especially St Thomas's, University College, St Bartholomew's and Barking Hospitals, our own work would not have started. We are therefore indebted to all those who are helping us with this study. We would particularly like to thank Miss K.S. Raju for her enthusiasm and continued support. We have had a number of illuminating discussions with Dr Tim Bishop.

REFERENCES

1. Fearon, E.R. and Vogelstein, B. (1990) A genetic model for colorectal tumourigenesis. *Cell*, **61**, 759–67.
2. Slamon, D.J., Godolphin, W., Jones, L.A. *et al.* (1989) Studies of the HER-2/*neu* proto-oncogene in human breast and ovarian cancer. *Science*, **244**, 707–12.
3. Smith, D.M., Groff, D.E., Pokul, R.K. *et al.* (1989) Determination of cellular oncogene rearrangement amplification in ovarian adenocarcinomas. *Am. J. Obstet. Gynecol.*, **161**, 911–15.
4. Yokota, J., Tsunetsugu-Yokota, Y., Battifora, H. *et al.* (1986) Alterations of *myc*, *myb* and *ras*^{HA} proto-oncogenes in cancers are frequent and show clinical correlation. *Science*, **231**, 261–5.
5. Knudson, A.G. (1971) Mutation and cancer: statistical study of retinoblastoma. *Proc. Natl. Acad. Sci.*, **68**, 820–3.
6. Knudson, A.G., Meadows, A.T., Nichols, W.W. and Hill, R. (1976) Chromosomal deletions and retinoblastoma. *N. Engl. J. Med.*, **295**, 1120–3.
7. Knudson, A.G. (1978) Retinoblastoma: a prototypic hereditary neoplasm. *Semin. Oncol.*, **5**, 57–60.
8. Cavenee, W.B., Dryja, T.P., Phillips, R.A. *et al.* (1983) Expression of recessive alleles by chromosomal mechanisms in retinoblastoma. *Nature*, **305**, 779–84.
9. Friend, S.H., Bernards, R., Rogelj, S. *et al.* (1986) A human DNA segment with properties of a gene that predisposes to retinoblastoma and osteosarcoma. *Nature*, **323**, 643–6.
10. Nigro, J.M., Baker, S.J., Preisinger, A.C. *et al.* (1989) Mutations in the p53 gene occur in diverse human tumour types. *Nature*, **342**, 705–8.
11. Malkin, D., Li, F.P., Strong, L.C. *et al.* (1990) Germ line p53 mutations in a familial syndrome of breast cancer, sarcomas and other neoplasms. *Science*, **250**, 1233–8.
12. Sager, R. (1989) Tumor suppressor genes: the puzzle and the promise. *Science*, **246**, 1406–12.
13. Harris, A.L., (1991) Telling changes of base. *Nature*, **350**, 377–8.
14. Wake, N., Hreshchyshyn, M.M., Piver, S.M. *et al.* (1980) Specific cytogenetic changes in ovarian cancer involving chromosome 6 and 14. *Cancer Res.*, **40**, 4512–18.
15. Hall, J.M., Lee, M.K., Newman, B. *et al.* (1990) Linkage of early onset familial breast cancer to chromosome 17q21. *Science*, **250**, 1684–9.
16. Lee, J.H., Kavanagh, J.J., Wharton, J.T. *et al.* (1989) Allele loss at the c-Ha-*ras*1 locus in human ovarian cancer. *Cancer Res.*, **49**, 1220–2.
17. Call, K.M., Glaser, T., Ito, C.Y. *et al.* (1990)

Isolation and characterization of a zinc finger polypeptide gene in the human chromosome 11 Wilms' tumor locus. *Cell*, **60**, 509–20.

18. Pritchard-Jones, K., Fleming, S., Davidson, D. *et al.* (1990) The candidate Wilms' tumour gene is involved in genitourinary development. *Nature*, **346**, 194–6.
19. Ehlen, T. and Dubeau, L. (1990) Loss of heterozygosity on chromosomal segments 3p, 6q and 11p in human ovarian cancer. *Oncogene*, **5**, 219–23.
20. Morita, R., Ishikawa, J., Tsutsumi, M. *et al.* (1991) Allelotype of renal cell carcinoma. *Cancer Res.*, **51**, 820–3.
21. Cohen, A.J., Li, F.P., Berg, S. *et al.* (1979) Hereditary renal cell carcinoma associated with a chromosomal translocation. *N. Engl. J. Med.*, **301**, 592–5.
22. Lee, J.H., Kavanagh, J.J., Wildrick, D.M. *et al.* (1990) Frequent loss of heterozygosity on chromosomes 6q, 11 and 17 in human ovarian carcinomas. *Cancer Res.*, **50**, 2724–8.
23. Koh, E.H., Ro, J. Wildrick, D.M. *et al.* (1989) Analysis of the estrogen receptor gene structure in human breast cancer. *Anticancer Res.*, **9**, 1841–6.
24. Sasano, H., Comerford, J., Silverberg, S.G. and Garrett, C.T. (1990) An analysis of abnormalities of the retinoblastoma gene in human ovarian and endometrial carcinoma. *Cancer*, **66**, 2150–4.
25. Li, S.B., Lee, E.Y.-H., Schwartz, P.E. *et al.* (1990) High frequency of allelic loss at retinoblastoma locus in human ovarian cancer. *Am. J. Hum. Genet.*, **3** (Suppl. 1), A11 1.15.
26. Russell, S.E.H., Hickey, G.I., Lowry, W.S. *et al.* (1990) Allele loss from chromosome 17 in ovarian cancer. *Oncogene*, **5**, 1581–3.
27. Eccles, D.M., Cranston, G., Steel, C.M. *et al.* (1990) Allele loss on chromosome 17 in human epithelial ovarian cancer. *Oncogene*, **5**, 1599–601.
28. Gullick, W.J. (1990) The role of oncogenes in ovarian cancer, in *Ovarian Cancer. Biological and Therapeutic Challenges* (eds F. Sharp, W.P. Mason and R.E. Leake), Chapman & Hall, London, pp. 63–8.
29. Haldane, J.S., Hird, V., Hughes, C.M. and Gullick, W.J. (1990) c-*erb*B2 oncogene expression in ovarian cancer. *J. Pathol.*, **162**, 231–7.
30. Schreiber, G. and Dubeau, L. (1990) C-*myc* proto-oncogene amplification detected by polymerase chain reaction in archival human ovarian carcinomas. *Am. J. Pathol.*, **137**, 653–8.
31. Enomoto, T., Inoue, M., Perantoni, A.O. *et al.* (1990) K-*ras* activation in neoplasms of the human female reproductive tract. *Cancer Res.*, **50**, 6139–45.
32. Ponder, B.A.J., Easton, D.F. and Peto, J. (1990) Risk of ovarian cancer associated with a family history: preliminary report of the OPCS study, in *Ovarian Cancer. Biological and Therapeutic Challenges* (eds F. Sharp, W.P. Mason and R.E. Leake), Chapman & Hall, London, pp. 3–6.
33. Lynch, H.T., Conway, T. and Lynch, J. (1990) Hereditary ovarian cancer, in *Ovarian Cancer. Biological and Therapeutic Challenges* (eds F. Sharp, W.P. Mason and R.E. Leake), Chapman & Hall, London, pp. 7–19.
34. Lynch, H.T., Watson, P., Bewtra, C. *et al.* (1991) Hereditary ovarian cancer. Heterogeneity in age at diagnosis. *Cancer*, **67**, 1460–6.

Chapter 5

Molecular pathogenesis of ovarian cancer

W.S. LOWRY, S.E.H. RUSSELL, G.I. HICKEY and R.J. ATKINSON

Cytogenetic and molecular analysis of ovarian tumour cell lines and fresh ovarian tumour specimens have shown a number of structural genetic abnormalities. These include amplification of the *erb*B2 oncogene on chromosome 17q in 30% of cases[1] and areas of tumour-specific allele loss on 3p, 6q, and 11p in small numbers of cases[2,3].

Investigations in our own unit began with characterization of several ovarian tumour-derived cell lines. Karyotype analysis suggested that abnormalities of chromosome 17 were common.

5.1 ALLELE LOSS

We used variable number tandem repeat (VNTR) probes to analyse DNA from a bank of over 50 primary ovarian tumours and peripheral blood lymphocyte matched control DNAs in a search for allele loss at loci on chromosomes 1, 11 and 17.

Approximately 38% of tumours showed allele loss for 17p13. This probably reflects inactivation of the tumour suppressor gene p53 and demonstrates that, as in other tumour types, loss of function of this gene also contributes to the pathogenesis of ovarian cancer.

Allele loss was also found on the long arm of chromosome 17, 17q23-25 in 63% mixed ovarian tumours[4]. When benign and borderline tumours were excluded, 81% of cases of ovarian cancer carried this lesion. This high rate of allele loss is well above expected 'background loss' and is strongly suggestive of inactivation of a new tumour suppressor gene.

Mapping to the short arm of chromosome 11 (11p15) identified allele loss in approximately 45% of cases. Initial mapping of chromosome 1p only demonstrated allele loss in approximately 20% of cases.

The findings on 17q have been confirmed in an independent investigation by colleagues in the ICRF Medical Oncology Unit and the MRC Human Genetics Unit in Edinburgh[5]. Similar results have been obtained by the Aberdeen group. Currently a collaborative study is underway between all three centres and the results from a tumour bank of over 150 tumour specimens will shortly be pooled enabling a more detailed analysis of pathological subtypes.

Gradually an allelotype is emerging for ovarian cancer but the predominant abnormality appears to be allele loss on 17q, and in some cases perhaps loss of the entire chromosome. By correlating the molecular data with the clinical, pathological and epidemiological findings, it is now possible to construct a working hypothesis for the pathogenesis of ovarian cancer.

5.2 INCESSANT OVULATION

In 1971 Fathalla[6] proposed that 'incessant ovulation' is a risk factor in ovarian neoplasia. In a study of comparative pathology, he noted that the human female is very extravagant with her ova compared with other mammals. Social conditions not only render the majority of ovulations purposeless, but allow relatively infrequent non-ovulatory rest periods. Fathalla suggested that the repeated minor trauma to the epithelial surface of the ovary caused by frequent ovulation results in increased risk of ovarian cancer. In contrast, other mammals have lower rates of ovulation and lower risk of ovarian neoplasm. Again, he noted that contrary to the situation in the ovary the related surface epithelium of the testis does not show any neoplastic potential.

5.3 OVARIAN PATHOPHYSIOLOGY

Most cases of ovarian cancer arise on the epithelial surface of the ovary, not the ovary itself. In the normal course of events, the female child has all her oocytes present at birth. These are quiescent for about 13 years. With the onset of menarche, the oocytes mature. The epithelial covering of the ovary is thereafter ruptured at regular intervals to release the ovum at ovulation throughout the reproductive years. Immediately prior to rupture, the normal ovarian follicle can be up to 2 cm diameter. Following rupture these epithelial tears are repaired on each occasion by the process of cell division. When repair is complete, growth is switched off, leaving the surface of the ovary pitted with minute epithelial scars. It is proposed that this repetitive proliferation allows promotion of cells already bearing allele loss and, by inference, those carrying inactivated tumour suppressor genes. This may lead to uncontrolled cell division and malignant transformation.

5.4 EPIDEMIOLOGICAL DATA

An increasing number of epidemiological studies are consistent with the hypothesis. Fathalla originally noted that pregnancy suppresses ovulation and protects against ovarian cancer. Even incomplete pregnancies protect, and the protective effect increases with increasing parity[7,8]. Oral contraception inhibits ovulation and is also associated with decreased risk of ovarian cancer[9,10]. This reduction is related to the length of use of oral contraception. Furthermore, the incidence of ovarian cancer peaks after the menopause and completion of maximum ovulatory cyclical damage[11]. The effect is demonstrated graphically: Figure 5.1 shows the typical straight-line graph of increasing incidence with age seen in most solid cancers; Figure 5.2 shows the same curve for ovarian cancer with a flattening around the time of menopause[12]. More recently, Cruickshank[13] predicted that as ovulation occurs more often in the right ovary, one might expect ovarian cancer to develop more commonly on the right side. In a study of women with epithelial ovarian cancer in the Grampian region of Scotland, he confirmed the prediction. Significantly more tumours were found in the right ovary than in the left ovary.

In an independent review of the risk factors in ovarian cancer, Pike[12] concluded that the major impetus to cell replication is repair of the epithelial surface after ovulation. He noted that each month cells divide to seal

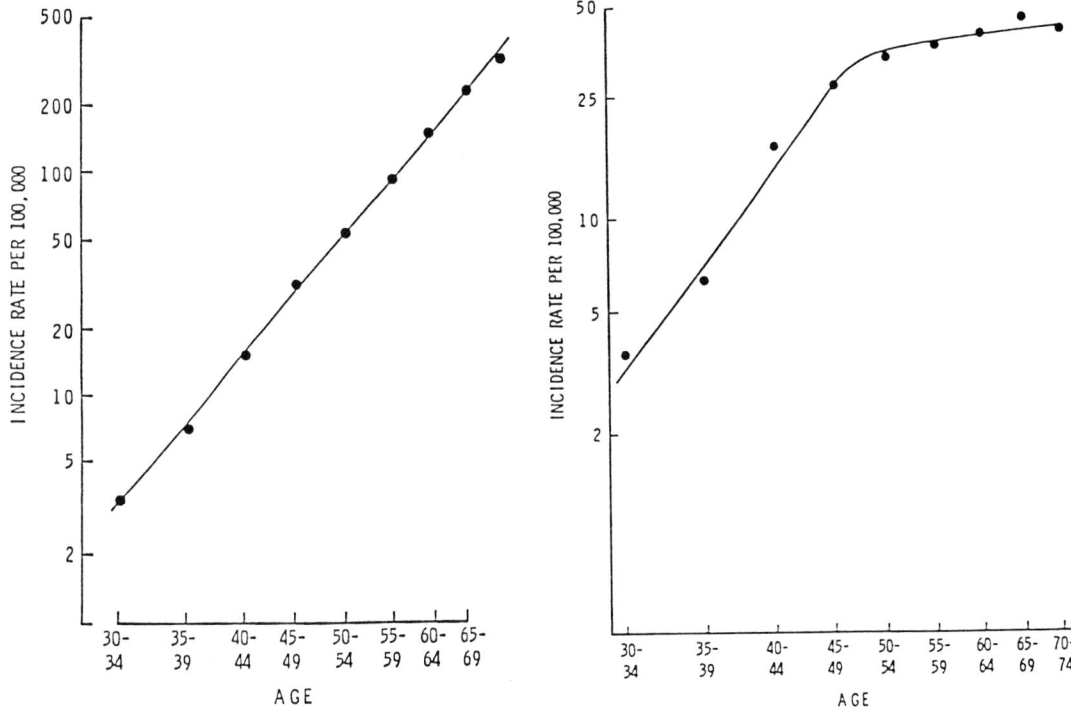

Figure 5.1 Age–incidence curve of colorectal cancer (after Pike [12], reproduced with permission).

Figure 5.2 Age–incidence curve of ovarian cancer (after Pike [12], reproduced with permission).

the hole: 'not that cell division itself causes cancer, but whatever does is made worse by cell division'. His colleague, Henderson[14] added that division of cells that would ordinarily not be replicating may be especially error prone, because efficient repair mechanisms are absent. The epidemiological evidence thus points to defective cellular repair following ovulation as a major risk factor in ovarian cancer. The molecular evidence suggests that this may be due to inactivated tumour suppressor genes. Taken together these factors may lead to malignant transformation.

5.5 CONCLUSION

Ovarian cancer embraces a variety of different tumours, even among the epithelial group, and one would not necessarily expect to find a single causative factor for all cases. Nevertheless the finding of allele loss in over 80% of cases is one of the highest reported for sporadic cancers. By itself allele loss is probably not sufficient to cause ovarian cancer. The underlying genetic changes are almost certainly more complex.

The finding of a genetic abnormality in a common tumour raises questions about the definition of malignancy. In our series of cases, allele losses on 17q were found in two cases described as borderline malignancy and in two others described as benign. This suggests that allele loss may be a very early change in the chain of events leading to malignancy. Alternatively, it may be that although the lesion is not recognizable morphologically as malignant, it may already be transformed in biochemical genetic terms. It will be of interest to follow the clinical course of such patients and correlate the molecular pathology with prognosis and other data as the study progresses.

Although the combined body of data is increasingly consistent with the Fathalla hypothesis, the evidence is not yet totally conclusive[15]. It is tempting therefore to look at alternative hormonal models. The ovary was phylogenetically developed for reproduction. The cycle of incessant ovulation, frequent menstruation, and intermittent breast stimulation is largely wasteful. Cell proliferation is continually being turned on and off in the target organs. It is therefore perhaps not surprising that malignant diseases of the reproductive organs (ovary, endometrium, breast) all occur more frequently in nulliparous women, and share other common aetiological endocrine features. Allele loss on chromosome 17 has been described in some cases of breast cancer, and we are currently looking for similar defects in endometrial cancer. As more information becomes available on breast and endometrium, the common epidemiological features may yet be linked together at a molecular level creating a unifying hypothesis for all three diseases.

REFERENCES

1. Slamon, D.J., Godolphin, W., Lovell, A.J. et al. (1989) Studies of the HER-2/neu proto-oncogene in human breast and ovarian cancer. *Science*, **244**, 707–12.
2. Lee, J.H., Kavanagh, J.J., Wharton, J.T. et al. (1989) Allele loss at the c-Ha-*ras* 1 locus in human ovarian cancer.
3. Ehlen, T. and Dubeau, L. (1990) Loss of heterozygosity on chromosomal segments 3p, 6q and 11p in human ovarian carcinomas. *Oncogene*, **5**, 219–23.
4. Russell, S.E.H., Hickey, G.I., Lowry, W.S. et al. (1990) Allele loss from chromosome 17 in ovarian cancer. *Oncogene*, **5**, 1581–3.
5. Eccles, D.M., Cranston, G., Steele, C.M. et al. (1990) Allele losses on chromosome 17 in human epithelial ovarian carcinoma. *Oncogene*, **5**, 1599–601.
6. Fathalla, M.F. (1971) Incessant ovulation – A factor in ovarian neoplasia? *Lancet*, **ii**, 163.
7. Beral, V., Fraser, P. and Chilvars, C. (1978) Does pregnancy protect against ovarian cancer? *Lancet*, **i**, 1083–7.
8. Cramer, D.W., Hutchinson, G.B., Welch, W.R. et al. (1983) Determinants of ovarian cancer risk. Reproductive experiences and family history. *J. Natl Cancer Inst.*, **71**, 711–16.
9. Casagrande, J.T., Pike, M.C., Ross, R.K. et al. (1979) Incessant ovulation and ovarian cancer. *Lancet*, **ii**, 170–3.
10. Cramer, D.W., Hutchinson, G.B., Welch, W.R. et al. (1982) Factors affecting the association of oral contraceptives and ovarian cancer. *N. Engl. J. Med.*, **307**, 1047–51.
11. Piver, M.S. (1987) Epidemiology of ovarian cancer, in *Ovarian Malignancies*, Churchill Livingstone, New York, p. 1.
12. Pike, M.C. (1990) in *Accomplishments in Cancer Research 1989* (eds J.G. Fortner and J.E. Rhodes), J.B. Lippincott, Philadelphia, pp. 327–56.
13. Cruickshank, D.J. (1990) Aetiological importance of ovulation in epithelial ovarian cancer: a population based study. *Br. Med. J.*, **301**, 524–5.
14. Henderson, B.E. and Preston-Martin, S. (1990) Increased cell proliferation as a cause of human cancer. *Proc. Am. Assoc. Cancer Res.*, **31**, 513–15.
15. Beral, V. (1987) The epidemiology of ovarian cancer, in *Ovarian Cancer – The Way Ahead* (eds F. Sharp and W.P. Soutter), Royal College of Obstetricians and Gynaecologists, London, p. 26.

Chapter 6

Molecular structure and clinical applications of a cancer-associated mucin

J. TAYLOR-PAPADIMITRIOU, D. ALLEN,
M. GRANOWSKA, N. PEAT, T. DUHIG,
A. SPICER, J. BURCHELL and S.J. GENDLER

6.1 INTRODUCTION

In the field of oncology the importance of identifying tumour-associated antigens for the diagnosis and therapy of cancer is well established. For some time, antibodies directed to these antigens have been used for detection of circulating antigen and for *in vivo* imaging of tumours, particularly ovarian cancer. More recently the potential of the antigens in therapy has been investigated, either using antibodies directed to them for targeting drugs, toxins or high dose radioactivity, or using the antigens themselves in the immunotherapy of cancer.

A tumour-associated antigen which has received much attention is a mucin produced by breast, ovarian and other adenocarcinomas. We have studied the mucin at the molecular level, and defined in detail the epitopes recognized by the monoclonal antibodies directed to it which have been and are being used for imaging of ovarian cancer. The mucin, which we have called polymorphic epithelial mucin (PEM) is the product of the *MUC*1 gene. The core protein of the mucin is a transmembrane protein with most of the extracellular domain being made up of tandem repeats, each of which has several potential glycosylation sites. The epitopes recognized by the HMFG-1, HMFG-2 and SM-3 antibodies, all of which have been used to image ovarian cancer, are core protein epitopes and fall within the tandem repeat between potential glycosylation sites (Table 6.1).

PEM is expressed on many simple epithelial cells, but the expression appears to be upregulated in adenocarcinomas. However, since the carbohydrate sidechains found on the tumour-associated mucin are shorter than those found on the normally processed mucin, core protein epitopes are unmasked and novel epitopes are found in the carbohydrate sidechains of the cancer-associated mucin. Thus, because of the differences in glycosylation, the tumour-associated mucin is antigenically different from the normal mucin, even though the core protein

Table 6.1 Epitopes recognized by antibodies reactive with the PEM (*MUC1*) tandem repeat†

Antibodies	Epitope
BrE-2	TRP
BrE-3	TRP
RINA 9/22	DTR
RINA 5/2	DTR
F36/22	RPAP
HMFG-1	PDTR
HMFG-2	DTR
SM-3	PDTRP

† Antibodies examined in the 1st International Workshop on Carcinoma-associated Mucins, San Francisco, November 1990.

```
  V T S A P D T R P A P G S T A P P A H G  (tandem
                                            repeat sequence)
      *       *         *     * *
```

* Potential glycosylation sites on *MUC1* tandem repeat sequence.

expressed by normal tissues and by cancer cells is the same. Since the cancer-associated mucin is antigenically distinct from the corresponding normal product, it should be possible (i) to develop antibodies which show selective reactivity with the mucin, and (ii) to use in immunotherapy the cancer-associated mucin, carrying novel, possibly non-self, epitopes. In this chapter we will discuss the molecular biology of PEM, and progress which is being made towards using the mucin, and antibodies to it, for therapy of breast and ovarian cancer.

6.2 GENERAL FEATURES OF MUCINS

Mucins are large molecular weight (250 000 to >1 million), heterogeneous glycoproteins which are composed of 50–90% carbohydrate O-linked through N-acetylgalactosamine to serine and/or threonine, characteristics which make the proteins difficult to analyse biochemically. They are present on the apical surfaces of highly polarized, secretory epithelial cells and are aberrantly expressed by most carcinomas, including ovarian. Largely because of their complexity, the detailed structure of the mucins has been difficult to analyse. These molecules are categorized mainly by the fact that they contain a high level of carbohydrate which is attached in O-linkage to serine and/or threonine via the linkage sugar N-acetylgalactosamine. The mucous secretions produced by some epithelial cells, particularly those lining the gastrointestinal tract and the lungs, contain mucins along with other products and these components have been studied for some time at the biochemical level. However, other glandular epithelial cells, such as the salivary gland, breast, ovary, endometrium and sweat glands, also produce mucins, and some of these simpler mucins have recently received much attention. This is because many antibodies selected for epithelial or tumour specificity have been found to react with high molecular weight glycoproteins which are produced by simple epithelial cells and have the properties of mucins.

Studies done in analysing the primary structure of the mucins from various animal species suggested that there is a great variety in the structure of the carbohydrate sidechains which may be composed of only one type of disaccharide, as in ovine submaxillary mucin. It may also be complex, containing several sugars and branched chains as in most gastric mucins[1]. The variability of the core protein structure in mucins obtained from different cell types in the same animal species is less clear[2]. However, recently, partial cDNA clones coding for the core proteins of several mucins have been produced, and three different genes have so far been identified in humans[3–5].

Full-length cDNA clones have been

The MUC1 gene and its products

Table 6.2 Sequence comprising the tandem repeat units of the seven mucin genes characterized. Sequences contain a high proportion of serines and/or threonines (to which the linkage sugar N-acetylgalactosamine may be attached) and prolines, which provide an extended core protein to allow for up to 90% of the molecule to consist of carbohydrate. The designation of *MUC1*, 2 and 3 is as proposed by Gum [5] with the numbers reflecting the order in which the human genes were cloned. The mouse homologue of *MUC1* is designated m*MUC1* and porcine submaxillary and *Xenopus* integumentary mucins are designated p*MUC4* and x*MUC5* and 6, respectively. The number of amino acids in the repeat is given in parentheses following the sequence

	immunogenic domain
MUC1	PPAHGVTSAPDTRPAPGSTA (20 AA)
m*MUC1*	SPVHSGTSSPATSAPEDSTS (20–21 AA)
MUC2	PTTTPITTTTVTPTPTPTGTQT (23 AA)
MUC3	HSTPSFTSSITTTETTS (17 AA)
p*MUC4*	GAGPGTTASSVGTETARPSVAGSGTTGTVSGASGSTGSSSGSPGATGASIGQPETSRIS VAGSSGAPAVSSGASQAAGTS (81 AA)
x*MUC5*	GESTPAPSETT (11 AA)
x*MUC6*	VPTTPETTT (9 AA)

obtained by a number of investigators for a mucin found abundantly in milk [2,6–11]. The gene coding for this mucin has recently been termed *MUC1*, and we have referred to the protein product as PEM[2,6–11]. As well as being referred to as PEM[6,7], the product of the *MUC1* gene has also been designated peanut-lectin binding urinary mucin (PUM) [12], episialin or MAM-6[8], DF3 antigen [9,10], H23 antigen[11], PAS-O[13], epithelial membrane antigen (EMA)[14], Ca1[15], non-penetrating glycoprotein (NPGP)[16] and NCRC11 antigen[17]. The multiplicity of names has led to confusion in the field regarding the identity of the protein. We propose to designate the mucin genes MUC followed by a number reflecting the order in which the genes were cloned as suggested by Gum and colleagues[5]. However, we will continue to refer to the product of the *MUC1* gene as PEM. Partial cDNAs for two human intestinal mucins (*MUC2* and *MUC3*) have also been reported[4,5], as well for the porcine submaxillary mucin[18] and two *Xenopus* integumentary mucins[19,20].

Tandem repeats appear to be a characteristic of mucins, being found in all seven mucins described so far. The repeat units of the different mucins show no similarity to each other in either sequence or number of amino acids in the repeat (Table 6.2), although in each case serines and/or threonines make up a high percentage of the amino acids, giving the molecules the potential to be highly glycosylated. Due to the presence of the tandem repeats, there is great variability observed in the size of the mucin molecules which suggests that length is not crucial to mucin function, but rather that the core protein exists in an extended form as a scaffold for O-linked carbohydrate.

6.3 THE *MUC1* GENE AND ITS PRODUCTS

6.3.1 *MUC1* CORE PROTEIN

The *MUC1* gene is the only mucin gene for which the full-length cDNA and genomic sequences have been determined [6,7,21,22], and the predicted structure of the *MUC1* product is shown diagrammatically in Figure 6.1. The deduced amino acid sequence shows the core protein to be a transmembrane protein, with most of the molecule extracellular (amino terminal), and a cytoplasmic tail of 69 amino acids. That the *MUC1* product is indeed a transmembrane protein has been confirmed by showing that an antibody directed to sequences in the cytoplasmic tail only stains permeabilized cells[23]. The most striking feature of the molecule is the extracellular domain which is made up largely of tandem repeats of 20 amino acids. The number of tandem repeats in the northern European population varies from 20 to 125[7] and it is this variability which accounts for the polymorphism seen at the DNA, RNA and protein level[24]. The tandem repeat domain in fact represents an expressed variable number tandem repeat (VNTR) unit. It is important to note that the genomic sequence obtained from a cosmid library made from normal DNA[22] agrees with the cDNA sequence obtained from breast cancer cell lines[7]. This means that the core protein of MUC1 is unaltered in breast cancers, and any differences seen in the normal and cancer-associated mucin are due to differences in glycosylation[25,26].

6.3.2 THE MOUSE HOMOLOGUE OF MUC1

Using the human 3' sequence as a probe, the mouse mucin has been cloned[27]. Interestingly, the gene has undergone substantial evolutionary changes while maintaining a high potential for O-glycosylation. The repeat domain of the mouse differs from that of the human in codon and amino acid sequence, the length and number of tandem repeats and the location of repeats within the protein. The mouse sequence contains a variable 20–21 amino acid tandem repeat, and although the amino acid composition is similar to that of the human, the homology is less than 40% at the protein level. It is notable that only two of the 16 repeats are identical in sequence, in contrast to the human gene which contains largely identical repeats. A comparison of the consensus amino acid sequence of the tandem repeat in

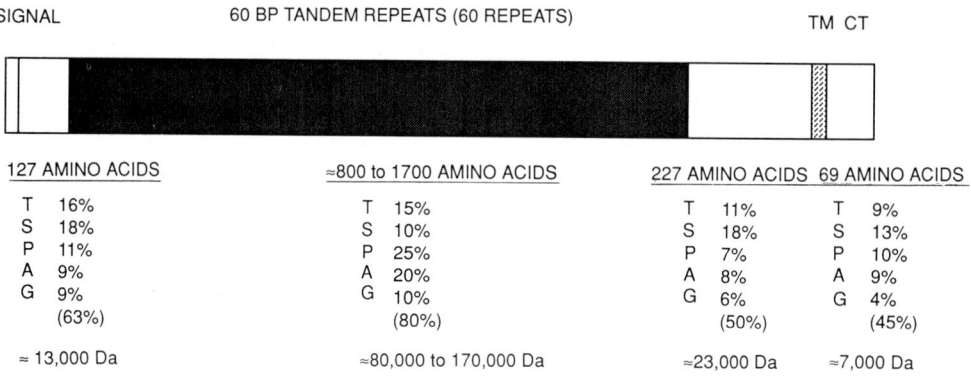

Figure 6.1 Diagram of MUC1 protein. The mucin core protein is drawn to scale for a molecule containing 60 tandem repeats to show that most of the protein consists of repeat units. The mucin is a transmembrane molecule which is expressed in the apical membrane of polarized secretory epithelial cells. The 31 amino acid transmembrane region is designated TM and the 69 amino acid cytoplasmic tail as CT.

Human: G S T A P P A H G V T S A P D T R P A P

Mouse: D S T S S P V H S G T S S P A T S A P V̲
                                                E

Figure 6.2 Alignment of consensus amino acids in tandem repeats of human and mouse MUC1 core proteins. Note that the region of the human repeat that is the epitope for many of the monoclonal antibodies, PDTRP, has become PATSA in the mouse. Presumably it is the presence of the two charged amino acids, aspartic acid and arginine, missing in the mouse homologue, that explain why the human mucin is so immunogenic in the mouse.

the mouse and human *MUC1* products shows differences in the domain in which the epitopes recognized by many monoclonal antibodies are found (see Table 6.1 and Figure 6.2), thus explaining why this domain is so immunogenic in the mouse[28].

6.4 POSSIBLE FUNCTIONS OF MUCINS

In spite of the differences in the tandem repeat sequences, a comparison of the sequences of transmembrane and cytoplasmic tail domains of the mouse and human MUC1 products shows 87% homology. This implies a conserved function, which may involve interactions with intracellular cytoskeletal components and associated proteins.

As shown in Table 6.2, tandem repeats are found to be present in all the seven mucins which have been cloned. Regarding the function of the mucin core protein to act as a scaffold for oligosaccharide attachment, a repetitive amino acid sequence with some glycosylation sites appears to be a necessary requirement. This would lead to a repetitive final structure, which in turn could lead to strong associations in oligomerization. Most mucins are found as aggregates and it may be that the property of self-association is related to the ability to form many similar, possibly weak, attachments. If this is so, we might expect to find that a repetitive structure is a key feature of most mucin core proteins, even though the individual sequences may show great divergence.

6.5 VARIATIONS IN GLYCOSYLATION OF THE SAME CORE PROTEIN IN DIFFERENT TISSUES

Unlike the situation with N-glycosylation, the factors governing O-glycosylation and extension of the oligosaccharide sidechains are not well understood. However, from the gene cloning studies it has become clear that the same core protein can be glycosylated variously in different tissues, resulting in the production of very different mucin products. This has been clearly demonstrated in the case of *MUC1*, which forms the core protein of a mucin present in pancreatic and ovarian cancer cells and normal and malignant breast cells. The sequence of the core protein of the pancreatic mucin[21] and the ovarian cancer mucin (M. Ford, personal communication) established by gene cloning, have been found to be identical to the published sequence of *MUC1*[7]. However, the native pancreatic mucin itself has a different profile of epitopes as compared to the native milk mucin or mucin isolated from breast cancer cells. Indeed, analysis of the carbohydrate sidechains on different mucins, derived by glycosylation of the same core protein, shows them to be very different[25,26,29] (Table 6.3). One interpretation of these results is that a different profile of glycosyltransferases is expressed or active in different tissues. As yet the carbohydrate sidechains of PEM isolated from ovarian cancers has not been directly analysed. However, the antigenic

Molecular structure and clinical applications of a cancer-associated mucin

Table 6.3 Oligosaccharide side chains added to MUC1 in normal and malignant breast cells, and in a pancreatic cancer cell line

Tissue or cell line	Oligosaccharide side chain	Reference
Normal milk	Galß(1→3)GalNAc→Ser/Thr *Galß(1–3/4)GlcNAcß[(1→6)Galß(1–4)GlcNAcß]n(1→6)GalNAc→Ser/Thr	[26]
BY20 (Breast cancer cell line)	Gal(ß1→3)GalNAc→Ser/Thr NANA(α2–3)Gal(ß1→3)GalNAc→Ser/Thr NANA(α2–3)Gal(ß1→3)→GalNAc→Ser/Thr with (α2–6)NANA branch	[25]
Pancreatic tumour	NANA(α2→3)Gal(ß1–4)GlcNAc(ß1→6) → Gal(ß1–4)GlcNAc–6–SO$_4$(1→3/4)GalNAc→Ser/Thr ← NANA(α2→3)Gal(ß1–4)GlcNAc(ß1–3)	[29]

* Neutral oligosaccharides.

profile of the mucin in these cancers appears to resemble that seen in breast cancers.

6.6 ABERRANT GLYCOSYLATION OF MUC1 IN NORMAL AND MALIGNANT EPITHELIAL CELLS

An examination of the structures in Table 6.3 shows that not only is there a difference in the glycosylation of MUC1 between pancreas and breast, but also between normal and malignant breast. The oligosaccharide sidechains found in the breast cancer mucin are shorter, probably terminated early by the addition of sialic acid. An important result of the aberrant glycosylation which occurs in breast cancer is that novel epitopes appear in the carbohydrate sidechains, and core protein epitopes, which are normally masked, are exposed. It is thus possible to find monoclonal antibodies which show some specificity in their reaction with breast cancers. SM-3 is such an antibody which reacts with a core protein epitope in the tandem repeat unit containing the core sequence PDTRP[28]. The antibody was developed using deglycosylated mucin as immunogen, and does not react with the fully glycosylated mucin produced by the lactating mammary gland[30]. The phenomenon of aberrant glycosylation of PEM is quite general in adenocarcinomas and SM-3 has also been found to react with lung, ovarian and colon carcinomas.

It should be noted that many monoclonal antibodies react with the core protein of the MUC1 product, and the epitopes of those that have been mapped fall within the amino acid sequence APDTRPAP (Table 6.1). Unlike

Use of the SM-3 antibody for imaging ovarian cancer

the SM-3 epitope, however, most of these epitopes are also available for binding in the normal mucin. An examination of a Chou–Fasman secondary structure plot with antigenicity according to Jameson–Wolf superimposed (Figure 6.3) reveals that these amino acids clearly comprise the antigenic region of the core protein. The repetitive motif and lack of this particular sequence in the mouse homologue of *MUC1* (see above and Figure 6.2) would make this region of the mucin particularly immunogenic.

Several novel carbohydrate epitopes have been found to be expressed by cancer-associated mucins. Examples of these are the T and Tn antigens[31] and sialylated Le^a[32].

Whether they are reactive with core protein epitopes or carbohydrate epitopes, antibodies showing a selective reaction with cancer-associated mucins could be very useful agents for targeting toxins, drugs or radioactivity to some adenocarcinomas.

6.7 USE OF THE SM-3 ANTIBODY FOR IMAGING OVARIAN CANCER

Antibodies HMFG-2 and HMFG-1, directed to the core protein of MUC1, were previously used to image ovarian cancer[33]. Since the antibody SM-3 was more specific in its reaction with malignant tumours in tissue sections, this antibody was also tested for its

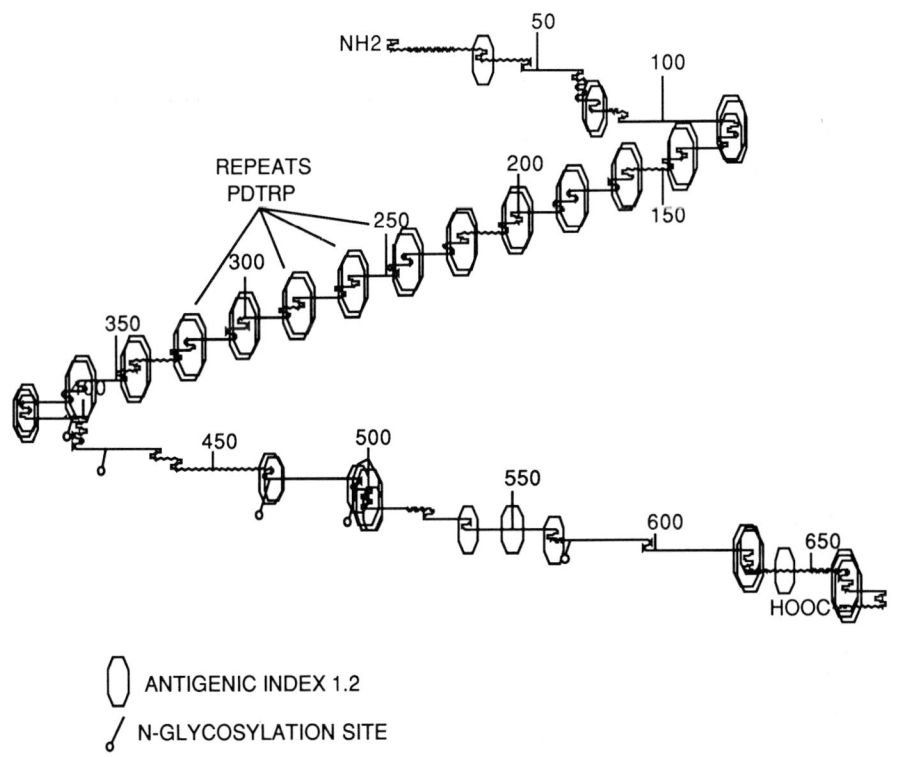

Figure 6.3 Diagram of Chou–Fasman secondary structure plot of MUC1 protein. The core protein used in the plot contains 12 tandem repeat units to show the repetitiveness of the antigenic region. The antigenicity is determined according to the Jameson–Wolf algorithm which is superimposed on the secondary structure plot. The computer analysis was performed on a VAX computer using the Wisconsin Genetics Computer Group (CGC) software.

Table 6.4 Imaging of ovarian tumours using radiolabelled SM-3 antibody

	^{111}In	^{123}I	^{99m}Tc	All cases
Number of studies	30	7	8	45
Number of patients	28	7	8	42
Positive images	27	5	6	38
True positive	18	4	6	28
False positive	7	1	0	8
No operation	2			2
Negative images	3	2	2	6
True negative	2	2	2	6
False negative	1	0	0	1

From Jobling et al. [34]

ability to image ovarian cancer and to distinguish malignant from benign tumours[34]. The antibody was labelled with ^{111}In, ^{123}I or ^{99m}Tc, and successful imaging of ovarian cancers was achieved with all types of label (for summary of data see Table 6.4). Although only one false negative was recorded out of 42 patients, some false positives were noted. These were either tumours, pathologically classified as benign, or non-ovarian carcinomas. The results were encouraging and it has been suggested that radioimmunoscintigraphy with SM-3 might be used as a second screen after ultrasound, which detects a large number of false positives.

The affinity of SM-3 is not very high and it would be desirable to increase the affinity of the antibody for use in conjugate therapy. It might also be advisable to use antibody fragments which could penetrate better into solid tumours. To this end the variable regions of the heavy and light chains of the SM-3 antibody have been cloned and sequenced, and functional 'domain antibodies' (DAbs) [35] expressed in E. coli (D. Allen, in preparation). After mutation and initial selection of variants in vitro, derivatives of SM-3 with improved affinity can be tested in a xenograft model, using the OS tumour derived originally from an ovarian cancer by Bruce Ward, which expresses the epitope for SM-3.

One isotope which should be extremely useful for targeting doses of irradiation to kill tumours is ^{32}P. As can be seen from Table 6.5, ^{32}P has an equivalent particle range to ^{90}Y, and since it emits β particles, patients do not have to be barrier-nursed after injection. Antibodies can be labelled with ^{32}P to a high specific activity without loss of binding capacity, by attaching a heptapeptide (Kemptide) which is a substrate for a protein kinase and phosphorylating with [^{32}P]-γ-ATP[36]. Kemptide has been successfully attached to the SM-3 antibody and in experiments using the OS xenograft model, effective localization of the isotope to the ovarian cancer cells has been shown (H.A. Band and A.M. Creighton, in preparation). The dose-limiting toxicity of these conjugates is to the bone marrow, mainly caused by the incorporation of ^{32}P-phosphate released following metabolism. To circumvent this problem, a current approach under investigation is the replacement of phosphate by the metabolically more stable thiophosphate (A.M. Creighton, personal communication).

6.8 T CELL EPITOPES IN PEM

Using a mouse model, it has been shown that the T and Tn antigens are carried on the mouse mucin epiglycanin and can induce a T cell response and induce tumour rejection[37]. More recently, cytotoxic T cells have

Table 6.5 Principal radionucleotides suitable for imaging and/or therapy

Isotope	Half-life (hours)	Particle	Max. energy (MeV)	Mean particle range (mm)
For imaging				
^{111}In*	68	γ	0.25	–
^{99m}Tc*	6	γ	0.14	–
^{67}Ga*	78	γ	0.30	–
^{123}I	13	γ	0.16	–
For therapy				
^{131}I	192	β	0.81	2.5
		γ	0.36	–
^{90}Y*	64	β	2.29	10
^{32}P	343	β	1.71	7.5
^{67}Cu*	62	β	0.58	–
		γ	0.19	–
^{212}Pb*	11	β	0.57	0.3
		γ	0.24	–
^{212}Bi*	1	α	6.09	0.007
		β	2.25	–
		γ	0.73	–

* Chelated by DTPA and related ligands.

been isolated from breast and pancreatic cancer patients which kill cells expressing a mucin with the MUC1 core protein[38,39]. Of great interest is the fact that the SM-3 antibody can block the cytotoxic effect of the T cells, suggesting that a T cell epitope lies in the tandem repeat unit of the core protein and overlaps the SM-3 epitope. The above studies suggest that because of the aberrant glycosylation, mucin antigens expressed by breast and other cancer cells carry T cell epitopes which may be considered 'non-self' and, therefore, could be potentially useful as antigens for immunotherapy of some carcinomas.

6.9 ANIMAL MODELS FOR TESTING ANTIGENS BASED ON *MUC1* IN TUMOUR REJECTION

As part of the study investigating the possibility of using the MUC1 gene and its products in immunotherapy of breast and ovarian cancer, we have begun to develop mouse models where cellular immune responses and tumour rejection can be studied. A suitable mouse ovarian cancer cell line has not been available, but a syngeneic mouse model has been produced by transfecting the MUC1 gene into a Balb/c mouse mammary tumour cell line. The human mucin is expressed both in the membrane and cytoplasmically in several transfectants and these have been tested for their ability to induce tumours in Balb/c mice[40]. Interestingly, at low inocula there is a dramatic reduction in tumour incidence seen in animals injected with a transfectant showing homogeneous expression of MUC1. Moreover, pre-immunization with a low inoculum of the transfectant could protect against subsequent challenge with a higher inoculum[40]. These results are encouraging and suggest that cellular immunity to tumours can be induced in mice by cell-associated MUC1.

A better model would be a transgenic

mouse expressing *MUC1* on the normal epithelial tissues on which it is expressed in humans as well as on a tumour. We have progressed with this model to the point of having developed a transgenic mouse expressing the *MUC1* product in a tissue-specific manner and glycosylated in a similar way as in humans (in preparation). We are presently introducing a mammary tumour into the transgenic mouse. In both the syngeneic and transgenic mouse model, a variety of immunogens based on the *MUC1* coded mucin can be tested. These include peptides and recombinant viruses containing *MUC1* sequences, as well as cells expressing the gene. Identification of the regulatory sequences of the *MUC1* gene will allow us to express an oncogene in the same tissue-specific manner as the mucin. This approach may lead to the development of a model for ovarian tumours in the transgenic mouse expressing the *MUC1* gene, which in turn would allow the selection of an effective antigen based on PEM for use in immunotherapy of ovarian tumours.

6.10 CONCLUSIONS

Although mucins have been studied at the biochemical and biophysical level for some time, attempts to define their structures in detail were only partially successful due to their size and complexity. The advent of monoclonal antibodies reactive with these molecules introduced a new approach to structural studies by defining antigenic epitopes, by allowing purification of the mucin molecules by affinity chromatography, and by providing a means to clone genes coding for the core proteins. By their profile of reactivity with the normal and cancer-associated mucin in particular tissue, the antibodies also defined a difference in the mucin derived from the two sources. It is now clear that this difference lies in the carbohydrate sidechains, as the core proteins are identical. Since the mucins are tumour-associated antigens and the cancer mucins can express epitopes which are relatively tumour specific, this family of molecules is now being intensively studied. In the next few years, we should expect to see the continued application of mucin-reactive antibodies in the clinic and the investigation of their use as agents for immunotherapy of ovarian and breast cancers.

REFERENCES

1. Hilkens, J. (1988) Biochemistry and function of mucins in malignant disease. *Cancer Rev.*, **11–12**, 25–54.
2. Taylor-Papadimitriou, J. and Gendler, S.J. (1988) Molecular aspects of mucins. *Cancer Rev.*, **11–12**, 1–24.
3. Gendler, S.J., Burchell, J.M., Duhig, T. *et al.* (1987) Cloning of partial cDNA encoding differentiation and tumour-associated mucin glycoproteins expressed by human mammary epithelium. *Proc. Natl Acad. Sci. USA*, **84**, 6060–4.
4. Gum, J.R., Byrd, J., Hicks, J. *et al.* (1989) Molecular cloning of human intestinal mucin cDNAs. *J. Biol. Chem.*, **264**, 6480–7.
5. Gum, J.R., Hicks, J., Swallow, D. *et al.* (1990) Molecular cloning of cDNAs derived from a novel human intestinal mucin gene. *Biochem. Biophys. Res. Commun.*, **171**, 407–15.
6. Gendler, S., Taylor-Papadimitriou, J., Duhig, T. *et al.* (1988) A highly immunogenic region of a human polymorphic epithelial mucin expressed by carcinomas is made up of tandem repeats. *J. Biol. Chem.*, **263**, 12820–3.
7. Gendler, S., Lancaster, C., Taylor-Papadimitriou, J. *et al.* (1990) Molecular cloning and expression of human tumour-associated polymorphic epithelial mucin. *J. Biol. Chem.*, **265**, 15286–93.
8. Ligtenberg, M., Vos, H., Gennissen, A. and Hilkens, J. (1990) Episialin, a carcinoma-associated mucin, is generated by a polymorphic gene encoding splice variants with alternative amino termini. *J. Biol. Chem.*, **265**, 5573–8.
9. Siddiqui, J., Abe, M., Hayes, D. *et al.* (1988)

Isolation and sequencing of a cDNA coding for the human DF3 breast carcinoma-associated antigen. *Proc. Natl Acad. Sci. USA*, **85**, 2320–3.

10. Abe, M. and Kufe, D. (1989) Sequence analysis of the 5′ flanking region of the human DF3 breast carcinoma-associated antigen gene. *Biochem. Biophys. Res. Commun.*, **165**, 644–9.
11. Wreschner, D.H., Hareuveni, M., Tsarfaty, I. et al. (1990) Human epithelial tumor antigen cDNA sequences. *Eur. J. Biochem.*, **189**, 463–73.
12. Swallow, D., Griffiths, B., Bramwell, M. et al. (1986) Detection of the urinary 'PUM' polymorphism by the tumour-binding monoclonal antibodies Ca1, Ca2, Ca3, HMFG-1 and HMFG-2. *Dis. Markers*, **4**, 247–54.
13. Shimizu, M. and Yamauchi, K. (1982) Isolation and characterisation of mucin-like glycoproteins in human milk fat globule membranes. *J. Biochem.*, **91**, 515–24.
14. Heyderman, E., Steele, K. and Ormerod, M.G. (1979) A new antigen on the epithelial membrane: its immunoperoxidase localization in normal and neoplastic tissues. *J. Clin. Pathol*, **32**, 35–9.
15. Bramwell, M.E., Bhavanandan, V.P., Wiseman, G. and Harris, H. (1983) Structure and function of the Ca antigen. *Br. J. Cancer*, **48**, 177–83.
16. Ceriani, R., Peterson, J., Lee, J. et al. (1983) Characterisation of cell surface antigens of human mammary epithelial cells with monoclonal antibodies prepared against human milk fat globule. *Somatic Cell Genet.*, **9**, 415–27.
17. Price, M., Edwards, Owainati, A. et al. (1985) Multiple epitopes on a human breast carcinoma associated antigen. *Int. J. Cancer*, **36**, 567–72.
18. Timpte, C., Eckhardt, A., Abernethy, J. and Hill, R. (1988) Porcine submaxillary gland apomucin contains tandemly repeated, identical sequences of 81 residues. *J. Biol. Chem.*, **263**, 1081–8.
19. Hoffmann, W. (1988) A new repetitive protein from *Xenopus laevis* skin highly homologous to pancreatic spasmolytic polypeptide. *J. Biol. Chem.*, **263**, 7686–90.
20. Probst, J.C., Gertzen, E.-M. and Hoffman, W. (1990) An integumentary mucin (FIM-B.1) from *Xenopus laevis* homologous with von Willebrand factor. *Biochemistry*, **29**, 6240–4.
21. Lan, M., Batra, S., Qi, W.-N. et al. (1990) Cloning and sequencing of a human pancreatic tumour mucin cDNA. *J. Biol. Chem.*, **265**, 15294–9.
22. Lancaster, C., Peat, N., Duhig, T. et al. (1990) Structure and expression of the human polymorphic epithelial mucin gene: an expressed VNTR unit. *Biochem. Biophys. Res. Commun.*, **173**, 1019–29.
23. Gendler, S.J., Spicer, A.P., Pemberton, L. et al. (1991) Molecular and evolutionary analysis of an expressed hypervariable gene for a tumor-associated mucin, PEM, in *Breast Cancer and Immunodiagnosis and Immunotherapy* (ed. R.L. Ceriani), Plenum Publishing Corp, New York, in press.
24. Gendler, S., Burchell, J., Girling, A. et al. (1988) in *Breast Cancer: Scientific and Clinical Progress*, Kluwer Academic, Boston, pp. 112–26.
25. Hull, S., Bright, A., Carraway, K. et al. (1989) Oligosaccharide differences in the DF3 sialomucin antigen from normal human milk and the BT20 human breast carcinoma cell line. *Cancer Comm.*, **1**, 261–7.
26. Hanisch, F.-G., Uhlenbruck, G., Peter-Katalinic, J. et al. (1989) Structures of neutral O-linked polylactosaminoglycans on human skim milk mucins. A novel type of linearly extended poly-N-acetyl-lactosamine backbones with Galβ(1–4) GlcNAcβ(1–6) repeating units. *J. Biol. Chem*, **264**, 872–83.
27. Spicer, A.P., Parry, G. and Gendler, S.J. Molecular cloning of the mouse homologue of the tumor-associated mucin, MUC1, reveals conservation of transmembrane and cytoplasmic domains but divergence of the external repeat domain. Manuscript submitted.
28. Burchell, J., Taylor-Papadimitriou, J., Boshell, M. et al. (1989) A short sequence, within the amino acid tandem repeat of a cancer-associated mucin, contains immunodominant epitopes. *Int. J. Cancer*, **44**, 691–6.
29. Khorrami, A., Lan, M.S., Metzgar, R.S. and Kaufman, B. (1989) Characteristics of a sulphated human pancreatic adenocarcinoma mucin glycoprotein. *Glycoconjugate J.*, **6**, 428.
30. Burchell, J., Gendler, S., Taylor-Papadimitriou,

30. ...J. et al. (1987) Development and characterisation of breast cancer reactive monoclonal antibodies directed to the core protein of the human milk mucin. *Cancer Res.*, **47**, 5476–82.
31. Itzkowitz, S., Yuan, M., Montgomery, C. et al. (1989) Expression of Tn, Sialosyl-Tn, and T antigens in human colon cancer. *Cancer Res.*, **49**, 197–204.
32. Magnani, J., Nilsson, B., Brockhaus, M. et al. (1982) A monoclonal antibody-defined antigen associated with gastrointestinal cancer is a ganglioside containing sialylated lacto-N-fucopentaose 11. *J. Biol. Chem.*, **257**, 14365–9.
33. Granowska, M., Shepherd, J., Britton, K.E. et al. (1984) Ovarian cancer: diagnosis using ^{123}I monoclonal antibody in comparison with surgical findings. *Nucl. Med. Commun.*, **5**, 485–99.
34. Jobling, T.W., Granowska, M., Britton, K.E. et al. (1990) Radioimmunoscintigraphy of ovarian tumors using a new monoclonal antibody, SM-3. *Gynecol. Oncol.*, **38**, 468–72.
35. Ward, E.S., Gussow, D., Griffiths, A.D. et al. (1989) Binding activities of a repertoire of single immunoglobulin variable domains secreted from *E. coli*. *Nature*, **341**, 544–6.
36. Foxwell, B.M.J., Band, H.A. and Long, J. (1988) Conjugation of monoclonal antibodies to a synthetic peptide substrate for protein kinase: A method for labelling antibodies with ^{32}P. *Br. J. Cancer*, **57**, 489–93.
37. Henningson, C., Selvaraj, S., Maclean, G. et al. (1987) T cell recognition of a tumour-associated glycoprotein and its synthetic carbohydrate epitopes: stimulation of anti-cancer T cell immunity *in vivo*. *Cancer Immunol. Immunother.*, **25**, 231–41.
38. Barnd, D., Lan, M., Metzgar, R. and Finn, O. (1989) Specific, MHC-unrestricted recognition of tumour-associated mucins by human cytotoxic T cells. *Proc. Natl Acad. Sci. USA*, **86**, 7159–63.
39. Jerome, K.R., Barnd, D.L., Boyer, C.M. et al. (1990) Adenocarcinoma reactive cytotoxic T lymphocytes recognize an epitope present on the protein core of epithelial mucin molecules, in *Cellular Immunity and the Immunotherapy of Cancer*, Wiley-Liss, New York, pp. 321–8.
40. Lalani, E.-N. Berdichevsky, F., Shearer, M. et al. Expression of the gene coding for a human mucin in mouse mammary tumour cells can affect their tumorigenicity. *J. Biol. Chem.* (submitted).

Part Two
Oncogenes, Growth Factors and Cytokines

Chapter 7

Expression of the epidermal growth factor receptor, HER-2/*neu* and p53 in ovarian cancer

A. BERCHUCK, J.R. MARKS and R.C. BAST JR

We now believe that most human cancers arise due to activation or inactivation of cellular genes that encode proteins normally involved in regulation of cellular proliferation. Two classes of cancer-causing genes have been described, proto-oncogenes[1] and tumour suppressor genes[2]. The proto-oncogenes, which were described first, encode diverse proteins that are believed to serve various roles in growth regulation. Examples of oncogenes include *sis*, which encodes a secreted growth factor (platelet-derived growth factor); *erb*B, which encodes a cell surface receptor (epidermal growth factor receptor); *ras*, which encodes a cytoplasmic protein; and *jun* and *fos*, which encode nuclear proteins. The proto-oncogenes act as dominant transforming genes when activated by either mutation, overexpression or translocation. The second class of cancer-causing genes are the tumour suppressor genes, which also are called anti-oncogenes. These genes encode proteins, most of which reside in the nucleus, that are thought to normally act by inhibiting proliferation. Because these genes inhibit, rather than stimulate, proliferation, inactivation of both copies of a tumour suppressor gene is required to elicit malignant transformation. Therefore, some have referred to these genes as recessive oncogenes.

For several years, our group at Duke University has examined expression of several cancer-causing genes in both normal ovarian epithelium and epithelial ovarian cancers. These studies have included large enough numbers of patients to allow us to begin to unravel the relationship between aberrant expression of various cancer-causing genes and biological behaviour and clinical outcome. Our initial studies focused on two closely related cell surface molecules, the epidermal growth factor (EGF) receptor and HER-2/*neu*. Both the EGF receptor and HER-2/*neu* are glycoproteins comprising a cysteine-rich extracellular domain, a hydrophobic membrane spanning region and an intracellular tyrosine kinase domain[3,4]. Binding of both

53

EGF and transforming growth factor-α to the EGF receptor results in activation of the inner tyrosine kinase domain, which is thought to be involved in transmission of the mitogenic signal to the nucleus. Due to the close structural similarity between the EGF receptor and HER-2/*neu*, it has been postulated that HER-2/*neu* also encodes a growth factor receptor. In this regard, recently, a 30-kDa ligand that binds to HER-2/*neu* has been described, but this putative growth factor has not yet been completely characterized[5].

Several studies have demonstrated that expression of the EGF receptor and HER-2/*neu* may be abnormal in some human cancers. Specifically, it has been shown that the EGF receptor gene is amplified in a proportion of squamous cancers, which results in marked overexpression of EGF receptor protein[6]. In contrast, Harris and coworkers have shown that a proportion of breast cancers appear to have lost EGF receptor expression[7]. In breast cancer, EGF receptor expression is inversely related to oestrogen receptor expression and is associated with poor prognosis. We examined expression of EGF receptor immunohistochemically in snap frozen tissue samples from normal ovaries and ovarian cancers. In this study, we used murine monoclonal antibody 528, which is specifically reactive with the external domain of the EGF receptor[8]. In the normal ovary, EGF receptor always was seen in both the surface epithelium and the underlying stroma. Among the advanced stage ovarian cancers, EGF receptor was seen in 77% of cases. Survival of patients with EGF receptor-negative cancers was significantly better than that of patients with EGF receptor positive cancers ($P < 0.05$)[9]. Thus, similar to breast cancers, a proportion of ovarian cancers appears to have lost EGF receptor and loss of receptor is associated with a favourable prognosis. It is possible that this favourable prognosis relates to loss of a growth regulatory pathway that promotes proliferation of both normal epithelium as well as some cancers.

Slamon has shown that the HER-2/*neu* oncogene is amplified in a proportion of breast cancers and that these cancers markedly overexpress HER-2/*neu* protein[10]. In addition, he and others have reported that overexpression of HER-2/*neu* is associated with poor prognosis in breast cancer. In surveying other human cancers, Slamon noted that epithelial ovarian cancers also frequently overexpressed HER-2/*neu* and that HER-2/*neu* overexpression in ovarian cancer was associated with poor prognosis, similar to breast cancer[11].

To investigate further the relationship between HER-2/*neu* expression and clinicopathological features and survival in ovarian cancer, we used an immunohistochemical technique to localize HER-2/*neu* in frozen sections of 73 advanced stage ovarian cancers [12]. Monoclonal antibody TA1, which is specifically reactive with the extracellular domain of HER-2/*neu*, was used as the primary antibody[13]. The intensity of staining in the cancers was compared to that seen in normal ovaries. In the normal ovary, light to moderate staining for HER-2/*neu* was seen in the epithelium. Staining in the underlying stroma was faint or absent. Similarly, 50 of 73 stage III/IV epithelial ovarian cancers that we examined also exhibited light to moderate staining for HER-2/*neu*. In contrast, 23 of the cancers (32%) exhibited strong staining for HER-2/*neu* consistent with overexpression. In several of the cancers, we examined HER-2/*neu* expression in the primary tumour and multiple metastases. We found that regardless of whether the cancer had normal or elevated HER-2/*neu* expression the intensity of staining was similar between various sites in a given patient. In addition, in a given patient, the intensity of staining for HER-2/*neu* did not change between initial surgery and second-look operation. These findings suggest that in patients with advanced disease the level of HER-2/*neu* expression can be determined using any viable fresh frozen tumour sample.

We also examined the relationship between HER-2/*neu* overexpression and other known prognostic factors in ovarian cancer [12]. There was no relationship between HER-2/*neu* overexpression and histological grade or the ability to perform optimal cytoreductive surgery. However, patients whose cancers had normal HER-2/*neu* expression were fivefold more likely to achieve a negative second-look laparotomy, compared to patients whose cancers overexpressed HER-2/*neu*. In addition, the median survival of patients whose cancers overexpressed HER-2/*neu* was strikingly worse (16 months) than that of patients whose cancers had normal HER-2/*neu* expression (32 months) (Figure 7.1).

In summary, we found that HER-2/*neu* is expressed by normal ovarian epithelial cells. Approximately 30% of advanced stage ovarian cancers overexpress HER-2/*neu* and HER-2/*neu* overexpression is associated with poor survival. Further studies are needed to determine the molecular mechanisms responsible for the association between overexpression of HER-2/*neu* and poor survival.

The p53 gene encodes a nuclear phospho-

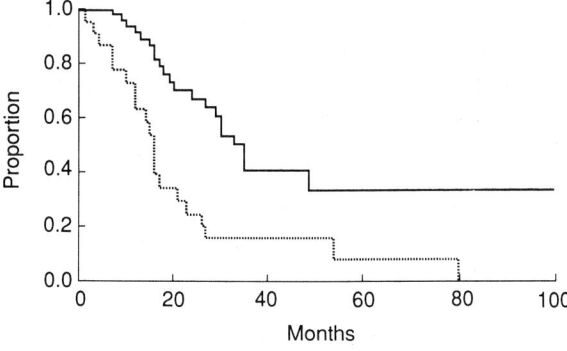

Figure 7.1 Relationship between the level of HER-2/*neu* expression and survival in advanced stage ovarian cancer ($P<0.001$).——, normal HER-2/*neu* expression; . . . ., HER-2/*neu* overexpression.

protein with a short half-life that is present in small amounts in most normal cells[14]. Several lines of evidence are suggestive that p53 normally acts as a tumour suppressor gene. First, transfection of normal wild-type p53 into some transformed cells suppresses the malignant phenotype. Conversely, deletion of both copies of the wild-type p53 gene is associated with transformation of some cells in culture. In addition, however, there is evidence that p53 can act as a dominant transforming oncogene. First, it has been shown that transfection of some cells with mutant p53 genes that encode proteins with increased stability can result in transformation. The increased stability of these mutant forms of p53 is due to decreased degradation of the mutant protein. In addition, it has been shown that mutant p53 protein complexes with wild-type p53, which may lead to inactivation of the suppressor function of wild-type p53 protein.

Mutation of the p53 gene is the most frequent genetic change to be described in human cancers thus far[15]. Mutant p53 genes have been noted in 60% of colon and lung cancers and 25% of breast cancers. In addition, p53 mutations often are associated with loss of the other wild-type p53 allele[16]. The p53 mutations have been shown to be diverse but most occur in highly conserved regions of the gene that presumably are of functional importance. Since mutant p53 genes encode proteins with a prolonged half-life, mutation of the p53 gene usually leads to relative overexpression of p53 protein.

Recently, we have begun to examine expression of the p53 gene in epithelial ovarian cancer[17]. For these studies, snap frozen tissue samples were available from 9 ovaries that were normal or had benign epithelial tumours and from 107 epithelial ovarian cancers. Among the 107 ovarian cancers, in 69 cases tissue was obtained at initial surgery while in 38 cases tissue was obtained at a second-look or subsequent operation. All the patients in this study underwent surgery at Duke University between 1985 and 1990. We used an avidin–biotin peroxidase immunohistochemical technique to localize p53 in frozen sections. The primary antibody, PAb1801, is a murine monoclonal antibody specifically reactive with p53[18]. In addition, the ploidy of the cancers was determined using a computerized image-analysis system to quantitate nuclear Feulgen staining in touch imprints of the tumours.

Similar to prior studies of normal and benign tissues, we did not find immunohistochemically detectable p53 in any of the 9 normal or benign ovaries because the small amounts of p53 present in most cells are below the threshold of detection. Similar to normal tissues, immunohistochemically detectable p53 was not seen in 53 (50%) of the ovarian cancers. In contrast, 54 of 107 cancers (50%), had strong nuclear staining in the majority of malignant cells, consistent with overexpression. Overexpression of p53 was seen with equal frequency in early stage cancers and advanced cancers. Among the stage I/II cancers, p53 overexpression was seen in 8 of 15 (53%) cases compared to 46 of 92 (50%) of the stage III/IV cases. In this study, however, many of the tissue samples from early stage cancers were obtained at the time of recurrence. Therefore, this group of early stage cancers is biased towards more virulent lesions. Further studies of larger numbers of early stage cancers are needed.

We also examined the relationship between p53 expression and other prognostic factors in patients with advanced stage disease. Among patients with stage III/IV disease in whom tumour samples were obtained at initial surgery, there was no relationship between histological grade and p53 overexpression. Overexpression of p53 was seen in 33% of well-differentiated cancers, 59% of moderately differentiated cancers and 47% of poorly differentiated cancers. There also was no difference in the incidence of p53 overexpression between patients who were optimally debulked such that the largest residual tumour nodule was less than 1 cm and those who were suboptimally debulked. We did, however, observe a significant relationship between ploidy and p53 overexpression. Overall, 23% of the ovarian cancers were diploid with a DNA index of 1.0 ± 0.1 while 77% were aneuploid. We found that p53 overexpression was significantly more common in aneuploid cancers (58%), than in diploid cancers (30%). Finally, we examined the relationship between p53 expression and survival.

To determine if overexpression of p53 in ovarian cancer is associated with mutation of the p53 gene, sequence analysis of the p53 gene was performed in several ovarian cancers[17]. All mutations that have been shown to increase the stability of p53 are located between exons 5–8, which encode amino acids 126–306. Within this area there are four 'hot spots' that correspond to highly conserved regions of the gene. Many, but not all, of the mutations have been found to occur in these hot spots. This region of the p53 message was sequenced in three cancers in which immunohistochemically detectable p53 was seen and in three cancers in which staining for p53 was not seen. Individual pieces of tumour tissue used for sequencing were assessed histologically to ensure that a high proportion of cancer cells was present. Total RNA was isolated and used as a template for p53 cDNA synthesis using reverse transcriptase and a p53 oligonucleotide primer. Exons 4–9 were then amplified using the polymerase chain reaction. Both cDNA strands then were sequenced using the dideoxy method.

The three cancers in which immunohistochemically detectable p53 was not seen were all found to have a normal wild-type p53 sequence. In contrast, the three cancers that demonstrated high levels of p53 protein all contained point mutations that altered the coding sequence. In the cancer shown in Figure 7.2, there is a mutation at codon 278 which changes the sequence from the wild-type CCT to CGT, with a resulting substitution of arginine for proline. No wild-type allele is seen. In another cancer, we observed a mutation at codon 282 (CGG to TGG), which changed the sequence from arginine to tryptophan. Again, no wild-type allele was seen. Both of these mutations in ovarian cancers which overexpress p53 fall within the area of 'hot spot' number 4 between codons 272 and 286 where mutations frequently have been described in other human cancers. In another ovarian cancer, which also over-expressed p53, we found a mutation at codon 216. The sequence changed from the wild-type GTG to ATG, which results in the substitution of methionine for valine. In this case a faint band corresponding to the wild-type sequence also was seen. The relatively faint intensity of this band was suggestive that it is due to contamination with a small amount of normal cells rather than due to the presence of a copy of the wild-type allele in the cancer. Although the mutation in this case is positioned outside of the four 'hot spots' of most consistent evolutionary conservation, this area is conserved between frog, rodent and man. Thus, it still is a relatively conserved region of the gene and mutations in this area have been reported in other cancers.

In summary, we found that normal and benign ovarian tissues do not express im-

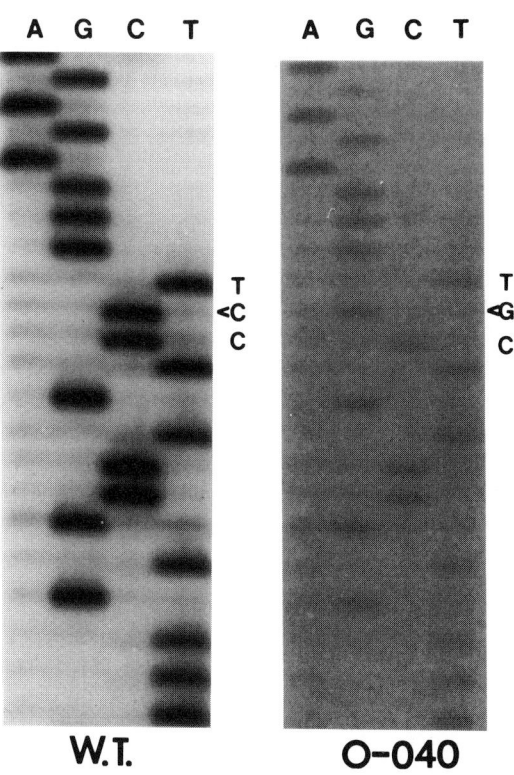

Figure 7.2 Sequencing of the p53 gene from an ovarian cancer that overexpresses p53 protein. Point mutation at codon 278 (C to G) changes proline to arginine.

munohistochemically detectable p53. In contrast, p53 is overexpressed in approximately 50% of epithelial ovarian cancers. We did not observe any correlation between p53 overexpression and stage, grade, debulking status or survival. In addition, we found that, similar to other human cancers, overexpression of p53 is associated with point mutations in the p53 gene.

Future considerations include the following. First is the question of whether mutation of p53 in ovarian cancer is associated with loss of the other wild-type p53 allele, as has been found in other cancers. The sequencing data that we have presented is suggestive that this also is the case in ovarian cancer, but additional studies of larger numbers of cases are needed to confirm this impression. In addition, however, our finding that p53 overexpression was more common in aneuploid cancers also is consistent with an association between p53 mutation and allelic loss. In this regard, studies by other investigators have shown that allelic loss on chromosome 17p, where the p53 gene resides, occurs in from one-third to three-quarters of ovarian cancers[18,19].

Although it is thought that wild-type p53 may normally act as a transcriptional regulator, further studies are needed to determine the molecular mechanisms by which wild-type p53 inhibits proliferation. In addition, the mechanism by which overexpression of mutant p53 participates in malignant transformation remains unknown. Finally, we must determine whether p53 mutations are early causative events in ovarian carcinogenesis or late events that occur due to genetic instability associated with the malignant phenotype. Although it appears that p53 mutation is a late event in colon carcinogenesis [16], there is some evidence that p53 mutation may be an early event in the development of some human cancers. Specifically, it has been shown that patients with Li–Fraumeni familial cancer syndrome, who develop multiple cancers at a young age, inherit mutant p53 genes[20]. It is possible then that p53 mutation may also be an early acquired somatic defect in some normal ovarian epithelial cells during the process of malignant transformation.

The discovery of cancer-causing genes has provided us with the exciting opportunity to begin to understand the molecular pathology of ovarian cancer. Aberrant expression of several of these genes including the EGF receptor, HER-2/*neu*, and p53 have been described in some ovarian cancers. Much remains to be learned, however. In the future, when we have a better understanding of the molecular mechanisms of ovarian carcinogenesis, hopefully this will allow us to better diagnose and treat and eventually to prevent ovarian cancer.

REFERENCES

1. Bishop, J.M. (1991) Molecular themes in oncogenesis. *Cell*, **64**, 235–48.
2. Marshall, C.J. (1991) Tumour suppressor genes. *Cell*, **64**, 313–26.
3. Downward, J., Yarden, Y., Mayes, E. *et al.* (1984) Close structural similarity of epidermal growth factor receptor and v-*erb*-B oncogene protein sequences. *Nature*, **307**, 521–7.
4. Bargmann, C.L., Hung, M.C. and Weinberg, R.A. (1986) The *neu* oncogene encodes an epidermal growth factor receptor-related protein. *Nature*, **319**, 226–9.
5. Lupu, R., Colmer, R., Zugmaier, G. *et al.* (1990) Direct interaction of a ligand for the *erb*B2 oncogene product with the EGF receptor and p^{185}. *Science*, **249**, 1552–5.
6. Ozanne, B., Richards, C.S., Hendler, F. *et al.* (1986) Over-expression of the EGF receptor is a hallmark of squamous cell carcinomas. *Am. J. Pathol.*, **149**, 9–14.
7. Sainsbury, J.R., Farndon, J.R., Needham, G.K. *et al.* (1987) Epidermal growth factor receptor status as predictor of early recurrence of and death from breast cancer. *Lancet*, **i**, 1398–402.

8. Gill, G., Kwamamoto, T., Cochet, C. et al. (1984) Monoclonal anti-epidermal growth factor receptor antibodies which are inhibitors of epidermal growth factor binding and antagonists of epidermal growth factor-stimulated tyrosine protein kinase activity. *J. Biol. Chem.*, **264**, 7755–60.
9. Berchuck, A., Rodriguez, G.C., Kamel, A. et al. (1991) Epidermal growth factor receptor expression in normal ovarian epithelium and ovarian cancer I – Correlation of receptor expression with prognostic factors in patients with ovarian cancer. *Am. J. Obstet. Gynecol.*, **164**, 669–74.
10. Slamon, D.J., Clark, G.M., Wong, S.G. et al. (1987) Human breast cancer: correlation of relapse and survival with amplification of the HER-2/*neu* oncogene. *Science*, **235**, 177–82.
11. Slamon, D.J., Godolphin, W., Jones, L.A. et al. (1989) Studies of HER-2/*neu* proto-oncogene in human breast and ovarian cancer. *Science*, **244**, 707–12.
12. Berchuck, A., Kamel, A., Whitaker, R. et al. (1990) Overexpression of HER-2/*neu* is associated with poor survival in advanced epithelial ovarian cancer. *Cancer Res.*, **50**, 4087–91.
13. McKenzie, S.J., Marks, P.J., Lam, T. et al. (1989) Generation and characterization of monoclonal antibodies specific for the human *neu* oncogene product, p185. *Oncogene*, **4**, 543–8.
14. Lane, D.P. and Benchimol, S. (1990) p53: oncogene or anti-oncogene? *Genes Dev.*, **4**, 1–8.
15. Nigro, J.M., Baker, S.J., Preisinger, A.C. et al. (1989) Mutations in the p53 gene occur in diverse human tumour types. *Nature*, **342**, 705–8.
16. Baker, S.J., Preisinger, A.C., Jessup, J.M. et al. (1990) p53 gene mutations occur in combination with 17p allelic deletions as late events in colorectal tumorigenesis. *Cancer Res.*, **50**, 7717–22.
17. Marks, J.R., Davidoff, A.M., Kerns, B.J. et al. (1991) Overexpression and mutation of p53 in epithelial ovarian cancer. *Cancer Res.*, (in press).
18. Eccles, D.M., Cranston, G., Steel, C.M. et al. (1990) Allele losses on chromosome 17 in human epithelial ovarian carcinoma. *Oncogene*, **5**, 1599–601.
19. Russell, S.E.H., Hickey, G.I., Lowery, W.S. et al. (1990) Allele loss from chromosome 17 in ovarian cancer. *Oncogene*, **5**, 1581–3.
20. Malkin, D., Li, F.P., Strong, et al. (1990) Germ line p53 mutations in a familial syndrome of breast cancer, sarcomas and other neoplasms. *Science*, **259**, 1233–8.

Chapter 8a

Factors regulating the growth of normal and malignant ovarian epithelium

R.C. BAST JR, G.C. RODRIGUEZ, S. WU,
C.M. BOYER, and A. BERCHUCK

Epithelial ovarian cancers are thought to arise from a single layer of cells that cover the ovarian surface or line inclusion cysts. Multiple steps are probably required for malignant transformation of ovarian epithelium. In other cell types, autocrine or paracrine growth have been shown to stimulate or inhibit normal epithelial growth or to sustain the proliferation of tumour cells[1]. Few studies have examined growth regulation in normal and malignant human ovarian epithelium[2].

In previous studies[3], we have examined the role of several known peptide factors in growth of epithelial ovarian carcinoma cell lines. These factors include epidermal growth factor (EGF), platelet-derived growth factor (PDGF), fibroblast growth factor (FGF), and transforming growth factor-β (TGF-β). Production of and response to these growth factors was found to vary between different ovarian cancer cell lines. In other types of normal epithelial cells, EGF and transforming growth factor-α (TGF-α) stimulated proliferation, whereas TGF-β inhibited proliferation. If similar regulation were observed in normal ovarian epithelium, increased autocrine/paracrine growth stimulation by EGF/TGF-α or decreased autocrine/paracrine growth inhibition by TGF-β could contribute to oncogenesis (Figure 8.1).

To facilitate comparison of normal and malignant ovarian epithelium, normal surface epithelial cells were obtained from apparently normal human ovaries removed at surgery. Epithelial cells were scraped from the ovarian surface and grown in culture media originally described by Auersperg and her colleagues[4]. Cells could be passaged from two to more than 20 times. The epithelial characteristics of cells isolated from normal ovaries included immunohistochemically detectable expression of cytokeratins and the presence of microvilli and desmosomes by electron microscopy.

EGF is a glycoprotein with a molecular mass of 6 kDa. The three-dimensional structure of EGF is similar to that of TGF-α which

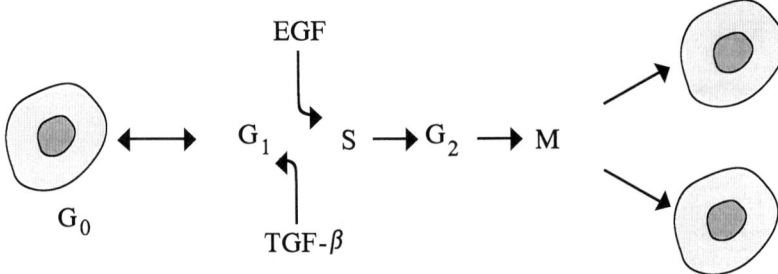

Figure 8.1 Autocrine/paracrine growth regulation by EGF and TGF-β. EGF stimulates, whereas TGF-β inhibits, epithelial cell proliferation.

also binds to the receptor for EGF. The EGF receptor is a transmembrane protein with an extracellular ligand binding domain, a membrane spanning region, and an intracellular tyrosine kinase domain[5]. EGF and TGF-α stimulate growth of many normal cells in culture including human foreskin fibroblasts, mammary epithelium and mesothelial cells. Proliferation of normal ovarian epithelial cells was measured in the presence of different concentrations of EGF. In each of five individuals, EGF stimulated uptake of [^{3}H] thymidine from 2.4- to 4-fold, albeit at different optimal concentrations of EGF ranging from 0.2 to 20 ng/ml[6]. In contrast, EGF in concentrations up to 20 ng/ml produced less than twofold stimulation in each of five epithelial ovarian cancer cell lines and one of these cancer cell lines was unresponsive to EGF.

Differences in responsiveness to EGF between normal and malignant epithelium might relate to a decrease in the expression of EGF receptor by the malignant epithelial cells. However, when Scatchard analysis of EGF receptors in benign and malignant ovarian epithelium was undertaken, similar numbers of EGF receptors of similar affinity were observed in benign and malignant epithelial cells[6]. Consequently, the decreased responsiveness of malignant epithelium to EGF does not appear to be due to decreased receptor expression. Alterations in signals downstream from the binding of ligand to its receptor are likely to be present. Whatever the source of this defect, malignant ovarian epithelium appears less, rather than more responsive to EGF relative to normal ovarian epithelial cells.

TGF-β is a glycoprotein dimer of molecular mass 24 kDa. To date, the TGF-β receptor has not been well characterized. TGF-β stimulates the growth of some mesenchymal cells in culture, but inhibits the growth of many epithelial cells including mammary epithelium and mesothelial cells[7]. TGF-β can suppress immune responses and inhibit differentiation of adipocytes or myoblasts. This growth factor can also stimulate matrix protein formation and the expression of receptors for matrix protein. Incubation of normal ovarian epithelial cells from five different donors with TGF-β consistently down-regulated growth. As little as 0.2 ng/ml of the growth factor significantly inhibited proliferation of normal ovarian epithelium. Similar dose–response effects were observed in each of five cases[8]. Again, in contrast, epithelial ovarian cancer cell lines varied in their response to TGF-β. OVCA 420 growth was strongly inhibited (95%) by 2 ng/ml of TGF-β, whereas OVCA 432 and 433 were slightly (15–20%), but significantly, inhibited by higher concentrations of TGF-β (10 ng/ml). Growth of OVCA cells was not affected by TGF-β.

In collaboration with Brad Arrick, expression of TGF-β mRNA by benign and malignant epithelial cells was characterized using the RNase protection assay[8]. Message for TGF-β1 and β2 was detected in normal ovarian epithelial cells from two women, but no TGF-β3 transcripts could be detected. When five different ovarian cancer cell lines were analysed, TGF-β and β2, but not β3, mRNA was present in four of the five cell lines. Each of four ovarian cancer cell lines and normal epithelial cell cultures was further analysed for the presence of latent and active TGF-β in serum-free culture supernatants. Although the normal epithelial cells produced active TGF-β, only one of four ovarian cancer cell lines, OVCA 433, secreted active TGF-β. This cancer cell line also retained responsiveness to exogenous TGF-β. When neutralizing monoclonal and polyclonal reagents were added to cultures of each of four epithelial ovarian cell lines and normal epithelial cells, only growth of OVCA 433 cells and normal epithelial cells was stimulated. These data are consistent with the possibility that TGF-β normally exerts an autocrine growth inhibitory effect in normal ovarian epithelial cells and OVCA 433 cells (Figure 8.2). Loss of this autocrine inhibitory loop could be one step in the development of some epithelial ovarian cancers.

In earlier studies[9], we had evaluated expression of several proto-oncogenes in a panel of snap frozen human epithelial ovarian cancer samples. Expression of l-*myc*, c-*myb* and c-*erb*B and c-*mos* was not detected. However, expression of c-*fos* was detected in 100%, c-*myc* in 77%, n-*myc* in 29%, c-H-*ras* in 50%, v-*fms* in 57% and c-*erb*B2 in 86% of cases tested. Expression of *myc* and *fos* is associated with rapidly proliferating cells. Similarly, H-*ras* is often expressed in nonmalignant cells. The expression of v-*fms* and of c-*erb*B2 in a majority of cases was of particular interest.

The c-*fms* proto-oncogene encodes a 150-kDa cell surface protein with tyrosine kinase activity. The *fms* gene product binds CSF-1 (M-CSF) and is probably identical to the receptor for this growth factor. In studies performed in collaboration with S. Ramakrishnan, M-CSF was found in culture media of each of six ovarian cancer cell lines tested[10]. Elevated levels of M-CSF also were found in sera of approximately 70% of patients with epithelial ovarian cancer. It is possible that M-CSF might participate in autocrine or paracrine growth regulation (Figure 8.3). Kacinski and colleagues demonstrated that incubation of cancer cell lines that express high levels of *fms* with M-CSF can stimulate invasiveness[11]. In addition, Man-

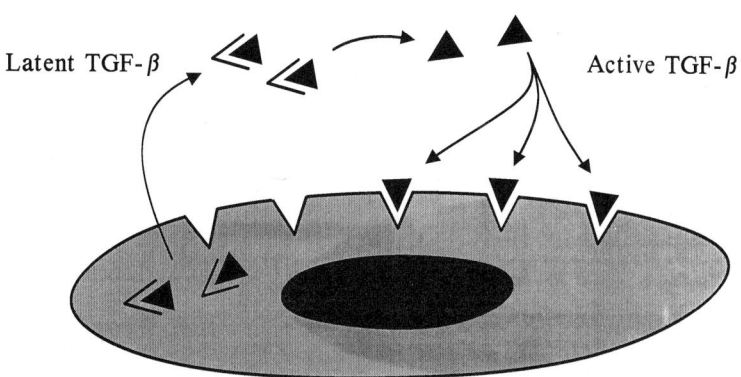

Figure 8.2 Autocrine growth inhibition by TGF-β

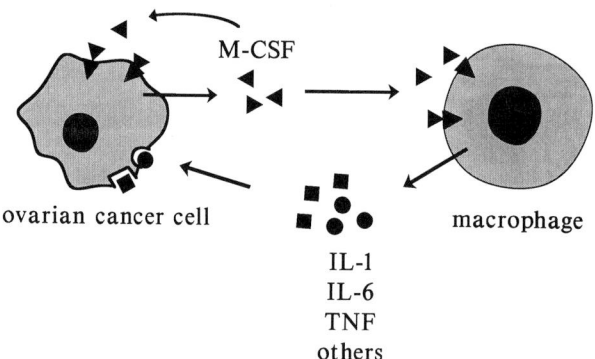

Figure 8.3 Autocrine/paracrine growth regulation by M-CSF.

tovani and his colleagues have demonstrated that M-CSF is a potent chemo-attractant for monocytes and macrophages[12].

Different numbers of monocytes and macrophages have been found associated with malignant ascites and within solid ovarian cancers. Although cancer immunologists have often assumed that activated macrophages inhibit tumour growth, Hamburger and colleagues had reported some time ago that macrophages can stimulate clonogenic growth of ovarian tumour cells[13]. Consequently, we have sought factors produced by monocytes and macrophages that would stimulate ovarian tumour proliferation[14]. TNF-α, interleukin (IL)-1 and IL-6 can all stimulate growth in three out of four ovarian cancer cell lines tested. In addition, TNF-α, IL-1 and IL-6 can all be produced by ovarian cancers (Chapters 11 and 12). Consequently, macrophages may not be the only source of these cytokines, but might support the growth of those tumours without high endogenous expression.

In summary (Table 8.1), our data argue against the importance of EGF/TGF-α in mediating malignant transformation, but by no means rule out this possibility. On the other hand, stronger evidence has been obtained that some ovarian cancers have lost the ability to produce active TGF-β or to

Table 8.1 Growth regulation of normal and malignant ovarian epithelium

	Normal	Malignant
Autocrine stimulation by TGF-α/EGF	+	+/− to −
Autocrine inhibition by TGF-β	+	+/− to −
Expression of M-CSF	+	+
Expression of c-*fms*	−	+
Stimulation by IL-1α	?	+
Stimulation by IL-6	?	+
Stimulation by TNF-α	?	+
Expression of c-*erb*B2 (*neu*)	+	+ to +++
Expression of p53	+/−	+/− to +++

respond to the inhibitory effects of this growth factor. Consequently, loss of this normal growth inhibitory pathway might be involved in the development of a small proportion of ovarian cancers. In addition, both normal and malignant ovarian epithelium express and release M-CSF[10,14–16]. In preliminary studies with Barry Kacinski (Yale University), it appears that normal epithelium fails to express *fms*, whereas approximately half of ovarian carcinomas studied *in situ* expressed this receptor. Consequently, the appearance of *fms* also might be an important step in oncogenesis. IL-1α, IL-6 and TNF-α could potentially exert autocrine or paracrine regulation, the latter via interaction with macrophages that are attracted by M-CSF. As outlined in a separate chapter, both normal and malignant epithelium express c-erbB2 (HER-2/neu), but the gene is overexpressed in approximately 30% of cases and is associated with poor prognosis[17]. In addition, p53 is mutated and overexpressed in 50% of epithelial ovarian carcinomas[18]. Loss of the inhibitory activity of the p53 tumour suppressor gene product could contribute to ovarian oncogenesis.

We are therefore beginning to determine which factors are involved in growth regulation of normal ovarian epithelium. In addition, several potentially important differences between normal and malignant ovarian epithelium have been identified. As we continue to add to this understanding in the future, hopefully these insights will lead to novel strategies for diagnosis, treatment, and prevention of ovarian cancer.

ACKNOWLEDGEMENTS

This work was supported through a grant, CA39930, from the National Cancer Institute, Department of Health and Human Services, USA.

REFERENCES

1. Knabbe, C., Lippman, M.E., Wakefield L.M. et al. (1987) Evidence that transforming growth factor-β is a hormonally regulated negative growth factor in human breast cancer cells. Cell, **48**, 417–28.
2. Siemens, C.H. and Auersperg, N. (1988) Serial propagation of human ovarian surface epithelium in tissue culture. J. Cell Physiol., **134**, 347–56.
3. Berchuck, A., Olt, G.J., Everitt, L. et al. (1990) The role of peptide growth factors in epithelial ovarian cancer. Obstet. Gynecol., **75**, 255–62.
4. Kruk, P.A., Maines-Bandiera, S.L. and Auersperg, N. (1990) A simplified method to culture human ovarian surface epithelium. Lab. Invest., **63**, 132–6.
5. Carpenter, G. (1984) Properties of the receptor for epidermal growth factor. Cell, **37**, 357–8.
6. Rodriguez, G.C., Berchuck, A., Whitaker, R.S. et al. (1991) Epidermal growth factor receptor expression in normal ovarian epithelium and ovarian cancer II. Relationship between receptor expression and response to epidermal growth factor. Am. J. Obstet. Gynecol., **164**, 745–50.
7. Wakefield, L.M. and Sporn, M.B. (1990) Suppression of carcinogenesis: A role for TGF-β and related molecules in prevention of cancer, in Tumour suppressor genes (ed. G. Klein), Marcel Dekker, New York, pp. 217–43.
8. Berchuck, A., Rodriguez, G., Olt, G. et al. Regulation of growth of normal ovarian epithelium and ovarian cancer cell lines by transforming growth factor-β. (Submitted for publication).
9. Tyson, F.L., Soper, J.T., Daly, L. et al. (1988) Overexpression and amplification of the c-erbB-2 (HER2/neu) proto-oncogene in epithelial ovarian tumours and cell lines. Proc. Am. Assoc. Cancer Res., **29**, A1872.
10. Ramakrishnan, S., Xu, F.J., Brandt, S.J. et al. (1989) Constitutive production of macrophage colony-stimulating factor by human ovarian and breast cancer cell lines. J. Clin. Invest., **83**, 921–6.
11. Kacinski, B.M., Carter, D., Mittal, K. et al. (1988) High level expression of fms proto-oncogene mRNA is observed in clinically aggressive human endometrial adenocarcinomas. Int. J. Radiat. Oncol. Biol. Phys., **15**, 823–9.
12. Bottazzi, B., Erba, E., Nobili, N. et al. (1990) A paracrine circuit in the regulation of the proliferation of macrophages infiltrating murine sarcomas. J. Immunol., **144**, 2409–12.
13. Hamburger, A.W., White, C.P. and Dunn, F.E. (1984) Modulation of tumour colony growth by irradiated accessory cells, in Fourth Conference on Tumour Cloning, Orlando, Florida, Grune and Stratton, pp. 53–65.
14. Wu, S., Rodabaugh, K., Martinez-Maza, O. et al. Stimulation of ovarian tumor cell proliferation with monocyte products including interleukin-1-alpha, interleukin-6 and tumor necrosis factor-alpha. (Submitted for publication).
15. Lidor, Y.J., Xu, F.J., Olt, G.J. et al. (1991)

Constitutive production of macrophage colony stimulating factor (M-CSF) and interleukin-6 (IL-6) by human ovarian surface epithelial cells. *Proc. Am. Assoc. Cancer Res.* **31**, A1431.
16. Kacinski, B.M., Carter, D., Kohorn, E.I. *et al.* (1989) Markedly elevated plasma levels of a tumor-produced cytokine CSF-1 (M-CSF), the macrophage colony-stimulating factor are seen in ovarian carcinoma patients with active gynecological neoplasms. *Soc. Gynecol. Invest.*, 182.
17. Berchuck, A., Kamel, A., Whitaker, R. *et al.* (1990) Overexpression of HER-2/*neu* is associated with poor survival in advanced epithelial ovarian cancer. *Cancer Res.*, **50**, 4087–91.
18. Marks, J.R., Davidoff, A.M., Kerns, B.J. *et al.* (1991) Overexpression and mutation of p53 in epithelial ovarian cancer. *Cancer Res.*, **51**, 2979–84.

Chapter 8b

Inhibition of breast and ovarian tumour cell growth by antibodies and immunotoxins reactive with distinct epitopes on the extracellular domain of HER-2/*neu* (c-*erb*B2)

R.C. BAST JR, F.J. XU, G.C. RODRIGUEZ,
R. WHITAKER, M. BOENTE, A. BERCHUCK,
S. McKENZIE, L.L. HOUSTON and C.M. BOYER

The *neu* (c-*erb*B2) oncogene was first identified in neuroblastomas that occur in rats exposed to ethyl nitosourea *in utero*[1]. Activation of *neu* in this model was found to be due to a single point mutation in the membrane spanning region[2]. The product of the human c-*erb*B2 gene (HER-2/*neu*) is a 185-kDa (p185) integral membrane glycoprotein with a cysteine-rich extracellular tyrosine kinase domain. The p185 molecule is highly homologous with the epidermal growth factor receptor and with the v-*erb*B oncogene, particularly in the tyrosine kinase region. Due to this close structural similarity, p185 is thought to act as a growth factor receptor.

Recently, Lupu and colleagues have identified a putative ligand of 30 kDa which binds to the extracellular domain of p185[3].

Approximately 30% of human breast and ovarian cancers overexpress structurally normal p185 protein. In most cases, overexpression results from amplification of the c-*erb*B2 gene (2–20-fold), but some breast and ovarian carcinomas overexpress c-*erb*B-2 RNA and protein despite having only a single gene copy. Overexpression of c-*erb*B2 has been associated with poor survival in both breast and ovarian cancer[4,5]. In a study performed by our group, the median survival of patients with ovarian cancers that had normal p185 expression was 32 months, whereas survival of patients whose tumours overexpressed p185 was only 16 months[5].

Green and co-workers reported down-regulation of rat tumour cell growth with antibodies directed against the extracellular domain of the rat *neu* gene product that contains a mutation in the intramembranous portion of the molecule[6]. Our group has considered the possibility that antibodies against the extracellular domain of human p185 might down-regulate the growth of breast and ovarian cancer cells. In addition, we have begun to compare inhibition of tumour cell growth with unconjugated antibodies to the inhibition produced by immunotoxins in which antibodies are conjugated with the A chain of ricin.

Eleven murine monoclonal antibodies reactive with the extracellular domain of p185 have been obtained from Applied Biotechnology (RC1, TA1, NB3, RC6, PB3, BD5, ID5, OD3)[7] or from Cetus (454C11, 520C9, and 741F8)[8,9]. Epitopes have been defined by cross-blocking experiments. Binding of ^{125}I-labelled anti-p185 antibodies to SKBr3 breast cancer cells that overexpress p185 was inhibited with different non-labelled antibodies [10]. Results of these studies are consistent with the map detailed in Figure 8.4. Ten antibodies recognize epitopes that are expressed in a linear array. One antibody, OD3, recognizes an epitope distinct from the others.

Binding of unconjugated antibodies to the extracellular domain of cells which overexpress p185 has inhibited their growth. Seven of the 11 antibodies inhibit anchorage-independent clonogenic growth of SKBr3 cells in soft agar by 40–80%. Four antibodies (TA1, 454C11, 520C9 and OD3) failed to inhibit anchorage-independent growth when used individually. However, when each of the antibodies that were inactive individually was used in combination, significant inhibition of anchorage-independent growth could be achieved. Consequently, each antibody appeared to recognize a functionally distinct determinant. In addition, multiple monoclonal antibodies were more effective than single reagents and some interactions appeared synergistic. Inhibition of anchorage-dependent tumour cell growth has been obtained with the SKOv3 ovarian cancer cell line that overexpresses p185 using the ID5 antibody[10].

The effect of different antibodies on anchorage-dependent growth of SKBr3 cells also was assessed by measuring incorporation of [^{3}H]-thymidine[7]. In this case, only BD5 and ID5 significantly inhibited growth in replicate experiments. Interestingly, preliminary work done in collaboration with Ruth

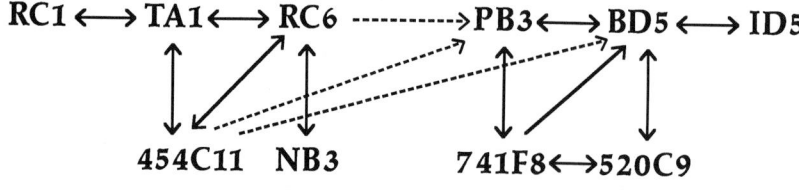

Figure 8.4 Epitope map of the extracellular domain of p185. Double arrows indicate reciprocal blocking of antibody binding. Single arrows indicate blocking only by one antibody. Solid lines indicate more intense blocking than broken lines.

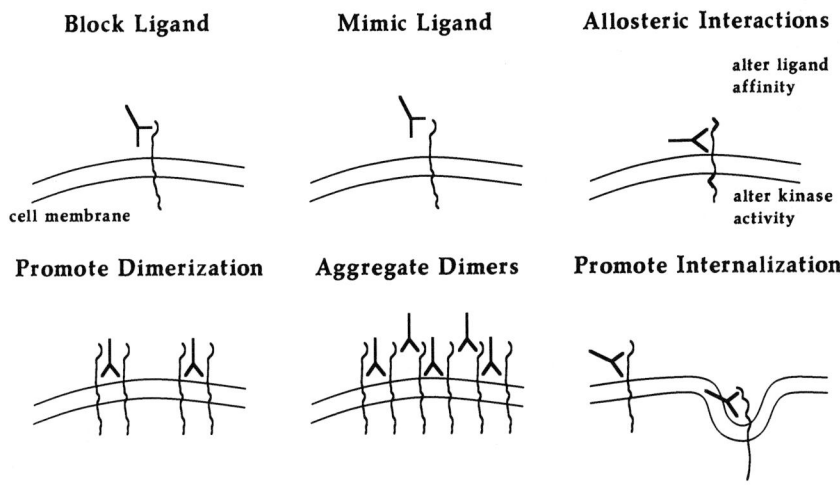

Figure 8.5 Possible mechanisms of inhibition of growth by antibodies that bind to the extracellular domain of p185.

Lupu suggests that these two antibodies inhibit the binding of the 30-kDa putative ligand to the extracellular domain of p185.

Inhibition of tumour growth might be due to several mechanisms (Figure 8.5). If binding of the ligand for p185 stimulated tumour growth, antibodies that block ligand binding might inhibit tumour growth. However, only two of the seven antibodies that inhibited anchorage-independent growth appear to block the ligand that has been described to date. Alternatively, if the ligand down-regulates growth of cells that overexpresses p185, as seems to be the case with SKBr3[3], binding of antibody to the extracellular domain of p185 might mimic ligand binding. Alternatively, binding of antibodies at sites adjacent to the ligand binding region could induce conformational changes in the extracellular domain of p185 that would alter the affinity of p185 for ligand binding. Another less likely possibility is that binding of antibody to the extracellular domain might cause a conformational change in the intracytoplasmic kinase, altering its activity.

The additive or synergistic effects of multiple monoclonal antibodies on growth might relate to aggregation of multiple receptors on the cell surface, creating a high local concentration of kinase activity. Alternatively, receptor aggregates may be more easily internalized, resulting in down-regulation of cell surface p185 following antibody binding. If the mechanism of activation for p185 resembles that for the p170 epidermal growth factor receptor, dimerization of p185 and autophosphorylation of the kinase regions may be particularly important. If antibody-induced dimerization of p185 is critical for growth inhibition, Fab fragments should not be as effective as intact antibody. However, in studies with ID5, RC1 and PB3 essentially similar degrees of inhibition of anchorage-independent growth have been obtained with intact IgG and Fab fragments[10].

As the kinase activity of p185 is likely to be important for cell growth regulation, additional studies have been performed to examine the effect of anti-p185 antibodies on phosphorylation of intracellular substrates. The ID5 antibody, which down-regulates growth of SKBr3 cells, increases phosphorylation of at least three unidentified intracellular substrates seen on two-dimensional gels. Phospholipase C_γ is known to be a substrate of the p185 kinase. The phosphorylated form of this enzyme acts to generate diacylglycerol from cell membrane phospholipids. In preliminary studies, diacylglycerol levels have been decreased by approximately 40% following incubation of SKBr3 cells with antibodies that down-regulate proliferation, but not in the presence of an antibody that does not affect anchorage-independent growth. However, the magnitude of the effect on diacylglycerol levels suggests that substrates other than phosphatidylinositol might be affected by the changes induced through the binding of antibody to the extracellular domain of p185.

The cytotoxic potency of monoclonal antibodies can be increased by conjugation with radionuclides, drugs and toxins. Each of the anti-p185 antibodies that has been used to inhibit growth of SKBr3 cells has been conjugated to the A chain of ricin (RTA). These ricin A chain conjugates require internalization with subsequent translocation of the RTA into the cytoplasmic compartment where there is catalytic inactivation of ribosomes. Using constructs which delete different portions of the intracellular domain, we have studied the requirements for internalization of the c-erbB2 gene product following binding of antibodies to the extracellular domain[11]. Membrane stripping and immunogold electron microscopy have been used to track the internalization of anti-p185 antibodies. In the presence of specific antibody, constructs are internalized that lack all but six amino acids of the intracellular domain. Mutants with a single codon mutation in the intramembranous domain, analogous to that observed in active rat neu, are internalized slightly more promptly than the wild type. However, each of the constructs serves equally well as a target for TRA-immunotoxins prepared with the TA1 antibody. Taken together, these data suggest that immunotoxins reactive with p185 do not require kinase activity for internalization, translocation of TRA, or tumour cell killing. However, preliminary data are consistent with the possibility that the intracellular domain and kinase activity are critical for inhibition of tumour cell growth with unconjugated anti-p185 antibodies. Consequently, the cytotoxicity exerted by conjugated and unconjugated antibody is likely to utilize fundamentally different mechanisms.

Further studies are needed to define which epitopes of p185 are the optimal targets for immunotoxins. In addition, more detailed studies of p185 expression in normal human tissues will be critically important prior to initiation of clinical trials, particularly if immunotoxins are to be employed. Despite expression by some normal tissues, overexpression of c-erbB2 by breast and ovarian carcinomas may provide an acceptable therapeutic index for treatment of patients with these neoplasms.

ACKNOWLEDGEMENTS

This work was supported through a grant, CA39930, from the National Cancer Institute, Department of Health and Human Services, USA.

REFERENCES

1. Padhy, L.C., Shih, C., Cowing, D. et al. (1982) Identification of a phosphoprotein specifically induced by the transforming DNA of rat neuroblastomas. Cell, **28**, 865–71.
2. Bargmann, C.I., Hung, M.C. and Weinberg, R.A. (1986) Multiple independent activations of the neu oncogene by a point mutation altering the transmembrane domain of p185. Cell, **45**, 649–57.
3. Lupu, R., Colomer, R., Zugmaier, G. et al. (1990) Direct interaction of a ligand for the erbB-2 oncogene product with the EGF receptor and $p185^{erbB2}$. Science, **249**, 1552–5.
4. Slamon, D.J., Godolphin, W., Jones, L.A. et al. (1989) Studies of HER-2/neu proto-

oncogene in human breast and ovarian cancer. *Science*, **244**, 707–12.
5. Berchuck, A., Kamel, A., Whitaker, R. *et al.* (1990) Overexpression of HER-2/*neu* is associated with poor survival in advanced epithelial ovarian cancer. *Cancer Res.*, **50**, 4087–91.
6. Drebin, J.A., Link, V.C., Stern, D.F. *et al.* (1985) Down-modulation of an oncogene protein product and reversion of the transformed phenotype by monocolonal antibodies. *Cell*, **41**, 697–706.
7. McKenzie, S.J., Marks, P.J., Lam, T. *et al.* (1989) Generation and characterization of monoclonal antibodies specific for the human *neu* oncogene product, p185. *Oncogene*, **4**, 543–8.
8. Bjorn, M.J., Ring, D.B. and Frankel, A.E. (1985) Evaluation of monoclonal antibodies for the development of breast cancer immunotoxins. *Cancer Res.*, **45**, 1214–21.
9. Ring, D.B., Clark, R. and Saxena, A. Reactivity of monoclonal antibodies against BCA200 with the recombinant extracellular region of c-*erb*B-2. *Mol. Immunol.* (in press).
10. Xu, F.J., Rodriguez, G.C., Whitaker, R. *et al.* (1991) Antibodies against immunochemically distinct epitopes on the extracellular domain of HER-2/*neu* (c-*erb*B-2) inhibit growth of breast and ovarian cancer cell lines. *Proc. Am. Asoc. Cancer Res.*, **32**, 260.
11. Rodriguez, G.C., Boente, M., Xu, J.F. *et al.* Cytotoxic effect of anti-*erb*B-2 antibodies and immunotoxins against epithelial ovarian cancer (in preparation).
12. Maier, L.A., Xu, F.J., Hester, S. *et al.* Requirements for the internalization of a murine monoclonal antibody directed against the HER-2/*neu* (c-*erb*B-2) gene product. *Cancer Res.* (in press).

Chapter 9

Cloning of genes encoding ovarian carcinoma-specific antigens

W.D. FOULKES, T.A. JONES
and J. TROWSDALE

9.1 INTRODUCTION

Ovarian cancer accounts for over half of the deaths due to gynaecological malignancy with a five-year survival rate of only 20–30% [1]. The poor prognosis in ovarian cancer is generally due to the advanced stage of the disease at the time of diagnosis. There is clear evidence that detection at an earlier stage results in a better prognosis and this has prompted the search for tumour-associated antigens which can be used as targets for monoclonal antibodies (MAb) in diagnosis and therapy.

One of the most useful MAbs to date is OC125 which recognizes approximately 80% of ovarian cancers and is used widely for diagnosis and for monitoring response to therapy[2]. CA-125, the antigen defined by OC125, is not truly specific to ovarian cancer as it can also be found in normal coelomic derived epithelium and elevations of CA-125 are also found in patients with other cancers and in those without malignant disease[3]. The reason for this is that many MAbs directed against human cancer-associated antigens recognize epitopes consisting of carbohydrate or glycolipid and are associated with high M_r glycoproteins or mucins of cell membranes[4]. These high M_r molecules are also expressed on normal tissue and MAbs directed against epitopes on them often react with a diverse range of normal and tumour tissue[5]. Recently new MAbs have been developed in an attempt to achieve better specificities, although many still react with non-ovarian neoplastic tissue as well as normal tissue. One of the most promising antibodies is MOv18 which recognizes a glycoprotein, CaMOv18, of M_r 38 000–40 000 present on approximately 90% of ovarian carcinomas[6–8].

Although many other tumour-associated markers have been described little is known about their molecular nature and few of the cDNA encoding them have been isolated. However, such information may provide insights into selective mechanisms involved in tumour growth and may also indicate a route for therapy or for the development of novel reagents for detection. The importance of a detailed knowledge of the molecular structure of these antigens has been demonstrated by the studies of Taylor-Papadimitriou[5] on the antibody SM-3. This MAb is reactive to an epitope of the polymorphic epithelial mucin (PEM) which is expressed in malignant tissue but not in the corresponding normal tissue, even though PEM is expressed at comparable levels in both. Once the cDNA for PEM was cloned a structural analysis of the core protein was possible which facilitated the mapping of the epitopes recognized by SM-3 and other PEM-reactive antibodies. They were able to demonstrate that SM-3 reacted to an epitope that was marginally expressed in normal tissue but as a result of aberrant glycosylation was selectively unmasked in a range of carcinomas[9]. While SM-3 is not particularly useful in detection of ovarian cancer it may be possible to engineer more discriminating and higher affinity antibodies based on the knowledge of its epitope. This approach may be relevant for many other ovarian tumour markers since, as discussed above, they too appear to be highly glycosylated and cross-reactivity with other types of tumours and with normal tissue is common. With this in mind we have undertaken to clone cDNA sequences encoding ovarian carcinoma-specific antigens.

9.2 cDNA CLONING STRATEGY

Traditional approaches to cloning mammalian cDNA have involved either expression of cDNA libraries in bacteria and detection of peptide with antibody or sequencing the peptide and then synthesizing a complementary oligonucleotide for subsequent DNA hybridization screening. The former method is rarely successful since bacteria lack most of the post-translational modification systems necessary for production of the correct epitope. This is a problem particularly when using MAbs given that they rely on a single epitope for peptide recognition. The latter method is generally more reliable but many proteins are difficult or impossible to purify in sufficient quantity for peptide sequencing.

Recently a new method for cDNA cloning has been described based on the transient expression of cDNA in mammalian cells and recovery of expressing cells with monoclonal antibody and has been used with great success for the cloning of cell differentiation antigens[10,11]. In this chapter we describe a modification of this procedure which we have used for the cloning of the cDNA encoding the antigen recognized by MAb MOv18.

9.3 METHODS AND RESULTS

9.3.1 cDNA CLONING BY 'PANNING'

The cDNA isolation protocol used was first described by Seed[10] and is colloquially known as 'panning' since it relies upon the adherence of expressing cells to the surface of antibody-coated petri dishes while non-expressing cells are washed off. We have modified this system in two ways. Firstly, all transfections were performed by electroporation[12], which is a simple and rapid procedure that achieves transient expression frequencies of 50–75%. Secondly, antibody-coated petri dishes were replaced by magnetic beads coated with rabbit anti-mouse IgG which are as efficient as petri dishes but enable the entire selection to be performed in a single tube.

The procedure is summarized in Figure 9.1. Briefly, cDNA libraries from HT29 colon and SKOV3 ovarian carcinoma lines were introduced separately into WOP cells[11] and 48–72 hours later CaMOv18 positive transfectants were selected with MOv18 antibody and sheep anti-mouse IgG antibody-coated magnetic beads. The CaMOv18 enriched

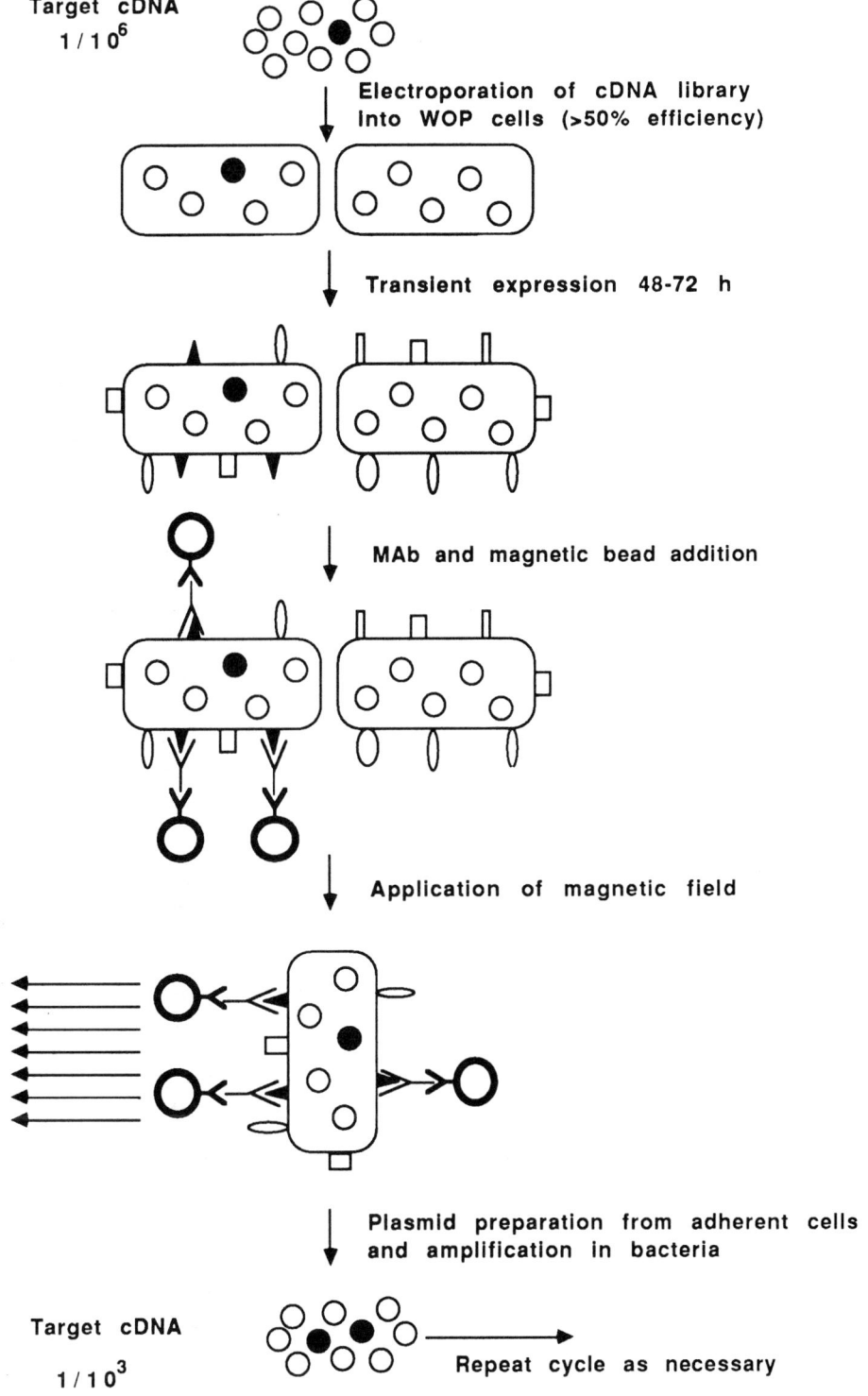

Figure 9.1 Summary of 'panning' protocol for cDNA cloning.

plasmid prepared from these cells was then amplified and the cycle repeated until the end of the third round. Plasmid DNA was then prepared from 10 individual bacterial transformants from each library and transfected separately into WOP cells. Three of the 10 HT29 clones and five of the SKOV3 clones showed surface expression of the MOv18 antigen as detected by indirect immunofluorescence.

9.3.2 MOv18 cDNA SEQUENCE ANALYSIS

The cDNA derived from SKOV3, cSKMOv18, was 1325 bp and contained a 771 bp open reading frame flanked by 5' and 3' non-coding sequences of 430 bp and 124 bp, respectively (Figure 9.2a). The sequence derived from HT29, cHTMOv18, consisted of 936 bp. Compared with cSKMOv18 the sequence of cHTMOv18 begins at nucleotide −11 and contains an additional 31 bp of 3' non-coding sequence as indicated. The cDNA sequences encoded identical peptides of 257 amino acids. A database search identified this sequence as an adult high-affinity folate binding protein (FBP) with an amino acid sequence identical to FBPs isolated by different strategies from normal placenta[13], KB cells[13,14] and Caco-2 cells[15].

Differences in the 5' non-coding region were evident between cSKMOv18 and the Caco-2 cell FBP and the KB cell FBP and appeared to be products of alternate RNA splicing at the junction of nucleotides −10 and −11 and between nucleotides −76 and −77 as illustrated in Figure 9.2b.

9.3.3 SOUTHERN ANALYSIS

Southern blots were carried out to determine the complexity of the FBP sequences at the genomic level. BamHI and EcoRI digests of SKOV3 and normal human DNA were hybridized with a ^{32}P-labelled cHTMOv18 probe and washed under stringent conditions (Figure 9.3). The large number of bands visible in Southern blots suggests that the FBP family may contain more than the two members so far described and/or the existence of pseudogenes.

9.3.4 DNA POLYMORPHISM

To identify restriction fragment length polymorphisms, DNA from 15–27 unrelated individuals were digested with a variety of enzymes and probed with ^{32}P-labelled cHTMOv18 probe. No polymorphisms were detected with BamHI, EcoRI, HindIII and PvuII.

Two PstI polymorphisms were detected among 27 unrelated individuals. An example blot is shown in Figure 9.4, panel A. The first was a four allele polymorphism with variable bands at 7.5 kb (allele A1), 6.8 kb (allele A2), 6.3 kb (allele A3) and 5.9 kb (allele A4) with frequencies of 0.04, 0.42, 0.5 and 0.04 respectively. The second was a three allele poly

---

Figure 9.2a Complete nucleotide sequence of the SKOV3 folate binding protein cDNA clone, cSKMOv18, and its predicted amino acid sequence. Polyadenylation signal [AATAAA] is underlined. Nucleotide positions based on the AUG start codon are numbered above the sequence and amino acid positions below. Putative RNA splice sites at nucleotides −10 and −76 are indicated by arrows (see also Figure 9.2b). The lower case sequences in the 3' non-coding region are present in cHTMOv18 but not cSKMOv18. Nucleotides indicated above the main sequence indicate substitutions found in cHTMOv18 and below substitutions in the Caco-2 cell [15] and KB cell [13] FBP sequences.

```
            -430                                                                    -351
            GGAAAGGATTTTCTCAGCCCCCATCTCCAGCACTGTGTGTTGGCCGCACCCATGAGAGCCTCAGCACTCTGAAGGTGCAG
                                                                                    -271
            GGGGCAAAGGCCAAAAGAGCTCTGGCCTGAACTTGGGTGGTCCCTACTGTGTGACTTGGGGCATGGCCTCATCTGTGCTG
                                                                                    -191
            AAATGATTCCACAAAGATTAAACTGGCTATCATTTGTTGATTTCCCCCTTCTTACATTTAATCCTTGCAGGAGAAAGCTA
                                                                                    -111
            AGCCTCAAGATAGTTTGCTTCTCTTTCCCCCAAGGCCAAGGAGAAGGTGGAGTGAGGGCTGGGGTCGGGACAGGTTGAAC
                                                                                    -31
            GGGAACCCTGTGCTCTAACAGTTAGGGCCCGCCGAGGAACTGAACCCAAAGGATCACCTGGTATTCCCTGAGAGTACAGA
                                                ↑
                                                1                                   36
            TTTCTCCGGCGTGGCCCTCAAGGGACAGAC  ATG GCT CAG CGG ATG ACA ACA CAG CTG CTG CTC CTT
                                          ↑ MET Ala Gln Arg MET Thr Thr Gln Leu Leu Leu Leu
                                            1                                       96
            CTA GTG TGG GTG GCT GTA GTA GGG GAG GCT CAG ACA AGG ATT GCA TGG GCC AGG ACT GAG
            Leu Val Trp Val Ala Val Val Gly Glu Ala Gln Thr Arg Ile Ala Trp Ala Arg Thr Glu
            13                                                                      156
            CTT CTC AAT GTC TGC ATG AAC GCC AAG CAC CAC AAG GAA AAG CCA GGC CCC GAG GAC AAG
            Leu Leu Asn Val Cys MET Asn Ala Lys His His Lys Glu Lys Pro Gly Pro Glu Asp Lys
            33                                                                      216
            TTG CAT GAG CAG TGT CGA CCC TGG AGG AAG AAT GCC TGC TGT TCT ACC AAC ACC AGC CAG
            Leu His Glu Gln Cys Arg Pro Trp Arg Lys Asn Ala Cys Cys Ser Thr Asn Thr Ser Gln
            53                                                                      276
            GAA GCC CAT AAG GAT GTT TCC TAC CTA TAT AGA TTC AAC TGG AAC CAC TGT GGA GAG ATG
            Glu Ala His Lys Asp Val Ser Tyr Leu Tyr Arg Phe Asn Trp Asn His Cys Gly Glu MET
            73                                                                      336
            GCA CCT GCC TGC AAA CGG CAT TTC ATC CAG GAC ACC TGC CTC TAC GAG TGC TCC CCC AAC
            Ala Pro Ala Cys Lys Arg His Phe Ile Gln Asp Thr Cys Leu Tyr Glu Cys Ser Pro Asn
            93                                                                      396
            TTG GGG CCC TGG ATC CAG CAG GTG GAT CAG AGC TGG CGC AAA GAG CGG GTA CTG AAC GTG
            Leu Gly Pro Trp Ile Gln Gln Val Asp Gln Ser Trp Arg Lys Glu Arg Val Leu Asn Val
            113                                                                     456
            CCC CTG TGC AAA GAG GAC TGT GAG CAA TGG TGG GAA GAT TGT CGC ACC TCC TAC ACC TGC
            Pro Leu Cys Lys Glu Asp Cys Glu Gln Trp Trp Glu Asp Cys Arg Thr Ser Tyr Thr Cys
            133                                                                     516
            AAG AGC AAC TGG CAC AAG GGC TGG AAC TGG ACT TCA GGG TTT AAC AAG TGC GCA GTG GGA
            Lys Ser Asn Trp His Lys Gly Trp Asn Trp Thr Ser Gly Phe Asn Lys Cys Ala Val Gly
            153                                                                     576
            GCT GCC TGC CAA CCT TTC CAT TTC TAC TTC CCC ACA CCC ACT GTT CTG TGC AAT GAA ATC
            Ala Ala Cys Gln Pro Phe His Phe Tyr Phe Pro Thr Pro Thr Val Leu Cys Asn Glu Ile
            173            T                                                        636
            TGG ACT CAC TCC TAC AAG GTC AGC AAC TAC AGC CGA GGG AGT GGC CGC TGC ATC CAG ATG
            Trp Thr His Ser Tyr Lys Val Ser Asn Tyr Ser Arg Gly Ser Gly Arg Cys Ile Gln MET
            193                                                                     696
            TGG TTC GAC CCA GCC CAG GGC AAC CCC AAT GAG GAG GTG GCG AGG TTC TAT GCT GCA GCC
            Trp Phe Asp Pro Ala Gln Gly Asn Pro Asn Glu Glu Val Ala Arg Phe Tyr Ala Ala Ala
            213                                                                     756
            ATG AGT GGG GCT GGG CCC TGG GCA GCC TGG CCT TTC CTG CTT AGC CTG GCC CTA ATG CTG
            MET Ser Gly Ala Gly Pro Trp Ala Ala Trp Pro Phe Leu Leu Ser Leu Ala Leu MET Leu
            233
                          771                       G                               829
            CTG TGG CTG CTC AGC TGA CCTCCTTTTACCTTCTGATACCTGAAAATCCCTGCCCTGTTCAGCCCCACAGCTC
            Leu Trp Leu Leu Ser End                   G
            253
                                                                                    895
            CCAACTATTTGGTTCCTGCTCCATGGTCGGGCCTCTGACAGCCACTTTGAATAAACCAGACACCGCacatgtgtcttgaga

            attatttggaaaaaaa
```

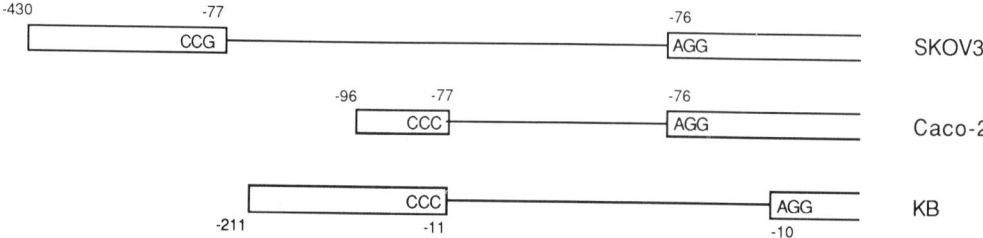

Figure 9.2b Schematic presentation (not to scale) of the proposed alternative RNA splicing in FBP transcripts based on sequences from SKOV3 (this chapter), Caco-2[15] and KB[13]. Cloned cDNA sequences are shown as shaded boxes and proposed intron regions as horizontal lines. Within the shaded boxes homologous sequences are aligned. Numbering is based on the AUG start codon.

morphism with variable bands at 4.2 kb (allele B1), 3.9 kb (allele B2) and 3.8 kb (allele B3) with frequencies of 0.61, 0.37 and 0.02, respectively.

*Msp*I detected a two allele polymorphism with fragment sizes of 4.5 kb (allele A1) or 3.2 kb (allele A2) as well as several constant bands (Figure 9.4, panel B). Among 22 unrelated individuals the allelic frequencies of A1 and A2 were 0.07 and 0.93, respectively. *Hae*III also revealed a two allele polymorphism (data not shown). The band sizes are approximately 3.8 kb and 2.8 kb, with frequencies of 0.63 and 0.37 respectively in 14 unrelated persons.

9.3.5 CHROMOSOME LOCALIZATION

Fluorescent *in situ* hybridizations[16,17] were performed on spreads of human metaphase chromosomes from peripheral blood lymphocytes using a cosmid clone containing part of the FBP region. Chromosome 11 was identified with a centromere probe. The positions of paired signals along chromosome arm 11q were measured on both chromatids of 14 copies of chromosome 11 from 6 metaphases. Figure 9.5 shows an example of a metaphase cell hybridized under these conditions. Analysis showed a mean FLcen-ter of 0.3024 with +/− standard deviation confidence limits of 0.2417–0.3631, corresponding to a location of 11q13.3-q14.1. No signals at other locations were observed.

9.3.6 EXPRESSION OF FBP IN CANCER CELL LINES

To confirm that an FBP was indeed the antigen to which MOv18 was reactive and to assess how widespread was its expression, various cell lines were examined by FACS using MOv18 and by northern blot analysis (Figure 9.6) using cHTMOv18 as probe. The results are summarized in Table 9.1. Surface expression of the MOv18 antigen among cell lines correlated with FBP mRNA levels in all but one case, HT29, where it appears that a post-translational modification results in a form of FBP which is not recognized by MOv18.

9.3.7 COPY NUMBER OF FBP GENES IN CELL LINES AND OVARIAN TUMOURS

To determine whether the wide variation in levels of FBP expression was due to difference in copy number of rearrangements, genomic DNA from 20 cell lines was digested with *Bam*HI and subjected to Southern analysis. No amplification of the FBP locus was evident in any cell line. However in HeLa and PE/01 deletion of some bands were observed.

We used the *Pst*I and *Hae*III polymorph-

Discussion

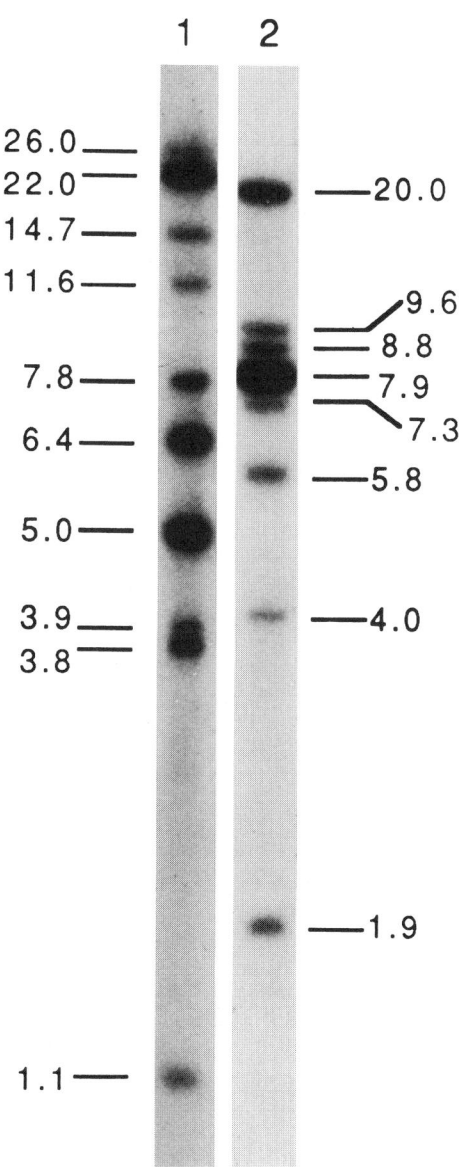

Figure 9.3 Hybridization of the cHTMOv18 probe with normal human DNA digested with *Bam*HI (lane 1) and *Eco*RI (lane 2) washed under stringent conditions. Sizes of fragments in kilobases are indicated. These were calculated from *Hin*dIII digested λ phage fragments run alongside.

isms noted above to look for allele loss at this locus with tumour/normal paired DNA samples on genomic Southern blots using cHTMOv18 as the probe. We have analysed 18 ovarian adenocarcinomas to date. Of the 15 that are polymorphic with the enzymes used, five show loss of one allele with reduplication. This is above the level of random loss seen on allelotypes[18]. Among these 15 samples a further three showed a 50% amplification making a total of 8/15 samples with alternations of the FBP locus. No amplification of this locus was observed in the remaining three samples which were non-informative for *Pst*I. The mapping of these polymorphisms will be important as the allele loss seen may have functional significance with regard to the FBP. The presence of gene amplification at this low level indicates that overexpression of FBP in ovarian cancer cannot be by this means.

9.4 DISCUSSION

We describe here the isolation of a cDNA sequence encoding the ovarian cancer-associated antigen recognized by MAb MOv18 and its identification as a high-affinity FBP. MOv18 does not appear to be recognizing an altered form of FBP unique to ovarian cancer since strong reactivity is observed with some normal tissue such as the renal distal and proximal tubules and fallopian tubes[7].

Folic acid plays an essential role in cellular biochemistry and FBPs are crucial to the assimilation, distribution and retention of this vitamin. Two classes of FBPs have been observed in a variety of normal and neoplastic tissue and body fluids (see reviews [19,20]). High-affinity FBPs constitute the major class and are characterized by their affinities for folate in the nanomolar range. They are found predominantly on the plasma membranes of tissues from placenta[13], kidney tubules[21], lactating mammary epithe-

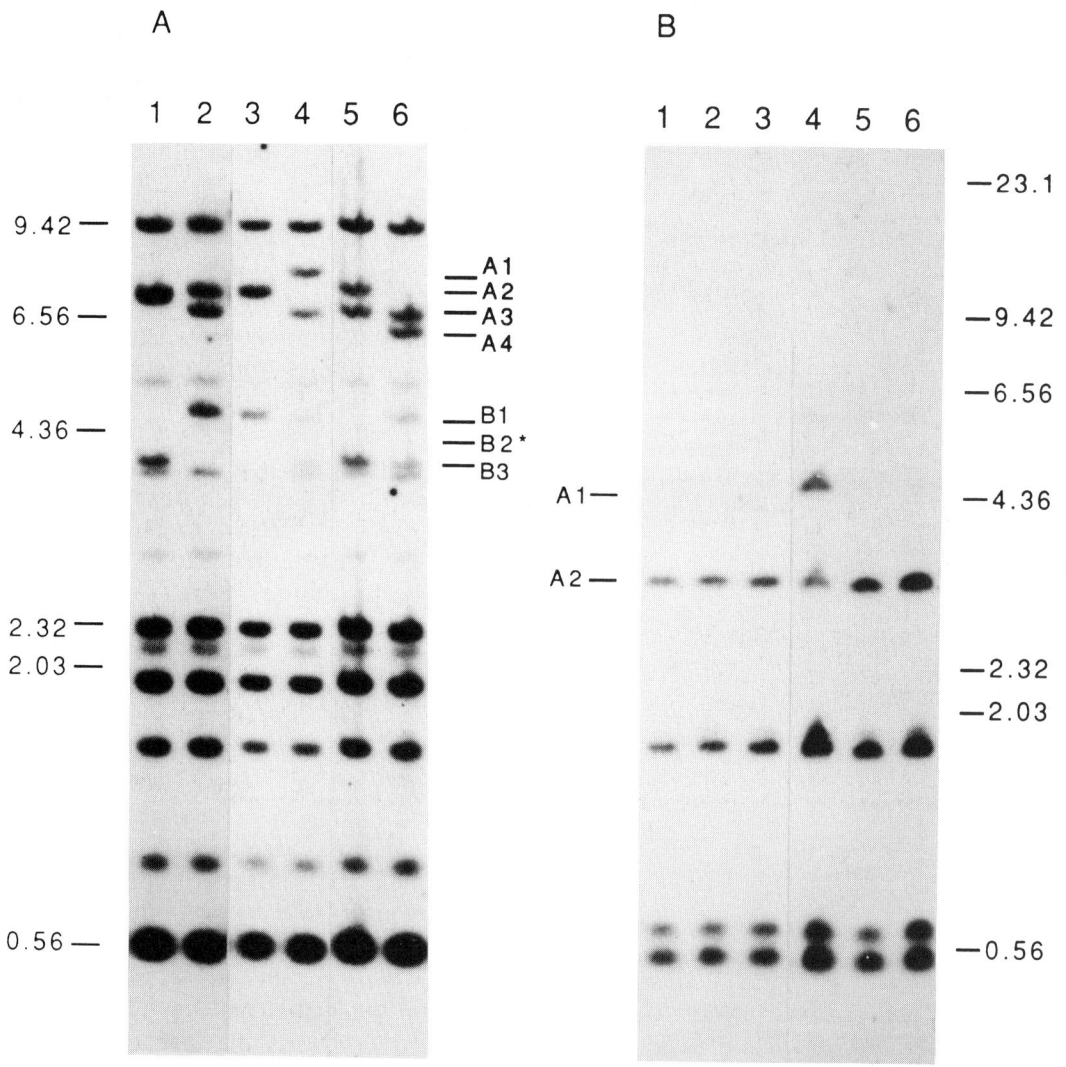

Figure 9.4 Southern blot of human genomic DNA from six unrelated individuals showing *Pst*I (Panel A) and *Msp*I (Panel B). Restriction fragment length polymorphisms detected by the cHTMOv18 probe and washed under stringent conditions. Sizes in kilobases of *Hind*III digested λ phage molecular weight markers are indicated. Fragments corresponding to the *Pst*I alleles are indicated on the right of panel A and to *Msp*I alleles on the left of panel B. *This allele is not present in any of these individuals but when present migrates to this position.

lium[19] and in some cultured cell lines [22–24] and mediate the transport of folates across plasma membranes via an endocytotic mechanism[25]. *In vitro* this class of FBP enables cells to proliferate in medium containing low folate concentrations[26–28]. The

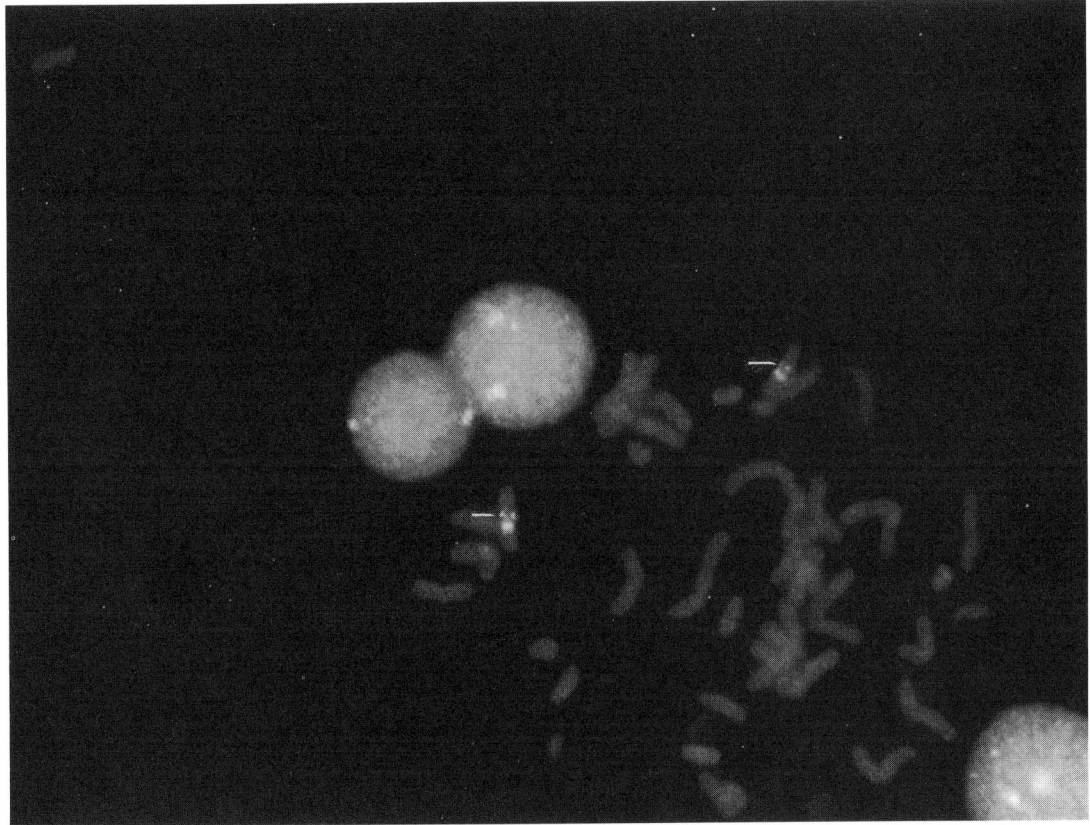

Figure 9.5 *In situ* hybridization of cosmid G53.3 and chromosome 11 centromere probe D11Z1 with metaphase chromosomes of a normal male. Markers indicate signals localized over 11q13.3–14.1.

membrane form of FBP is anchored to the cell membrane via a glycosyl-phosphatylinositol (GPI) linkage [15,29] and a soluble form is generated by cleavage of this GPI link, as is the case for alkaline phosphatase[30].

The role of high affinity FBPs in ovarian cancer is intriguing. It is possible that the attachment of FBP by a GPI link may act not only as a means of controlling cell surface expression[31] but may also be involved in cell activation or communication. Simultaneous with protein release by specific phospholipases would be the production of biologically active lipids such as 1,2-diacylglycerol and phosphatidic acid that could cross the plasma membrane and affect intracellular metabolism[30]. A number of lymphocyte proteins that use GPI anchors have been shown to be capable of mediating mitogenic responses and have been implicated as a second messenger of insulin action[32]. Alternatively, its involvement in ovarian cancer may be purely nutritional, reflecting the increased folate requirement of rapidly growing cells, although why this should be specific to ovarian cancer is unclear.

The mechanism by which FBP expression is elevated *in vitro* and *in vivo* does not in general involve gene amplification although

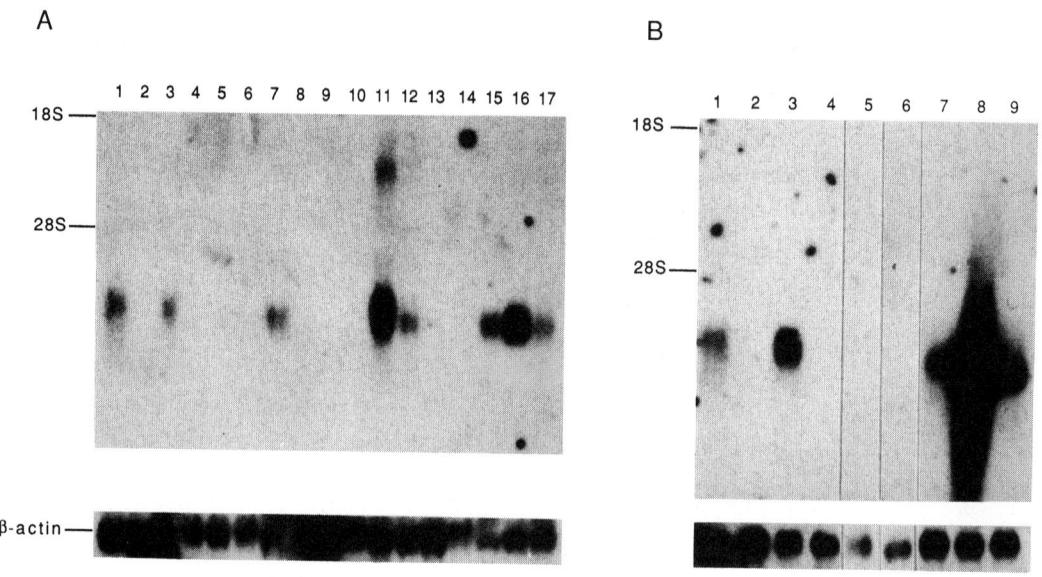

Figure 9.6 Northern hybridization analysis of cell lines with cHTMOv18 probe. 0.5 μg of poly [A]⁺RNA or 10 μg of total RNA were loaded onto 1% agarose/formaldehyde gels and electrophoresed at 25 V for 20 hours. The gel was blotted onto Hybond N⁺ and probed with ^{32}P-labelled cHTMOv18 (top panel) or β-actin as a loading control (bottom panel) and washed under stringent conditions. (A) Track 1, SKOV3; track 2, PE/01; track 3, DX3; track 4, SV80; track 5, PAF; track 6, MRC5; track 7, HCT-116; track 8, MANN; track 9, CC20; track 10, SW1222; track 11, SW620; track 12, DLD-1; track 13, LS174T; track 14, SW1417; track 15, HT29; track 16, OVCA432; track 17, OVCA433. (B) Track 1, SKOV3; track 2, PE/01; track 3, PE/04; track 4, JAMA-2; track 5, TR175; track 6, JA-1; track 7, LoVo; track 8, HeLa; track 9, HEp2.

the rearrangements evident in HeLa and PE/01 cell lines may involve regions containing regulatory elements. It is noteworthy that the 5′ non-coding regions of all the FBP cDNA sequences so far reported are different and may have functional significance. Alternative processing of RNA is a mechanism used by eukaryotic cells to generate multiple transcripts from a single gene and allows several possibilities for modulation of gene expression. Untranslated 5′ regions may affect mRNA stability or the translational capacity of the transcript allowing for expression in a tissue-specific manner[33,34]. A control mechanism that may have particular relevance to FBP expression in ovarian cancer is that of the rat insulin-like growth factor where transcripts containing one of three alternate 5′ untranslated regions are differentially regulated by growth hormone[35].

Although the role of FBP in ovarian cancer is unclear it nevertheless provides a unique opportunity to target such cells not only by specific MAb but also by virtue of the high affinity and high specificity of FBP for folates and folate analogues. Indeed the small molecular size of such analogues compared to MAbs should also be of considerable advantage. Alternative therapies involving antifolate drugs which can utilize or interfere with the activity of high affinity FBP may be useful in killing ovarian cancer cells which rely solely on these for folate acquisition. One such folate analogue, homofolate, has been

Table 9.1 Expression of CaMOv 18 among cell lines

Cell line	Source	FACS*	FBP mRNA†
SKOV3	Ovarian carcinoma	10	++
OVCA433	Ovarian carcinoma	5	++
PE/01	Ovarian carcinoma	0	−
PE/04	Ovarian carcinoma	9	++
JA-1	Ovarian carcinoma	0	−
TR175	Ovarian carcinoma	0	−
JAMA	Ovarian carcinoma	0	−
SW620	Colon adenocarcinoma	30	+++
HT29	Colon adenocarcinoma	1	++
SW1222	Colon adenocarcinoma	0	−
LS-174T	Colon adenocarcinoma	0	−
LoVo	Colon adenocarcinoma	15	+++
SW1417	Colon adenocarcinoma	nt	−
DLD-1	Colon adenocarcinoma	nt	++
HCT-116	Colon adenocarcinoma	nt	++
CC20	Colon adenocarcinoma	nt	−
MCF-7	Breast adenocarcinoma	3	nt
SkBR3	Breast metastatic pleural effusion	0	nt
NB100	Neuroblastoma	0	nt
EJ	Bladder carcinoma	0	nt
HT1080	Fibrosarcoma	0	nt
HeLa S3	Epitheloid carcinoma of cervix	60	++++
HEp-2	Epidermoid carcinoma—larynx	18	+++
MOLT-4	Lymphoblastic leukaemia	0	nt
DX3	Melanoma	nt	+
Mann	B-cell	nt	−
MRC5	Fetal lung—normal	nt	−
PAF	Fibroblast	nt	−
SV80	Fibroblast	nt	−

* Relative fluorescence with MOv18 measured in arbitrary units.
† Summary of northern hybridization with cHTMOv18 probe shown in Figure 9.6.
Indication of intensity of signal above background: −, no signal; +, 2–5×; ++, 5–10×; +++, 10–100×; ++++, >100×. nt, not tested.

used successfully *in vitro* and specifically inhibits the growth of cells utilizing the high affinity FBPs but not those using other folate transport systems[36].

The location of the FBP locus on 11q13.3-14.1 places it near *INT2*, the human homologue of the murine mammary virus integration site[37]. Other oncogenes and putative tumour suppressor genes have been mapped to this region, including the *HST1* (heparin binding secretory transforming factor) gene [38], and *PP1* (protein phosphatase)[39]. Translocations involving 11q13 and 14q32.3 are frequently observed in several human pathologies and a rare folate-sensitive fragile site is located at 11q13.3[40]. Studies of this important region may benefit from the *Pst*I, *Msp*I and *Hae*III polymorphisms in the FBP loci we have identified.

The isolation of cDNA clones for the MOv18 antigen represents a significant advance in the understanding of the nature and function of tumour markers and is one of only a few reports where the genes for such antigens have been cloned[41–43]. This information should assist in elucidating the mechanism(s) involved in tumour growth and also provide insight into devising alternative therapies. Further study into the genomic organization of the FBP genes is warranted in order to identify the factors that regulate FBP gene expression and to assess the possible involvement of alternative transcripts in the elevation of FBP expression in ovarian cancer.

REFERENCES

1. Bast, R.C. Jr, Boyer, C.M., Olt, G.J. *et al.* (1990) Identification of markers for early detection of epithelial ovarian cancers, in *Ovarian Cancer. Biological and therapeutic challenges* (eds F. Sharp, W.P. Mason and R.E. Leake), Chapman & Hall, London, pp. 265–75.
2. Bast, R.C. Jr, Feeney, M., Lazarus, H. *et al.* (1981) Reactivity of a monoclonal antibody with human ovarian carcinoma. *J. Clin. Invest.*, **68**, 1331–7.
3. Bast, R.C. Jr, Hunter, V. and Knapp, R.C. (1987) Pros and cons of gynecological tumor markers. *Cancer*, **60**, 1984–92.
4. Hakomori, S. (1984) Tumor-associated carbohydrate antigens. *Annu. Rev. Immunol.*, **2**, 103–26.
5. Girling, A., Bartkova, J., Burchell, J. *et al.* (1989) A core protein epitope of the polymorphic epithelial mucin detected by the monoclonal antibody SM-3 is selectively exposed in a range of primary carcinomas. *Int. J. Cancer*, **43**, 1072–6.
6. Miotti, S., Canevari, S., Menard, S. *et al.* (1987) Characterization of human ovarian carcinoma-associated antigens defined by novel monoclonal antibodies with tumor-restricted specificity. *Int. J. Cancer*, **39**, 297–303.
7. Veggian, R., Fasolato, S., Menard, S. *et al.* (1989) Immunohistochemical reactivity of a monoclonal antibody prepared against human ovarian carcinoma on normal and pathological female genital tissues. *Tumori*, **75**, 510–13.
8. Boerman, O.C., van Niekerk, C.C., Makkink, K. *et al.* (1989) Comparative immunohistochemical study of four monoclonal antibodies directed against ovarian carcinoma-associated antigens. *Int. J. Gynecol. Pathol.* (in press).
9. Gendler, S.L., Burchell, J.M., Duhig, T. *et al.* (1987) Cloning of partial cDNA encoding differentiation and tumor-associated mucin glycoproteins expressed by human mammary epithelium. *Proc. Natl Acad. Sci. USA*, **84**, 6060–4.
10. Seed, B. (1987) An LFA-3 cDNA encodes a phospholipid membrane protein homologous to its receptor CD2. *Nature*, **329**, 840–2.
11. Aruffo, A. and Seed, B. (1987) Molecular cloning of a CD28 cDNA by a high-efficiency COS cell expression system. *Proc. Natl Acad. Sci. USA*, **84**, 8573–7.
12. Chu, G., Hayakawa, H. and Berg, P. (1987) Electroporation for the efficient transfection of mammalian cells with DNA. *Nucl. Acid Res.*, **15**, 1311–26.
13. Elwood, P.C. (1989) Molecular cloning and characterization of the human folate-binding protein cDNA from placenta and malignant tissue culture (KB) cells. *J. Biol. Chem.*, **264**, 14893–901.
14. Sadasivan, E. and Rothenberg, S.P. (1989) The complete amino acid sequence of a human folate binding protein from KB cells determined from the cDNA. *J. Biol. Chem.*, **264**, 5806–11.
15. Lacey, S.W., Sanders, J.M., Rothberg, K.G. *et al.* (1990) Complementary DNA for the folate binding protein correctly predicts anchoring to the membrane by glycosyl-phosphatidylinositol. *J. Clin. Invest.*, **84**, 715–20.
16. Pinkel, D., Landegent, J., Collins, C. *et al.* (1985) Fluorescence *in situ* hybridization with human chromosome-specific libraries: Detection of trisomy 21 and translocations of chromosome 4. *Proc. Natl Acad. Sci. USA*, **85**, 9138–42.
17. Lichter, P., Chang, Tang, C., *et al.* (1989) High-resolution mapping of human chromosome 11 by *in situ* hybridisation with cosmid clones. *Science*, **247**, 264–8.
18. Vogelstein, B., Fearon, E.R., Kern, S.E. *et al.* (1989) Allelotype of colorectal carcinomas. *Science*, **244**, 207–11.

19. Henderson, G.B. (1990) Folate binding proteins. *Annu. Rev. Nutr.*, **10**, 319–35.
20. Kane, M.A. and Waxman, S. (1989) Biology of disease; Role of folate binding proteins in folate metabolism. *Lab. Invest.*, **60**, 737–46.
21. Selhub, J., Nakamura, S. and Carone, F.A. (1987) Renal folate absorption and the kidney folate binding protein: II. Microinfusion studies. *Am. J. Physiol.*, **252**, F757–F760.
22. McHugh, M. and Cheng, Y.C. (1979) Demonstration of a high-affinity folate binder in human cell membranes and its characterization in human KB cells. *J. Biol. Chem.*, **254**, 11312–18.
23. Jansen, G., Westerhof, G.R., Kathmann, I., et al. (1989) Identification of a membrane-associated folate-binding protein in human leukemic CCRF-CEM cells with transport-related methotrexate. *Cancer Res.*, **49**, 2455–9.
24. Henderson, G.B., Tsuji, J.M. and Kumar, H.P. (1988) Mediated uptake of folate by a high affinity binding protein in sub-lines of L1210 cells adapted to nanomolar concentrations of folate. *J. Membr. Biol.*, **101**, 247–58.
25. Rothberg, K.G., Ying, Y., Kolhouse, J.F. et al. (1990) The glycophospholipid-linked folate receptor internalizes folate without entering the clathrin-coated endocytic pathway. *J. Cell Biol.*, **110**, 637–49.
26. Kamen, B.A. and Capdevila, A. (1986) Receptor-mediated folate accumulation is regulated by cellular folate content. *Proc. Natl Acad. Sci. USA*, **83**, 5983–7.
27. Jansen, G., Kathmann, I., Rademaker, B.C. et al. (1989) Expression of a folate binding protein in L1210 cells grown in low folate medium. *Cancer Res.*, **49**, 1959–63.
28. Kane, M.A., Elwood, P.C., Portillo, R.M. et al. (1988) Influence on immunoreactive folate binding proteins of extracellular folate concentration in cultured human cells. *J. Clin. Invest.*, **81**, 1398–406.
29. Alberti, S., Miotti, S., Fornaro, M. et al. (1990) The Ca-MOv18 molecule, a cell-surface marker of human ovarian carcinomas, is anchored to the cell membrane by phosphatidylinositol. *Biochem. Biophys. Res. Comm.*, **171**, 1051–5.
30. Low, M.G. and Saltiel, A.R. (1988) Structural and functional roles of glycosyl-phosphatidylinositol in membranes. *Science*, **239**, 268–75.
31. Low, M.G. and Prasad, A.R.S. (1988) A phospholipase D specific for the phosphatidylinositol anchor of cell-surface proteins is abundant in plasma. *Proc. Natl Acad. Sci. USA*, **85**, 980–4.
32. Low, M.G. (1990) Glycosyl-phosphatidylinositol: a versatile anchor for cell surface proteins. *FASEB J.*, **3**, 1600–8.
33. Leff, S.E. and Rosenfeld, M.G. (1986) Complex transcriptional units: diversity in gene expression by alternative RNA processing. *Annu. Rev. Biochem.*, **55**, 1091–117.
34. Chobert, M.N., Lahuna, O., Lebargy, F. et al (1990) Tissue specific expression of two γ-glutamyl transpeptidase mRNAs with alternative 5′ ends encoded by a single copy gene in the rat. *J. Biol. Chem.*, **265**, 2352–7.
35. Low, W.L. Jr, Roberts, C.T. Jr, Lasky, S.R. and LeRoith, D. (1987) Differential expression of alternative 5′ untranslated regions in mRNAs encoding rat insulin-like growth factor. *Proc. Natl Acad. Sci. USA*, **84**, 8946–50.
36. Henderson, G.B. and Strauss, B.P. (1990) Growth inhibition by homofolate in tumor cells utilizing a high-affinity folate binding protein as a means for folate internationalization. *Biochem. Pharmacol.*, **39**, 2019–25.
37. Casey, G., Smith, R., McGillvray, D. et al. (1986) Characterization and chromosome assignment of the homolog of int-2, a potential proto-oncogene. *Mol. Cell. Biol.*, **6**, 502–10.
38. Yoshida, M.C., Wada, M., Satoh, H. et al. (1988) Human HST1 [HSTF1] gene maps to chromosome band 11q and co-amplifies with the INT2 gene in human cancer. *Proc. Natl Acad. Sci. USA*, **85**, 4861–4.
39. Barker, H.M., Jones, T.A., Da Cruze Silva, E.F. et al. (1990) Localization of the gene encoding a type I protein phosphatase catalytic subunit to human chromosome band 11q13. *Genomics*, **7**, 159–66.
40. Julier, C., Nakamura, Y., Lathrop, M. et al. (1990) A detailed genetic map of the long arm of chromosome 11. *Genomics*, **7**, 335–45.
41. Zimmerman, W., Ortlieb, B., Friedrich, R. and von Kleist, S. (1987) Isolation and characteriza-

tion of cDNA clones encoding the human carcinoembryonic antigen reveal a highly conserved repeating structure. *Proc. Natl Acad. Sci. USA*, **84**, 2960–4.

42. Hotta, H., Ross, A.H., Huebner, K. *et al.* (1988) Molecular cloning and characterization of an antigen associated with early stages of melanoma tumor progression. *Cancer Res.*, **48**, 2955–62.

43. Strnad, J., Hamilton, A.E., Beavers, L.S. *et al.* (1989) Molecular cloning and characterization of a human adenocarcinoma/epithelial cell surface antigen complementary DNA. *Cancer Res.*, **49**, 314–17.

Chapter 10
Cytokines and ovarian cancer

S.T.A. MALIK, M.S. NAYLOR and F.R. BALKWILL

10.1 INTRODUCTION

Cytokines are low molecular weight regulatory proteins that control many physiological and pathological processes. Cytokines are of importance to ovarian cancer for two main reasons: first, some cytokines may have therapeutic value in this disease, either by acting directly on the tumour, or by stimulating a host reaction to the disease; second, and paradoxically, some cytokines may contribute to the progression of the disease by acting as autocrine growth factors or stimulating tumour cell spread. In this chapter we describe our recent findings concerning cytokines and ovarian cancer in animal models and in biopsy samples from patients with the disease. Our work has mainly focused on the cytokines tumour necrosis factor (TNF), interleukin 1 (IL-1), and interferon-γ (IFN-γ).

10.2 TUMOUR NECROSIS FACTOR AND INTRAPERITONEAL TUMOURS IN ANIMALS

We have studied the effect of intraperitoneal (i.p.) TNF administration on human ascitic ovarian cancer xenograft models[1]. TNF therapy prolonged survival in two out of three xenograft lines, but, paradoxically, restored the ability of all three non-TNF producing ovarian tumours to form peritoneal tumour implants. Exogenous TNF administration therefore led to a pathological picture reminiscent of metastatic human ovarian cancer. The cytokine IL-1 also promoted invasion of the peritoneal surface, but no other cytokine that we have studied, including IFN-α, IFN-γ, IL-6 and IL-8, had this property (Malik et al., in preparation). To study the possible role of tumour-produced TNF on peritoneal implantation and invasion, the behaviour of Chinese hamster ovary (CHO) cells that had been transfected with the human TNF gene (CHO/TNF cells) was compared to CHO cells that contained the transfection vector alone (CHO/NEO cells)[2]. CHO/TNF cells showed enhanced implantation to the surface of the peritoneum and liver, and also metastasized to the lungs. Furthermore, the metastatic capability of CHO/TNF cells could be specifically abrogated by injection of antibodies to TNF. Properties of TNF that could potentially contribute to promotion of metastasis include

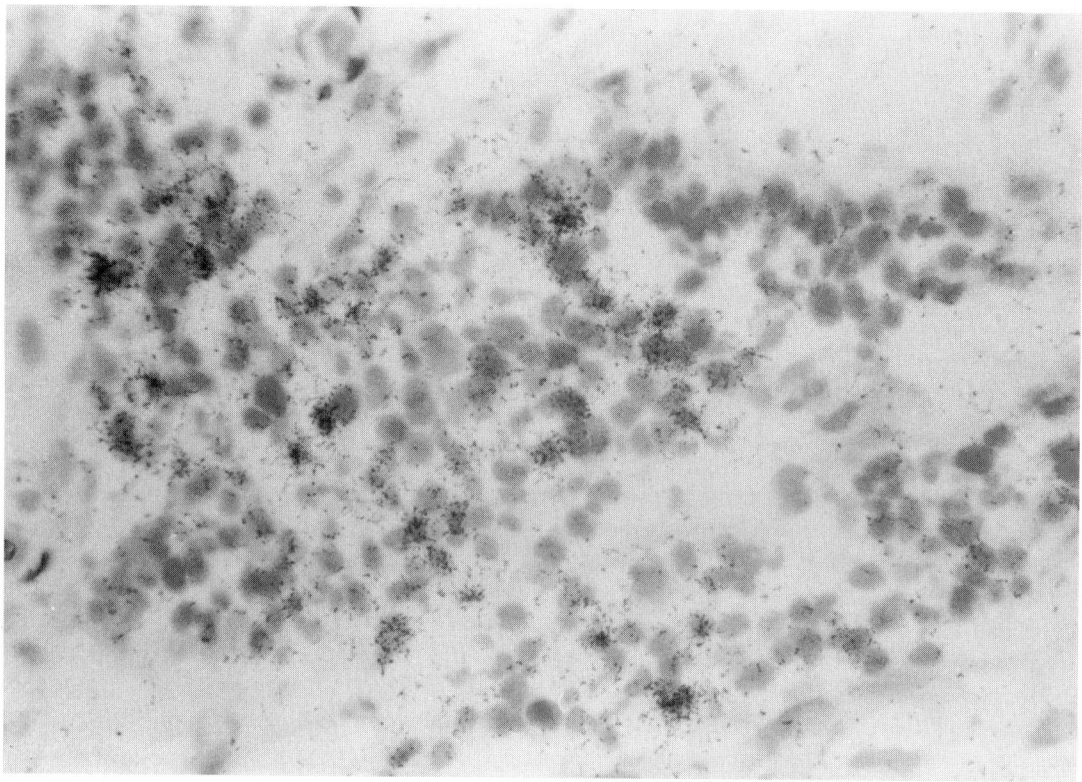

Figure 10.1 *In situ* expression of mRNA for TNF in a biopsy of human ovarian cancer.

stimulation of angiogenesis[3], enhancement of tumour cell adhesion to host cells[4], and induction of tumour cell metalloproteinases[5]. The known procoagulant activities of TNF[6], and the induction of platelet-derived growth factor[7], could contribute to the generation of tumour stroma, an important step in the establishment of tumours in metastatic sites[8].

10.3 CYTOKINE PRODUCTION IN HUMAN OVARIAN CANCER

Our observations in the i.p. animal models led us to look for cytokine production in primary tumour material from patients with ovarian cancer. Using *in situ* hybridization and immunohistochemistry ([9] and unpublished data) we have studied 50 biopsies taken at laparotomy for the presence of mRNA and protein for TNF and IL-1. In over 50% of cases a minority of cells in the epithelial areas of the tumour contained TNF mRNA. In individual high power fields as many as 40% of cells were positive for TNF message. Immunohistochemical studies on sequential sections, and the morphology of the positive cells, led us to the conclusion that the ovarian tumour cells were transcribing the TNF gene. The distribution of the TNF-producing cells within the tumours was of interest. Cells expressing TNF mRNA were not randomly scattered throughout the tumour, but occurred in clusters of high expression (Figure 10.1) whereas other areas of the tumour had no TNF expressing cells at all. Over 99.9% of cells expressing TNF mRNA were of epithelial origin as assessed by morphology and immunostaining for tumour-associated antigens on sequential sections. We also found immunohistochemical evidence of the production of TNF protein by the tumour cells and TNF protein in a tumour lysate. TNF protein was found in tumour cells and also, in contradiction to the mRNA localization, in some stromal areas. The possible reasons for this discrepancy are several: for instance, TNF mRNA may be more stable in the tumour cells than in stromal cells; the level of mRNA in the stromal cells may be below the level of detection of *in situ* hybridization; or the protein found in stromal areas may have been produced by the tumour cells. Whatever the explanation, the presence of TNF in human ovarian cancer tissues may influence the biology of the disease and alter response to therapy.

We are currently correlating TNF production with response to therapy and disease course in our series of 50 patients. By extrapolation from our data in the animal models, we would predict that tumour cells expressing TNF may have an increased ability to spread. We are therefore studying biopsies from the advanced stages of ovarian cancer, and cells from ascitic tumours.

We have also found evidence for IL-1 production in the tumour. Cells expressing IL-1 mRNA and its protein product are not so frequent as those producing TNF, and are found in both tumour and stromal areas.

In a limited number of samples we have looked for IL-6 and IFN-γ expression using *in situ* hybridization. Cells expressing mRNA for these cytokines are rarely found (<0.01% of cells in the tumour) in our primary untreated biopsies.

There are reports of constitutive production of several other cytokines by human ovarian cancer cell lines and in fresh tumour biopsy material. Macrophage colony stimu-

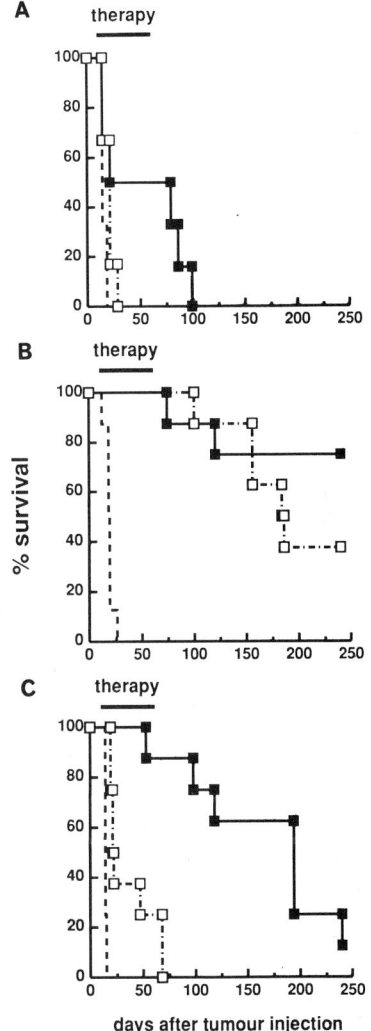

Figure 10.2 Survival of mice bearing ascitic xenografts of human ovarian cancer. Mice were injected i.p. with tumour cells. There were 6–8 mice in each experimental group, rHu IFN-γ therapy started i.p. 7 days later for 8 weeks. A, OS xenograft; B, HU xenograft; C, LA xenograft. –·–, diluent treated mice; –·□·–, rHuIFN-γ 1.5×10^5 U twice weekly; –■–, rHu IFN-γ 1.5×10^5 U daily. There was a significant increase in survival in all treated mice in B and C ($P < 0.001$). Twice weekly therapy was less effective than daily therapy in C ($P < 0.002$).

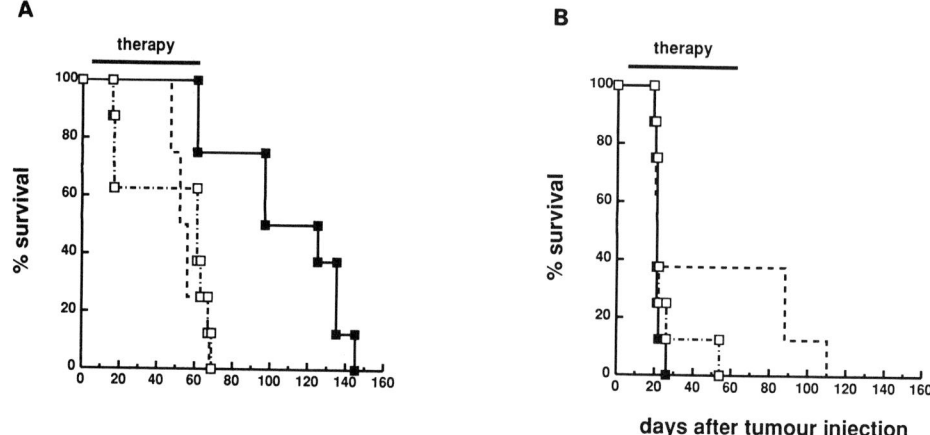

Figure 10.3 Survival of mice bearing solid peritoneal tumour xenografts of human ovarian cancer. Mice were injected i.p. with tumour cells. There were 8 mice in each experimental group. After 7 days, mice received rHuTNF 1 μg/mouse i.p. for 7 days. After 1 day's rest, 8 weeks i.p. rHu IFN-γ therapy began. A, HU xenograft; B, LA xenograft. –·–·, diluent treated mice; –·□·–, rHu IFN-γ 1.5×10^5 U twice weekly; –■–, rHu IFN-γ 1.5×10^5 U daily. In A, daily IFN therapy gave a significant increase in survival ($P < 0.005$).

lating factor (M-CSF) and expression of the c-*fms* oncogene (encoding the M-CSF receptor) has been demonstrated in human ovarian cancers and ovarian cancer cell lines[10,11], but not in normal ovarian epithelium. Indeed, serial measurements of serum M-CSF in patients with ovarian cancer may be useful in monitoring disease activity[12]. A monocyte chemotactic factor produced by ovarian cancer cell lines[13], is now known to be the TNF and IL-1 inducible cytokine monocyte chemotactic protein (MCP-1)[14]. IL-6 production is transiently induced in gonadotrophin-primed hyperstimulated ovaries [15], can be detected in the ascitic fluid of ovarian cancer patients[16], and is constitutively produced by human ovarian cancers and ovarian cancer cell lines[17].

Thus epithelial ovarian cancers and ovarian cancer cell lines produce inappropriately a number of cytokines. These cytokines can potentially perpetuate their own biological effects by recruiting cytokine-producing host cells and inducing further cytokine production. Experimental evidence indicates a potential role for at least two cytokines, TNF and IL-1, in the promotion of ovarian cancer metastasis, and two others (M-CSF and the ligand for the c-*erb*B2 proto-oncogene) as putative autocrine growth factors in ovarian cancer.

10.4 THERAPEUTIC POTENTIAL OF IFN-γ

One result of our findings in the ovarian cancer xenograft models was the ability to generate two stages of the disease in these mice, ascitic tumour, and, after a short course of TNF therapy, bulky solid tumour. Recent trials of i.p. IFN-γ in human ovarian cancer have yielded promising results[18] and we have attempted to model this. We have thus investigated the action of recombinant human IFN-γ, rHuIFN-γ, against human ovarian cancer xenografts growing as ascites or as bulky solid i.p. tumours in nude mice (Figures 10.2 and 10.3). Both forms of the disease responded to i.p. IFN-γ with significant increases in mouse survival time, and in two of three ascitic models the mice were cured of their peritoneal disease (Figure 10.2). The activity of IFN-γ was dose and schedule dependent and xenografts derived from three different patients showed a heterogeneity of response. Peak i.p. levels of rHuIFN-γ in nude mice bearing multiple i.p. solid tumours were similar to those found in ovarian cancer patients receiving i.p. rHuIFN-γ, but clearance was more rapid in the mice. The rHuIFN-γ appeared to have direct effects on the tumour cells, because rat IFN-γ had no antitumour activity at the same doses and schedules, although it had some biological activity in the nude mice. Histological examination of rHuIFN-γ treated tumours revealed increased necrosis and loss of cellular organization with large areas of hypocellular epithelial mucin (Malik et al., submitted for publication).

We are currently using these models to investigate mechanisms of action and markers predictive of response of ovarian cancer to IFN-γ.

10.5 CONCLUSION

An understanding of the *in situ* production of cytokines by the different stages of ovarian cancer may allow the development of more specific therapies. Cytokines may be acting to create a hormone-modulated tumour-promoting environment in this disease. If so, future therapies for ovarian cancer may involve negating the effects of these cytokines, either by inhibition of cytokine secretion, inhibition of the recruitment of cytokine-producing cells, or treatment with specific cytokines or their antagonists.

REFERENCES

1. Malik, S.T.A., Griffin, D.B., Fiers, W. and Balkwill, F.R. (1989) Paradoxical effects of tumour necrosis factor in experimental ovarian cancer. *Int. J. Cancer*, **44**, 918–25.
2. Malik, S.T.A., Naylor, M.S., Oliff, A. and Balkwill, F.R. (1990) Cells secreting tumour necrosis factor show enhanced metastasis in nude mice. *Eur. J. Cancer*, **26**, 1031–4.
3. Frater-Schroder, M., Risau, W., Hallmann, R. et al. (1987) Tumor necrosis factor type alpha, a potent inhibitor of endothelial cell growth *in vitro*, is angiogenic *in vivo*. *Proc. Natl Acad. Sci. USA*, **84**, 5277–81.
4. Rice, G.E., Gimborne, M.A. and Bevilacqua, M.P. (1988) Tumour cell endothelial interactions. Increased adhesion of human melanoma cells to activated vascular endothelium. *Am. J. Pathol.*, **133**, 204–10.
5. Ito, A., Sato, T., Iga, T. and Mori, Y. (1990) Tumour necrosis factor bifunctionally regulates matrix metalloproteinases and tissue inhibitor of metalloproteinases (TIMP) production by human fibroblasts. *FEBS Lett.*, **269**, 93–5.
6. Pober, J.S. (1987) Effects of tumour necrosis factor and related cytokines on vascular endothelial cells. *Ciba Found. Symp.*, **131**, 170–84.
7. Hajjar, K.A., Hajjar, D.P., Silverstein, R.L. and Nackman, R.L. (1987) Tumour necrosis factor-mediated release of platelet derived growth factor from cultured endothelial cells. *J. Exp. Med.*, **166**, 235–45.
8. Dvorak, H.F. (1986) Tumours: Wounds that do not heal. *N. Engl. J. Med.*, **315**, 1650–9.
9. Naylor, M.S., Malik, S.T.A., Jobling, T. et al. (1990) Demonstration of tumour necrosis factor in human ovarian cancer by *in situ* hybridisation. *Eur. J. Cancer*, **26**, 1027–30.
10. Ramakrishnan, S., Xu, F.L., Brandt, S.J. et al. (1989) Constitutive production of macrophage-colony stimulating factor by human breast and ovarian cancer cell lines. *J. Clin. Invest.*, **83**, 921–6.
11. Kacinski, B.M., Carter, D., Mittal, K. et al. (1990) Ovarian adenocarcinomas express *fms*-complementary transcripts and *fms* antigen, often with coexpression of CSF-1. *Am. J. Pathol.*, **137**, 135–47.
12. Kacinski, B.M., Stanley, E.R., Carter, D. et al. (1989) Circulating levels of CSF-1 (M-CSF) a lymphohaemopoietic cytokine may be a useful

marker of disease status in patients with malignant ovarian neoplasms. *Int. J. Radiat. Oncol. Biol. Phys.*, **17**, 159–64.
13. Bottazzi, B., Ghezzi, P., Tarabolletti, G. *et al.* (1985) Tumor-derived chemotactic factors from human ovarian carcinoma: evidence for a role in the regulation of macrophage content of neoplastic tissues. *Int. J. Cancer*, **36**, 167–73.
14. Bottazzi, B., Collotta, F., Nobili, N. *et al.* (1990) A chemoattractant expressed in human sarcoma cells (tumour derived chemotactic factor, TDCF) is identical to monocyte chemoattractant protein 1/monocyte chemotactic and activating factor (MCP-1/MCAF). *Int. J. Cancer*, **45**, 795–7.
15. Motro, B., Itin, A., Sachs, L. and Keshet, E. (1990) Pattern of interleukin 6 gene expression *in vivo* suggests a role for this cytokine in angiogenesis. *Proc. Natl Acad. Sci. USA*, **87**, 3092–6.
16. Erroi, A., Sironi, M., Chiaffarino, F. *et al.* (1989) IL-1 and IL-6 release by tumor associated macrophages from human ovarian carcinoma. *Int. J. Cancer*, **44**, 795–801.
17. Watson, J.M., Sensintaffar, J.L., Berek, J.S. and Martinez-Maza, O. (1990) Constitutive production of interleukin-6 by ovarian cancer cell lines and by primary ovarian tumour cultures. *Cancer Res.*, **50**, 6959–65.
18. Pujade-Lauraine, E., Guastella, J.P., Colombo, N. *et al.* (1991) Intraperitoneal human r-IFN gamma as treatment of residual carcinoma (OC) and laparotomy (SLL). *Proc. ASCO*, **10**, 195.

Chapter 11

Characterization of cytokines produced by ovarian cancer cells

L.F. CARSON, M.M. MORADI, B.-Y. LI,
M.C. OLSON, D. MOHANRAJ, S.A. ELG
and S. RAMAKRISHNAN

11.1 INTRODUCTION

Ovarian carcinoma is one of the few intra-abdominal malignancies associated with the production of ascitic fluid. The composition of ascitic fluid has not been completely characterized. Gross analyses of the cellular contents of the fluid have shown the presence of tumour cells admixed with varying numbers of leucocytes. It is conceivable that the factors secreted by the tumour cells and the infiltrating leucocyte populations would play an important role in the progression of the disease. The biological relevance of some of the factors secreted by the tumour cells which accumulate in ascitic fluid will be discussed below.

11.2 INTERACTION BETWEEN THE TUMOUR CELLS AND THE HOST

The mechanism by which cancer cells escape from homeostatic control leading to uninhibited growth is yet to be fully understood. Some of the contributory factors have been well established. They include viral infection, genetic rearrangements, abnormal activation of cellular genes/oncogenes and inappropriate production of growth factors. In ovarian cancer no clear evidence has been thus far provided for specific genetic lesions or association of viral infection. However, several recent investigations have indicated the overexpression of certain proto-oncogenes and growth factor genes in the tumour cells. In the light of this new information, the interactions between tumour cell and host cells are being reassessed.

The immune cells present at the tumour site are believed to restrain the growth of cancer cells. However, cells from the host may also inadvertently facilitate tumour cell growth by paracrine mechanisms. Some of the earlier work demonstrated the existence of a paracrine loop between the macrophages present at the tumour site and the cancer cells. For example, when macrophage conditioned media were added to cultures of ovarian cancer cell lines *in vitro* it was found to increase the clonogenic growth. The increased cloning efficiency was not restricted to autologous macrophages but even a heterologous source of macrophage conditioned media was able to support the growth[1]. Additional evidence stems from depletion experiments wherein the removal of macrophages from ascitic fluid reduced the clonogenic potential of the ovarian cancer cells. While these studies indicate a positive role for the macrophages in the growth of tumour cells, other investigators have noticed a direct influence of the tumour cells on the production of myeloid cells or its activation. Tumour cell transplantation in experimental animals, for example, has been found to increase the number of myeloid cells in circulation. Similar changes in granulo-myelopoiesis have been noted in some cancer patients also. Besides increasing the production of myeloid/lymphoid cells, tumour cell-derived growth factors or substances could be chemotactic in attracting the immune cells to the tumour site[1].

11.3 EXPRESSION OF GROWTH FACTORS/GROWTH FACTOR RECEPTORS

In many tumour systems, cellular homologues of retroviral oncogenes (proto-oncogenes) are activated. Many of these proto-oncogenes code for either a growth factor or growth factor receptor. One of the best examples is the proto-oncogene c-*sis* that produces the B-chain of the platelet-derived growth factor[2]. Similarly increased levels of c-*erb*B2 oncogene were seen in a variety of cancers including ovarian, breast, glial and squamous cell carcinomas. c-*erb*B2 codes for the truncated homologue of EGF receptor lacking the extracellular domain[3,4]. Recent studies have shown that the proto-oncogene HER-2/*neu* is amplified in a significant fraction of ovarian cancers. The expression of HER-2/*neu* is correlated with poor prognosis[5]. Another group of proto-oncogenes such as c-*fos* and c-*jun* code for nuclear binding proteins which are involved in the transcriptional regulation of various cellular genes. Activation of these genes could indirectly lead to tumour cell proliferation. Yet another group of oncogenes code for proteins which are involved in signal transduction pathways. These include GTP binding proteins and tyrosine kinases.

11.4 STUDIES ON MACROPHAGE COLONY STIMULATING FACTOR (M-CSF)

Since a greater number of macrophages are noted in association with ovarian cancer we investigated whether the tumour cells could be producing colony stimulating factors related to the ontogeny of myeloid lineage. Initially, our attention was focused on six different ovarian cancer cell lines. Using a specific radioimmunoassay we demonstrated that all cancer cell lines constitutively produced M-CSF. In addition to ovarian cancer cell lines, three out of five breast cancer cell lines also secreted detectable but low levels of this growth factor. Expression of M-CSF gene was determined initially by estimating the steady-state levels of the mRNA. All the ovarian cancer cell lines showed multiple transcripts for M-CSF[6]. The relative amounts of these transcripts were however varied between cell lines. Most of them showed an abundance of larger transcripts (4.2 kb). Presence of transcripts alone is not sufficient to conclude that ovarian cancer cells produce M-CSF. Therefore, experiments were carried out to find out whether M-CSF protein could be detected in these cells. In competitive inhibition assays (radioimmunoassay and radioreceptor assay) culture supernatants of ovarian cancer cell lines were found to contain a molecule which was similar to the authentic M-CSF. Finally, the biological activity of this molecule was estab-

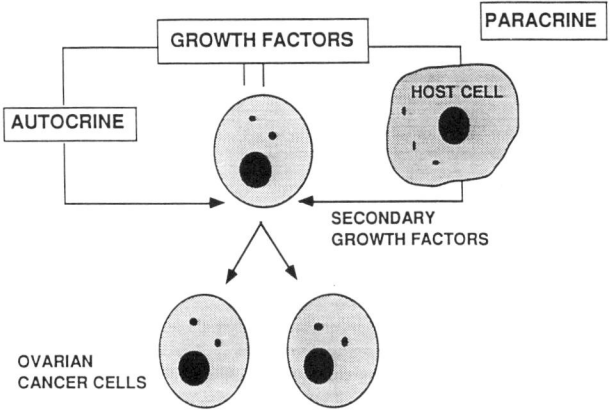

Figure 11.1 Ovarian cancer cell microenvironment.

lished in a bone marrow colony forming assay, wherein inclusion of culture supernatants from ovarian cancer cell lines induced the formation of macrophage colonies. The increased production of macrophage colonies was blocked by the pre-incubation with a polyclonal antiserum generated against recombinant M-CSF which confirmed the nature of the colony stimulating factor.

Parallel studies in primary tumours obtained from patients showed positive transcripts for M-CSF and M-CSF receptor. When soluble proteins were extracted from tumour tissues, there were significant amounts of immunoreactive M-CSF (unpublished work). The biological relevance of M-CSF production by ovarian tumour cells is yet to be fully understood. However, the constitutive production could lead to eventual accumulation in circulation. Therefore, it was important to investigate whether ovarian cancer patients would have an increased serum level of this factor. Sera obtained from primary ovarian cancer patients did show an increased level of M-CSF[7]. Normal control subjects had a mean level of 1.16 ng/ml of M-CSF, whereas the cancer patients had almost three-fold higher amounts of M-CSF in circulation. These studies led to the evaluation of this factor as a possible tumour marker. It must be noted at this point that M-CSF levels could be increased above the basal levels in non-malignant states such as infection. Our studies along with others clearly indicate that M-CSF may have a potential use in the detection of ovarian cancer when used in conjunction with other tumour markers such as CA-125. Further studies are needed to evaluate critically the clinical utility of M-CSF in ovarian cancer.

We also investigated whether the levels of M-CSF would have a predictive value in patients having second-look laparotomy. The majority of these patients had normal or below normal levels of the established tumour marker CA-125. The sera from 33 ovarian cancer patients undergoing second-look laparotomy were analysed for M-CSF and showed significantly higher levels compared to normal subjects[8].

Ovarian cancer cell lines not only produced M-CSF but also coexpressed its receptor[9]. After serum starvation, significant amounts of transcripts related to M-CSF receptor could be found in the majority of these cells. The proto-oncogene c-*fms* is homologous to the M-CSF receptor with some minor differences at the cytoplasmic domain. M-CSF receptor belongs to the tyrosine kinase receptor family and its coexpression along with the ligand alluded to the possibility of an autocrine loop.

11.5 PROPERTIES OF M-CSF

M-CSF is a homodimeric glycoprotein of 90 kDa and is involved in the ontogeny of macrophages. Analyses of cDNA clones indicate that M-CSF is produced with a 32 amino acid long signal peptide, followed by a polypeptide of about 224 residues long. Several studies suggest that at least one-third of the molecule at the C-terminal end is removed prior to its secretion[10,11]. Pluripotent stem cells in bone marrow differentiate to various lineages of functionally distinct populations under the influence of hormone-like colony stimulating factors. The proliferative effect of M-CSF is strictly restricted to mononuclear phagocytic lineage [11–14].

Some studies suggest that M-CSF could also function as a chemotactic factor in attracting tissue macrophages[15]. Treatment of macrophages with M-CSF activates bacteriocidal properties of the phagocytic cells by the increased production of hydrogen peroxide. Other studies have shown that exposure to M-CSF could enhance the tumoricidal properties of macrophages[16–20]. M-CSF is produced by many cell types including endothelial, fibroblast, stromal, some cells of epithelial origin as well as monocytes and lymphocytes[21–26].

The levels of constitutive production of M-CSF varies among these cell populations. Exogenous inductive signals, such as mitogen stimulation, treatment with bacterial lypopolysaccharides, or other related growth factors, significantly induce its production [27–30]. Recent reports have shown an interplay between many of the CSFs. For example, macrophage proliferation is synergized by TNF-α and CSF-1[28]. TNF could increase the production of GM-CSF, M-CSF, and IL-1; GM-CSF could increase IL-1 α, β, and TNF-α by mononuclear cells[29]. TNF could also increase the secretion of GM-CSF. Similarly, IL-1 can induce GM-CSF[31] and M-CSF could stimulate macrophages to produce interferon, TNF, and other CSF activities[32].

11.6 STUDIES ON OTHER CYTOKINES

We have recently extended these studies to other growth factors/cytokines. Total cellular RNA and poly A selected RNA isolated from ovarian cancer cells were analysed for the presence of a number of cytokines. While all the ovarian cancer cell lines were positive for M-CSF production, only some of them coexpressed a related cytokine, GM-CSF, which is involved in the production of granulocyte/macrophage lineage. None of the cancer cell lines had detectable amounts of granulocyte colony stimulating factor (G-CSF). Interestingly, ovarian cancer cells seem to produce both forms of IL-1 (unpublished work). There were no detectable levels of transcripts related to other lymphokines such as IL-2 and IL-3. However, significant amounts of transcripts could be found for IL-6[33]. It is interesting to note that ovarian cancer cells produce such a variety of haemopoietic growth factors some of which are involved in the inflammatory process. While one can speculate on the possible roles for these factors in the progression of ovarian cancers it is suggested that at least IL-6 may not be an autocrine factor[33]. IL-1 and TNF-α secreted by the tumour cells could be important in the manifestation of cachexia in ovarian cancer patients. In animal model systems using human xenografts, production of IL-1 and TNF-α has been seen in association with cachexia. Injection of neutralizing antibodies to either of these factors resulted in the

Table 11.1 Growth factors secreted by ovarian cancer cells

Growth factor/cytokine	Biological activity	Reference
M-CSF	Production/differentiation of macrophages	[6]
GM-CSF	Production/differentiation of granulocytes/macrophages	unpublished
IL-1α	Acts on T cells, B cells	unpublished
IL-1β	Acts on T cells, B cells	unpublished
TNF-α	Induces cachexia	unpublished
IL-6	Pleotropic. Acts on stem cells, hepatocytes, T cells, B cells, keratinocytes	[33]
TGF-β	Negative growth regulatory factor. Immunosuppressive	unpublished

reversal of weight loss in tumour-bearing athymic mice[34]. Production of multiple cytokines could lead to synergistic action on host cells to produce secondary growth factors which may play an important role in the proliferation of tumour cells. Such factors have been found in ovarian ascitic fluid[35]. Injection of human ovarian ascitic fluid and not benign peritoneal fluid was able to support the tumorigenic properties of an established ovarian cell line in mice. While the nature of these molecules is not yet identified, it is likely that novel therapeutic strategies could evolve by understanding the nature and composition of various growth factors secreted by ovarian cancer cells *in vivo*. A summary of growth factors secreted by ovarian cancer cells is shown in Table 11.1.

11.7 CONCLUSIONS

Tumour cells produce various growth factors/cytokines. Some of these factors stimulate the cells from which they are derived while others act on a distinct population of host cells and induce the production of secondary growth factors (paracrine loop). Understanding the mechanisms of the interrelationship between host cells and the cancer cells could be important in developing novel therapeutic strategies. These factors could also be used in early detection of ovarian cancer. In addition, growth factors could be utilized in targeted therapy. For example, by modulating the type of growth factors secreted by the tumour cells, their chemo/radiosensitivity could be increased. If we could identify the type of receptor molecules that are abundantly expressed in ovarian cancer cells, then we could target toxin polypeptides via the respective ligand. Such methods could be useful in selectively eliminating microscopic nodules of the tumour without affecting the normal cells.

ACKNOWLEDGEMENTS

This work was supported in part by a Grant from the National Cancer Institute, CA-48068, Women's Cancer Center, University of Minnesota, and Cancer Research Foundation of America.

REFERENCES

1. Buick, R.N., Fry, S.E. and Salmon, S.E. (1980) Effect of host-cell interactions on clonogenic

1. carcinoma cells in human malignant effusions. *J. Cancer*, **41**, 695–704.
2. Waterfield, M.D., Scarce, G.T., Whittle, H. et al. (1983) Platelet-derived growth factor is structurally related to the putative transforming protein p28sis of simian sarcoma virus. *Nature*, **304**, 35–9.
3. King, C.R., Kraus, M.H. and Aaronson, S.A. (1985) Amplification of a novel v-*erb*-B-related gene in a human mammary carcinoma. *Science*, **229**, 974–6.
4. Varmus, H.E. (1984) The molecular genetics of cellular oncogenes. *Annu. Rev. Genet.*, **18**, 553–612.
5. Slamon, D.J., Godolphin, W., Jones, L.A. et al. (1989) Studies of the HER-2/*neu* proto-oncogene in human breast and ovarian cancer. *Science*, **244**, 707–10.
6. Ramakrishnan, S., Xu, F.J., Brandt, S.J. et al. (1989) Constitutive production of macrophage colony-stimulating factor by human ovarian and breast cancer cell lines. *J. Clin. Invest.*, **83**, 921–6.
7. Xu, F.J., Ramakrishnan, S., Daly, L. et al. (1991) Increased serum levels of macrophage colony-stimulating factor in ovarian cancer. *Obstet. Gynecol.* (in press).
8. Elg, S.A., Yin, Y.Y, Carson, L.F. et al. (1991) Serum levels of macrophage colony-stimulating factor in ovarian cancer patients undergoing second look laparotomy. *Obstet. Gynecol.* (in press).
9. Kacinski, B.M., Carter, D., Khushbakhat, M. et al. (1990) Ovarian adenocarcinomas express *fms*-complementary transcripts and *fms* antigen, often with coexpression of CSF-1. *Am. J. Pathol.*, **137**, 135–47.
10. Stanley, E.R., and Heard, P.M. (1977) Factors regulating macrophage production and growth. Purification and some properties of the colony-stimulating factor from medium conditioned by mouse L cells. *J. Biol. Chem.*, **252**, 4305–12.
11. Stanley, E.R. (1985) The macrophage colony-stimulating factor, CSF-1. *Methods Enzymol.*, **116**, 564–87.
12. Metcalf, D. (1989) The molecular control of cell division, differentiation commitment and maturation in haematopoietic cells. *Nature*, **339**, 27–30.
13. Ralph, P., Warren, M.K. and Nakoinz, I. et al. (1986) Biological properties and molecular biology of the human macrophage growth factor, CSF-1. *Immunobiology*, **172**, 194–204.
14. Hamburger, A.W. and White, C.P. (1986) Growth factors for human tumour clonogenic cells elaborated by macrophages isolated from human malignant effusions. *Cancer Immunol. Immunother.*, **22**, 186–90.
15. Wang, J.M., Griffin, J.D., Rambaldi, A. et al. (1988) Induction of monocyte migration by recombinant macrophage colony-stimulating factor. *J. Immunol.*, **141**, 575–9.
16. Ralph, P. and Nakoinz, I. (1986) CSF-1 stimulates macrophage tumoricidal activity.
17. Wing, E.J., Magee, D.M., Whiteside, T.L. et al. (1989) Recombinant human granulocyte/macrophage colony-stimulating factor enhances monocyte cytotoxicity and secretion of tumour necrosis factor alpha and interferon in cancer patients. *Blood*, **15**, 643.
18. Li, H., Schwinzer, R., Baccarini, M. and Lohmann-Matthes, M.L. (1989) Cooperative effects of colony-stimulating factor-1 and recombinant interleukin 2 on proliferation and induction of cytotoxicity of macrophage precursors generated from mouse bone marrow cell cultures. *J. Exp. Med.*, **169**, 973–86.
19. Mufson, R.A., Aghajaninan, J., Wong, G. et al. (1989) Macrophage colony-stimulating factor enhances monocyte and macrophage antibody-dependent cell-mediated cytotoxicity. *Cell. Immunol.*, **119**, 182–92.
20. Hume, D.A., Donahue, R.E. and Fieler, I.J. (1989) The therapeutic effect of human recombinant macrophage colony-stimulating factor (CSF-1) in experimental murine metastatic melanoma. *Lymphokine Res.*, **8**, 69–77.
21. Herrmann, F., Cannistra, S.A. and Griffin, J.D. (1986) T cell monocyte interactions in the production of humoral factors regulating granulopoiesis *in vitro*. *J. Immunol.*, **136**, 2858.
22. Felix, R., Fleish, H. and Elford, P.R. (1989) *Calcif. Tissue Int.*, **44**, 356–60.
23. Becker, S., Devlin, R.B. and Haskill, J.S. (1989) Differential production of tumour necrosis factor, macrophage colony stimulating factor and interleukin 1 by human alveolar macrophages. *J. Leukoc. Biol.*, **45**, 353–61.

24. Sieff, C.A., Niemeyer, C.M., Mentzer, S.J. and Faller, D.V. (1988) Interleukin-1, tumour necrosis factor, and the production of colony-stimulating factors by cultured mesenchymal cells. *Blood*, **72**, 1316–23.
25. Quesenberry, P.J. and Gimbrone, M.A. (1980) Vascular endothelium as a regulator of granulopoiesis: Production of colony stimulating activity by cultured human endothelial cells. *Blood*, **56**, 1060.
26. Tsai, S., Emerson, S.G., Sieff, C.A. and Nathan, D.G. (1986) Isolation of a human stromal cell secreting hemopoietic growth factors. *J. Cell. Physiol.*, **127**, 137.
27. Aizawa, S., Tsurusawa, M. and Mori, K.J. (1986) Effect of hydrocortisone and bacterial lipopolysaccharide on colony-stimulating activity production from mouse narrow adherent cells, spleen cells and peritoneal macrophages in vitro. *Int. J. Cell Cloning*, **4**, 415–23.
28. Branch, D.R., Turner, A.R. and Guilbert, L.J. (1989) Synergistic stimulation of macrophage proliferation by the monokines tumour necrosis factor-alpha and colony-stimulating factor 1. *Blood*, **73**, 307–11.
29. Kaushansky, K., Broudy, V.C., Harlan, J.M. and Adamson, J.W. (1988) *J. Immunol.*, **141**, 3410.
30. Sisson, S.D. and Dinarello, C.A. (1988) Production of interleukin-1 alpha, interleukin-1 beta and tumour necrosis factor by human mononuclear cells stimulated with granulocyte-macrophage colony-stimulating factor. *Blood*, **72**, 1368–74.
31. Lu, L., Walker, D. and Graham, C.D. et al. (1988) Enhancement of hematopoietic colony-stimulating factors CSF-1 and G.CSF by recombinant human tumour necrosis factor-alpha: synergism with recombinant human interferon-gamma. *Blood*, **72**, 34–41.
32. Warren, M.K. and Ralph, P. (1986) Macrophage growth factor CSF-1 stimulates human monocyte production of interferon, tumour necrosis factor and colony stimulating activity. *J. Immunol.*, **137**, 2281–5.
33. Watson, J.M., Sensintaffar, J.L., Berek, J. and Martinez-Maza, O. (1990) Constitutive production of interleukin 6 by ovarian cancer cell lines and by primary ovarian tumour cultures. *Cancer Res.*, **50**, 6959–65.
34. Gelin, J., Moldawer, L.L., Lonnroth, C. et al. (1991) Role of endogenous tumour necrosis factor alpha and interleukin 1 for experimental tumour growth and the development of cancer cachexia. *Cancer Res.*, **51**, 415–21.
35. Mills, G.B., May, C., Hill, M. et al. (1990) Ascitic fluid from ovarian cancer patients contains growth factors necessary for intraperitoneal growth of human ovarian adenocarcinoma cells. *J. Clin. Invest.*, **86**, 851–5.

Chapter 12
Role of interleukin-6 in ovarian cancer

J.S. BEREK, J.M. WATSON
and O. MARTINEZ-MAZA

12.1 INTRODUCTION

Of all gynaecological malignancies, ovarian cancer has the lowest overall survival. Most patients have advanced stage disease at the time of diagnosis[1]. Failure to detect the disease in the early stages is partly due to the lack of suitable tumour markers. The ovarian cancer-associated marker, CA-125, has been the most well-characterized tumour marker in ovarian cancer, and it has been useful in helping to monitor the course of disease in many patients with non-mucinous epithelial malignancies[2]. Also, ovarian malignancies frequently metastasize throughout the peritoneal cavity prior to the development of symptoms[1], and thus better methods of early detection are needed.

Tumour development has often been associated with either oncogene expression and amplification or inappropriate growth factor regulation. With the exception of the recently identified HER-2/*neu* oncogene[3], which was found to be overexpressed in approximately 30% of ovarian malignancies and was usually indicative of poor clinical prognosis, there has been little success in correlating significant oncogene expression and amplification with ovarian tumours[4]. The HER-2/*neu* oncogene[3] fits into both categories, since HER-2/*neu* is both an oncogene and appears to encode for an epidermal growth factor (EGF) receptor-like module. Macrophage colony stimulating factor (M-CSF, also known as CSF-1) is a cytokine that is constitutively produced by ovarian cancer cells[5]. Levels of serum M-CSF have been reported to correlate with serum CA-125 levels[6]. Interestingly, the receptor for M-CSF (M-CSF-R) is encoded by the cellular form of an oncogene, c-*fms*[7]. These observations support the concept that inappropriate growth factor production, regulation and/or responsiveness plays an important role in ovarian tumour development and progression.

A cytokine that might play a role in the development and progression of ovarian cancer is interleukin 6 (IL-6). IL-6 (also known as BSF-2, IFN-β2, and hybridoma/plasmacytoma growth facctor) is a 26-kDa glycoprotein consisting of 212 amino acids, with a wide range of biological effects on a variety of cell types[8,9]. The gene for IL-6 has been cloned and sequenced, and the structure of the IL-6 gene described[10]. Also,

101

the receptor for IL-6 (IL-6R) has been cloned [9,11]. It is composed of two polypeptides: an 80-kDa IL-6-binding protein, which is associated on IL-6 binding to a second, 130-kDa signal-transducing molecule. Recently, NF-IL6, a nuclear factor that binds to a cytokine-responsive (IL-1) element in the IL-6 regulatory region, was described[9,12]. The NF-IL6 protein has significant homology with the *fos* and *myc* oncogene products, and interacts with the NF-IL6-binding motif (AGATTGCACAATCT) within the IL-6 gene regulatory region, inducing IL-6 gene expression. Normally, NF-IL6 is not expressed, but its expression is induced by exposure of IL-6-producing cells to various inducers of IL-6 gene expression, including lipopolysaccharide (LPS) or cytokines.

The biological activities that have been described for IL-6 include: induction of differentiation of activated B cells, support of plasmacytoma and myeloma growth, induction of acute phase reactants and stimulation of hepatocytes, nerve growth factor-like activity, induction of cytotoxic T cells, and support of haemopoietic differentiation[8,9]. IL-6 can be produced by several types of cells, including T lymphocytes, monocyte/macrophages, fibroblasts, certain tumour cells, epithelial cells, and endometrial cells. Clearly, IL-6 is a pleiotropic factor, with both growth and differentiation inducing properties.

IL-6 is produced by cardiac myxoma tumour cells[13], by human epidermal cells and epidermoid carcinoma cells[15]. In fact, IL-6 has been shown to act as an autocrine or paracrine growth factor for several types of human tumours, including multiple myeloma[16,17] and renal cancer[18]. IL-6 transgenic mice develop plasmacytomas[19]. In our own recent work, we found that IL-6 acts as an autocrine growth factor for AIDS-associated Kaposi's sarcoma (KS)[20]. However, the role of IL-6 in the host response to cancer and as a tumour growth promoting or suppressing factor has not been completely elucidated. In a recent report, IL-6 was shown to act as an IL-1 inducible, autocrine growth-suppressive factor for malignant melanoma[21]. Therefore, IL-6 can potentially have both growth-inducing and suppressing effects. IL-6 also could have profound effects on the activation of the immune system in tumour-bearing hosts, leading to either the augmentation or suppression of antitumour responses. Recent studies showed that IL-6 is rapidly produced following implantation of chemically induced sarcomas in mice[22], and that IL-6 is induced *in vitro* and *in vivo* by exposure to tumour necrosis factor (TNF), IL-1, or IL-2[23]. It is not clear whether IL-6 is being produced in response to tumour growth, or is being produced by the tumour itself and is acting as a growth factor.

Since the majority of ovarian carcinomas arise from the coelomic epithelium that covers the ovarian surface[1] and epithelial cells produce IL-6[24], we examined the production of IL-6 from several established ovarian carcinoma cell lines and from freshly excised epithelial ovarian tumours. We found that ovarian cancer cell lines, as well as primary cultures of ovarian tumour cells, produce IL-6[25]. Also, elevated levels of IL-6 were detected in the ascitic fluid[25] and serum[26] of women with ovarian cancer.

Follicular development involves many cytokines in complex interaction. Recent reports indicate that ovarian stroma and follicles contain IL-6[27,28]. This IL-6 gene expression appears to be related to angiogenesis[27]. Also normal ovarian germinal epithelial cells produce detectable levels of IL-6[29]. Thus, IL-6 may normally function as an ovarian paracrine growth factor. In tumorigenesis, neo-angiogenesis is a critical step [30]. Therefore, the overexpression of IL-6 might play a role in the genesis of ovarian neoplasia in a manner that helps to support the growth of the tumour. With tumour growth, ovarian cancer-produced IL-6 might be secreted into the peritoneal cavity and then into the systemic circulation.

In summary, the factors involved in the generation and maintenance of ovarian cancer are not completely understood. However, it is clear that cytokines, including IL-6, can play an important role in tumour growth; cytokines can function as growth factors for tumour cells, and also can affect the host immune response to cancer.

12.2 EPITHELIAL OVARIAN CANCER CELLS PRODUCE IL-6

Since ovarian cancer cells derive from the ovarian epithelium and epithelial cells are known to produce IL-6, we examined the production of IL-6 from several established ovarian cell lines, as well as from excised epithelial ovarian tumours[25]. IL-6, measured by bioassay or enzyme-linked immunosorbent assay (ELISA), was detected in culture supernatants of several established ovarian tumour cell lines, as well as supernatants from primary cultures of ovarian cancer. In initial experiments, culture supernatants from human ovarian cancer cell lines, CAOV-3, OVCAR-3 and SKOV-3 were found to contain IL-6, while PA-1 culture supernatants did not contain any detectable IL-6 (Figure 12.1, Table 12.1). Subsequently, we tested several other ovarian cancer cell lines for IL-6 production, and found that OC436, A2780, OC8, and C30 produced IL-6, while OC222 and an adriamycin-resistant A2780 subline did not produce detectable IL-6 (not shown). Therefore, seven of ten ovarian cancer cell lines tested secreted detectable levels of IL-6 into their culture supernatant.

Supernatant IL-6 levels were dependent on cell concentration and increased with time in culture (Figure 12.1). Interestingly, the levels of IL-6 produced by each of these ovarian cancer cell lines is different: CAOV-3 produces significantly more IL-6 than SKOV-3, while OVCAR-3 produces several fold less IL-6 than SKOV-3 (Figure 12.1, Table 12.1). The levels of IL-6 in culture supernatants continued to increase with time, even after the cultures had reached cellular confluency, although IL-6 levels did eventually plateau (Figure 12.1b). The serum content of the medium used to culture ovarian cancer cell lines had different effects on IL-6 secretion, depending on the cell line tested[25]. For instance, while CAOV-3 and SKOV-3 cells produced decreasing levels of IL-6 when cultured in decreasing fetal bovine serum (FBS) concentrations (ranging from 10% FBS to 2% FBS, to serum-free Iscove's medium), decreasing FBS concentration had little effect on IL-6 production by OVCAR-3 cells (not shown). We concluded from these experiments that IL-6 producing ovarian cancer cell lines constitutively produce IL-6.

IL-6 production and activity was measured in several ways[25]. IL-6 activity in culture supernatants was measured using an IL-6 bioassay based on an IL-6 dependent murine hybridoma cell line (MH60.BSF-2, kindly provided for these studies by Tadamitsu Kishimoto, Osaka, Japan), as previously described [31]. Essentially, these IL-6 dependent cells were exposed to serially diluted supernatants (or hrIL-6 standards), cultured for 42 hours, pulsed with ^{3}H-thymidine, harvested, and the radioactivity counted. The levels of IL-6 in culture supernatants were determined by comparison with a standard curve derived using hrIL-6. This MH60.BSF-2 cell line has been shown to be solely responsive to IL-6; other cytokines, including IL-1-α, IL-1-β, IL-2, IL-3, IL-4, IL-5, interferon (IFN)-β, IFN-γ, and granulocyte colony stimulating factor (G-CSF) do not induce cell growth[31]. The specificity of IL-6 produced by these ovarian cancer cell lines in stimulating the proliferation of the MH60.BSF-2 cells was shown by the ability of a rabbit anti-human IL-6 polyclonal serum (Ig fraction) to significantly reduce the IL-6 activity in the culture

Role of interleukin-6 in ovarian cancer

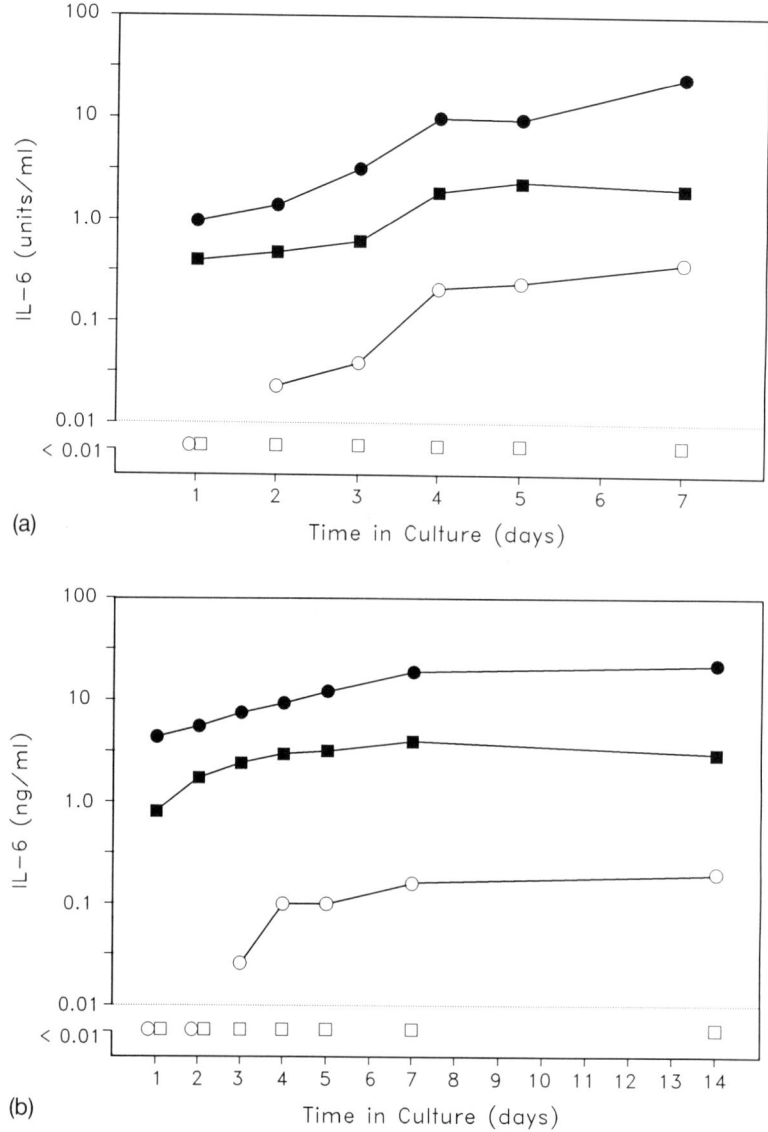

Figure 12.1 Epithelial ovarian cancer cell lines produce IL-6. Supernatants from ovarian cancer cell lines collected 5 days after passage at 10×10^4 cells/ml were tested for IL-6 biological activity by the proliferative response of IL-6 dependent MH60.BSF2 hybridoma cells (a), and for the presence of antigenically related IL-6 by an IL-6 specific ELISA (b). Data points represent mean value of duplicate supernatants. ●, CAOV-3; ■, SKOV-3; ○, OVCAR-3; □, PA-1 culture supernatants.

supernatant. When the antiserum was used at a final concentration of 0.2 µg/µl, the IL-6 activity of CAOV-3 was markedly decreased by >90% of the control CAOV-3 supernatant. IL-6 activity in SKOV-3 and OVCAR-3 supernatants was reduced by >90% and >80% respectively at a final antibody concentration of 0.1 µg/µl.

Table 12.1 Ovarian cancer cell lines produce IL-6

Cell line	IL-6 activity (U/ml)*	IL-6 concentration (ng/ml)†
CAOV-3	8.4 ± 0.9‡	9.1 ± 0.4
OVCAR-3	0.2 ± 0.02	0.2 ± 0.1
PA-1	≤0.03	≤0.05
SKOV-3	2.5 ± 0.3	2.7 ± 0.5
Medium	≤0.03	≤0.05

* Results determined by MH60.BSF-2 bioassay.
† Results determined by ELISA.
‡ All results expressed as mean ± SEM from triplicate experiments using 5-day supernatants harvested from cells plated at 10×10^4 cells/ml.

IL-6 concentration in culture supernatants was also determined using an IL-6 specific sandwich ELISA[20]. This ELISA employed a monoclonal anti-human IL-6 primary antibody (kindly provided by F. Takatsuki, Ajinomoto, Kawasaki, Japan) coated onto plastic microplates. Supernatants or hrIL-6 standards were added, incubated and washed, then rabbit anti-IL-6 polyclonal serum (Genzyme) was added. The plates were then incubated with an anti-rabbit horseradish peroxidase coupled goat serum (Tago) prior to development with the o-phenylenediamine (OPD) substrate. Again, IL-6 activity was determined by comparison with a curve generated from hrIL-6 standards. Generally, IL-6 levels determined by bioassay and ELISA correlated well.

The CAOV-3, NIH-OVCAR-3 (OVCAR-3), SKOV-3, and PA-1 epithelial ovarian cancer cell lines were obtained from and maintained according to directions from American Type Culture Collection. All cells were grown in tissue culture flasks and split after reaching confluency. In cultures of primary ovarian cancer cell isolates, the cells were washed and fed prior to reaching confluency (2–3 weeks) and the supernatant collected 14 days later. In kinetics experiments, cell lines were passaged after reaching confluency and supernatants were collected at various time points after passage (24, 48, 72, 96, 120 hours, 7 and 14 days).

It is unlikely that the IL-6 produced by these ovarian cancer cells is a result of some exogenous stimulus, such as endotoxin contamination of the medium used to culture these cells. These ovarian cancer cell lines produced IL-6 when cultured in medium known not to induce IL-6 production by unstimulated peripheral blood mononuclear cells, which produce IL-6 when exposed to very low levels of endotoxin. This medium had a LPS content of 0.16 ng/ml, as determined by the *Limulus* assay. Also, the medium used failed to support the growth of the IL-6-responsive test cell line (MH60.BSF-2 cells), indicating that the culture medium itself did not contain any IL-6 activity. Finally, while LPS (0.1–1 μg/ml) did induce IL-6 production by some ovarian cancer lines (CAOV-3 and OVCAR-3), LPS did not induce increase IL-6 production by OC436 cells, which constitutively secrete IL-6 (not shown).

Fibroblasts and monocytes have been shown to produce multiple forms of IL-6[32]. To determine the molecular mass forms of IL-6 synthesized and secreted by ovarian cancer cell lines, we immunoprecipitated IL-6 from the culture supernatants of ovarian cancer cell lines biosynthetically labelled with ^{35}S-*trans*-methionine. The supernatant was collected and precipitated with anti-IL-6 polyclonal antisera (Genzyme), followed by the addition of protein-A conjugated to sepharose beads (Pharmacia). Antigen-antibody complexes were released from protein-A–sepharose by boiling, separated by polyacrylamide gel electrophoresis, and analysed by autoradiography. All four ovarian cancer lines tested produced a polyclonal anti-IL-6 antibody, including the line PA-1, which does not appear to secrete biologically active IL-6[25]. The molecular mass of the IL-6 secreted by these ovarian cancer cell lines was 24 kDa, with little or no heterogeneity in the molecular mass of ovarian cancer-produced IL-6, as has been shown for IL-6 produced by epidermal cells and epidermoid carcinoma cell lines[33].

IL-6 expression by ovarian cancer cells also was examined by immunoperoxidase staining; all ovarian cancer cell lines tested stained positive for cytoplasmic IL-6 (Table 12.2)[25]. All of the CAOV-3, SKOV-3 and OVCAR-3 cells were immunoperoxidase positive using a monoclonal anti-IL-6 antibody. There was slight heterogeneity in the intensity of the staining for cytoplasmic IL-6. However, this staining heterogeneity was also seen in positive control slides, when cells were stained for MHC class I (not shown). The PA-1 cell line, which does not appear to secrete IL-6 as determined by bioassay and ELISA, was lightly positive for cytoplasmic IL-6[25]. Immunoperoxidase staining was done by culturing ovarian cancer cells on slides, which were then washed in PBS, fixed in acetone, and air dried. Prior to staining, cells were washed in PBS and incubated with blocking reagent (mouse serum). Primary antibody (murine monoclonal anti-IL-6), or isotype control antibody (IgG$_1$), was then added, followed by biotinylated antibody (horse anti-mouse IgG), ABC (Vector), and AEC substrate solution (freshly prepared), with colour development checked by light microscopy.

Slot blot mRNA analysis revealed the presence of IL-6 specific mRNA in SKOV-3 and CAOV-3 cell lines, but failed to detect OVCAR-3 and PA-1 mRNA specific for IL-6[25]. The levels of IL-6 mRNA did not correlate with the amount of secreted IL-6, suggesting that mRNA stability or translation efficiency could differ among the cell lines tested[25]. There is some precedent for the detection of a biologically active protein without the detection of associated mRNA, as in the case of OVCAR-3 produced IL-6: Ramakrishnan and co-workers[5] have shown detectable M-CSF activity in culture supernatants of ovarian cancer cell lines, including OVCAR-3, while failing to detect corresponding specific cytokine mRNA.

The receptor for IL-6 has been defined and cloned[11]. It is composed of an 80-kDa IL-6 binding polypeptide that associates with a second, 130-kDa signal-transducing polypeptide on binding to IL-6. Recent preliminary experiments indicate that ovarian cancer cell lines constitutively express high levels of IL-6 receptor (80 kDa) mRNA, and suggest that IL-6 could act as an autocrine or paracrine growth factor for ovarian cancer cells (Watson, in preparation).

Isolation and analysis of mRNA from the ovarian cancer cell lines was performed as previously described[20,34,35]. Briefly, cells were trypsinized, washed, and lysed with 4 M guanidium isothiocyanate to extract the total cellular RNA. RNA was pelleted by caesium chloride ultracentrifugation, resuspended in water, quantified by spectrophotometry, blotted onto nylon membranes, and probed with the ^{32}P-labelled 440-bp *Taq*1-*Ban*II fragment of BSF-2/IL6 cDNA (pBSF-2.38, kindly provided by Toshio Hirano, Osaka, Japan)[10].

Table 12.2 Ovarian cancer cell lines and primary tumour cells stain positive for cytoplasmic IL-6 by immunoperoxidase staining

	% Positive cells	Staining intensity
Established lines		
CAOV-3	>98	++/+++
OVCAR-3	>98	+++
PA-1	>75	±
SKOV-3	>98	++
Primary tumour cells		
449s	>75	±
451	>80	±
453	>80	±
455	>80	±

±, faint, cytoplasmic staining; +, majority of cells lightly stained; ++, areas of dark staining; +++, areas of intense staining.

Table 12.3 Modulation of IL-6 secretion from ovarian cancer cells by TNF-α, IL-1β, IFN-γ

	IL-6 concentration (ng/ml)*		
	CAOV-3	OVCAR-3	SKOV-3
Control	35.3 ± 0.7	0.1 ± 0.03	0.6 ± 0.4
TNF-α			
10 U/ml	51.4 ± 0.1	0.3 ± 0.09	1.00 ± 0.00
100 U/ml	54.9 ± 3.0	0.9 ± 0.2	1.7 ± 0.2
1000 U/ml	52.1 ± 1.9	0.8 ± 0.3	2.8 ± 0.2
IL-1β			
1 U/ml	35.0 ± 0.3	0.04 ± 0.0	0.5 ± 0.4
10 U/ml	37.2 ± 9.3	1.1 ± 0.07	1.7 ± 0.5
100 U/ml	33.4 ± 1.2	10.3 ± 1.3	6.2 ± 1.8
IFN-γ			
1 U/ml	45.6 ± 10.6	0.1 ± 0.02	0.9 ± 0.04
100 U/ml	50.6 ± 6.7	0.1 ± 0.05	1.4 ± 0.05
10 000 U/ml	52.8 ± 1.7	0.1 ± 0.03	1.5 ± 0.06

* Values represent mean ± SD of triplicate samples from one representative experiment ($n=4$).

12.3 IL-6 PRODUCTION BY OVARIAN CANCER CELLS CAN BE MODULATED BY EXPOSURE TO CYTOKINES

The production of IL-6 by various cells can be induced or modified by various cytokines [9]. We examined the effects of TNF-α, IL-1β, G-CSF, granulocyte-macrophage colony stimulating factor (GM-CSF), and IFN-γ on IL-6 production by ovarian cancer cell lines [25]. IL-6 production by CAOV-3 cells was enhanced by TNF-α and by IFN-γ, but not by IL-1β (Table 12.3). Both IL-1β and TNF-α significantly increased IL-6 production by OVCAR-3 cells, while exposure of these cells to IFN-γ did not result in an increase in IL-6 secretion (Table 12.3). IL-1β, TNF-α, and IFN-γ all increased IL-6 production by SKOV-3 cells (Table 12.3), while none of these cytokines induced IL-6 production by PA-1 or OC222 cells (not shown). G-CSF and GM-CSF had little effect on IL-6 secretion by ovarian cancer cell lines (not shown). Clearly, exposure of ovarian cancer cells to various cytokines can lead to increased IL-6 secretion.

12.4 IL-6 IS PRODUCED BY FRESHLY ISOLATED PRIMARY OVARIAN CANCER CELLS

Since ovarian cancer cell lines represent fully transformed cells which have been selected for *in vitro*, the abnormal constitutive IL-6 production seen in our studies could have been a consequence of such selection. To determine whether ovarian cancer cells produce IL-6 *in vivo*, we examined freshly isolated ovarian tumours for the production of IL-6. Every primary culture supernatant ($n = 19$) from freshly excised ovarian tumours had detectable IL-6 activity (Table 12.4). This activity ranged from 12 to 36 ng/ml and was detectable in both the IL-6 ELISA and bioassay (not shown). Primary cultures of ovarian tumour cells also were positive for cytoplasmic IL-6 expression when assessed by immunoperoxidase staining (Table 12.2). To culture ovarian tumour specimens, small portions of fresh primary ovarian tumours were minced with sterile scissors in medium; the resultant supernatant was seeded into tissue culture flasks, and cultured in complete media (RPMI-1640, 10% FCS, 1% L-glutamine and 1% pencillin/streptomycin) as described earlier for established ovarian cancer cell lines.

12.5 ELEVATED LEVELS OF IL-6 IN THE ASCITIC FLUID OF WOMEN WITH OVARIAN CANCER

Knowing that ovarian cancer cells, both established ovarian cancer cell lines and primary cultures of freshly excised ovarian tumour cells, produced IL-6, we examined *in vivo* IL-6 levels in ovarian cancer. Ascites were collected from women with ovarian cancer. After collection, these ascites specimens were centrifuged to remove cells and debris, and IL-6 content was determined by

Table 12.4 Primary culture supernatants from freshly excised ovarian tumours and ascitic fluid from ovarian cancer patients contain high levels of IL-6

	IL-6 concentration (ng/ml)	
	Range	Mean ± SEM
Ovarian cancer patients		
Ascitic fluids (n=23)	3.2 – 42.0	20.0 ± 2.6
Primary culture supernatants (n=19)	11.6 – 35.9	24.4 ± 1.8
Non-cancer patients		
Peritoneal fluids (n=108)	0.2 – 2.6	1.5 ± 0.4

both ELISA (Table 12.4) and bioassay (not shown), as described. All ovarian cancer ascitic fluids tested ($n = 24$) had detectable levels of IL-6 (Table 12.4), with IL-6 levels ranging from 3.2 to 42 ng/ml[25]. While the high IL-6 levels seen in the ascitic fluid of women with ovarian cancer cannot be directly linked to IL-6 produced by ovarian tumour cells *in vivo*, it is important to note that ascitic fluids isolated from women without ovarian cancer ($n = 108$, patients with gynaecological conditions other than ovarian cancer, or those undergoing voluntary tubal ligations) all contained much lower (<3.0 ng/ml) levels of IL-6 (Table 12.4).

12.6 ELEVATED SERUM LEVELS OF IL-6 ASSOCIATED WITH OVARIAN CANCER

Since epithelial ovarian cancer cells were seen to produce IL-6, and since elevated levels of IL-6 were detected in ascites from women with ovarian cancer, we examined serum IL-6 levels in epithelial ovarian cancer. The levels of serum IL-6 in 36 patients (90 separate serum specimens) were measured by bioassay, as described. Elevated IL-6 levels were detected in many patients with ovarian cancer (Table 12.5)[26].

Mean IL-6 serum levels were correlated with the extent of disease at the time of

Table 12.5 IL-6 serum levels in epithelial ovarian cancer patients

Disease status	No. of patients	No. of specimens	IL-6 levels	
			Elevated†	Mean ± SEM
Macroscopic	21	57	16 (76%)	0.26 ± 0.04**
Bulky (>2 cm)	9	27	8 (89%)	0.31 ± 0.05*
MRD‡ (>2 cm)	12	30	8 (67%)	0.23 ± 0.04**
Microscopic	15	33	2 (13%)	0.09 ± 0.03
Total	36	90	18 (50%)	0.24 ± 0.04**
Control	12	12	2	0.12 ± 0.03

* $P < 0.01$, ** $P < 0.05$, versus control.
† sensitivity (macroscopic) = 0.76, microscopic + macroscopic = 0.50.
‡ MRD, minimal residual disease.

Table 12.6 Correlation of serum IL-6 levels and CA-125 levels in epithelial ovarian cancer patients

CA-125 Level	Patients	IL-6 level elevated
Normal (<35 U/ml)	22	4 (18%)
Elevated (>35 U/ml)	14	12 (86%)
Correlation coefficient = 0.84		

exploratory laparotomy[26]. The mean serum IL-6 level of those ovarian cancer patients with macroscopic disease ($n = 57$) was 0.26 units/ml. Of 21 patients in this group 16 (76%) had elevated (>0.20 units/ml IL-6) levels of serum IL-6, with serum IL-6 levels approaching 1 unit/ml in some patients. Of the 9 patients with bulky tumour (residual >2 cm maximum tumour dimension), 8 (89%) had an elevated serum IL-6 level (mean = 0.31 units/ml, $P<0.01$), while 8 of 12 (66%) patients with minimal residual disease (<2 cm maximum tumour dimension) showed elevated serum IL-6 levels ($P <0.05$). Only 2 of 15 (13%) patients who were in clinical remission, and who had microscopic disease only, had elevated serum IL-6 values. Based on a mean serum IL-6 value of 0.12 ± 0.03 units/ml in healthy adult female volunteer donors ($n = 12$), IL-6 serum levels in patients with ovarian cancer were elevated significantly.

Serum CA-125 levels were correlated with serum IL-6 levels in ovarian cancer[26]. Of the 36 patients tested, 22 were CA-125 negative (<35 U/ml), and of these 4 (18%) had elevated IL-6 levels. Of the 14 patients with elevated CA-125 levels, 12 (86%) had elevated IL-6 levels (Table 12.6). Of the 36 patients tested 15 had microscopic disease only at the time of measurement, and all but one of these patients had a negative CA-125.

12.7 CONCLUSIONS

We have reported that several well-characterized human ovarian carcinoma cell lines constitutively produce IL-6. These ovarian cancer cell lines, as well as primary cultured ovarian tumour cells, secrete biologically active and immunochemically reactive IL-6, as determined by IL-6 bioassay, ELISA, and immunoperoxidase staining. By the methods employed (biosynthetic labelling and radio-immunoprecipitation), this ovarian cancer-produced IL-6 had the biological and antigenic characteristics of IL-6 produced by lymphocytes or monocytes; the IL-6 produced by these ovarian cancer cell lines had a molecular mass of approximately 24 kDa, and exhibited little of the molecular mass heterogeneity often associated with IL-6 produced by epithelial cells. Also, several ovarian cancer cell lines were seen to constitutively display IL-6 gene expression, with moderate to high levels of IL-6 mRNA, as well as IL-6 receptor gene expression.

Most of the primary ovarian tumours and ovarian cell lines that produced IL-6 activity were classified as adenocarcinomas. One of the ovarian cancer cell lines that failed to produce any IL-6 was PA-1, which was originally derived from a teratocarcinoma [36]. This germ cell tumour is derived from undifferentiated cells exhibiting characteristics of embryonal carcinoma cells, as opposed to the adenocarcinomas, which characteristically contain more differentiated cells.

Significant levels of IL-6 were detected not only in the supernatants of primary ovarian tumour cultures, but also in ovarian cancer patients' ascitic fluid. Although the source of the IL-6 detected in these ascitic fluids cannot be ascribed to IL-6 solely produced by the ovarian tumours, IL-6 has been detected in ovarian tumour sections by immunoperoxidase staining[37]. Despite the uncertain source of the IL-6 in ascitic fluid, it is apparent that there are extremely high levels of this cytokine within the peritoneal cavity of patients with ovarian cancer. The high concentrations of IL-6 within a confined and localized region reported here are not unique, as high levels (>1000 U) of IL-6 have been reported to be present in amniotic fluid[38], in the synovial fluid of rheumatoid arthritis patients[39], and in localized regions of acute bacterial infections[4]. It remains to be determined what immune function these high levels of IL-6 might play in such localized areas, and whether in high, sustained concentrations IL-6 actively suppresses or enhances immune function directly or induces a state of non-responsiveness through indirect means.

Our results indicate that the serum of many patients with epithelial ovarian cancer contains detectable levels of IL-6, levels that often were significantly higher than those found in normal controls. The levels of serum IL-6 seen in patients with ovarian cancer correlated with the extent of residual disease, such that those patients who had a large tumour burden also tended to have high IL-6 serum levels. Furthermore, in patients who developed evidence of clinical progression, IL-6 levels became elevated, and the survival of patients from the time of surgery was seen to correlate with the level of serum IL-6, suggesting that overexpression of this growth factor might be prognostic in epithelial ovarian cancer. Together, these results suggest that serum IL-6 may be a useful biomarker for epithelial ovarian cancer.

The presence of high levels of IL-6 in several patients who had a low level of the surface antigen, CA-125, suggests that the assay might contribute to the monitoring of such patients. However, because the correlation coefficient between the two markers is high, the lack of complementarity might diminish the utility of IL-6. The relevance of tumour differentiation to IL-6 production cannot be drawn from our studies to date, as there are too few patients with each tumour type. However, it would be interesting to examine this in future studies, to determine if such a correlation exists. The ovarian carcinoma marker CA-125 has been demonstrated to be associated with approximately 85% of serous ovarian tumours and 70% of undifferentiated tumours, and 30–40% of mucinous tumours[4].

IL-6 has been implicated as an autocrine growth factor in several human tumours, including myeloma, renal cell carcinoma, and AIDS-associated Kaposi's sarcoma[16–18,20]. The possible role of ovarian cancer cell-produced IL-6 as an autocrine, exocrine or paracrine growth factor is intriguing, especially if considered in light of the high levels of IL-6 that are produced and retained within the peritoneal cavity and the rarity of ovarian cancer metastasis outside of the peritoneal cavity. Certainly, our observation that ovarian cancer cells express high levels of IL-6 receptor mRNA constitutively suggests that IL-6 is a potential growth-modulating factor for ovarian cancer cells. Interestingly, recent work indicates that the supernatants of activated peripheral blood mononuclear cells, or monocyte cell lines, could stimulate the growth of ovarian carcinoma cell lines, and that part of this monocyte-derived, growth-stimulating activity may have been due to IL-6[42]. Experiments in which IL-6 translation in ovarian cancer cells is arrested using antisense oligonucleotides specific for a sequence in the second exon of the IL-6 gene[20] are currently underway in our laboratory. Very preliminary results indicate that the specific inhibition of IL-6 production by OC436 or OVCAR-3 ovarian cancer cell lines, induced by this IL-6 antisense oligonucleotide, results in greatly decreased cellular proliferation (Watson *et al.*, unpublished observation). In any case, this approach should allow us to better understand the growth-enhancing or suppressive role of IL-6 in ovarian cancer.

The secretion of IL-6 into the blood of patients with epithelial ovarian cancer might result from the up-regulated production of IL-6 by the ovarian epithelial and stromal cells of these malignancies. A recent report by Motro and co-workers[27] suggests that IL-6 is a paracrine growth factor secreted by epithelial and stromal cells of normal ovaries and by the ovarian follicles, and that IL-6 gene expression is associated with angiogenesis within the stroma. It has been proposed by Tamm *et al.*[43] that IL-6 may promote tumour metastasis and invasiveness because exogenous IL-6 has been shown to increase motility and decrease adherence junctions in breast carcinoma cell lines. It is not known whether ovarian cancer cell, autonomously produced, IL-6 would affect the motility and adherence of these cells, but this phenomenon could explain the extensive peritoneal carcinomatosis seen in patients with ovarian carcinoma. Also, IL-6 could play an important role in the establishment of autonomous tumour by acting as an angiogenesis-inducing factor[27]. Based on these observations, we postulate that, during the process of tumorigenesis, IL-6 gene expression helps to promote tumour growth, and that tumour-produced IL-6 is secreted into the peritoneal cavity and blood of patients with ovarian cancer.

The results of these studies suggest that ovarian tumour progression may be related to abnormal growth factor regulation, as well as inappropriate oncogene expression and amplification[3]. Evidence relating abnormal growth factor regulation to ovarian tumour progression has been demonstrated by the amplification and overexpression[3] of an epidermal growth factor-like receptor[44] encoded by the HER-2/*neu* oncogene. Also, recent reports have shown that M-CSF, the receptor for which is a product of the c-*fms* proto-oncogene, also is secreted by many ovarian epithelial tumours[5]. While the unregulated expression of IL-6 has been implicated in the pathogenesis of numerous diseases, including various malignancies, it remains to be determined what role such deregulated expression of IL-6 has in the development and progression of ovarian cancer.

ACKNOWLEDGEMENTS

This material is based upon work supported by grants from the University of California Cancer Research Coordinating Committee (CRCC), the California Institute for Cancer Research (CICR), NIH Tumor Immunology Institutional Training Grant CA09120, the American Cancer Society (JFRA-165), the Ramona Moskovitz Memorial Cancer Research Fund, and the Brindell and Milton Gottlieb Gynecologic Oncology Research Laboratory. The authors would like to thank Catherine Chung, Klara Kaldi, Reba Knox, Ahmad Rezai, Mary Salke, John Sensintaffer, for their assistance with these studies; Dr Tadamitsu Kishimoto and Dr Toshio Hirano of Osaka University, Osaka, Japan, for providing IL-6, IL-6 cDNA and the IL-6 responsive cell line, MH60.BSF2; Dr Fumihiko Takatsuki of Ajinomoto Co., Inc, Kawasaki, Japan, for providing monoclonal antibodies to IL-6; Dr Robert C. Bast for advice and additional specimens; and Dr Donna Vredevoe for the use of laboratory facilities. Some of the results presented in this communication were published previously[25,26].

REFERENCES

1. Berek, J.S. (1989) Epithelial ovarian cancer, in *Practical Gynecologic Oncology* (eds J.S. Berek and N.F. Hacker), Williams and Wilkins, Baltimore, p. 327.
2. Neuntenfel, W. and Breitenecker, G. (1989) Tissue expression of CA 125 in benign and malignant lesions of ovary and fallopian tube: A comparison with CA 19–9 and CEA. *Gynecol. Oncol.*, **32**, 297.
3. Slamon, D.J., Godolphin, W., Jones L.A. *et al.* (1989) Studies of HER-2/*neu* proto-oncogene in human breast and ovarian cancer. *Science*, **244**, 707.
4. Boltz, E.M., Kefford, R.F., Leary, J.A. *et al.*

(1989) Amplification of c-*ras*-Ki oncogene in human ovarian tumours. *Int. J. Cancer*, **43**, 428.
5. Ramakrishnan, S., Xu, F.J., Brandt, S.J. *et al.* (1989) Constitutive production of macrophage colony-stimulating factor by human ovarian and breast cancer cell lines. *J. Clin. Invest.*, **83**, 921.
6. Kacinski, B.M., Carter, D., Kohorn, E.I. *et al.* (1989) Markedly elevated plasma levels of a tumor-produced cytokine CSF-1 (M-CSF), the macrophage colony stimulating factor are seen in ovarian carcinoma patients with active disease and may be a useful circulating tumor market for ovarian and other gynecological neoplasms. *Soc. Gynecol. Invest.*, Abstract 202.
7. Sherr, C.J., Rettenmeir, C.W., Sacca, R. *et al.* (1985) The c-*fms* proto-oncogene product is related to the receptor for the mononuclear phagocyte growth factor, CSF-1. *Cell*, **41**, 665.
8. Kishimoto, T. and Hirano, T. (1988) Molecular regulation of B lymphocyte response. *Annu. Rev. Immunol.*, **6**, 485.
9. Hirano, T., Akira, S., Taga, T. and Kishimoto, T. (1990) Biological and clinical aspects of interleukin 6. *Immunol. Today*, **11**, 443.
10. Hirano, T., Yasukawa, K., Harada, H. *et al.* (1986) Complementary DNA for a novel human interleukin (BSF-2) that induces B lymphocytes to produce immunoglobulin. *Nature*, **324**, 73.
11. Yamasaki, K., Taga, T., Hirata, Y. *et al.* (1988) Cloning and expression of human interleukin-6 (BSF-2/IFNb$_2$) receptor. *Science*, **241**, 825.
12. Akira, S., Isshiki, H., Sugita, T. *et al.* (1990) A nuclear factor for IL-6 expression (NF-IL6) is a member of a C/EBP family. *EMBO J.*, **9**, 1897.
13. Hirano, T., Taga, T., Yasukawa, K. *et al.* (1987) Human B cell differentiation factor defined by an anti-peptide antibody and its possible role in autoantibody production. *Proc. Natl Acad. Sci. USA*, **84**, 228.
14. Krutmann, W., Borth, D., Damm, G. *et al.* (1989) IFN-b2, B cell differentiation factor 2, or hybridoma growth factor (IL-6) is expressed and released by human epidermal cells and epidermoid carcinoma cell lines. *J. Immunol.*, **142**, 1922.
15. Rawle, F.C., Shields, J., Smith, S.H. *et al.* (1986) B cell growth and differentiation induced by supernatants of transformed epithelial cell lines. *Eur. J. Immunol.*, **16**, 1017.
16. Kawano, M., Hirano, T., Matsuda, T. *et al.* (1988) Autocrine generation and requirement of BSF-2/IL-6 for human multiple myelomas. *Nature*, **332**, 83.
17. Klein, B., Zhang, X.-G., Jourdan, M. *et al.* (1989) Paracrine rather than autocrine regulation of myeloma-cell growth and differentiation by interleukin-6. *Blood*, **73**, 517.
18. Miki, S., Iwano, M., Miki, Y. *et al.* (1989) Interleukin-6 (IL-6) functions as an *in vitro* autocrine growth factor in renal cell carcinomas. *FEBS Lett.*, **250**, 607.
19. Matsuda, T., Suematsu, S., Kawano, M. *et al.* (1989) IL-6/BSF2 in normal and abnormal regulation of immune responses. *Ann. N. Y. Acad. Sci.*, **557**, 466.
20. Miles, S.A., Rezai, A.R., Salazar-Gonzalez, J.F., *et al.* (1990) AIDS Kaposi's sarcoma-derived cells produce and respond to interleukin-6. *Proc. Natl Acad. Sci. USA*, **87**, 4068.
21. Morinaga, Y., Suzuki, H., Takatsuki, F. *et al.* (1989) Contribution of IL-6 to the antiproliferative effect of IL-1 and tumor necrosis factor on tumor cell lines. *J. Immunol.*, **143**, 3538.
22. McIntosh, J.K., Jablons, D.M., Mule, J.J. *et al.* (1989) *In vivo* induction of IL-6 by administration of exogenous cytokines and detection of *de novo* serum levels of IL-6 in tumor-bearing mice. *J. Immunol.*, **143**, 162.
23. Jablons, D.M., Mule, J.J., McIntosh, J.K. *et al.* (1989) IL-6/IFN-b-2 as a circulating hormone: Induction by cytokine administration in humans. *J. Immunol.*, 142, 1542.
24. Kupper, T.S., Min, K., Sehgal, P. *et al.* (1989) Production of IL-6 by keratinocytes: implications for epidermal inflamation and immunity. *Ann. N. Y. Acad. Sci.*, **557**, 454.
25. Watson, J.M., Sensintaffar, J.L., Berek, J.S. and Martínez-Maza, O. (1990) Epithelial ovarian cancer cells constitutively produce interleukin-6 (IL6). *Cancer Res.*, **50**, 6959.
26. Berek, J.S., Chung, C., Kaldi, K. *et al.* Serum IL-6 levels correlate with disease status in epithelial ovarian cancer patients. *Am. J. Obstet. Gynecol.* (in press).
27. Motro B., Itin, A., Sachs, L. and Keshet, E. (1990) Pattern of interleukin 6 gene expression

in vivo suggests a role for this cytokine in angiogenesis. *Proc. Natl Acad. Sci. USA*, **87**, 3092.

28. Buyalos, R., Watson, J.M. and Martínez-Maza, O. (1990) Detection of interleukin-6 in human follicular fluid. American Fertility Society, Meeting Abstracts.
29. Lidor, Y.J., Xu, F.J., Olt, G.J. *et al.* (1990) Constitutive production of macrophage colony stimulating factor (M-CSF) and interleukin-6 (IL-6) by human ovarian surface epithelial cells. American Association for Cancer Research, Meeting Abstracts.
30. Folkman, J. (1982) Tumor invasion and metastasis, in *Pathogenesis of Cancer* (eds L.A. Liotta and I.R. Hart), Martinus Nijhoff, The Hague, p. 167.
31. Matsuda, T., Hirano, T. and Kishimoto, T. (1988) Establishment of an interleukin 6 (IL-6)/B cell stimulatory factor 2-dependent cell line and preparation of anti-IL-6 monoclonal antibodies. *Eur. J. Immunol.*, **18**, 951.
32. May, L.T., Ghrayeb, J., Santhanam, U. *et al.* (1988) Synthesis and secretion of multiple forms of b2-interferon/B-cell differentiation factor 2/hepatocyte-stimulating factor by human fibroblasts and monocytes. *J. Biol. Chem.*, **263**, 7760.
33. Kirnbauer, R., Kock, A., Schwartz, T. *et al.* (1989) IFN-b2, B cell differentiation factor 2, or hybridoma growth factor (IL-6) is expressed and released by human epidermal cells and epidermoid carcinoma cell lines. *J. Immunol.*, **142**, 1922.
34. Maniatis, T., Fritsch, E.F. and Sambrook, J. (1982) Extraction, purification, and analysis of mRNA from eukaryotic cells, in *Molecular Cloning*, Cold Spring Harbor Laboratory, Cold Spring Harbor, NY, p. 187.
35. Nakajima, K., Martínez-Maza, O., Hirano, T. *et al.* (1989) Induction of IL-6 (B cell stimulatory factor-2/IFN-b2) production by HIV. *J. Immunol.*, **142**, 531.
36. Zeuthen, J., Norgaard, J.D.R., Avner, P. *et al.* (1980) Characterization of a human ovarian teratocarcinoma-derived cell line. *Int. J. Cancer*, **25**, 19.
37. Tabibzadeh, S.S., Poubouridis, D., May, L.T. and Sehgal, P.B. (1989) Interleukin-6 immunoreactivity in human tumors. *Am. J. Pathol.*, **135**, 427.
38. Anuradha, R., Tatter, S.B., Santhanam, V. *et al.* (1989) Regulation of expression of interleukin-6: Molecular and clinical studies. *Ann. N. Y. Acad. Sci.*, **557**, 353.
39. Hirano, T., Matsuda, T., Turner, M. *et al.* (1988) Excessive production of interleukin 6/B cell stimulatory factor-2 in rheumatoid arthritis. *Eur. J. Immunol.*, **18**, 1792.
40. Helfgott, D.C., Tatter, S.B., Santhanam, U. *et al.* (1989) Multiple forms of IFN-b2/IL-6 in plasma and body fluids during acute phase bacterial infection. *J. Immunol.*, **142**, 948.
41. Lavin, P.T., Knapp, R.C., Malkasian, G. *et al.* (1987) CA 125 for the monitoring of ovarian carcinoma during primary therapy. *Obstet. Gynecol.*, **69**, 223.
42. Wu, S., Rodabaugh, K., Watson, J.M. *et al.* Stimulation of ovarian tumor cell proliferation with monocyte products including inter-leukin-1-alpha, interleukin-6 and tumor necrosis factor-alpha (submitted for publication).
43. Tamm, I., Cardinale, I., Krueger, J. *et al.* (1989) Interleukin-6 decreases cell–cell association and increases motility of ductal breast carcinoma cells. *J. Exp. Med.*, **170**, 1649.
44. Kokai, Y., Dobashi, K., Weiner, D.B. *et al.* (1988) Phosphorylation process induced by epidermal growth factor alters the oncogenic and cellular *neu* (NGL) gene products. *Proc. Natl Acad. Sci. USA*, **85**, 5389.

Chapter 13

CSF-1 and its receptor in ovarian and other gynaecological neoplasms

B.M. KACINSKI

13.1 INTRODUCTION

13.1.1 BREAST, OVARIAN AND ENDOMETRIAL CARCINOMA: CLINICAL BIOLOGY

Taken together, carcinomas of the ovary, endometrium and breast together will account for nearly half of all new diagnoses of cancer and one-quarter of all cancer deaths in women in North America and western Europe in 1991[1–8]. For all three of these common female neoplasms, therapy of disease confined to the organ of origin (ovary, uterine corpus, or breast) or with minimal locoregional spread is quite effective, with overall five-year cure rates of 60–80% or better, in striking contrast to disease-free survivals of less than 25% at five years for patients with extensive locoregional spread or distant metastases even after the most aggressive combinations of radical surgery, chemotherapy and therapeutic irradiation [1–8].

Also, despite major advances in surgical and radiotherapeutic techniques, combination chemotherapy, and hormonal manipulation, overall cure rates for these three common female malignancies are not significantly better now than they were 30 years ago[1–8]. Substantial improvements in therapeutic results are extremely unlikely without a better understanding of the molecular events which underlie ovarian, endometrial and mammary carcinogenesis, disease progression, and responsiveness to therapy, especially for those locally advanced and metastatic tumours which remain incurable in 1991.

13.1.2 ONCOGENES, GROWTH FACTORS AND RECEPTORS IN OVARIAN, ENDOMETRIAL AND BREAST CARCINOMA

During the 1980s, when the role of oncogenes was first investigated in human cancers, abnormalities in the expression and/or function of a variety of oncogenes (c-*myc*, *int*-2, *hst*, *neu*, *erb*B, *ras*) were observed in breast, ovarian and endometrial carcinoma specimens and/or cell lines. To date, however, no single abnormality or set of alterations in proto-oncogene expression or function is clearly pathognomonic of clinically aggressive behaviour or poor outcome. Nor are any of the protein products encoded by the proto-oncogenes themselves obviously implicated either in the physiology of invasion by normal or malignant 'invasive' cells (e.g. macrophages, granulocytes, trophoblast, carcinoma cells) or in the regulation of host immune antitumour responses[9–12]. Even abnormalities in the expression and/or gene copy number of the (wild-type) HER-2/*neu* oncogene, associated by some investigators [9,10] with poorer than average prognosis in breast and ovarian carcinoma, are not clearly pathognomonic of invasive and/or metastatic behaviour, since similar overexpression of *neu* transcripts and protein is observed in such minimally invasive neoplasms as comedo carcinomas[10] of the breast and borderline carcinomas of the ovary[11]. Also paradoxically, the putative gp30 *neu*-ligand [12] can inhibit as well as stimulate the proliferation of breast carcinoma cell lines known to express the HER-2/*neu* receptor.

13.1.3 IS THERE A ROLE FOR OTHER GENES INVOLVED IN 'NORMAL' INVASION AND IMMUNOMODULATION?

Such results strongly suggest that other genes must play important roles in tumourigenesis and dissemination. Thus it seems inherently reasonable that genes (such as CSF-1 and its receptor) already implicated in such normal non-neoplastic invasive and immunomodulatory processes as placental implantation, which involves trophoblastic invasion into the uterine wall and evasion of maternal immune defences, and macrophage activation during infection and wound healing, which also requires cell migration, invasion through tissue barriers, and modulation of the function of other immune cells, might also regulate the expression of similar traits by invasive carcinoma cells.

13.2 OUR OWN INVESTIGATIONS

13.2.1 EXPRESSION OF *fms* (CSF-1R) *IN VIVO* BY OVARIAN, ENDOMETRIAL AND BREAST CARCINOMAS: *IN SITU* HYBRIDIZATION AND IMMUNOHISTOCHEMICAL STUDIES

In our earlier studies of tumour cell-specific expression of oncogenes, growth factor and growth factor receptor transcripts in human tissue specimens (initially from ovarian and endometrial carcinomas eventually extended to include adenocarcinomas of the breast), we naively attempted to determine how the levels of the expression of transcripts of a large number of different oncogenes, growth factors, and growth factor receptors (for which cDNA probes were available to us in 1985) correlated with clinical and pathological features prognostic of poor outcome[13–15]. Since we were intrigued by the parallel phenotypes expressed by implanting trophoblast, activated immune cells, and carcinoma cells, we also included a probe for the only oncogene known to be involved in both processes, *fms*; this was first characterized in a feline leukaemia virus and later shown to encode the receptor for the macrophage colony-stimulating factor, CSF-1.

Our initial studies of ovarian, endometrial and breast carcinomas[13–15] revealed many correlations between levels of expression of different pairs of oncogenes. In breast, ovarian and endometrial carcinomas, these included strong correlations of the levels of expression of c-*fos* and c-*myc*, whose transcript levels are elevated by many mitogenic stimuli, with a variety of growth factor receptors including *fms*, EGF-R and HER-*neu*. We also observed statistically significant correlations of levels of *neu* (HER-2, *erb*B2) or *erb*B1 (EGF-R) and levels of *fms* (CSF-1R) transcripts suggesting that breast and ovarian carcinomas often coexpress CSF-1R and either HER-2/*neu* or EGF-R. However only levels of *fms* transcripts correlated strongly with those of high histological grade and advanced clinical stage presentations which are strongly prognostic of poor outcome [13–15] in ovarian and endometrial carcinoma patients. Since quantitated *in situ* hybridization techniques did not permit us to discriminate *fms* transcripts of abundant tumour infiltrating macrophages from those expressed by the malignant epithelial cells, immunohistochemical studies were needed to more precisely localize the cellular origin of tumour *fms* transcripts.

Confirmatory imunohistochemical studies with an anti-CSF-1 receptor monoclonal antibody (specific for a tyrosine kinase insert epitope unique to c-*fms*[16]) revealed expression of *fms* antigen by tumour infiltrating macrophages as well as by the malignant epithelial cells of ovarian, endometrial, and breast adenocarcinomas. At the time of our initial observations, CSF-1R gene expression had not been reported in cell types other than those of monocyte–macrophage lineage, placental trophoblast, and their neoplastic derivatives (certain myelomonocytic leukaemias and choriocarcinoma cells[17–19]); but for these it was clear that this receptor/ligand pair played important roles in:

1. the control of macrophage differentiation and phenotypic 'activation' and interactions with other host immune cells[17]; and
2. the regulation of normal placental implantation and development[19].

Therefore, as we suggest above, it was not unreasonable to investigate further whether such a haemopoietic ligand/receptor pair important in two normal invasive/immunomodulatory processes might also control the expression of similar phenotypes in ovarian, endometrial and breast carcinomas.

13.2.2 EXPRESSION OF CSF-1 *IN VIVO* BY OVARIAN, ENDOMETRIAL AND BREAST CARCINOMAS: IMMUNOHISTOCHEMICAL (IHC) STUDIES

IHC studies with anti-CSF-1 antibodies revealed that at least one-third of CSF-1R-positive breast, ovarian and endometrial carcinoma specimens also coexpress CSF-1 [13,14,17]. In a collaborative study with investigators at the Institut Curie[20], IHC evidence of tumour cell CSF-1 antigen expression was strongly correlated with dense (macrophage and T-cell) lymphocytic infiltration of human breast carcinoma – a not unexpected finding for CSF-1, which is a known potent chemo-attractant and phenotypic activator of human macrophages [17,21]. In addition, breast tumour cell CSF-1 antigen expression was strongly correlated with amplification of other oncogenes (HER-2/*neu*, *int*-2) associated with poor prognosis[20]. The latter clinicopathological association relates less obviously to the known effects of CSF-1 as a colony-stimulating factor and phenotypic activator of macrophages [17,21] and may reflect other actions either as an autocrine growth factor for CSF-1R-positive carcinoma cells (see below) or as a paracrine, but adverse, modifier of macrophage function (see below).

13.2.3 ELEVATED CIRCULATING CSF-1 LEVELS IN OVARIAN, ENDOMETRIAL AND BREAST CARCINOMA PATIENTS WITH ACTIVE NEOPLASTIC DISEASE

In vivo, we have observed such markedly elevated plasma levels of CSF-1 in a large percentage (>80%) of ovarian, endometrial ([22–25] and Figure 13.1) and breast carcinoma patients (unpublished observations) with active neoplastic disease that determinations of circulating CSF-1 levels can be exploited as 'tumour markers' of tumour burden and disease activity. In breast and

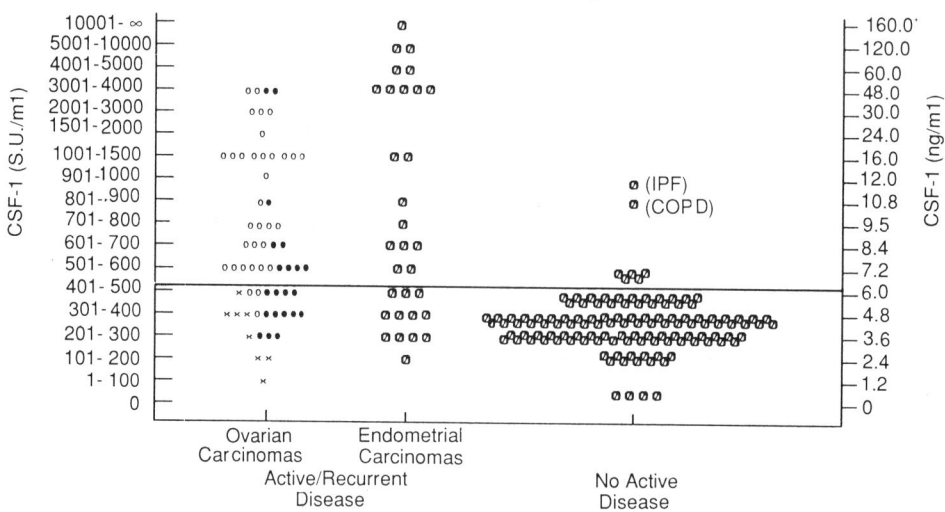

Figure 13.1 Circulating CSF-1 levels in patients with ovarian and endometrial carcinoma as a function of disease status. As we have described in detail elsewhere[22–25], frozen plasma samples from patients of the Yale University Gynecologic Oncology clinic were stored at −70°C until use: 62 were from patients with active or recurrent ovarian carcinoma and 31 were from patients with active or recurrent endometrial carcinoma; 122 were from patients without evidence of active or recurrent ovarian or endometrial carcinoma. In the latter group, two samples were from patients with active chronic obstructive pulmonary disease (COPD) and idiopathic pulmonary fibrosis (IPF) and are indicated as such on the plot of the distribution of CSF-1 levels. RIA was carried out with iodinated recombinant CSF-1 (provided by Cetus Corp.) and anti-CSF-1 PAbs as described elsewhere[22–25] and measured in S.U./ml where 1 S.U./ml = 0.44 fmol = 12 pg/ml CSF-1. Significantly higher CSF-1 levels were observed in patients with active or recurrent ovarian or endometrial carcinoma than in patients without evidence of active neoplastic disease.

lung carcinomas[25] and unpublished observations, elevated circulating levels of CSF-1 correlate very strongly ($P<0.0001$) with the presence of metastatic, as opposed to local, disease; while high circulating CSF-1 levels at time of (breast carcinoma) relapse appear to be associated with a rapidly progressive clinical course (unpublished observations). While elevated CSF-1 levels themselves may be the consequence of 'tumour burden' and discharge into the circulation of a copiously shed tumour antigen, their association with an aggressive clinical course suggests that tumour-produced CSF-1 may exert adverse effects on tumour/host interactions as an autocrine cytokine for CSF-1R-positive tumour cells, a paracrine adverse modifier (see below) of host macrophage function, or both.

13.2.4 EXPRESSION OF FUNCTIONAL CSF-1R BY HUMAN OVARIAN, ENDOMETRIAL AND BREAST CARCINOMA CELL LINES

In vitro, we have been able to demonstrate low but detectable levels of CSF-1 receptor (*fms*)-complementary transcripts, CSF-1R protein (Figure 13.2, panel F) and functional CSF-1-specific binding sites in several generally available breast, endometrial and early passage ovarian adenocarcinoma cell lines (Table 13.1 and [14,22,26,27]). As expected, high levels of *fms* transcripts were observed

Our own investigations

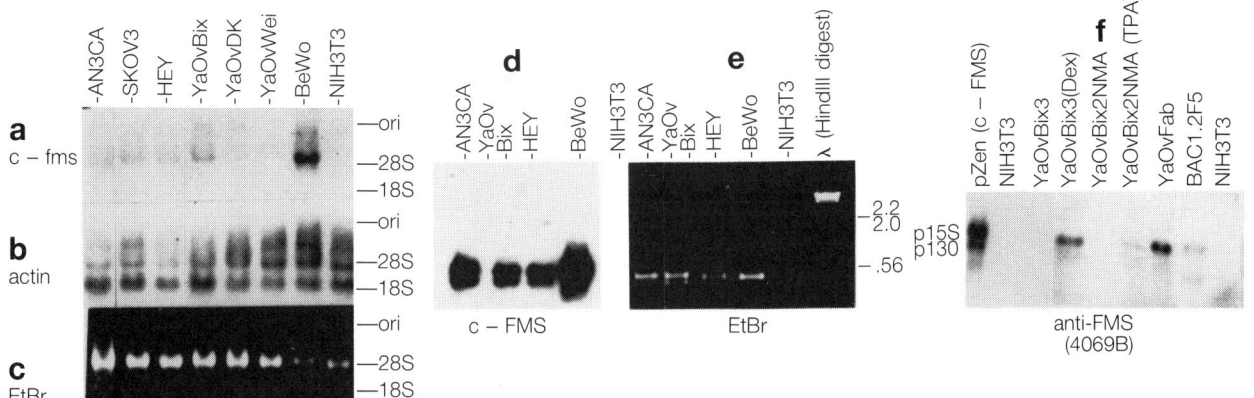

Figure 13.2 CSF-1R (*fms*) transcript and protein expression in ovarian and endometrial carcinoma cell lines.

Panels a,b,c: As described elsewhere[26] total RNAs were extracted from several generally available ovarian (SKOV3, HEY) and endometrial (AN3CA) cell lines as well as from several (YaOvBix, YaOvDK, YaOvWei) adenocarcinoma cell lines[26] which we cultured from the malignant ascites of several Yale ovarian carcinoma patients. BeWo choriocarcinoma RNA and NIH3T3 total RNA were included as positive and negative controls, respectively, for *fms* transcript expression. Total RNA was electrophoresed in formaldehyde-agarose gels, stained with ethidium bromide (panel c) blotted to Gene Screen Plus membranes which were hybridized with beta-actin (panel b) and human c-*fms* probes (panel a). Low but detectable levels of normal length *fms* transcripts were observed in AN3CA, SKOV3, HEY, YaOvBix and YaOvDk cells relative to the BeWo choriocarcinoma positive control.

Panels d,e: Reverse transcription and polymerase chain reaction (PCR) amplification was carried out with oligonucleotides complementary to nucleotides 996–1016 (sense) and 1514–1494 (antisense) on total RNA isolated from AN3CA endometrial carcinoma, YaOvBix and HEY ovarian carcinoma and BeWo choriocarcinoma and NIH3T3 total RNA was determined as we have described elsewhere[26]. PCR products were analysed by agarose gel electrophoresis and ethidium bromide staining (panel e) and Southern blotting with a human *fms* probe (panel d). Amplification of a 0.52-bp DNA fragment as expected from the *fms* transcript sequence (with no evidence of the ~8.0-kb fragment which would have been amplified from chromosomal DNA contaminants) was observed (as expected) for BeWo choriocarcinoma cell RNA and for RNA extracted from AN3CA endometrial carcinoma and HEY and YaOvDK ovarian carcinoma cells and not observed for the NIH3T3 negative control.

Panel f: Western blotting with an anti-*fms* polyclonal antibody was carried out as described elsewhere [27] with protein extracts of NIH3T3 (pZen-c-*fms*), a cell line engineered to express very high levels of CSF-1 receptors, BAC12F5 macrophages, NIH3T3 cells, and several ovarian (YaOvBix3), YaOv-Bix2NMA, YaOvFab) carcinoma cell lines. Some of these extracts were carried out on cells exposed to phorbol esters (TPA) or dexamethasone (dex). As expected, high levels of *fms* expression were observed in NIH3T39 (pZen-c-*fms*) and BAC1.2F5 cells and significant levels were detected in YaOvBix3 cells treated with dexamethasone, YaOvBix2NMA cells treated with phorbol esters as well as in untreated YaOvFab ovarian carcinoma cells.

on BAC1.2F5 murine macrophages and low but detectable numbers of ^{125}I CSF-1 binding sites were observed on ovarian (YaOvBix) and breast carcinoma (BT20) cell lines in which low levels of *fms* transcripts were observed. These levels were significantly increased by treatment of cells with 1 μM dexamethasone (dex). No significant level of

Table 13.1

Cell line	Cell type	^{125}I-CSF-1 binding sites per cell
1. BAC1.2F5	Murine macrophage	40 000–80 000
2. YaOvBix	Ovarian carcinoma	500–1000
3. BT20	Breast carcinoma	1000–2000
4. BT20 (dex)	Breast carcinoma	15 000–20 000
5. MCF-7	Breast carcinoma	<100/cell

CSF-1R binding was found on MCF-7 cells in which we have observed little or no expression of *fms* transcripts[27]. Partial sequence analysis of breast (SKBR3, approximately 75% complete), ovarian and endometrial carcinoma cell line-derived *fms* transcripts have thus far failed to reveal any of the previously described[28,29] 'activating' mutations at codon 301 or 374, or for that matter any functionally significant differences from the published wild-type c-*fms* sequence[30]. However, the extreme 5′ termini of the ovarian and breast carcinoma cell *fms* transcripts are significantly different from those expressed by macrophages and appear very similar (if not identical) to those of choriocarcinoma cell lines.

More complete sequence analysis of *fms* transcripts of ovarian, endometrial and breast carcinoma cell lines is underway to better define the relationship of these transcripts to those expressed by macrophages and trophoblast.

Breast, ovarian and endometrial carcinoma CSF-1R p140 and p160 also appear very similar, if not identical, to the CSF-1R proteins expressed by choriocarcinoma and macrophage cell line positive controls (Figure 13.1).

13.2.5 EXPRESSION OF CSF-1 BY HUMAN OVARIAN, ENDOMETRIAL AND BREAST CARCINOMA CELL LINES

Nearly one-half of these same ovarian, endometrial and breast carcinoma cell lines can be shown to coexpress CSF-1 *in vitro*, a result which has been reported by ourselves and others[24,25,31,32]. The highest levels of CSF-1 production were observed in several ovarian carcinoma-derived cell lines (SKOV3, HEY, YaOvBix1 [24,25]) while lower but still detectable levels were observed in several endometrial (HEC1A, AN3CA [24]) and breast (SKBR3 [31,32]) carcinoma-derived lines. At least in the cell lines studied thus far, the predominant CSF-1 transcript is full-length (4.2 kb transcript[33]) without evidence for those smaller (e.g. 2.0 kb) CSF-1 transcripts which have been described in the pregnant uterus[19] or in a pancreatic carcinoma-derived cell line, PANC-1[33].

CSF-1 produced by ovarian, endometrial and breast carcinoma cell lines is released into culture supernatants in biologically active form[24,25,32], similar if not identical to that which is observed at elevated levels in the sera of ovarian, endometrial and breast carcinoma patients *in vivo* (see above).

13.2.6 LACK OF OBVIOUS GENETIC REARRANGEMENTS OR AMPLIFICATIONS IN THE STRUCTURAL GENES FOR CSF-1R OR CSF-1 TO EXPLAIN THEIR EXPRESSION IN CARCINOMAS OR TUMOUR-DERIVED CELL LINES

We have sought but have failed thus far to find genomic DNA amplifications and/or rearrangements[26] or consistent karyotypic abnormalities of human chromosome 5q consistent with the rearrangement or amplification of either the CSF-1 or *fms* loci in breast, ovarian and endometrial carcinoma cell lines or tissue specimens. Such negative results suggest that the expression of either or both of these genes is the consequence of mutations or more subtle rearrangements in promoter sequences or of mutations in other loci (including so-called tumour suppressor genes) which affect the rate of *fms* transcription or *fms* transcript stability, hypotheses which we are actively investigating.

13.2.7 EFFECTS OF CSF-1 ON CSF-1R-POSITIVE CARCINOMA CELLS

Exposure of a non-autocrine CSF-1 receptor-positive carcinoma cell line to recombinant human CSF-1 (rhM-CSF) leads to the phosphorylation on tyrosine of a variety of proteins including one 160 kDa protein whose M_r and rapid phosphorylation kinetics (maximal by 3 minutes) are consistent with that of macrophage and trophoblast CSF-1 receptors [27]. More recently, we have also been able to demonstrate that CSF-1 treatment stimulates several-fold the invasion of CSF-1R-positive (BT20) breast carcinoma cells through an amnionic membrane support[34] and that this invasion is blocked by anti-urokinase antibodies. We are currently investigating in somewhat greater detail these and other phenotypic effects of exogenous CSF-1 on CSF-1R-positive ovarian and breast cancer cell lines (discussed in section 13.3).

13.2.8 CONTROL OF EXPRESSION OF CSF-1R IN BREAST AND OVARIAN CARCINOMA-DERIVED CULTURED CELL LINES: UP-REGULATION BY GLUCOCORTICOIDS

In ongoing studies of steroid and cytokine effects on *fms* expression, we have observed that levels of expression of breast, ovarian and choriocarcinoma cell line *fms* transcript and protein are dramatically augmented by glucocorticoids (Figure 13.2, Table 13.1;[27]) and (to a lesser degree) by progestins in cell lines which express glucocorticoid receptors. This effect is completely blocked by RU-486, (RU-38486, mifepristone), a potent competitive inhibitor of both glucocorticoids and progestins. Nuclear run-off and ActD/mRNA half-life studies (unpublished observations) strongly suggest that glucocorticoids increase CSF-1R gene expression by inducing the synthesis of a protein which dramatically stabilizes pre-existing *fms* transcripts and not by stimulating new *fms* gene transcription. We are currently attempting to isolate and clone this glucocorticoid inducible protein as well as define those sequences necessary for its interaction with carcinoma cell *fms* transcripts.

13.2.9 INTERACTIONS OF CSF-1R WITH OTHER BIOLOGICALLY RELEVANT TYROSINE KINASE GROWTH FACTOR RECEPTORS: EGF-R, HER-2/*neu*

As mentioned above, overexpression of EGF-R and/or HER-2/*neu* has already been linked to aggressive biological behaviour and adverse prognosis in some studies for breast and ovarian carcinoma[9,36,37], but the molecular physiology underlying these clinicopathological associations remains unclear. It is clear, however, that ligand activation of either one of these two tyrosine kinase growth factor receptors can result in the transphosphorylation and cross-activation of the other when both are expressed in the same cell, suggesting that receptor transphosphorylation cascades and inter-receptor 'cross-talk'[38,39] may permit one ligand to activate the expression of phenotypes normally under the control of a second distinct ligand/receptor pair.

In earlier clinicopathological studies[14,15], we reported statistically significant correlations of expression of CSF-1R and either EGF-R or HER-2/*neu* in breast and ovarian carcinomas, but the biological relevance of these observations remained unclear. Based upon these observations, we are currently investigating whether functional CSF-1R is ever coexpressed together with either EGF-R and/or HER-2/*neu* in at least some ovarian and breast carcinoma cell lines; and if so, whether ligand activation of CSF-1R leads to cross-activation of either or both of these receptors.

13.2.10 POTENTIAL EFFECTS OF TUMOUR-PRODUCED CSF-1 ON MACROPHAGE FUNCTION AND INTERACTIONS WITH TUMOUR CELLS AND OTHER HOST IMMUNE CELLS

CSF-1-stimulated macrophages synthesize a wide variety of cytokines including interleukin (IL)-1, IL-6, tumour necrosis factor (TNF) and probably a wide variety of other cytokines (TGF-α, TGF-β, ?gp30) active both on tumour cells and other host immune cells. It is quite possible that the net effect of this local production of cytokines by CSF-1 stimulated macrophages may foster, rather than inhibit, tumour cell proliferation and dissemination and adversely modify host immune anti-tumour responses. We are actively investigating the possibility that therapeutic strategies designed to interrupt this paracrine tumour/stromal macrophage interaction may disrupt tumour proliferation, dissemination and evasion of host immune anti-tumour defences.

With regard to the latter, it has also recently been recognized[39] that CSF-1 can exert paradoxical 'immunosuppressive' effects on macrophages in striking contrast to its known stimulatory effects on macrophage oxidative burst and antimicrobial activity [17,21]. Specifically, CSF-1 (i) down-regulates macrophage MHC II antigen expression (to thereby interfere with macrophage antigen presentation to $CD4^+$ T cells) and (ii) is also able to block the increases in macrophage MHC II antigen expression induced by other theoretically 'immunostimulant' cytokines including interferon-γ and GM-CSF[38], without measurable effects on MHC I antigen expression. If such events occur *in vivo*, then tumour production of CSF-1 could interfere with host responses to otherwise foreign tumour antigens and thereby aid tumour cell 'evasion' of host immune defences. It is this latter very intriguing possibility that we are now investigating, employing a strongly antigenic mouse tumour cell line[40] which we have 'engineered' to express high levels of CSF-1 prior to subcutaneous implantation in wild-type and immunodeficient mice.

13.3 ONGOING RESEARCH AND FUTURE DIRECTIONS

Based on the observations which we have summarized above, we now hypothesize that local and circulating CSF-1 in ovarian, endometrial and breast carcinoma patients should be available to act both as (i) an autocrine growth and differentiation factor for CSF-1R-positive carcinoma cells by itself and through interaction with other tyrosine kinase receptors and (ii) a paracrine effector of monocyte/macrophage function locally and at metastatic sites. We are currently carrying out further studies to investigate *in vitro*:

1. the effects of CSF-1 on CSF-1R-positive breast and ovarian carcinoma cell lines;
2. biochemical and physiological interactions involving CSF-1R and other tyrosine kinase receptors or recognized importance in ovarian and breast carcinomas (e.g. EGF-R, HER-2/*neu*);
3. molecular controls of CSF-1 and CSF-1R expression focusing specifically on:
   (a) control of CSF-1R expression by glucocorticoids and progestins;
   (b) control of CSF-1R expression by a

tumour suppressor locus on chromosome 11.

We are attempting to study *in vivo*:

4. the effects that tumour cell production of CSF-1 may exert on:
   (a) macrophage production of tumour-stimulating and immunosuppressive cytokines;
   (b) macrophage presentation of antigen and modulation of antitumour activity of other host immune cells.

We are also investigating:

5. the prognostic significance of tumour expression of other antigenic markers shared by trophoblast, activated macrophages, and carcinoma cells;
6. the utility of measurements of circulating CSF-1 levels as a 'tumour marker' complementary to CA-125 in the screening, diagnosis, and monitoring of response to therapy of ovarian, endometrial and breast carcinoma patients.

ACKNOWLEDGEMENTS

Dr Kacinski is supported by the PHS-NIH Award R29-CA47292 and a Bristol-Myers Cancer Research Award R100-063.

He acknowledges the contributions of his collaborators at Yale, Dr Darryl Carter in the Department of Pathology and Dr Setsuko K. Chambers and other members of the Gynecologic Oncology Section headed by Dr Peter E. Schwartz in the Department of Obstetrics and Gynecology.

He also acknowledges the contributions of three postdoctoral fellows, Dr Eva Sapi, Dr Bonnie L. King and Dr Lisa D. Yee and a student Kimberly A. Scata and that of external collaborators Dr E.R. Stanley at Albert Einstein College of Medicine and Dr Larry R. Rohrschneider and his former postdoctoral fellow Dr Victoria M. Rothwell at the Fred Hutchinson Cancer Research Center.

He also graciously acknowledges the gift of recombinant CSF-1 for RIA studies from Cetus and the generous assistance of Dr Eugene Brown of Genetics Institutes Cambridge who graciously provided rhM-CSF and other cytokines, antibodies, and PCR oligonucleotides necessary for many of the studies described in this communication.

REFERENCES

1. Boring, C.B. (1991) Cancer statistics, *CA*, **41**, 19–36.
2. Young, R.C., Knapp, R.C., Fuks, Z. and DeSaia, P.J. (1985) Cancer of the ovary, in *Cancer: Principles and Practice of Oncology*, (eds V.T. DeVita, S. Hellman and S.A. Rosenberg), J.P. Lippincott, Philadelphia, pp. 1083–109.
3. Day, T.G., Gallager, H.S. and Rutledge, F. (1975) Epithelial carcinoma of the ovary: prognostic importance of histologic grade. *Natl Cancer Inst. Monogr.*, **42**, 15–18.
4. Perez, C.A., Knapp, R.C., DiSaia, P.J. and Young, R.C. (1985) Gynecologic tumors, in *Cancer: Principles and Practice of Oncology*, (eds V.T. DeVita, S. Hellman and S.A. Rosenberg), J.B. Lippincott, Philadelphia, pp. 1013–81.
5. Berman, M.L., Ballon, S.C., Lagasse, L.D. and Watring, W.G. (1990) Prognosis and treatment of endometrial cancer. *Am. J. Obstet. Gynecol.*, **136**, 679–88.
6. Halsted, W.S. (1970) The results of radical operations for the cure of cancer of the breast. *Ann. Surg.*, **46**; 1–19.
7. Fisher, B. and Slack, N.H. (1970) Number of lymph nodes examined and the prognosis of breast cancer. *Surg. Gynecol. Obstet.*, **131**, 79–88.
8. O'Reilly, S.M., Camplejohn, R.S., Barnes, D.M. *et al.* (1990) Node negative breast cancer: prognostic subgroups defined by tumor size and flow cytometry. *J. Clin. Oncol.*, **8**, 2040–6.
9. Slamon, D.J., Godolphin, W., Jones, L.A. *et al.* (1989) Studies of the HER-2/*neu* proto-oncogene in human breast and ovarian cancer. *Science*, **244**, 707–12.
10. Van de Vijver, M.J., Peterse, J.L., Mooi, W.J. *et al.* (1988) Neu protein overexpression in breast cancer: Association with comedo-type ductal carcinoma *in situ* and limited prognostic

value in stage II breast cancer. *N. Engl. J. Med.*, **319**, 1239–46.
11. Kacinski, B.M., King, B.L., Chambers, S.K. and Schwartz, P.E. (1991) Expression of the *neu* oncogene in benign, borderline, and invasive ovarian neoplasms and carcinoma-derived cell lines. Abstract, Proceedings of the 22nd Annual Meeting of the Society of Gynecologic Oncologists, Orlando, FL, 17–20 February (manuscript submitted).
12. Lupu, R., Colmer, R., Zugmaier, G. *et al.* (1990) Direct interaction of a ligand for the *erb*B2 oncogene product with the EGF receptor and p185-*erb*B2. *Science*, **249**, 1552–5.
13. Kacinski, B.M., Carter, D., Mittal, K. *et al.* (1988) High level expression of *fms* proto-oncogene mRNA is observed in clinically aggressive human endometrial adenocarcinomas. *Int. J. Radiat. Oncol. Biol. Phys.*, **15**, 823–9.
14. Kacinski, B.M., Carter, D., Kohorn, E.I. *et al.* (1989) Oncogene expression *in vivo* by ovarian adenocarcinomas and mixed-müllerian tumors. *Yale J. Biol. Med.*, **62**, 379–92.
15. Yee, L., Kacinski, B.M. and Carter, D.C. (1989) Oncogene structure, function and expression in breast cancer. *Semin. Diagn. Pathol.*, **6**, 110–25.
16. Taylor, G.R., Reedijk, M., Rothwell, V. *et al.* (1989) The unique insert of cellular and viral *fms* protein tyrosine kinase is dispensable for enzymatic and transforming activity. *EMBO J.*, **8**, 2029–37.
17. Clark, S.C. and Kamen, R. (1987) The human hematopoietic colony stimulating factors. *Science*, **236**, 1229–37.
18. Rettenmier, C.W., Sacca, R., Furman, W.L. *et al.* (1986) Expression of the human c-*fms* proto-oncogene product (colony-stimulating factor-1 receptor) on peripheral blood mononuclear cells and choriocarcinoma cells lines. *J. Clin. Invest.*, **77**, 1740–6.
19. Pollard, J.W., Bartocci, A., Arceci, R. *et al.* (1987) Apparent role of the macrophage growth factor CSF-1 in placental development. *Nature*, **330**, 484–6.
20. Tang, R., Kacinski, B.M., Validire, P. *et al.* (1990) Oncogene amplification correlates with dense lymphocyte infiltration in human breast cancers: a role for hematopoieitic growth factor release by tumor cells? *J. Cell. Biochem.*, **45**, 1–10.
21. Stanley, E.R. (1986) The action of the colony stimulating factor CSF-1. *Ciba Found. Symp.*, **118**, 29–41.
22. Kacinski, B.M., Chambers, S.K., Stanley, E.R. *et al.* (1990) The cytokine CSF-1 (M-CSF), whose receptors are expressed by endometrial carcinomas *in vivo* and *in vitro*, may also be a circulating tumor marker of neoplastic disease activity in endometrial carcinoma patients. *Int. J. Radiat. Oncol. Biol. Phys.*, **19**, 619–26.
23. Kacinski, B.M., Stanley, E.R., Carter, D. *et al.* (1989) Circulating levels of CSF-1 (M-CSF), a lymphohematopoietic cytokine, may be a useful marker of disease status in patients with malignant ovarian neoplasms. *Int. J. Radiat. Oncol. Biol. Phys.*, **17**, 159–64.
24. Kacinski, B.M., Bloodgood, R.S., Schwartz, P.E. *et al.* (1989) The macrophage colony stimulating factor CSF-1 is produced by human ovarian and endometrial adenocarcinoma-derived cell lines and is present at abnormally high levels in the plasma of ovarian carcinoma patients with active disease, in *Molecular Diagnostics of Human Cancer*, Cold Spring Harbor Press, Cold Spring Harbor, NY, pp. 333–7.
25. Kacinski, B.M., Chambers, S.K., Carter, D. *et al.* (1990) The macrophage colony stimulating factor CSF-1, an auto- and paracrine tumor cytokine, is also a circulating 'tumor marker' in patients with ovarian, endometrial, and pulmonary neoplasms. *Prog. Leuk. Biol.*, **10B**, 393–400.
26. Kacinski, B.M., Carter, D., Yee, L.D. *et al.* (1990) Ovarian adenocarcinomas express *FMS*-complementary transcripts and *FMS* antigen *in vivo* and *in vitro*, often with co-expression of CSF-1. *Am. J. Pathol.*, **137**, 135–47.
27. Kacinski, B.M., Scata, K.A., Carter, D. *et al.* (1991) *FMS* (CSF-1 receptor) and CSF-1 transcripts and protein are expressed by human breast carcinomas *in vivo* and *in vitro*. *Oncogene* (in press).
28. Woolford, J., McAuliffe, A. and Rohrschneider, L.R. (1988) Activation of the feline c-*fms* proto-oncogene: multiple alterations are required to generate a fully transformed phenotype. *Cell*, **55**, 965–77.

29. Roussel, M.F., Downing, J.R., Rettenmier, C.W. and Sherr, C.J. (1988) A point mutation in the extracellular domain of the human CSF-1 receptor (c-*fms* proto-oncogene product) activates its transforming potential. *Cell*, **55**, 979–88.
30. Coussens, L., Van Beveren, C., Smith, D. *et al.* (1986) Structural alteration of viral homologue of receptor proto-oncogene *fms* at carboxyl terminus. *Nature*, **320**, 277–80.
31. Horiguchi, J., Sherman, M.L., Sampson-Johannes, A. *et al.* (1988) CSF-1 and c-*fms* gene expression in human carcinoma cell lines. *Biochem. Biophys. Res. Commun.*, **157**, 395–401.
32. Ramakrishnan, S., Xu, F.J., Brandt, S.J. *et al.* (1989) Constitutive production of macrophage colony-stimulating factor by human ovarian and breast cancer cell lines. *J. Clin. Invest.*, **83**, 921–6.
33. Wong, G.G., Temple, P., Leary, A.C. *et al.* (1987) Human CSF-1: Molecular cloning and expression of a 4 kb cDNA encoding the human urinary protein. *Science*, **235**, 1504–8.
34. Filderman, A.E., Bruckner, A., Kacinski, B. and Remold, H. (1991) Macrophage colony-stimulating factor (CSF-1) enhances invasiveness in CSF-1 receptor (CSF-1R)-positive lung cancer cell lines. *J. Cell. Biochem.* (Suppl. 15E) (in press).
35. Harris, A.L. and Nicholson, S. (1988) Epidermal growth factor receptors in human breast cancer. *Cancer Treat. Resp.*, **40**, 93–118.
36. Stanbrige, E.J. and Nowell, P.C. (1990) Origins of human cancer revisited. *Cell*, 867–74.
37. Stern, D.F. and Kamps, M.P. (1988) EGF-stimulated tyrosine phosphorylation of $p185^{neu}$: a potential model for receptor interactions. *EMBO J.*, **7**, 995–1001.
38. Connelly, P.A. and Stern, D.F. (1990) The epidermal growth factor receptor and the product of the *NEU* protooncogene are members of a receptor tyrosine phosphorylation cascade. *Proc. Natl Acad. Sci. USA*, **87**, 6054–7.
39. Willman, C.L., Stewart, C.C., Miller, V. *et al.* (1989) Regulation of MHC class II gene expression in macrophages by hematopoietic colony-stimulating factors (CSF). *J. Exp. Med.*, **170**, 1559–67.
40. Torre-Amione, G., Beauchamp, R.D., Koeppen, H. *et al.* (1990) A highly immunogenic tumor transfected with a murine transforming growth factor type beta cDNA escapes immune surveillance. *Proc. Natl Acad. Sci. USA*, **87**, 1486–90.

Chapter 14
Regulation of growth of human ovarian cancer cells

G.B. MILLS, S. HASHIMOTO, J. HURTEAU,
R. SCHMANDT, S. CAMPBELL, C. MAY,
M. HILL, P. SHAW, R. BUCKMAN and D. HOGG

14.1 INTRODUCTION

Adenocarcinoma of the ovary is the leading cause of death from gynaecological malignancy and the fourth leading cause of death from cancer in females[1–3]. Current therapy, which combines surgery, radiation and chemotherapy, cures less than 40% of patients[1–3]. While the cure rate of patients with stage III or IV cancer of the ovary is very low, that of stage I is approaching 80%. The high mortality rate of ovarian cancer is due predominantly to occult progression of the tumour within the peritoneal cavity with initial diagnosis usually made at an advanced stage[1–3]. Modifications in chemotherapy, radiation therapy, or surgery are unlikely in the near future to improve the dismal prognosis that is associated with this disease. Since patients diagnosed at an early stage have a reasonable prognosis, techniques for early diagnosis are essential[1,2]. An improved understanding of the mechanisms regulating the growth of ovarian cancer cells may eventually lead to techniques which facilitate early diagnosis, establish prognosis or enable determination of response to therapy. Eventually, it may even be possible to design effective therapeutic approaches based on interfering with the biochemical processes which allow the growth of ovarian cancer cells.

Our working hypothesis is that normal ovarian epithelial cells proliferate in response to growth factors and that malignant ovarian epithelial cells may proliferate abnormally due to aberrations in the production, action or signalling processes activated by growth factors. In addition, ovarian cancer cells may have lost the ability to produce or respond to factors which normally limit proliferation of benign ovarian epithelial cells. By comparing the responses of benign and malignant ovarian epithelial cells at the molecular level, we may be able to determine the abnormalities which result in the proliferation of malignant ovarian cancer cells.

The natural history and frequency of adenocarcinoma of the ovary provides a unique opportunity to study the growth regulatory processes that allow the proliferation of malignant cells of many different lineages. Initial debulking of ovarian cancer provides as much as several kilograms of tumour for study[1,2]. Epithelial ovarian cancer is frequently associated with the accumulation of large quantities of fluid in the abdomen (3–4 litres). This ascites is removed at initial surgery and provides: (i) large numbers of tumour cells (10^{10}) in single cell suspension or in small clumps; (ii) tumour conditioned medium; and (iii) tumour infiltrating cells (10^9) composed of monocytes, neutrophils and lymphocytes[4]. Removal of ascites for palliation after tumour recurrence allows for repeated studies of freshly isolated tumour cells from the same patient. The short interval between therapy and recurrence combined with the availability of tumour cells both at initial diagnosis and at recurrence allows for the study of drug resistance in tumour cells from patients before and after therapy.

Studies of malignant transformation have been facilitated by the comparison of benign and malignant cells of the same tissue type. In some cancers such as bowel and cervix, a spectrum of normal, dysplastic, neoplastic and malignant cells is readily available. The benign counterpart of epithelial ovarian cancer consists of a single layer of cells on the surface of the ovary which until recently could not be propagated in culture[5]. The lack of sufficient numbers of normal ovarian epithelial cells has prevented comparisons of the physiology of normal and malignant ovarian epithelial cells. In addition, at the time of diagnosis, in most cases, both ovaries are completely replaced by malignant tumour cells. However, these problems have been ameliorated somewhat by newer techniques requiring fewer cells for immunohistochemical or molecular genetic analysis, combined with the development by Auersperg (Vancouver, Canada) of techniques to culture normal human ovarian epithelial cells, as well as to immortalize normal ovarian epithelial cells [5–7].

This chapter describes model systems which have been used to evaluate the mechanisms regulating the growth of epithelial ovarian cancer cells and to compare these mechanisms to those regulating the proliferation of normal cells. Although some of the interpretations presented herein are based on preliminary data, they provide a framework on which to plan and analyse future studies.

14.2 OVULATION AND OVARIAN CANCER

A number of epidemiological studies have established that frequency of ovulation is associated with the risk of developing ovarian cancer[1,2]. The physiological mechanism leading to this association is not clear. However, we have attempted to test some of the hypotheses which could explain this observation.

Ovulation is associated with a rupture of the tunica albuginea and the surface ovarian epithelium. The ovarian epithelium must migrate and proliferate to heal this wound. Ovulation is associated with the production of a number of growth factors for ovarian epithelium[8–10]. The concentrations of growth factors for ovarian epithelial cells, including transforming growth factor-α (TGFα), are high in follicular fluid[8–10]. Also both granulosa and theca cells have been demonstrated to produce a variety of different growth factors which can alter proliferation of ovarian epithelial cells[8–10]. Therefore following ovulation, a transient wave of proliferation of ovarian epithelial cells occurs. This proliferation must be tightly regulated to ensure that it occurs at the correct time and location, and also that it terminates following healing of the ovulation site. However, each time a cell divides there is the potential for an error in replication of DNA[11]. These mutations could result in the activation of an oncogene (those genes associated with the induction of cancer, also called dominant oncogenes), or in the inactivation of a tumour suppressor gene (those genes associated with inhibition of tumour growth, also called anti-oncogenes or recessive oncogenes)[11–13]. Therefore, the requirement to heal the wound left by ovulation results in ovarian surface epithelium having a high proliferative potential and this may, at least in part, play a role in the high frequency of ovarian cancer.

Epithelial ovarian cancer frequently appears to start from inclusion bodies within the ovary. These inclusion bodies may occur as the result of ovulation, followed by migration of epithelial cells into the defect and entrappment of these cells within the healed ovary. Alternatively, the inclusion bodies may represent embryonal rests. Why ovarian cancer starts so frequently from the small numbers of ovarian epithelial cells entrapped within the ovary is unknown. However, it may relate to the local production and action of growth factors.

Local action (autocrine or paracrine) of the growth factors is enforced by rapid diffusion and a resultant decrease in concentration due to dilution in interstitial fluid[14]. In some cases, the growth factor is not released at all, but rather remains on the cell surface and only stimulates cells that are in direct contact with the cell expressing the growth factor (juxtacrine). A role for growth factors in tumorigenesis is supported by the observation that many tumour cells, including ovarian cancer cells, produce and respond to a number of different growth factors[11–13]. Furthermore, many of the known oncogenes are either growth factors, growth factor receptors, or components of the pathways stimulated by growth factors[11–13]. Therefore, at least one of the components of malignant transformation appears to be the aberrant activation of the growth factor pathways which normally regulate the proliferation of normal cells.

The local concentration of growth factors produced by ovarian epithelial cells on the surface of the ovary rapidly decreases due to diffusion into the peritoneal cavity. However, growth factors produced by ovarian epithelial cells in inclusion bodies cannot diffuse into the peritoneal cavity and may attain high local concentrations. This may lead to the proliferation of ovarian epithelial cells entrapped within the ovary. Indeed, we and others have demonstrated that fluid from benign cysts contains growth promoting activity which can increase the proliferation of ovarian epithelial cancer cell lines[10].

Ovarian cancer normally breaks through the surface of the ovary and then seeds throughout the peritoneal cavity. Although distant metastases do occur, they are usually a relatively late event and the majority of the tumour remains within the peritoneal cavity[1,2]. The intraperitoneal growth of ovarian cancer cells is associated with the accumulation of ascitic fluid. This ascitic fluid contains the growth factors to which ovarian cancer cells would be exposed *in vivo*. Ascitic fluid from ovarian cancer patients increases proliferation of ovarian epithelial cells *in vitro* (Figure 14.1)[4] and *in vivo*[15], indicating that it is indeed a potent source of growth promoting activity. As demonstrated in Figure 14.1, the growth-inducing activity of ascitic fluid from patients with ovarian cancer does not return to baseline levels until greater than a 1/600 dilution is reached, as compared to 1/20 for ascites from patients with benign disease. Therefore, it is possible that ascitic fluid plays a direct role in the proliferation of ovarian cancer cells in the peritoneal cavity.

The pattern of growth of human ovarian cancer cells in immuno-incompetent nude mice supports the contentions described above. In the human, ovarian cancer cells normally proliferate within the peritoneal cavity and spontaneous subcutaneous metastases are rare. In contrast, human ovarian cancer cells grow readily in the subcutaneous tissues of nude mice. However, in the majority of cases, the same cells will not grow intraperitoneally. This suggests that local

Regulation of growth of human ovarian cancer cells

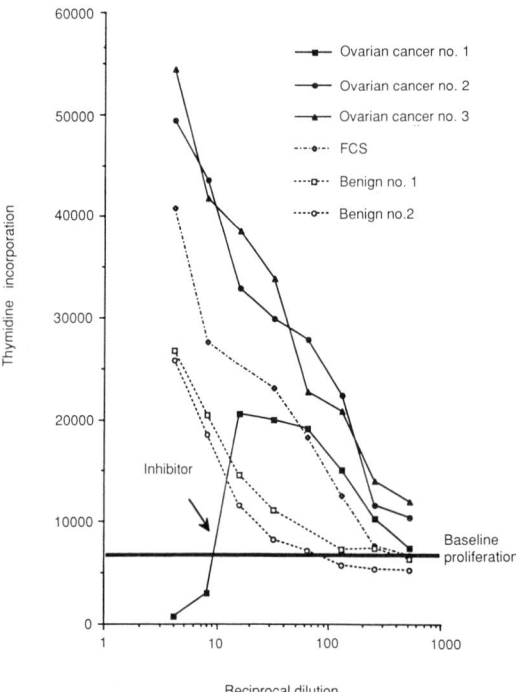

Figure 14.1 The HEY ovarian cancer cell line was carried in RPMI 1640 with 10% fetal calf serum (FCS). Cells were washed and incubated without FCS for 48 hours. Cells were harvested, and plated at 2×10^4/ml in 96-well microtitre plates in the presence of serial dilutions of FCS or ascites from patients with ovarian carcinoma or with benign hepatic disease. Cells were cultured for 48 hours, pulsed with [^{3}H]thymidine, harvested and counted in a beta counter. Cells cultured in RPMI 1640 alone incorporated 7000 c.p.m. Note that proliferation induced by the two benign samples dilutes to baseline levels at approximately 1 in 20, whereas all three ascitic fluids from malignant patients dilute to baseline at 1/600 or more, representing approximately 30 times more growth factor activity. Ovarian cancer no.1 represents a sample which contains growth inhibitory activity.

cancer cells injected into the peritoneal cavity of nude mice would diffuse away rapidly, resulting in low local concentrations. We have tested this hypothesis by determining the effect of ascitic fluid from ovarian cancer patients (as an exogenous source of growth factors to which the ovarian cancer cells are exposed in patients) on the intraperitoneal growth of human ovarian cancer cells[15]. In the majority of cases (23/28 mice), injection of semi-purified cell-free ascitic fluid from ovarian cancer patients, but not from patients with benign hepatic disease (1/15), resulted in the intraperitoneal growth of cells from the human ovarian cancer cell line HEY[15]. In contrast, intraperitoneal injection of HEY cells alone or with buffer did not result in intraperitoneal growth (0/21)[15]. In addition, through mechanisms which are not clear at this moment, many (approximately 50%) of the mice injected with HEY cells and human ovarian cancer ascitic fluid developed distant metastases, whereas mice injected with HEY cells either subcutaneously or intraperitoneally, without additional injection of ascitic fluid from human ovarian cancer patients, did not. Therefore, ascitic fluid from ovarian cancer patients contains potent growth factor activity which alters the growth of human ovarian cancer cells *in vitro* and *in vivo*. Therefore, it is likely that ascitic fluid also plays a major role in the intraperitoneal proliferation of ovarian cancer cells in patients. As a corollary, growth factors present in ascitic fluid are potential targets for therapy. In addition, measurement of levels of these growth factors in ascitic fluid or in serum may aid in early diagnosis or screening or possibly in determining prognosis.

14.3 GROWTH FACTORS AND THEIR RECEPTORS IN OVARIAN CANCER

Interleukin 1 IL-1), IL-6, platelet-derived growth factor (PDGF), and macrophage colony stimulating factor (M-CSF) have been

concentrations of growth factors in the subcutaneous tissues of nude mice (perhaps produced by the ovarian cancer cells themselves) may be sufficient to allow establishment of the human tumour. In contrast, growth factors produced by human ovarian

Table 14.1 Benign and malignant ovarian epithelial cell lines produce M-CSF. Short-term normal ovarian epithelial cells (nos. 1, 2) and malignant cells (nos. 4–7) as well as malignant cell lines (nos. 1–3) were established. Cells were cultured at 1×10^6/ml for 24 hours in RPMI 1640, supernatants collected and assessed for M-CSF levels by Cetus Corporation with a specific RIA as described by Shadduck et al.[23]. Lymphoid cells produced less than 0.25 ng/ml, and fibroblasts produced 1–5 ng/ml; 1 ng/ml is equivalent to 20 U/ml

	$ng/ml/10^6 cells/$ 24 hours
Malignant ovarian epithelial cells ($n=7$)	2.47 ± 1.6 (range 1–6)
Normal ovarian epithelial cells ($n=2$)	0.83 (range 0.25–1.4)

identified in either ascites or in extracts from solid ovarian tumours[10,16–22]. Controversy still exists as to whether epidermal growth factor (EGF) or transforming growth factor alpha (TGFα), which bind to and activate the EGF receptor, are present in ascitic fluid (see below;[10,16,17]). Furthermore, both benign and malignant ovarian epithelial cells can produce M-CSF (Table 14.1), EGF, TGFα, granulocyte colony stimulating factor (G-CSF), PDGF, TGFβ, IL-1, or IL-6, suggesting that malignant cells may be the *in vivo* source of these factors[16,18–20]. As described above, ascitic fluid also contains large numbers of mononuclear cells including lymphocytes. Using a sensitive and quantitative polymerase chain assay, we have demonstrated that these cells contain mRNA for a number of growth factors including the interleukins, interferons and colony stimulating factors. In addition to these known factors, we have demonstrated that ascitic fluid from all ovarian cancer patients contains what appears biochemically and by mechanism of action to be a putative unique growth factor, designated ovarian cancer growth factor (OCGF).

14.3.1 OVARIAN CANCER GROWTH FACTOR

Ascitic fluid from ovarian cancer patients is at least 10 times more effective at inducing the growth of ovarian epithelial cancer cells than are combinations of EGF, TGFα, M-CSF, insulin and transferrin in the presence of bovine serum albumin as a protein source. This suggested that ascitic fluid from ovarian cancer cells is likely to contain a number of other growth factors.

We decided to determine if ascitic fluid from ovarian cancer patients could activate any of the immediate/early biochemical events which have been associated with the action of growth factors in other systems [21,22,25,26]. It is likely, by analogy to these other growth factor systems, that a single growth factor could activate a specific biochemical process independent of the action of other growth factors. That specific biochemical process could then be used to compare the activity of growth factors in ascitic fluid to known factors, and also to purify unique growth factors from ascitic fluid.

Growth promoting activity of various growth factors receptors is dependent on tyrosine phosphorylation as shown by site-directed mutagenesis, intracellular injection of anti-phosphotyrosine antibodies and studies with tyrosine kinase inhibitors[26–28]. Most oncogenes are either mutated growth factors, growth factor receptors, or components of the signalling cascades activated by growth factor receptors[11–13]. In addition, many of the known oncogenes are tyrosine kinases, components of pathways which regulate tyrosine kinases, or are regulated by tyrosine kinases[11–13,26]. For example, mutated forms of the EGF, M-CSF, stem cell growth factor (SCF), and *neu* receptors constitute the *erb*B, *fms*, *kit*, and *erb*B2 tyrosine kinase oncogenes respectively and the *sis* oncogene produces a ligand which stimulates the PDGF receptor tyrosine kinase

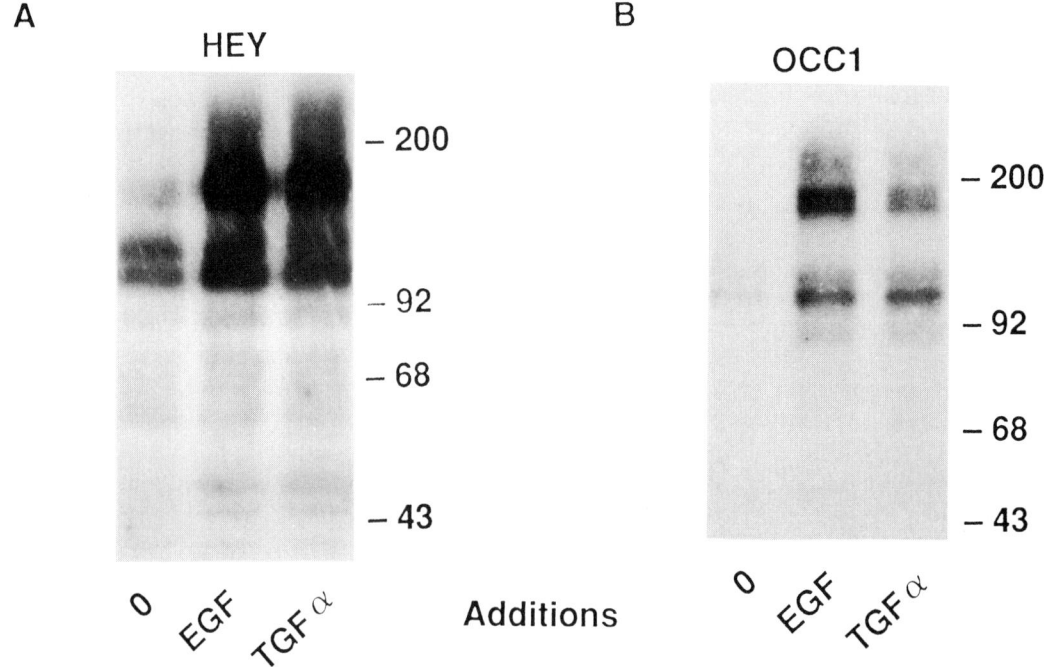

Figure 14.2 HEY (A) or OCC1 (B) ovarian cancer cell lines in logarithmic growth phase were starved of serum for 24 hours. Cells were then incubated for 15 minutes with a 10 ng/ml of EGF or TGF-α as indicated. Cells were lysed in SDS Laemlli buffer and proteins separated on SDS-PAGE as previously described. Proteins were electrophoretically transferred to nitrocellulose and western blotted with anti-phosphotyrosine antibodies detected with ^{125}I-protein A and autoradiography. Note that incubation with EGF or TGF-α results in an increase in tyrosine phosphorylation of a number of proteins in both HEY and OCCI. The most prominent band in activated cells is in the area of 175–185 kDa. In HEY this appears to be a single band (see Figure 14.3) and represents mainly the EGF receptor. In OCC1 cells, this band appears to be a doublet and represents tyrosine phosphorylation of both the EGF and *neu* receptors.

activity. Tyrosine kinase activity is necessary for transformation induced by these oncogenes, further emphasizing the importance of tyrosine phosphorylation in regulating the growth of both benign and malignant cells[11–13,26].

Despite inducing proliferation, ascitic fluid from 20 different ovarian cancer patients did not induce tyrosine phosphorylation in the ovarian cancer cell lines HEY or OCC1, as measured by western blotting with specific monoclonal anti-phosphotyrosine antibodies [29]. In each case, EGF or TGFα which stimulate the EGF receptor induced a marked and readily detectable increase in tyrosine phosphorylation in these cells (Figure 14.2)[29]. Therefore, at least at the level of sensitivity of this assay, ascitic fluid from ovarian cancer patients does not induce tyrosine phosphorylation, and ascitic fluid-induced proliferation is unlikely to be the consequence of increased tyrosine kinase activity.

In addition to stimulating tyrosine kinases, many growth factor receptors activate phospholipase C (PLC) which hydrolyses membrane phospholipids producing inositol phosphates and diacylglycerols[21,22,25]. These intracellular second messengers activate the serine/threonine specific protein kinase C (PKC) and increase cytosolic free calcium ($[Ca^{2+}]_i$). In many cases, activation of phospholipases appears to be a prerequisite for growth factor- and oncogene-induced cell proliferation[21,22,25].

We have demonstrated that ascites from ovarian cancer patients activates PLC[24]. This is reflected by increases in $[Ca^{2+}]_i$, as well as by increased phospholipid turnover and inositol phosphate production[24]. We have utilized this ability to activate PLC to purify and characterize OCGF, which is clearly different from other growth factors identified as present in ascites. EGF, TGFα, TGFβ, M-CSF, insulin, insulin-like growth factor (IGF)-1, fibroblast growth factor (FGF), PDGF, the interleukins, interferons, or colony stimulating factors do not alter $[Ca^{2+}]_i$ in ovarian cancer cells and do not alter the effect of OCGF on $[Ca^{2+}]_i$ in Jurkat cells (T cell leukaemia) which do not express *neu*, eliminating the possibility that it is a ligand for *neu*.

Using a combination of protein separation columns, we have been able to purify OCGF, as assayed by ability to induce changes in cytosolic calcium, over 6000-fold. The most purified material available retains the ability to induce cell growth as well as to increase $[Ca^{2+}]_i$. OCGF is a highly hydrophobic glycoprotein of approximately 30 kDa, which is stable to acid, base and reduction. These biochemical characteristics are different from those of other characterized growth factors and, combined with the ability of OCGF to increase $[Ca^{2+}]_i$, suggest that it is a unique growth factor.

Since OCGF can be absorbed to ovarian cancer cells, washed extensively with medium and then eluted from cells at a low pH, it is likely that a high affinity receptor for OCGF is present on ovarian cancer cells[24].

OCGF cannot be absorbed and eluted from freshly isolated peripheral blood lymphocytes, suggesting that the receptor for OCGF is not expressed on all cells[24].

Although OCGF concentrations are high in ascitic fluid from ovarian cancer patients, ascitic fluid from patients with other cancers and with some benign diseases also contain low but significant levels of a similar or the same factor (Figure 14.1)[24]. Indeed, a similar activity is present in serum from normal individuals as well as in fetal calf serum[24]. Although our studies have concentrated on the effect of OCGF on ovarian cancer cell lines, ascitic fluid from ovarian cancer patients will induce changes in $[Ca^{2+}]_i$ in murine NIH3T3 fibroblasts, rat mesangial cells, fresh human breast cancer cells, and the Jurkat T cell leukaemia line[24]. This suggests that OCGF may stimulate a wide variety of cells. It is possible that OCGF is produced by ovarian cancer cells and that its activity is restricted to the activation of ovarian cancer cells. If this is the case, then serum and benign ascitic fluid would have to contain a different factor which cross-reacts in the calcium assay. Alternatively, OCGF may be a more generalized growth factor, such as EGF or TGFα, which is produced by many cell types and plays a pleiomorphic role in cell activation. Resolution of these possibilities awaits the production of specific antibodies or cloning and characterization of the factor itself.

14.3.2 EGF AND TGFα

EGF and TGFα are small, approximately 7-kDa proteins which bind to and activate the EGF receptor. One of the ligands for *neu* (gp36, see below)[30], as well as a unique 22-kDa macrophage product[31] also bind to and stimulate the EGF receptor. Therefore, the EGF receptor is regulated by at least four different growth factors allowing for a number of potential levels of regulation. The

EGF receptor contains an intrinsic tyrosine kinase domain and the presence of a functional tyrosine kinase is required for the growth promoting activity of EGF once again underlining the importance of tyrosine phosphorylation in growth regulation[32,33].

EGF alters the growth of normal ovarian epithelial cells as well as some ovarian carcinoma cell lines[8,9,20,34]. This suggests that functional EGF receptors are expressed on both benign and malignant ovarian epithelial cells and that EGF or TGFα plays a role in the normal growth regulation of these cells. In most circumstances, EGF and TGFα are growth promoting factors. However, as is demonstrated in Table 14.2, some ovarian cancer cell lines are growth stimulated by these factors, whereas others are significantly growth inhibited[29]. It must be noted that OCC1, 2, and 3 had been in culture for a limited period of time when they were assessed for their response to EGF and TGFα. Thus they may reflect the normal physiological response of ovarian cancer cells to the ligands better than previous studies with lines that had been in culture for longer periods of time.

The mechanism leading to increase in proliferation by EGF and TGFα in some lines and inhibition of proliferation in other lines is unknown. Despite differences in their responses to EGF and TGFα, HEY and OCC1 both express similar numbers of high affinity receptors for EGF (500) on their cell surface with a similar dissociation constant (10^{-2} M) as assessed by Scatchard analysis[29]. In addition, all four cell lines described in Table 14.2 have been assessed for tyrosine phosphorylation induced by EGF and TGFα, a *sine qua non* for activation of the EGF receptor. In each case, EGF and TGFα induced an increase in tyrosine phosphorylation which is similar in amount and pattern to that seen with HEY and OCC1 (Figure 14.2). As will be discussed below, the EGF receptor and *neu* interact to regulate signalling in breast cancer cells[36–38] and possibly in ovarian cancer cell lines. Therefore, it is possible that differences in responses to EGF and TGFα may reflect differences in the expression of *neu* or differences in interaction between *neu* and the EGF receptor. This possibility is supported by the observation that the ovarian cancer cell lines in Table 14.2 which overexpress (as assessed by western blotting with anti-*neu* antibodies) the *neu* protein (OCC1 and OCC3) are growth stimulated by EGF and TGFα, whereas those that express

Table 14.2 Effect of EGF, TGF-α, and TFG-β on proliferation of ovarian cancer cells. OCC1, 2, and 3 are ovarian cancer cell lines. The epithelial nature of the cells has been confirmed by immunohistochemistry with 2G3[35]. As assessed by Scatchard analysis, both OCC1 and HEY cells express approximately 500 high affinity EGF receptors with a dissociation constant of approximately 10^{-12} M[29]. Note that OCC2 and HEY express similar levels of *neu* as do benign ovarian epithelial cell lines measured by western blotting with specific anti-*neu* antibodies[29]. OCC1 and OCC3 overexpress *neu* protein by approximately 5–10 fold[29]. The effect of all factors were determined over a dilution ranging from 0.2 ng/ml to 25 ng/ml. Data are presented for 10 ng/ml which was maximal in each case. For studies with EGF and TFG-α, cell lines were grown and starved as described in Figure 14.1. For studies with TGF-β, OCC2 cells were cultured in 199/105 medium supplemented with 0.5% FCS and 3 mg/ml bovine serum albumin, whereas HEY was cultured in complete medium. Note that the cell lines which over express *neu* are growth stimulated by EGF and TGF-α, while lines which express normal levels of *neu* are growth inhibited by the same reagents

	Increase over baseline response			
	OCC1	OCC3	OCC2	HEY
EGF	1.7×	2.0×	0.43×	0.7×
TGF-α	2.0×	2.6×	0.46×	0.43×
TGF-β	nd	nd	0.12×	1.75×

nd, not done.

normal levels (normal is defined as similar levels to those expressed by benign ovarian epithelial cell lines in culture) of *neu* (OCC2 and HEY) are growth inhibited.

Production of TGFα has been suggested to be an indication of tumour burden and to correlate with a bad prognosis in ovarian cancer[17]. In one study, immunoreactive TGFα was detected in greater than 90% of biopsies of ovarian cancer patients[17]. Low levels of immunoreactive EGF were also found in approximately one-third of ovarian cancer biopsies[39]. However, recent studies have failed to detect immunoreactive TGFα in ascitic fluid from ovarian cancer patients [10]. Both TGFα and EGF require proteolytic cleavage to become fully active. Therefore, measuring levels of EGF and TGFα by immunological techniques may either overestimate or underestimate the levels of functional factor. We have therefore used a functional assay to determine whether biologically active EGF or TGFα is present in ascitic fluid. As demonstrated in Figure 14.2, addition of EGF or TGFα to the human ovarian cancer cell lines HEY or OCC1 results in a rapid increase in tyrosine phosphorylation of a number of proteins including the EGF receptor itself (175 kDa), as well as *neu* (185 kDa) (see below). With this assay, we can detect 1 ng/ml of EGF or TGFα which is similar to the level of EGF or TGFα which alters cell proliferation in these lines[29].

In contrast to immunoassays which detect antigenic material, the phosphorylation assay detects the presence of functional TGFα or EGF. Even with this sensitive and specific assay we have been unable to detect biologically active TGFα or EGF in ascitic fluid from 20 different ovarian cancer patients. The failure to detect functional EGF or TGFα in ascitic fluid was not due to the presence of inhibitors of the assay in ascitic fluid, as the addition of ascitic fluid to exogenous TGFα or EGF does not alter the ability to induce changes in tyrosine phosphorylation[29]. This is somewhat surprising, considering that ovarian cancer cells from the majority of patients contain immunoreactive TGFα either in the cytosol or on the cell surface as measured by immunohistochemistry (S. Campbell and G.B. Mills, unpublished data), and that supernatants from the HEY line contain functional EGF or TGFα. Reports with other ovarian cancer cell lines have failed to detect secretion of biologically active EGF, as measured by the ability to induce proliferation of EGF-sensitive cells[20]. Taken together, the data suggest that ascitic fluid is not a rich source of functional EGF or TGFα activity.

The level of tyrosine phosphorylation in cells reflects the balance between the activities of tyrosine kinases and of tyrosine phosphatases. Tyrosine phosphatases appear to be ubiquitous and may play a role in preventing transformation, as incubation of cells with vanadate, which inhibits tyrosine phosphatases, can cause cells to exhibit several of the characteristics of malignant transformation[28]. EGF-induced tyrosine phosphorylation is transient suggesting that tyrosine phosphatases are activated in ovarian epithelial cells[29]. We have observed that inhibition of tyrosine phosphates with vanadate induces a similar pattern of tyrosine phosphorylation to that induced by EGF[29]. This suggests that tyrosine phosphatases play a major regulatory role in ovarian epithelial cells, and that the dominant tyrosine kinase in ovarian cancer cells is the EGF receptor.

Expression of the EGF receptor has been demonstrated to be indicative of bad prognosis in several types of malignancy[41,42]. Several groups have examined the EGF receptor in ovarian cancer and have not found evidence for elevation at the DNA, RNA or protein level[17,18,43,44]. Surprisingly, several reports suggest that expression of the EGF receptor is an indicator of a good prognosis in ovarian cancer[40].

14.3.3 NEU

The *neu* proto-oncogene is expressed by normal ovarian epithelial cells suggesting that it plays a role in the growth of these cells [29,45,46]. Expression of elevated levels of *neu* by ovarian cancer cells is observed in approximately 30% of patients[45,46]. Overexpression of *neu* is highly correlated with poor prognosis. *Neu* appears to be overexpressed in malignant cells as a consequence of gene amplification in most patients. However, increased transcription or stability of mRNA, or increased translation or stability of the protein accounts for elevated levels in approximately 10% of patients[45]. Flow cytometric analysis of *neu* expression suggests that *neu* is overexpressed in only a subset of malignant ovarian epithelial cells. This would suggest that amplification of *neu* is a late event related to tumour progression rather than initiation. This is in agreement with the evidence that overexpression of *neu* is associated with a bad prognosis rather than a characteristic of all ovarian cancer patients.

Two reports have identified growth factor activity released by transformed breast or fibroblast cell lines which stimulate *neu* tyrosine kinase activity and alter proliferation of cells expressing *neu*[30,47]. Surprisingly, one of these factors (gp36) binds to both *neu* and the EGF receptor with similar affinities[30]. High levels of gp36 inhibit the proliferation of cell lines which overexpress *neu*[30]. Preliminary data suggest that low levels of gp36 induce the growth of cell lines which express *neu*[30].

14.3.4 EGF AND TGFα ACTIVATE THE EGF AND *NEU* RECEPTORS

Following activation of the EGF receptor, *neu* can be cross-linked to the EGF receptor, suggesting that heteroconjugates between the EGF and *neu* receptors are formed[38].

Western blotting of total cell lysates with anti-phosphotyrosine antibodies represents a major technological advance in the study of signal transduction through tyrosine kinases [38–52]. This assay is limited by the amount of protein that can be separated by SDS-PAGE prior to western blotting [48–52]. The sensitivity and specificity of the assay can be increased by immunoprecipitation with specific antibodies against the substrate of interest such as the EGF receptor or *neu* followed by western blotting with anti-phosphotyrosine antibodies (Figure 14.3)[29]. This technique does not measure the levels of the immunoprecipitated EGF receptor or *neu* (or any other molecule), but rather determines the level of tyrosine phosphorylation of the EGF receptor or *neu*.

We have demonstrated that addition of EGF or TGFα to benign and malignant ovarian epithelial cell lines induces a rapid increase in tyrosine phosphorylation of a number of substrates (Figure 14.2)[29]. In particular, following activation of the EGF receptor, a 175 kDa band is rapidly tyrosine phosphorylated in HEY cells, whereas in OCC1 cells, which overexpress *neu*, EGF and TGFα induce tyrosine phosphorylation of a doublet of approximately 175 and 185 kDa (Figure 14.2)[29]. Following immunoprecipitation with anti-EGF receptor antibodies to identify the EGF receptor and western blotting with anti-phosphotyrosine antibodies to determine if the receptor is tyrosine phosphorylated (Figure 14.3[29], the 175-kDa band can be readily identified as the EGF receptor in both HEY and OCC1 cells. However, immunoprecipitation with anti-*neu* antibodies failed to detect any bands in HEY cells which were reactive with the anti-phosphotyrosine antibodies irrespective of whether the cells were incubated in the presence or absence of EGF or TGFα. This indicates that *neu* is not detectably tyrosine phosphorylated in these cells. In contrast, in OCC1 a low level of *neu* tyrosine phosphorylation is present in resting cells and this is markedly increased following activation of

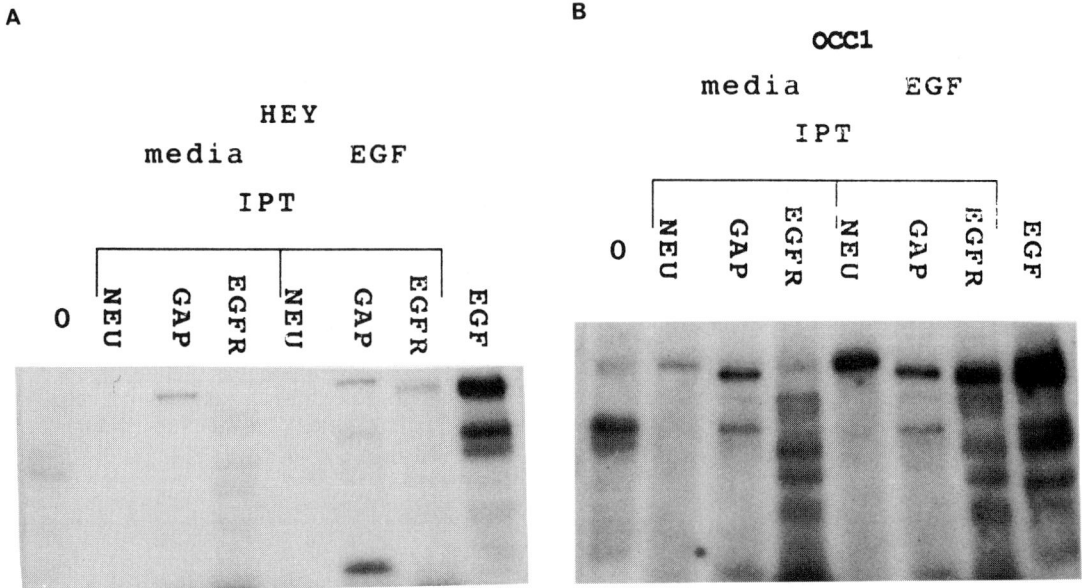

Figure 14.3 Identification of specific substrates following activation of the EGF receptor. HEY (A) or OCC1 (B) ovarian cancer cell lines in logarithmic growth phase were starved of serum for 24 hours. Cells were then incubated for 15 minutes with 10 ng/ml of EGF or TGF-α as indicated. Cells were lysed in RIPA buffer supplemented with 500 μM vanadate to decrease tyrosine dephosphorylation. Lysates were immunoprecipitated as described with antibodies to the EGFR receptor (J. Kudlow, Birmingham, AL, USA), *neu* (T. Pawson, Toronto, ON, Canada) or GAP (T. Pawson, Toronto, ON, Canada). Immunoprecipitates were washed and separated on SDS-PAGE. Proteins were western blotted for the presence of phosphotyrosine, as described in Figure 14.2. Proteins in the outer two lanes were not immunoprecipitated and represent total cell lysates similar to those described in Figure 14.2. The four left lanes are from resting cells, whereas the four right lanes are from cells incubated for 15 minutes with 10 ng/ml EGF.

This assay does not detect the levels of specific proteins but rather determines the level of tyrosine phosphorylation of these proteins. In particular, it is clear that the band observed in Figure 14.2 at 175–185 kDa can be resolved by this technique into at least three different tyrosine phosphorylated proteins. It includes the EGF receptor (175 kDa), the *neu* receptor (185 kDa) and a 180-kDa protein associated with the GAP protein. The *neu* antibody immunoprecipitates tyrosine phosphorylated *neu* in OCC1 cells which overexpress *neu* and does not immunoprecipitate the EGF receptor. Surprisingly, in HEY cells, a smaller protein is present in *neu* immunoprecipitates which is a likely co-immunoprecipitation of a tyrosine phosphorylated EGF receptor. This protein is only immunoprecipitated in mild detergent buffers such as NP40, and is lost in RIPA buffer, demonstrating that this is indeed co-immunoprecipitation of the EGF receptor. This indicates in HEY cells, that the EGF and *neu* receptors are closely associated, and that these molecules can be co-immunoprecipitated. The data indicate that activation of the EGF receptor in both HEY and OCC1, which are growth inhibited and growth stimulated by EGF and TGF-α respectively, results in increased tyrosine phosphorylation of a number of similar substrates. Strikingly, in the GAP immunoprecipitates of HEY cells, the 120-kDa GAP protein and a 62-kDa associated molecule are inducibly phosphorylated after activation of the EGF receptor, whereas they are constitutively phosphorylated in OCC1 cells. Whether this plays a role in the differential proliferative response to EGF or TGF-α is currently unknown.

the EGF receptor. This suggests that the *neu* and EGF receptor interact in ovarian cancer cells and that one of the possible mechanisms regulating *neu* is through the EGF receptor.

Why is overexpression of *neu* associated with a poor prognosis?

The strong correlation with a poor prognosis suggests that overexpression of *neu* plays a direct role in altering cell proliferation[45,46]. We have measured the effect of ascites from a number of different patients on tyrosine phosphorylation of *neu* in ovarian cancer cells as well as murine cell lines engineered to overexpress the human *neu* protein[29,14]. This is one of the assays used to identify and characterize the gp36 *neu* ligand as well as the *neu* ligand produced by transformed fibroblasts[30,47]. We have also concentrated ascitic fluid over 10-fold as well as followed the published protocol for partial purification of the gp36 *neu* ligand. In each case we have failed to detect the presence of the *neu* ligand in ascitic fluid as assessed by the ability to induce tyrosine phosphorylation. However, we have only assessed ascitic fluid from four individuals and it is possible that when larger numbers are studied the *neu* ligand will be present in the ascitic fluid of a subpopulation of patients.

High concentrations of the gp36 *neu* ligand inhibit proliferation of cells which overexpress *neu*[30]. We have demonstrated that low concentrations of ascites from all ovarian cancer patients stimulate proliferation of ovarian cancer cells[24]. However, high concentrations of ascites from some ovarian cancer patients inhibit proliferation of ovarian cancer cells both *in vitro* (Figure 14.1)[4] and *in vivo* (unpublished). The cell line tested expressed *neu*, albeit at low levels[29], so it is possible that many of the effects were secondary to binding of the *neu* ligand to *neu*. This would be in agreement with the studies mentioned above which suggest that if the *neu* ligand is present in ascitic fluid it is only present in a limited number of patients.

Since the studies described above have failed to detect the presence of functional ligands for *neu* or the EGF receptor in ascitic fluid, it is possible that the basal level of *neu* activity in unstimulated cells which overexpress *neu* is sufficient to result in a poor prognosis. As indicted in both Figures 14.2 and 14.3, low levels of tyrosine phosphorylation of *neu* are present in OCC1 cells independent of stimulation of the EGF receptor by exogenous ligand. Furthermore, when *neu* kinase activity was measured in a resting ovarian cancer cell line overexpressing *neu*, the *neu* kinase was active[29]. However, it remains to be determined whether this cell line (OCC1) secretes the *neu* ligand, or whether *neu* is activated in this cell line by the action of other kinases and ligands. We have not been able to detect *neu* tyrosine kinase activity in cells which do not overexpress *neu*, whereas *neu* kinase activity is readily detected in cells overexpressing *neu*[29]. This is most likely due to the low level of *neu* present on cells which do not overexpress *neu*; however, it may represent expression of an activated kinase in cells which overexpress *neu*.

Clinical applications

It may well be that overexpression of *neu* will correlate with responsiveness to particular forms of therapy and may identify those patients who should or should not receive alkylating agents of cisplatinum. However, 20% of patients with early disease recur. Measurement of prognostic indicators in this population may be particularly important and may suggest which patients do or do not require adjuvant therapy.

The levels of *neu* in some patients are significantly higher than those in normal tissues, thus giving a potential therapeutic advantage for therapy with reagents which target *neu*. This could consist of therapy with the *neu* ligand itself which decreases the growth of breast cancer cell lines which overexpress *neu*[30]. Remarkably, some antibodies to *neu* directly inhibit the growth of cell lines which overexpress *neu* while having limited effects on cell lines with normal levels of *neu* on their surface[53]. This suggests that anti-*neu* antibodies may provide effective modes of tumour therapy. It is likely that the efficiency of these antibodies or perhaps even of the growth factor could be increased if they were conjugated to radiolabels, cytotoxic drugs or to bacterial toxins or plant toxins. These modes of therapy are theoretically possible, have shown efficacy in nude mouse models in which murine cells transfected with human *neu* are growth inhibited by anti-*neu* antibodies, and appear to be nearing or in phase I human trials in several centres.

14.3.5 MACROPHAGE COLONY STIMULATING FACTOR

M-CSF, also known as CSF-1, is a 70–90 kDa glycoprotein homodimer[18]. In addition to stimulating mature monocytes and macrophages, M-CSF appears to play a role in the normal differentiation of the placenta as well as in the proliferation of choriocarcinoma cells[18].

Preliminary data suggest that serum and ascitic fluid M-CSF levels correlate with tumour bulk supporting the possibility that M-CSF is produced by the tumour cells and may provide a useful tumour marker[54]. Increased levels of M-CSF have been reported to be present in the serum of 70-80% of ovarian cancer patients[16,54]. Depending on the M-CSF level which is considered to be elevated, the sensitivity and specificity of the assay vary significantly. When 7 ng/ml as measured by RIA[23] is used as a cut-off, 94% of sera from ovarian cancer patients ($n = 35$) and 40% ($n = 29$) of benign samples have elevated levels[16]. When 8 ng/ml is used as the upper limit, 80% of ovarian cancer patients and 25% of benign samples have elevated levels[16]. Some ovarian cancer patients with low levels of serum CA-125 have elevated levels of M-CSF[54]. The number of patients in whom both assays are positive is over 90% suggesting that combining the two assays may have superior sensitivity to using either assay alone. It remains to be determined if M-CSF levels are elevated in the 50% of patients with stage I ovarian carcinoma who have normal CA-125 levels. The levels of M-CSF in serum frequently decrease after therapy and, at least in some patients rise prior to detectable clinical recurrence of ovarian cancer or increase in CA-125[54]. Therefore measurement of M-CSF serum levels complement CA-125 levels for determining response to therapy as well as identifying tumour recurrence.

Low levels of mRNA for M-CSFR, also known as c-*fms*, had been demonstrated to be present in approximately 60% of samples of ovarian carcinoma by northern blot analysis and immunohistochemistry[43]. However, the low level of expression observed could have represented a contribution by contaminating monocytes or other cells. More specific techniques using *in situ* hybridization have suggested that the M-CSFR is elevated in malignant cells from patients with ovarian carcinoma as compared to benign or borderline tumours and that the level of expression correlates with the grade of the tumour[55]. Very low levels of M-CSFR are detected in ovarian cancer cell lines; however, these can be increased by incubation with steroids or by culture in serum-free medium (Chapter 13). The presence of the M-CSFR on the cell surface combined with production of M-CSFR by ovarian cancer cells (Table 14.1;[16,19]) suggests that an autocrine loop may regulate proliferation of ovarian cancer cells.

14.4 GROWTH INHIBITORS

In addition to growth factors positively regulating growth, other polypeptide hormones such as TGFβ have been identified which decrease the growth of some cell types such as epithelia but stimulate others such as stroma. Significant concentrations of TGFβ-like activity are present in the ascitic fluid from many if not all ovarian cancer patients[56]. One study measured production of TGFβ as well as growth inhibition by TGFβ in four ovarian cancer cell lines to determine the role of production and action of TGFβ in proliferation of ovarian cancer cells[20]. One of the four lines tested produced TGFβ (Bast) and was growth inhibited by TGFβ[20]. This suggests that production of TGFβ by ovarian cancer cells in the peritoneal cavity may contribute to the amount of TGFβ present in ascitic fluid. The growth of benign ovarian epithelial cells is inhibited by TGFβ and 3/4 ovarian cancer lines tested were reported to be sensitive to the effects of TGFβ[20]. However, only one of the three ovarian cancer lines was highly sensitive to the effects of TGFβ; the proliferation of the other two lines was only slightly decreased by incubation with TGFβ. One of the ovarian cancer cell lines tested did not respond to TGFβ[20]. As demonstrated in Table 14.2, the ovarian cancer cell line OCC2 is quite sensitive to the inhibitory action of TGFβ whereas the proliferation of HEY is actually stimulated by TGFβ. The increase in proliferation of HEY induced by TGFβ is in contrast to reports with other ovarian cancer cell lines. The inhibition of growth of OCC2 suggests that at least a portion of ovarian cancer cell lines have retained their responsiveness to TGFβ. One of the components of malignant transformation, in some but not all cases, may be the loss of responsiveness to factors which normally inhibit cell growth.

As demonstrated in Figure 14.1, high concentrations of ascitic fluid from some, but not all, patients contain a potent inhibitory activity which decreases the growth of ovarian cancer cells[4]. The effects of the growth inhibitor appear to be dominant *in vitro* (Figure 14.1)[4], that is, at high concentrations of ascitic fluid growth is inhibited, while at low concentrations the growth promoting activity is observed. Indeed ascitic fluid from patients with high levels of growth inhibitory activity *in vitro* does not support the growth of HEY cells in the peritoneal cavity of nude mice and will actually decrease growth induced by ascitic fluid from ovarian cancer patients, which contains primarily growth promoting activity.

The nature of the growth inhibitory activity contained in ascitic fluid of some ovarian cancer patients is unknown. As demonstrated in Table 14.2, the HEY ovarian cancer cells used to assess the effect of ascitic fluid (Figure 14.1), are growth stimulated by the action of TGFβ suggesting that growth inhibitory activities in addition to TGFβ exist in ascitic fluid. Although HEY cells can be growth inhibited by EGF or TGFα, as we have discussed above, we have been unable to demonstrate the presence of bioactive EGF or TGFα in ascitic fluid. This suggests that EGF or TGFα and TGFβ do not account for all the growth inhibitory activities present in ascitic fluid and that additional growth inhibitory factors are likely to be present.

In addition to growth inhibitory factors present in ascitic fluid, mononuclear cells within the peritoneal cavity can inhibit the growth of ovarian cancer cells. As we reported in the previous proceedings of the Helene Harris Symposium[4], addition of mononuclear cells isolated from the peritoneal cavity of ovarian cancer patients to semi-purified ovarian cancer cells from the same patient decreased the colony forming activity of the ovarian cancer cells. This suggests that mononuclear cells in the peritoneal cavity are activated and that they can decrease tumour growth.

It may be possible to use the phenomenon

of lymphocyte activation in ovarian cancer as a marker of the presence of disease or of response to therapy. Following activation, T lymphocytes express high levels of the 55 kDa receptor for IL-2 (also commonly known as the T activation antigen or TAC)[14]. TAC is released from activated T lymphocytes likely through a process of enzymatic cleavage[57]. Resting T cells do not express or release TAC making measurement of soluble TAC in serum in excellent marker for T cell activation[57]. We have demonstrated high levels of TAC in ascites in 66/66 ovarian cancer samples[16]. Elevated levels were observed in only 1/10 patients with ascites as the result of benign disease[16]. This indicates that a significant number of T lymphocytes in the peritoneal cavity of patients with ovarian carcinoma are activated and that they may be responding to the presence of malignant cells. In addition, it suggests that measuring levels of the TAC antigen in ascitic fluid or in peripheral blood may provide an indicator of the presence of tumour cells or of the response to therapy.

14.5 TTK

As described herein, tyrosine phosphorylation plays a major role in growth regulation of both benign and malignant cells[11–13,26–28]. With this in mind, we have begun a systematic study to clone and characterize new tyrosine kinases. We have used two major approaches. The first relies on the polymerase chain reaction with degenerate oligonucleotide probes against conserved sequences in tyrosine kinases. The second is based on the observation that E. coli does not contain phosphotyrosine or tyrosine kinase [58]. Introduction of known tyrosine kinases into E. coli results in marked increases in tyrosine phosphorylation of a number of proteins indicating that E. coli contains tyrosine kinase substrates and likely contains low or absent levels of tyrosine phosphatases.

Therefore, probing of mammalian expression libraries in E. coli with anti-phosphotyrosine antibodies provides a sensitive functional approach to detecting mammalian tyrosine kinases. We will describe some of our studies with TTK, which, although it was initially identified by expression cloning from a human T lymphocyte library, appears to play a broader role in the proliferation of a number of cell types including ovarian cancer cells[58].

The complete cDNA sequence of TTK has been derived from cDNAs isolated from an ovarian cancer cell library from HEY cells as well as from lymphoid cells (Figure 14.1)[58]. The cDNA sequence predicts a mature protein with a molecular weight of 97 000 and this is compatible to the size of proteins immunoprecipitated with anti-TTK antibodies from ^{35}S-methionine-labelled cells. As illustrated in Figure 14.4, human TTK has two potential start sites. The upstream CTG start site (CTG is used as a start site in a number of genes including *myc*, *int*-2, and basic FGF) would give a protein with a potential leader peptide, whereas the ATG site would not. Downstream of these potential start sites is a domain which contains two sets of cysteine repeats, all 10–15 amino acids apart, as well as a number of potential N- and O-linked glycosylation sites. This structural motif is similar to that present in a number of growth factor receptors particularly those of the erythropoietin family[58]. The murine form of TTK appears to have a transmembrane domain which is expressed through alternate splicing in some cell types. This potential transmembrane domain combined with the putative leader peptides suggests that in some cells TTK may function as a growth factor receptor. However, a transmembrane domain has not yet been identified in human cells.

A kinase domain is located in the carboxy-terminal half of TTK[58]. This kinase domain exhibits only a low level of homology with

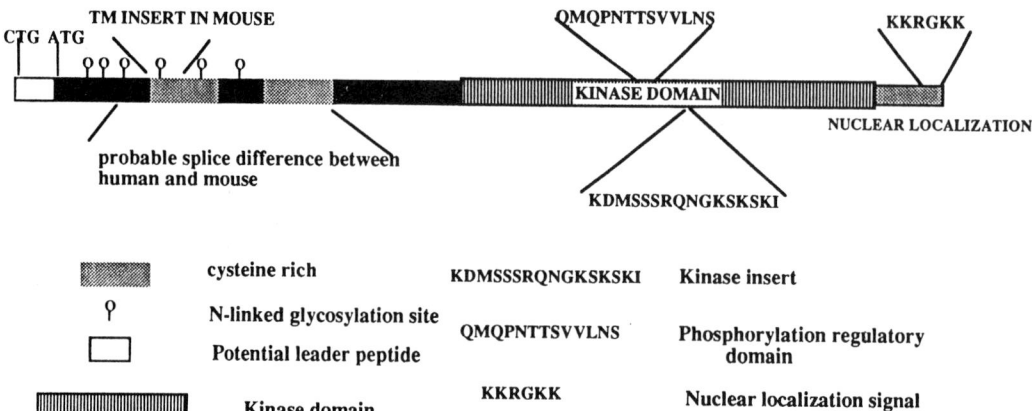

Figure 14.4 Structure of TTK. This structural model is derived from analysis of the predicted amino acid sequence of TTK as derived from the cDNA sequence.

known kinases suggesting that it is a member of a new family of kinases. Within the kinase domain are two potential regulatory domains (Figure 14.4) which contain numerous phosphorylation sites for a number of different serine and threonine kinases. This suggests that TTK may be a downstream mediator which integrates signals from a number of signalling pathways. TTK terminates with a potential nuclear localization signal similar to that present in the *abl* oncogene. Immunohistochemistry of HEY cells with affinity purified antibodies raised against a TTK fusion protein demonstrates the presence of TTK in the nucleus as well as in the cytoplasm. The cDNA for TTK contains a long 5' untranslated sequence suggesting that expression of TTK is regulated by this domain[58]. TTK probably has intracellular and nuclear forms. Indeed, immunological evidence for all forms have been obtained in different cell types. Ovarian cancer cells apparently express solely the intracellular and nuclear forms.

Previously kinases were thought to either phosphorylate tyrosine or serine and threonine exclusively[26]. The primary activity of TTK, both as expressed in *E. coli* and in human cells, appears to be phosphorylation of serine and threonine residues[58]. However, it also appears to be able to phosphorylate tyrosine, albeit at low levels. This suggests that TTK is member of a new class of kinase which phosphorylates tyrosine as well as serine and threonine[58].

When normal human tissues were probed for the presence of TTK mRNA, only testes and thymus expressed high levels of TTK [58]. Testes and thymus contain high numbers of proliferating cells. Indeed TTK was found to be expressed at high levels in proliferating cell lines in culture including a number of haemopoietic cell types as well as ovarian cancer cells (HEY), choriocarcinoma cells (BEWO), and cervical carcinoma cells (HELA)[58]. A 3.5-kb message was present in all cell types except for HELA cells which had a 3.0-kb message. Freshly isolated tumour cells from ascitic fluid obtained from 4 out of 5 ovarian carcinoma patients expressed high levels of the TTK message. TTK message was not elevated in mononuclear cells isolated from the ascitic fluid of the same ovarian cancer patients indicating that expression of TTK was restricted to the tumour cells[58]. In addition, antisense constructs directed against TTK decrease the proliferation of HEY ovarian cancer cells. This suggests that

TTK may play a role in regulating proliferation of ovarian cancer cells.

14.6 SUMMARY, CONCLUSIONS AND FUTURE DIRECTIONS

Over the last decade and in particular the last several years, remarkable progress has been made in understanding the mechanisms regulating the growth of ovarian cancer cells. This is very evident if one compares the previous proceedings of the Helene Harris Ovarian Cancer Symposium with the current. It is clear that we have made significant progress in determining the role of growth factors, oncogenes, and tumour suppressors in the growth of ovarian cancer cells. Furthermore several of these observations are beginning to be of significance in the clinical management of patients. Measurement of serum M-CSF levels in combination with CA-125 levels may be useful in screening or in management of treatment. Levels of *neu* on the surface of cells provide a prognostic indicator and perhaps a target for therapy.

REFERENCES

1. Richardson, G.S., Scully, R.E., Nikrui, N. and Nelson, J.H. (1985) Common epithelial cancer of the ovary. Part 1. *N. Engl. J. Med.*, **312**, 415–24.
2. Richardson, G.S., Scully, R.E., Nikrui, N. and Nelson, J.H. (1985) Common epithelial cancer of the ovary. Part 2. *N. Engl. J. Med.*, **312**, 474–83.
3. *Cancer in Ontario* (1989) Ontario Cancer Treatment and Research Foundation.
4. Mills, G.B. and May, C. (1990) Regulatory mechanisms in ascitic fluid, in *Ovarian Cancer: Biologic and Therapeutic Challenges* (eds F. Sharp, W.P. Mason and R.E. Leake), Chapman & Hall, London, pp. 55–62.
5. Kruk, P.A., Maines-Bandiera, S.L. and Auersperg, N. (1990) A simplified method to culture human ovarian surface epithelium. *Lab. Invest.*, **63**, 132–41.
6. Siemens, C.H., and Auersperg, N. (1988) Serial propagation of human ovarian surface epithelium in culture. *J. Cell. Physiol.*, **134**, 347–56.
7. Auersperg, N. and Roskelley, C. (1991) Retroviral oncogenes: interrelationships between neoplastic transformation and cell differentiation *Crit. Rev. Oncol. Hematol.*, **2**, 125–60.
8. Lobb, D.K., Kobrin, M.S., Kudlow, J.E. and Dorrington, J.H. (1989) Transforming growth factor-alpha in the adult bovine ovary: Identification in growing ovarian follicles. *Biol. Reprod.*, **40**, 1087–93.
9. Carson, R.S., Zhang, Z., Hutchinson, L.A. *et al.* (1989) Growth factors in ovarian function. *J. Reprod. Fertil.*, **85**, 735–46.
10. Wilson, A., Fox, H., Scott, I. *et al.* (1991) A comparison of the growth promoting properties of ascitic fluid, cyst fluids and peritoneal fluid from patients with ovarian tumours. *Br. J. Cancer*, **63**, 102–8.
11. Sporn, M.B. and Roberts A.B. (1985) Autocrine growth factors and cancer. *Nature*, **313**, 745–7.
12. Ullrich, A. and Schlessinger, J. (1990) Signal transduction by receptors with tyrosine kinase activity. *Cell*, **61**, 203–15.
13. Bishop, J.M. (1985) Viral oncogenes. *Cell*, **42**, 23–6.
14. Mills, G.B., Zhang, N., Schmandt, R. *et al.* (1991) Transmembrane signalling by interleukin 2. *Biochem. Trans.*, **19**, 277–85.
15. Mills, G.B., May, C., Hill, M. *et al.* (1990) Ascitic fluid from human ovarian cancer patients contains growth factors necessary for intraperitoneal tumor growth. *J. Clin. Invest.*, **86**, 851–5.
16. Mills, G.B., Hashimoto, S., Hurteau, J. *et al.* (1991) Role of growth factors, their receptors, and oncogenes in the diagnosis, prognosis, followup and therapy of ovarian cancer. *Diagn. Oncol.*, (in press).
17. Artega, C. Hanauske, A., Clark, G. *et al.* (1988) Immunoreactive α transforming growth factor activity in effusions from cancer patients as a marker of tumor burden and patient prognosis. *Cancer Res.*, **48**, 5023–8.
18. Stanley, E.R. (1986) CSF-1 and its receptor. *Ciba Found. Symp.*, **118**, 29–53.
19. Ramakrishnan, S., Xu, F.J., Brandt, S.J. *et al.*

(1989) Constitutive production of macrophage colony-stimulating factor by human ovarian and breast cancer cell lines. *J. Clin. Invest.*, **83**, 921–6.
20. Berchuk, A., Olt, G., Everritt, L. et al. (1990) The role of peptide growth factors in epithelial ovarian cancer. *Obstet. Gynecol.*, **75**, 255–62.
21. Moolenar, W.H., Defize, L.H.K., Van Der Saag, P.T. and De Latt, S.W. (1986) The generation of ionic signals by growth factors. *Curr. Top. Membranes Transport*, **26**, 137–54.
22. Berridge, M.J. (1987) Inositol trisphosphate and diacylglycerol: Two interacting second messengers. *Annu. Rev. Biochem.*, **56**, 159–67.
23. Shadduck, R. and Waheed, A. (1989) Development of a radioimmunoassay for human macrophage colony stimulating factor (CSF-1) *Ann. N.Y. Acad Sci.*, **554**, 156–66.
24. Mills, G.B., May, C., McGill, M. et al. (1988) Ascitic fluid from ovarian cancer patients contains a putative new growth factory: Identification, characterization and mechanism of action. *Cancer Res.*, **48**, 1066.
25. Kikkawa, U. and Nishizuka, Y. (1986) The role of protein kinase C in transmembrane signalling. *Annu. Rev. Cell Biol.*, **2**, 149–62.
26. Hanks, S.K., Quinn, A.M. and Hunter, T. (1988) The protein kinase family: Conserved features and deduced phylogeny of the catalytic domains. *Science*, **241**, 42–8.
27. Yaish, P., Gazit, A., Gilon, C. and Levitzki, A. (1988) Blocking of EGF-dependent cell proliferation of EGF receptor kinase inhibitors. *Science*, **242**, 933–7.
28. Klarlund, J.K. (1985) Transformation of cells by an inhibitor of phosphatases acting on phosphotyrosine in proteins. *Cell*, **41**, 707–17.
29. Hashimoto, S., Campbell, S., Hurteau, J. and Mills, G.B. (1991) EGF and TGFα induce tyrosine phosphorylation of the epidermal growth factor receptor and NEU in ovarian cancer cells: A possible explanation for the poor prognosis associated with overexpression of NEU (submitted).
30. Lupu, R., Colomer, R. Zugmaier, G. et al. (1990) Direct interaction of a ligand for the *erb*B2 oncogene product with the EGF receptor and P185erbB2. *Science*, **249**, 1552–5.
31. Higashiyama, S., Abraham, J., Miller, J. et al. (1990) A herapin-binding growth factor secreted by macrophage-like cells that is related to EGF. *Science*, **251**, 936–9.
32. Glenney, J.R. Jr, Chen, W.S., Lazar, C.S. et al. (1988) Ligand-induced endocytosis of the EGF receptor is blocked by mutational inactivation and by microinjection of anti-phosphotyrosine antibodies. *Cell*, **52**, 675–671.
33. Lyall, R.M., Zilberstein, A., Gazit, A. et al. (1989) Tyrphostins inhibit epidermal growth factor-receptor tyrosine kinase activity in living cells and EGF-stimulated cell proliferation. *J. Biol. Chem.*, **264**, 14503–12.
34. Sariban, E. Sitaras, N., Antoniades, H. et al. (1988) Expression of platelet-derived growth factor (PDGF)-related transcripts and synthesis of biologically active PDGF-like proteins by human malignant epithelial cell lines. *J. Clin. Invest.*, **82**, 1157–64.
35. Shaw, P., Buckman, R., Law, J. et al. (1988) Reactivity of tumor cells in malignant effusions with a panel of monoclonal and polyclonal antibodies. *Tumour Biol.*, **9**, 101–9.
36. Stern, D.F. and Kamps, M.P. (1988) EGF-stimulated tyrosine phosphorylation of p185^{neu}: a potential model for receptor interactions. *EMBO J.*, **7**, 995–1001.
37. Kokai, Y., Dobashi, K., Myers, J.N. et al. (1988) Phosphorylation process induced by epidermal growth factor alters cellular and oncogenic *neu* gene products. *Proc. Natl Acad. Sci. USA*, **85**, 5389–93.
38. Wada, T., Qian, X. and Greene, M. (1990) Intermolecular association of the p185^{neu} protein and EGF receptor modulates EGF receptor function. *Cell*, **61**, 1339–47.
39. Leake, R. and Owens, O. (1989) The prognostic value of steroid receptors, growth factors and growth factor receptors in ovarian cancer, in *Ovarian Cancer: Biologic and therapeutic challenges* (eds F. Sharp, W.P. Mason and R.E. Leake), Chapman & Hall, London, pp. 69–75.
40. Bauknecht, T., Runge, M., Schwall, M. and Pfleiderer, A. (1988) Occurrence of epidermal growth factor receptors in human adnexal tumors and their prognostic value in advanced ovarian cancers. *Gynecol. Oncol.*, **29**, 147–57.
41. Nicholson, S., Halcrow, P., Sainsbury, J.R.C. et al. (1988) Epidermal growth factor receptor (EGFr) status associated with failure of primary endocrine therapy in elderly postmeno-

pausal patients with breast cancer. *Br. J. Cancer*, **58**, 810–14.
42. Sainsbury, J., Farndon, J., Needham, G. *et al.* (1987) Epidermal growth-factor receptor status as a predictor of early recurrence of and death from breast cancer. *Lancet*, **i**, 1398–402.
43. Slamon, D., deKernion, J., Verma, I. and Cline, M. (1984) Expression of cellular oncogenes in human malignancies. *Science*, **224**, 256–62.
44. Gullick, W. (1989) The role of oncogenes in ovarian cancer, in *Ovarian Cancer: Biologic and Therapeutic Challenges* (eds F. Sharp, W.P. Mason and R.E. Leake), Chapman & Hall, London, pp. 63–8.
45. Berchuk, A., Kamel, A., Whitaker, R. *et al.* (1990) Overexpression of HER-2/*neu* is associated with poor survival in advanced epithelial ovarian cancer. *Cancer Res.*, **50**, 4087–91.
46. Slamon, D.J., Godolphin, W., Jones, L.A. *et al.* (1989) Studies of the HER-2/*neu* proto-oncogene in human breast and ovarian cancer. *Science*, **244**, 707–12.
47. Yarden, Y., and Weinberg, R.A. (1989) Experimental approaches to hypothetical hormones: Detection of a candidate ligand of the NEU protooncogene. *Proc. Natl Acad. Sci. USA*, **86**, 3179–83.
48. Kamps, M.P. and Sefton, B. (1989) Acid and base hydrolysis of phosphoproteins bound to immobilon facilitates analysis of phospho-aminoacids in gel-fractionated proteins. *Anal. Biochem.*, **176**, 22–34.
49. Kamps, M.P. and Sefton, B. (1988) Identification of multiple novel polypeptide substrates of the v-*src*, v-*yes*, v-*fps*, v-*ros*, and v-*erb*-B oncogenic tyrosine protein kinases utilizing antisera against phosphotyrosine. *Oncogene*, **2**, 305–12.
50. Mills, G.B., May, C., McGill, M. *et al.* (1990) Interleukin 2 receptor β is tyrosine phosphorylated. *J. Biol. Chem.*, **265**, 3561–7.
51. Mills, G.B., Stanley, J., Stewart, D. *et al.* (1990) Interrelationship between signals transduced by phytohemagglutinin and interleukin 1. *J. Cell. Physiol.*, **142**, 539–51.
52. Stanley, J., Huang, C.-K., Love, J. *et al.* (1990) Tyrosine phosphorylation is an obligatory event in IL2 production. *J. Immunol.*, **145**, 2189–98.
53. Drebin, J., Link, V., Stern, D. *et al.* (1985) Down-modulation of an oncogene protein product and reversion of the transformed phenotype by monoclonal antibodies. *Cell*, **41**, 695–706.
54. Kacinski, B., Stanley, R., Carter, D. *et al.* (1989) Circulating levels of CSF-1 (m-CSF) a lymphohematopoietic cytokine may be a useful marker of disease status in patients with malignant ovarian neoplasms. *Int. J. Radiat. Oncol. Biol. Phys.*, **17**, 159–64.
55. Kacinski, B., Carter, D., Kohorn, E. (1989) Oncogene expression *in vivo* by ovarian adenocarcinomas and mixed-mullerian tumors. *Yale J. Biol. Med.*, **62**, 379–92.
56. Hirte, H. and Clark, D.A. (1991) Generation of lymphokine-activated killer cells in human ovarian carcinoma ascitic fluid: Identification of transforming growth factor β as a suppressive factor. *Cancer Immunol. Immunother.*, **32**, 296–302.
57. Rubin, L. and Nelson, D. (1990) The soluble interleukin-2 receptor: biology, function and clinical application. *Ann. Intern. Med.*, **113**, 619–45.
58. Mills, G.B., Schmandt, R., McGill, M. *et al.* (1991) Cloning and characterization of TTK, a novel protein kinase which is associated with cell proliferation. *J. Biol. Chem.* (in press).

Part Three
Drug Resistance and Experimental Therapeutics

Chapter 15

Signal transduction pathway regulation of cisplatin sensitivity

S.B. HOWELL, R.D. CHRISTEN, P.A. ANDREWS, S.C. MANN and D. HOM

15.1 INTRODUCTION

Ovarian carcinoma cells still retain responsiveness to a variety of growth factors and cytokines that are capable of altering both proliferation and other phenotypic characteristics. We have been investigating the hypothesis that growth factor-induced changes in phenotype are also reflected by changes in drug sensitivity. Cisplatin has been the focus of attention because of the important role that this drug plays in the treatment of ovarian carcinoma.

While some ovarian carcinomas are intrinsically diamminedichloroplatinum (DDP) resistant and fail to respond to chemotherapy at all, the more common pattern is that the tumour responds well initially but subsequently either regrows during therapy or during the first few years after the end of treatment. The 'acquired' resistance that develops during treatment is generally reflected by a two to four fold change in the sensitivity of the cells when they are tested *in vitro*[1–4]. *In vitro* studies have confirmed that cell line variants demonstrating low level resistance are easily selected from parental cells, but that it is difficult to obtain sublines with more than 10- to 15-fold 'acquired' resistance[5]. This is in contrast to the situation with drugs that participate in the multiple drug resistance phenotype, or to the antimetabolites, where very high levels of resistance can be obtained. We have recently shown that low level 'acquired' resistance to DDP also emerges rapidly *in vivo*[6], and that ovarian carcinoma xenograft responsiveness to DDP is diminished when the cells are just 2.5-fold resistant *in vitro*. We have now demonstrated that several signal transduction pathways, including those mediated by protein kinase A (PKA) and the epidermal growth factor (EGF) receptor, are capable of modulating sensitivity to DDP in mammalian cells.

15.2 PROTEIN KINASE A PATHWAY

Forskolin is a direct activator of adenyl cyclase[7], and causes a rapid increase in intracellular cyclic AMP, resulting in the activation of PKA. We used the human

Signal transduction pathway regulation of cisplatin sensitivity

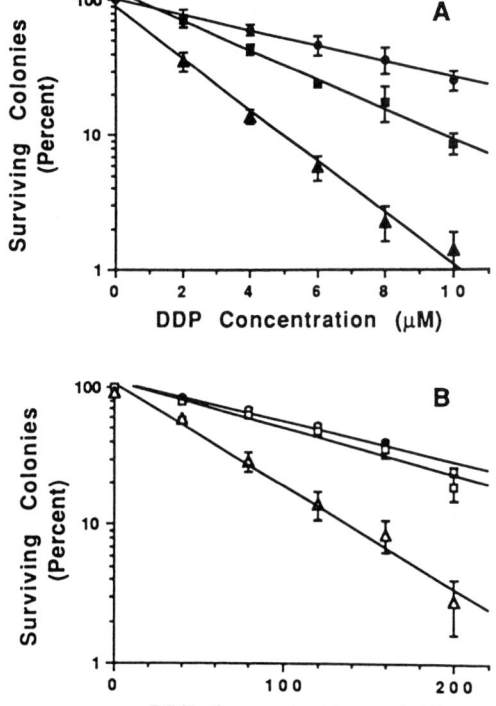

Figure 15.1 Dose–response curves for 2008 (A) and C13*5.25 (B) cells exposed to DDP for 1 hour either alone (circles), or in combination with 50 μU forskolin (squares) or 3 mM IBMX (triangles). Points are means of three experiments performed in triplicate; bars indicate standard error.

ovarian carcinoma cell line 2008[8], and a subline, 2008/C13*5.25, selected from 2008 cells by repeated *in vitro* exposure to DDP, to determine how PKA activation influenced sensitivity to a 1-hour exposure to DDP as quantitated by clonogenic assays[9]. Panel A of Figure 15.1 demonstrates that a concurrent 1-hour exposure to 50 μM forskolin and DDP increased the slope of the DDP dose–response curve for 2008 cells by a factor of 1.9-fold. A simultaneous 1-hour exposure to DDP and 3-isobutyl-1-methylxanthine (IBMX), which increases intracellular cyclic AMP by both blocking phosphodiesterase and stimu-

lating adenyl cyclase, increased the slope of the dose–response curve by 3.3-fold. When the nature of the interaction between forskolin and DDP was investigated by median effect analysis[10] using continuous drug exposures, it was found to be highly synergistic producing a combination index of 0.54 ± 0.02 (SE, $n = 4$) at 50% cell kill. Interestingly, panel B in Figure 15.1 shows that forskolin was ineffective at modulating DDP sensitivity in the 10-fold resistant C13*5.25 cells, and that the effect of IBMX was muted. Thus it appears that selection for DDP resistance resulted in a defect in the PKA pathway.

In DDP-sensitive 2008 cells, forskolin and IBMX caused a concentration-dependent increase in DDP accumulation at 10 minutes that reached a maximum of 2.1-fold and 2.3-fold, respectively. The inactive analogue, 1,9-dideoxyforskolin, decreased DDP accumulation relative to control. Neither forskolin nor IBMX had any effect on DDP accumulation in DDP-resistant C13*5.25 cells. The effect of both forskolin and IBMX on DDP uptake was detectable as early as 1 minute after the start of drug exposure. Half-maximal stimulation of DDP accumulation occurred at 0.2 μM for forskolin and 0.2 mM for IBMX.

The fact that both forskolin and IBMX were able to enhance sensitivity to DDP and increase DDP accumulation suggested that the defect in the PKA pathway in DDP-resistant cells was downstream of adenyl cyclase. However, no difference was detected in the ability of cyclic AMP to stimulate PKA activity in either cell line using the filter assay of Roskoski[11]. Thus we concluded that the lesion in the PKA pathway was also distal to PKA itself. Our current hypothesis is that enhanced DDP uptake is mediated by a substrate of the phosphorylation cascade activated by PKA, and that some component of this phosphorylation cascade or substrate is abnormal in DDP-resistant cells.

15.3 EPIDERMAL GROWTH FACTOR RECEPTOR PATHWAY

Binding of EGF to its receptor induces tyrosine phosphorylation of various cellular proteins including the EGF receptor itself. This is associated with a variety of changes in intracellular physiology, including activation of the Na^+/H^+ transporter and activation of PKA[12]. When 2008 cells were exposed to 10 nM EGF for 1 hour, and then to both EGF and DDP during the second hour, EGF increased sensitivity to DDP by a factor of 3.1 ± 0.9 (SD) as quantified by the ratio of the IC_{50} values[13]. EGF had a similar effect on another human ovarian carcinoma cell line, Colo 316, enhancing sensitivity by a factor of 2.4 ± 0.1. In neither cell line was the modulation of drug sensitivity due to an EGF-induced change in growth rate. Neither a 2-hour nor a continuous exposure to 10 nM EGF had a demonstrable impact on the doubling time of the cell lines.

The ability of EGF to enhance DDP sensitivity was found to be a function of both EGF concentration and EGF receptor number. A 2-hour exposure to EGF enhanced sensitivity to DDP at EGF concentrations as low as 0.4 nM, and the effect was maximal at concentrations of approximately 10 nM; a further increase in EGF concentration up to 100 nM produced no additional change in sensitivity. Mouse C127 fibroblasts stably transfected with a plasmid construct containing the human EGF receptor gene under the control of the transferrin receptor 3'-inducible regulator were used to investigate the effect of the number of EGF receptors. In this system, induction of EGF receptor expression increased human EGF receptor number by approximately twofold, as compared with uninduced cells. In the presence of 10 nM EGF, the induced cells were approximately twofold more sensitive to DDP than control cells.

The time course of the effect of EGF on the sensitivity of 2008 cells indicated that sensitivity to DDP was maximal at the end of 1 hour, and that enhanced sensitivity persisted for at least 5 hours after the end of a 1-hour EGF exposure, but had largely disappeared by 24 hours. EGF slightly increased glutathione content, had no discernible effect on glutathione-S-transferase activity, and no effect on DDP accumulation in 2008 cells.

As was the case for the PKA pathway, EGF was unable to modulate DDP sensitivity in 10-fold DDP-resistant C13*5.25 cells. Whether the defect in this pathway is the same as the defect in the PKA pathway that eliminates the effect of forskolin is not currently known.

15.4 CONCLUSION

These studies demonstrate that the DDP sensitivity of an ovarian carcinoma cell line can be influenced by the PKA and EGF receptor signal transduction pathways.

The biochemical and molecular basis for DDP resistance is not well defined, but it is clear that cell killing is a function of how much DDP gets into the cell, how much actually reacts with DNA, how tolerant the cell is of lesions in its DNA, and how effectively it removes DDP lesions from DNA. Cells defend themselves through the up- or down-regulation of biochemical processes such as DDP uptake, DNA repair[5], or the level of glutathione and metallothioneins to produce the phenotype of 'acquired' DDP resistance[5]. Also, at the present time it is not clear whether these observations can be generalized to other ovarian cell lines, or whether the effect can be produced under *in vivo* conditions. Nevertheless, the results do establish that a biochemical mechanism exists that allows sensitivity to be enhanced by growth factors.

One concern is that the magnitude of the change in DDP sensitivity produced by signal transduction pathways appears to be relatively small. However, it must be remembered that it only takes a small change in DDP sensitivity to convert 2008 xenografts from being sensitive to being clinically resistant to DDP therapy *in vivo*[6]. Likewise, what data do exist comparing the DDP sensitivity of tumours, assayed *in vitro*, before and after *in vivo* DDP therapy, suggest that 'acquired' DDP resistance is also of modest magnitude[3–6]. Several drugs that alter the activity of the PKA pathway are currently available for clinical use, and EGF itself may also eventually become available, offering the opportunity to undertake clinical investigations of whether DDP sensitivity can be modulated by either of these signal transduction pathways *in vivo*.

REFERENCES

1. Simmonds, A.P. and McDonald, E.C. (1984) Ovarian carcinoma cells in culture: assessment of drug sensitivity by clonogenic assay. *Br. J. Cancer*, **50**, 317–26.
2. Inoue, K., Mukaiyama, T. and Ogawa, M. (1985) *In vitro* evaluation of anticancer drugs in relation to development of drug resistance in the human tumor clonogenic assay. *Cancer Chemother. Pharmacol.*, **15**, 208–13.
3. Wilson, A.P., Ford, C.H.J., Newman, C.E. and Howell, A. (1987) Cisplatinum and ovarian carcinoma. *In vitro* chemosensitivity of cultured tumor cells from patients receiving high dose cisplatinum as first line treatment. *Br. J. Cancer*, **56**, 763–73.
4. Wolf, C.R., Hayward, I.P., Lawrie, S.S. *et al.* (1987) Cellular heterogeneity and drug resistance in two ovarian adenocarcinoma cell lines derived from a single patient. *Int. J. Cancer*, **39**, 695–702.
5. Andrews, P.A. and Howell, S.B. (1990) Cellular pharmacology of cisplatin: perspectives on mechanisms of acquired resistance. *Cancer Cells*, **2**, 35–43.
6. Andrews, P.A., Jones, J.A., Varki, N.M. and Howell, S.B. (1990) Rapid emergence of acquired cis-diamminedichloroplatinum (II) resistance in an *in vivo* model of human ovarian carcinoma. *Cancer Comm.*, **2**, 93–100.
7. Seamon, K.B., Padgett, W. and Daly, J.W. (1981) Forskolin: a unique diterpene activator of adenylate cyclase in membranes and in intact cells. *Proc. Natl Acad. Sci. USA*, **78**, 3363–7.
8. DiSaia, P.J., Sinkovics, J.G., Rutlege, F.N. and Smith, J.P. (1972) Cell-mediated immunity to human malignant cells. *Am. J. Obstet. Gynecol.*, **114**, 979–89.
9. Mann, S.C., Andrews, P.A. and Howell, S.B. (1991) Modulation of cis-diamminedichloroplatinum (II) accumulation and sensitivity by forskolin and 3-isobutul-1-methylxanthine in sensitive and resistant human ovarian carcinoma cells. *Int. J. Cancer* (in press).
10. Chou, T.C. and Talalay, P. (1984) Quantitative analysis of dose–effect relationships: the combined effects of multiple drugs or enzyme inhibitors. *Adv. Enzyme Regul.*, **22**, 27–55.
11. Roskoski, R. (1983) Assays of protein kinase. *Methods Enzymol.*, **99**, 3–6.
12. Bell, R.M. (1986) Protein kinase C activation by diacylglycerol second messengers. *Cell*, **45**, 631–2.
13. Christen, R.D., Hom, D.K., Porter, D.C. *et al.* (1990) Epidermal growth factor regulates the *in vitro* sensitivity of human ovarian carcinoma cells to cisplatin. *J. Clin. Invest.*, **86**, 1632–40.

Chapter 16
Immunotoxin therapy in ovarian cancer

M.A. BOOKMAN

16.1 INTRODUCTION

Most antitumour monoclonal antibodies fail to directly inhibit the growth of ovarian adenocarcinoma *in vitro* or *in vivo*. Thus, conjugation of monoclonal antibodies to toxins, radionuclides, chemotherapeutic agents, or lymphokines has been utilized to focus cytotoxic effects against tumour cells expressing the appropriate surface antigens. Potent toxins of plant, fungal, or bacterial origin have been identified which can irreversibly inhibit ribosomal protein synthesis. Conjugation of toxins to monoclonal antibodies directed against tumour-associated antigens has been used to create immunotoxins with preferential cytotoxicity for tumour cells. Two collaborative clinical trials have been completed using intraperitoneal immunotoxins for treatment of ovarian adenocarcinoma, and will be reviewed.

16.2 SELECTION OF IMMUNOTOXINS

Clinical development of immunotoxins requires identification of an appropriate antibody and toxin for conjugation, verification of activity in a preclinical model, and choosing the route of administration. The choice among toxins includes ricin, originally obtained from castor bean (*Ricinis communis*), and *Pseudomonas* exotoxin (PE), obtained from bacteria (*Pseudomonas aeruginosa*). Both of these toxins are available in recombinant as well as natural forms, and have been evaluated as immunotoxins in clinical trials against various malignancies. Ricin is composed of an A-chain (RA) with toxin activity, and a B-chain that facilitates cellular binding and translocation. PE consists of a single polypeptide chain with specific functional domains[1] and truncated molecules have been constructed that maintain toxin activity in the absence of cellular binding activity. RA is usually conjugated through a reducible disulphide bond to facilitate cytoplasmic translocation of free RA following cellular internalization, as there is no cleavage site within RA itself. In contrast, PE has an intrinsic cleavage site, and can be conjugated using a non-reducible thioether linkage, which is associated with improved *in vivo* stability.

Both toxins are immunogenic, resulting in the development of toxin-neutralizing antibodies within 10 days that prevent additional therapy. Approximately 10–20% of cancer patients will have pre-existing neutralizing antibodies directed against PE, which may be acquired following infection or multiple surgical procedures. Pre-existing antibodies against ricin are uncommon.

The choice among potential antibodies has been guided by targeting specificity and efficacy. For immunotoxins to be effective, they must be internalized into each target cell. There is no mechanism for bystander killing of antigen-negative cells admixed with antigen-positive cells. The marked heterogeneity of antigen expression among individual tumours from individual patients has limited the choice of candidate target antigens. In addition, specific antibodies vary greatly with regard to the efficiency of killing when conjugated to various toxins, which requires empiric evaluation in each case. Variable efficacy has been related to the rapidity and degree of antibody internalization after binding, as well as the transport of internalized immunotoxin among subcellular compartments that result either in toxin translocation or degradation.

Many antibodies have been described as 'tumour specific'. However, the vast majority of these also react with antigens expressed by normal host tissues and are not truly tumour specific, raising concerns about host toxicity. Routine immunohistochemical screening may not be sufficiently sensitive to identify potential sites of toxicity due to low-level antigen expression, which is nonetheless sufficient for immunotoxin binding and internalization. Xenogeneic treatment models using human tumours transplanted into athymic (nude) or scid mice also fail to predict host toxicity, as there is no expression of human antigen on the normal murine host tissues. Even non-human primates may not accurately predict toxicity due to species antigenic variation.

Unanticipated toxicity has resulted in closure of phase-I clinical trials. For example, a clinical trial of an immunotoxin (260F9-rRA, Cetus) directed against an adenocarcinoma-associated antigen was terminated as a result of peripheral neuropathy most likely due to previously undetected targeting of Schwann cells[2].

The failure to identify truly tumour-specific reagents with uniform reactivity against tumour cells has led to consideration of immunotoxins directed against receptors for endogenous growth factors. For example, some tumours express high levels of epidermal growth factor receptor, while all tumours require iron and express variable amounts of transferrin receptor. The transferrin receptor is efficiently internalized and recycled back to the cell surface during binding and transport of the iron–transferrin complex. Rapid internalization and utilization of specific subcellular pathways that favour toxin translocation have led to development of highly efficient immunotoxins. Tumours with a high growth fraction, abnormal growth regulation and increased receptor expression may exhibit greater sensitivity to immunotoxins than normal host tissues. Although some degree of host toxicity would be expected with these broadly reactive compounds, toxicity might be dose related and predictable from patient to patient, in a manner similar to that observed with conventional chemotherapy.

Two monoclonal antibodies were selected for clinical evaluation based on their efficacy in treatment of human ovarian adenocarcinoma xenografts in athymic (nude) mice and their expression on the majority of ovarian tumour specimens. OVB3 is a murine IgG2b antibody derived following immunization with the OVCAR3 cell line, and is directed against a tumour-associated antigen uniformly expressed on the majority of ovarian carcinoma cells[3]. OVB3 was linked through a non-reducible thioether bond with PE to form the immunotoxin OVB3-PE (Laboratory of Molecular Biology, National Cancer Institute, NIH). 454A12 is a murine IgG1 derived following immunization with the SKBR3 breast cancer line, and is directed against the human transferrin receptor[4]. 454A12 was linked through a reducible disulphide bond to recombinant non-glycosylated RA to form 454A12-rRA (Cetus Corporation, Emeryville, CA). Both immunotoxins were found to have 50% inhibitory concentrations (IC_{50}) *in vitro* of less than 10 ng/ml against ovarian carcinoma cell lines.

16.3 REGIONAL VERSUS SYSTEMIC THERAPY

Both immunotoxins reviewed in this chapter were administered by the intraperitoneal route, although debate continues regarding the relative merits and potential efficacy of intraperitoneal versus intravenous therapy. Regional therapy offers the possibility for sustained contact between high levels of antibody and tumour cells, particularly in patients with malignant ascites. In addition, peritoneal fluid can be serially monitored for antitumour and inflammatory effects, and can be removed in the event of toxicity. Absorption of macromolecules from the peritoneal cavity utilizes the same lymphatic pathways that are subject to metastatic tumour implantation, which may maximize tumour exposure in those sites. However, the ability of peritoneal immunotoxins to penetrate intraparenchymal or bulk tumour masses sequestered by adhesions and fibrosis appears limited. Initial studies suggested that small molecules, such as conventional chemotherapy drugs, behave quite differently from macromolecules. Specifically, depth of tissue penetration is governed by the passive diffusion rate, which is inversely related to size, and the tissue clearance rate, which is also inversely related to size. Small molecules may freely and rapidly cross the peritoneal surface, but are then efficiently cleared from normal tissues and tumours by capillaries, resulting in limited direct penetration. These molecules will then recirculate and have the potential to penetrate larger masses by the intravascular route. In contrast, macromolecules initially diffuse more slowly, but require lymphatics, rather than capillaries, for tissue clearance. Thus, penetration of parietal tissues, such as the diaphragm or abdominal wall, can eventually reach a significant depth[5]. Successful penetration of serosal tumour implants is more problematic, due to compressibility of visceral structures and the presence of nascent leaky capillaries, resulting in a net outward flow of serum proteins that directly opposes the inward diffusion of macromolecules from the peritoneal cavity.

Tissue penetration has been evaluated with monoclonal antibodies in nude mice bearing OVCAR3 tumours[6]. After 24 hours, the majority of solid tumour staining occurred within a few cell diameters of the peripheral stroma, consistent with minimal direct antibody penetration. However, within 3 days there was good overall penetration, most consistent with absorption from the peritoneal cavity followed by systemic redistribution. Distribution of iodinated HMFG2 antibody was evaluated in nude mice bearing human ovarian cancer xenografts[7], and in a series of ovarian cancer patients scheduled for laparotomy[8]. In the murine model, regional antibody achieved large concentration advantages on ascites cells, but failed to penetrate bulk peritoneal tumour masses any better than intravenous antibody. Results from the human clinical study were similar, with an advantage for intraperitoneal administration demonstrated only against malignant ascites cells, and with a disadvantage in absolute solid tumour uptake. In another study, uptake of B72.3 by small peritoneal implants was twofold greater following intraperitoneal versus intravenous injection, whereas tumour uptake in lymph node metastases and local recurrences was twofold greater following intravenous versus intraperitoneal injection[9].

Thus, although microscopic or small volume residual disease may be optimally

Table 16.1 Intraperitoneal therapy with OVB3-PE*

Dose level	μg/kg/dose	Days	Patients entered	Comments
I	1	1,4	5	
II	2	1,4	3	
III-A	5	1,4	3	
IV	10	1,4	5	2 neurological toxicity
III-B	5	1,4	2	
V	5	1,4,7	3	
VI	5	1,3,5,7	2	1 neurological toxicity

* Patients treated at the Medicine Branch and Biological Response Modifiers Program, National Cancer Institute, Duke University Medical Center, and the University of California at San Diego.

treated by regional therapy, an advantage for intraperitoneal therapy of solid tumours has not yet been demonstrated. Differences in regional scaling between mice and humans are associated with more rapid absorption in mice, and prolongation of peritoneal dwell in humans, which may accentuate differences between each route of administration.

16.4 CLINICAL TRIALS WITH INTRAPERITONEAL IMMUNOTOXINS

16.4.1 OVB3-PE

A collaborative phase-I dose-escalating trial of intraperitoneal OVB3-PE was conducted for patients with refractory ovarian adenocarcinoma[10]. Each patient received one cycle of therapy with a fixed dose of immunotoxin on multiple days (Table 16.1) in a final total volume of 1000–2000 ml of normal saline administered through an indwelling peritoneal Tenckhoff catheter. Patients were excluded if they were found to have pre-existing toxin neutralizing antibodies. Spontaneous ascites was drained prior to each infusion, and a 20-μg test dose was administered prior to the actual treatment dose. A total of 23 patients received doses at 1, 2, 5, or 10 μg/kg. Two episodes of encephalopathy occurred at the highest dose level of 10 μg/kg, and a second group of patients was treated with three or four doses at 5 μg/kg. However, the study was terminated when a third patient developed encephalopathy after three doses at 5 μg/kg.

Peak intraperitoneal concentrations of immunotoxin exceeded the IC_{50} at all dose levels tested, ranging from 40 to 600 ng/ml, with prolongation of clearance at higher dosage levels. Serum levels were undetectable (i.e. <4 ng/ml) after 1 and 2 μg/kg, and ranged from 7 to 40 ng/ml 24 hours after 10 μg/kg. All patients analysed, ($n = 12$) were found to develop anti-PE antibodies within 14 days and anti-murine antibodies within 28 days.

Patients were preferentially recruited with small-volume residual disease limited to the peritoneal cavity, and were staged by cytology, laparoscopy or laparotomy before and after therapy. No objective antitumour responses were documented, including patients with small-volume disease. Several patients had transient reductions in malignant cytology lasting less than 4 weeks.

Toxicity was not dose limiting with the exception of three episodes of encephalopathy, including one fatality (Table 16.2). Encephalopathy was characterized by con-

Table 16.2 Toxicity associated with intraperitoneal OVB3-PE

Toxicity	Number (total=23)	Comments
Abdominal pain	19(83%)	7 required narcotics
Nausea and vomiting	12(52%)	
Fever	5(22%)	
Elevated transaminases	8(34%)	
Elevated LDH	4(12%)	
Elevated alkaline phosphatase	9(28%)	
Central neurological toxicity	3(13%)	1 fatal

fusion, apraxia, and dysarthria with focal inflammatory abnormalities in the pons and midbrain on gadolinium-enhanced magnetic resonance imaging in one patient. Neurological toxicity was largely reversible in two patients, but progressed to status epilepticus and coma with widespread structural abnormalities in the third patient, who eventually died without recovery of neurological function. Cerebrospinal fluid protein levels were markedly elevated (700–800 mg/dl) without detectable levels of immunotoxin. Toxicity in that patient became apparent 6 hours after the third intraperitoneal dose of immunotoxin at 5 µg/kg.

Although subsequent immunohistochemical studies with OVB3 antibody have demonstrated low-level staining within the molecular layer of the cerebellum in some fresh brain specimens, it is uncertain if this explains any of the reported toxicity. Shared expression of antigens between neural tissue and ovarian cancer has been previously described[11], and may predispose to paraneoplastic cerebellar degeneration associated with development of anti-Purkinje cell antibodies. Treatment of non-human primates with OVB3-PE has not produced neurological toxicity, although there is no evidence that monkeys express an antigen recognized by the OVB3 antibody. Thus, the aetiology of encephalopathy following small doses of intraperitoneal immunotoxin remains unclear.

16.4.2 454A12-rRA

A collaborative phase-I dose-escalating trial was also conducted with 454A12-rRA anti-transferrin receptor immunotoxin for patients with carcinoma involving the peritoneal cavity[12]. Each patient received one cycle of therapy with a fixed dose of immunotoxin daily for five consecutive days in a final total volume of 1000–2000 ml of peritoneal dialysate administered through an indwelling peritoneal Tenckhoff catheter. Spontaneous ascites and residual peritoneal fluid was drained prior to each daily infusion, and a 20-µg test dose was administered prior to the first treatment dose. A total of 19 patients were treated at dose levels of 5, 10, 25, or 50 µg/kg/day, including 10 women with ovarian cancer. The study was terminated after one reversible and one fatal episode of encephalopathy at the highest dose level (Table 16.3).

Peak intraperitoneal levels exceeded the IC_{50} at all dose levels tested, ranging from 100 to 2000 ng/ml. In view of the daily dosage regimen, intraperitoneal levels were sustained in most patients throughout the duration of treatment. Serum levels were not routinely detected at any dose level, within the limits of assay sensitivity.

No objective antitumour responses were noted, although several patients had minor responses categorized by reductions in serum CA-125 antigen, transient clearance of malig-

Table 16.3 Intraperitoneal therapy with 454A12-rRA*

Dose level	μg/kg/dose	Days	Patients entered	Comments
I	5	1–5	3	2 ovarian
II	10	1–5	4	1 ovarian
III	25	1–5	7	4 ovarian
IV	50	1–5	5	3 ovarian; 2 neurological toxicity

* Patients treated at Fox Chase Cancer Center and the University of Massachusetts Medical Center.

nant cytology, and a palliative decrease in accumulation of ascites. Systemic toxicity, with the exception of dose-limiting encephalopathy, was mild, consisting of asymptomatic hypoalbuminaemia, two episodes of superficial mucositis, and abdominal discomfort. Specifically, there was no hepatic, renal, cardiovascular, pulmonary, haematological, or peripheral neurological toxicity.

Encephalopathy occurred in two patients after four doses at the 50 μg/kg dose level, and was characterized by aphasia, obtundation, agitation, and multifocal myoclonus. Toxicity was reversible in one patient after support in the intensive care unit. A second patient became comatose with evidence of massive oedema within the basal ganglia on cranial computed tomographic scanning, and died within 36 hours. Post-mortem examination revealed diffuse haemorrhagic necrosis of capillaries within the basal ganglia, with a normal appearance in the remainder of the brain. Localization of 454A12-rRA could not be demonstrated on the formalin-fixed brain tissue. Transferrin receptor is expressed on capillaries in the brain, but is also expressed on many other host tissues, and free transferrin receptor has been identified in the serum. Thus, the explanation for localized capillary necrosis within the brain is unknown, and may relate to either a specific antibody–receptor interaction, or reflect some unique property of the blood–brain barrier. Reversible aphasia and confusion have been encountered in other clinical trials with ricin A-chain immunotoxins administered by intravenous bolus, suggesting that some elements of neurological toxicity may not be antigen specific. In addition, the sustained low serum levels of immunotoxin achieved following intraperitoneal administration may be more likely to cause neurological toxicity than the transient peak serum levels that follow an intravenous bolus.

16.4.3 CONCLUSIONS FROM INTRAPERITONEAL IMMUNOTOXIN TRIALS

Immunotoxins are potent antitumour reagents with the ability to cause significant host toxicity at low serum levels. These two phase-I studies have demonstrated that high regional levels of active immunotoxins can be achieved following intraperitoneal infusion, and minor antitumour responses have been observed. Unfortunately, preclinical immunohistochemical and animal studies have thus far failed to predict serious human toxicities. In addition, most new antibodies that emerge as candidates for further testing will also share cross-reactivity with normal host antigens, reinforcing concerns about toxicity.

Immunogenicity of murine antibodies and xenobiotic toxins is associated with development of host neutralizing antibodies that limit the period of time patients can be treated to less than 2 weeks. Approaches currently available to reduce host antibody formation include the use of cyclosporin A[13] and modification of the murine antibodies to create smaller antigen-binding fragments with or without incorporation of sequences from human antibodies. Although these approaches may limit or slow down the formation of anti-murine antibodies, additional strategies may be required to blunt the antitoxin response, such as conjugation to polyethylene glycol[14]. At least one immunosuppressive reagent, 15-deoxyspergalin, has been shown to suppress the antibody response to PE and PE-containing immunotoxins in mice[15]. Thus, options exist that should allow extension of treatment beyond 14 days, if safe and effective immunotoxins can be identified.

16.5 NEW IMMUNOTOXIN TARGETS

There has been considerable interest in new targets for immunotoxin therapy. The c-*erb*B2 gene product is a transmembrane molecule that shares homology with the receptor for epidermal growth factor, and is expressed on a proportion of ovarian tumours and adenocarcinomas from other sites, as well as some normal host tissues[16]. A number of antibodies has emerged with reactivity against different domains on c-*erb*B2, and has been evaluated as potential immunotoxins. Unfortunately, the c-*erb*B2 antigen is not readily internalized following antibody binding[17], and may not be an efficient target for immunotoxin delivery to cells that do not have extremely high levels of antigen expression.

Efforts to prepare chimeric single-chain proteins with antigen binding and toxin activity are in progress[18]. Application of these recombinant techniques may permit refinements in binding specificity, antigen modulation, and subcellular localization that will yield safe and effective candidate immunotoxins for clinical evaluation in the treatment of ovarian cancer.

REFERENCES

1. Hwang, J., FitzGerald, D.J., Adhya, S. *et al.* (1987) Functional domains of *Pseudomonas* exotoxin identified by deletion analysis of the gene expressed in *E. coli. Cell*, **48**, 129–36.
2. Gould, B.J., Borowitz, M.J., Groves, E.S. *et al.* (1989) Phase I study of an anti-breast cancer immunotoxin by continuous infusion: report of a targeted toxic effect not predicted by animal studies. *J. Natl Cancer Inst.*, **81**, 775–81.
3. Willingham, M.C., FitzGerald, D.J. and Pastan, I. (1987) *Pseudomonas* exotoxin coupled to a monoclonal antibody against ovarian cancer inhibits the growth of human ovarian cancer cells in a mouse model. *Proc. Natl Acad. Sci. USA*, **84**, 2474–8.
4. Frankel, A.E., Ring, D.B., Tringale, F. and Hsieh-Ma, S.T. (1985) Tissue distribution of breast cancer-associated antigen defined by monoclonal antibodies. *J. Biol. Response Mod.*, **4**, 273–86.
5. Dedrick, R.L. (1985) Theoretical and experimental bases of intraperitoneal chemotherapy. *Semin. Oncol.*, **12** (Suppl. 4), 1–6.
6. Ong, G.L. and Mattes, M.J. (1989) Penetration and binding of antibodies in experimental human solid tumors grown in mice. *Cancer Res.*, **49**, 4264–73.
7. Ward, B.J. and Wallace, K. (1987) Localization of the monoclonal antibody HMFG2 after intravenous and intraperitoneal injection into nude mice bearing subcutaneous and intraperitoneal human ovarian cancer xenografts. *Cancer Res.*, **47**, 4714–18.
8. Ward, B.G., Mather, S.J., Hawkins, L.R. *et al.* (1987) Localization of radioiodine conjugated to the monoclonal antibody HMFG2 in human ovarian carcinoma: Assessment of intravenous and intraperitoneal routes of administration. *Cancer Res.*, **47**, 4719–23.
9. Colcher, D., Esteban, J., Carrasquillo, J.A. *et al.* (1987) Complementation of intracavitary and intravenous administration of a monoclonal antibody B72.3 in patients with carcinoma. *Cancer Res.*, **47**, 4218–24.
10. Pai, L.H., Bookman, M.A., Ozols, R.F. *et al.*

Clinical evaluation of intraperitoneal *Pseudomonas* exotoxin immunoconjugate OVB3-PE in patients with ovarian cancer. *J. Clin. Oncol.* (in press).
11. Furneaux, H.M., Rosenblum, M.K., Dalmau, J. *et al.* (1990) Selective expression of Purkinje-cell antigens in tumor tissue from patients with paraneoplastic cerebellar degeneration. *N. Engl. J. Med.*, **322**, 1844–51.
12. Bookman, M.A., Godfrey, S., Padavic, K. *et al.* (1990) Anti-transferrin receptor immunotoxin (IT) therapy: Phase-I intraperitoneal (i.p.) trial. *Proc. Am. Soc. Clin. Oncol.*, **9**, 187 (A722).
13. Ledermann, J.A., Begent, R.H.J., Bagshawe, K.D. *et al.* (1988) Repeated antitumour antibody therapy in man with suppression of the host response by Cyclosporin A. *Br. J. Cancer,* **58**, 654–7.
14. Hershfield, M.S., Buckley, R.H., Greenberg, M.L. *et al.* (1987) Treatment of adenosine deaminase deficiency with polyethylene glycol-modified adenosine deaminase. *N. Engl. J. Med.*, **316**, 589–96.
15. Pai, L.H., FitzGerald, D.J., Tepper, M. *et al.* (1990) Inhibition of antibody response to *Pseudomonas* exotoxin and an immunotoxin containing *Pseudomonas* exotoxin by 15-deoxyspergualin in mice. *Cancer Res.*, **50**, 7750–3.
16. Slamon, D.J., Godolphin, W., Jones, L.A. *et al.* (1989) Studies of the HER-2/*neu* proto-oncogene in human breast and ovarian cancer. *Science*, **244**, 707–12.
17. van Leeuwen, F., van de Vijver, M.J., Lomans, J. *et al.* (1990) Mutation of the human *neu* protein facilitates down-modulation by monoclonal antibodies. *Oncogene*, **5**, 497–503.
18. Chaudhary, V.K., Queen, C., Junghans, R.P. *et al.* (1989) A recombinant immunotoxin consisting of two antibody variable domains fused to *Pseudomonas* exotoxin. *Nature*, **339**, 394–7.

Chapter 17

Effects of granulocyte macrophage colony stimulating factor in cyclophosphamide- and carboplatin-treated patients

J.H. EDMONSON, G. COLON-OTERO,
H.J. LONG, T.R. FITCH, L.C. HARTMANN,
J.A. JEFFERIES and T.A. BRAICH

17.1 INTRODUCTION

In order to reduce the nausea and vomiting and to avoid the neurological and renal toxicities of cisplatin (CDDP), we and others have attempted to replace this agent in the treatment of ovarian carcinoma with carboplatin (CBDCA). Our initial dose-seeking study identified 225 mg/m^2 as the maximum dose of CBDCA which could be combined safely with 1000 mg/m^2 doses of cyclophosphamide (CYCLO)[1]. Even this moderate dose of CBDCA in the two-drug combination could not be continued at 4-week intervals for many treatment cycles without dose reductions in most of our patients, because of myelosuppression. Another more recent study demonstrated the apparent inferiority of CBDCA in comparison with CDDP, when the two were administered to patients with ovarian carcinoma in combination with the same 1000 mg/m^2 doses of CYCLO utilizing 150 mg/m^2 doses of CBDCA, which were equally as myelosuppressive as 60 mg/m^2 doses of CDDP[2]. We concluded that the substantial myelosuppression of CBDCA must be ameliorated if larger doses are to be utilized in this regimen, in a search for more effective therapy for ovarian carcinoma. Thus, we began a study attempting to extend the tolerable dose range for CBDCA plus CYCLO supported by the haemopoietic stimulant GM-CSF[3], intending to develop a more intensive dose CBDCA-based regimen for use in comparative studies of ovarian carcinoma.

17.2 MATERIALS AND METHODS

Between June 1988 and March 1991, patients with histologically proven advanced ovarian cancer and other similar tumours were enrolled in a pilot study of monthly intravenous doses of CYCLO 1 g/m², plus various doses of CBDCA supported by GM–CSF. Requirements for entry included Eastern Cooperative Oncology Group performance status of two or better, total leucocyte count of $\geq 4 \times 10^9$/l, platelet count of $\geq 130 \times 10^9$/l, and normal levels of serum creatinine (≤ 1.2 mg/dl) and direct reacting serum bilirubin (≤ 0.3 mg/dl). All patients had recovered from surgery and were free of significant infection or other serious medical problems which might contraindicate this treatment. At the initial 225 mg/m² dose level of CBDCA, patients previously treated with chemotherapy and limited irradiation were accepted. However, patients receiving higher doses were required to be free of such exposure.

Both leucocyte and platelet counts were performed three times a week during treatment, and cytotoxic drug doses were reduced by 20% if nadir leucocyte or platelet counts fell below 2×10^9/l or 50×10^9/l, respectively. If leucocytes or platelets receded below 1×10^9/l or 25×10^9/l, respectively, CYCLO and CBDCA doses were reduced by 40%. We attempted to maintain in each patient the assigned schedule and dosage of GM-CSF. However, severe and life-threatening toxicity sometimes required early cessation of treatment or dose reduction.

Beginning at 225 mg/m², CBDCA doses were to be escalated in consecutively treated groups of patients to 300, 400, 500, 600 and 700 mg/m² and further, if tolerable. No dosage escalation was practised in individual patients. GM-CSF (rh-GM-CSF, *E. coli*, non-glycosylated, Schering-Plough/Sandoz) was given initially intravenously as a daily 30-minute bolus beginning 2 days after chemotherapy and continuing for 20 days, in doses of 10 or 20 µg/kg. Other three-patient groups received these same doses on the same schedule by subcutaneous injection, all at the 225 mg/m² level of CBDCA. After CBDCA doses were escalated to the 300 mg/m² level, all GM-CSF treatment utilized the subcutaneous route. The doses and treatment schedules were adapted six more times (adaptations A through F, Table 17.1). Observational comparisons were made of leucocyte and platelet response curves and of their respective post-chemotherapy nadirs (simply the lowest values traversed by the leucocyte and platelet curves, respectively, between the initial chemotherapy and the beginning of cycle two or the completion of 30 days' observation, whichever came first).

Serum GM-CSF levels have been studied for 24 hours in two patients receiving 10 µg/kg every 12 hours subcutaneously and in two receiving 5 µg/kg according to the same schedule (Table 17.2). Toxic effects of GM-CSF have been described with efforts to correlate these with dose, route of administration and treatment schedule.

17.3 RESULTS

Among the 57 patients who entered our study, 28 had ovarian carcinoma, three had fallopian tube carcinoma, and three had cancers of the pelvic peritoneum. All 34 patients had been diagnosed by exploratory laparotomy, and all were thought to be surgically incurable. With median age 60 years (range 38–83 years), all except three of these women had definite stage III–IV ovarian cancer or its fallopian tube or peritoneal equivalent. One had possible mesothelioma; another was thought to have had stage IC ovarian cancer; and the third had apparently localized clear cell carcinoma of the ovary. All except six had performance status of zero or one; none had received irradiation, and only two had previously received cytotoxic drugs. Each of these patients entered the study with the intent to complete six cycles of chemotherapy utilizing GM-CSF according to the schedule active at date of entry. However, only 10 completed six cycles per protocol

Table 17.1 Adaptations of the GM-CSF regimen

GM-CSF regimen	Initial CBDCA dose (mg/m²)*	No. of patients treated	First cycle nadirs	
			WBC mean (×10⁹/l)	Platelet mean (×10⁹/l)
10 µg/kg i.v. (30-min bolus) daily, days 2–21	225	3	2.2	193
20 µg/kg i.v. (30-min bolus) daily, days 2–21	225	3	1.8	136
10 µg/kg s.c. daily, days 2–21	225	3	3.3	83
	300	3	1.8	19
20 µg/kg s.c. daily, days 2–21	225	3	4.1	122
Adaptation A 20 µg/kg daily s.c., days 1–7, then 10 µg/kg daily, days 8–14	300	3	1.7	55
Adaptation B 20 µg/kg daily s.c., days 1–14	300	3	2.7	84
Adaptation C 10 µg/kg every 12 hours s.c., days 1–14	300	3	2.8	102
	400	3	3.3	104
	500	3	1.8	69
	600	6	2.6	40
	700	3	0.5	32
Adaptation D 10 µg/kg every 12 hours s.c., days −6 to −3 and days 1–14	300	3	4.7	121
Adaptation E 5 µg/kg every 12 hours s.c., days −6 to −3 and days 1–14	300	3	6.1	143
	400	3	4.3	115
Adaptation F 5 µg/kg every 12 hours s.c., days 1–14	400	3	2.2	58
	600	6	2.6	62

* All patients also received CYCLO 1 g/m².
i.v., intravenous; s.c., subcutaneous.

Effects of GM-CSF in cyclophosphamide- and carboplatin-treated patients

Table 17.2 Serum GM-CSF during twice daily subcutaneous treatment*

Time (hours)	Serum GM-CSF (ng/ml)			
	10 µg/kg every 12 hours		5 µg/kg every 12 hours	
0	0	0	0	0
4	11.72	4.77	7.1	5.4
8	8.69	6.61	2.1	0.9
12	3.93	3.57	0	0
16	12.29	10.72	10.8	6.1
20	5.21	9.58	2.3	1.0
24	2.21	6.94	0	0

* Two patients received GM-CSF subcutaneously every 12 hours at 10 µg/kg and two received GM-CSF every 12 hours at 5 µg/kg. Assays were performed by Dr Sheila Jacobs, Schering Plough Research, Bloomfield, NJ.

among the 30 who have finished study participation so far (four currently remain on protocol treatment). Ten other patients among the 30 finished 'ovarian-type' cases have utilized non-protocol treatment to complete their prescribed six cycles of chemotherapy (usually CYCLO/CDDP or CYCLO/CBDCA without GM-CSF). Five additional patients have satisfactorily completed four or five cycles of chemotherapy, and three others have experienced disease progression during the initial four cycles of chemotherapy. One patient refused any further treatment for ovarian carcinoma following successful completion of three cycles of protocol treatment, and one other patient had stable disease after withdrawal from the study following one treatment cycle, because of a GM-CSF induced pulmonary reaction.

The initial 20-day GM-CSF subcutaneous treatments in patients receiving 225 mg/m^2 of CBDCA can produce significant leukaemoid reactions and leukaemoid infiltrates, as we observed in one of our ovarian cancer patients who developed tonsillar infiltrates when total leucocyte count reached 191×10^9/l. Another of our 57 patients also experienced this same complication at 107×10^9/l. Toxic effects attributed to GM-CSF in our study also included fatigue, lethargy, myalgia, fever, chills, arthralgia, bone pain, anorexia, and occasionally syncope, nausea and vomiting, or diarrhoea. More serious effects included pleuritis, pericarditis, atrial fibrillation, and pulmonary reactions and lung infiltrates with cough, dyspnoea, and tachypnoea. A variety of skin eruptions were observed including erythroderma; maculopapular and morbilliform rashes; and, frequently, pruritic reactions at cutaneous injection sites. Exfoliative dermatitis occurred in four of our 34 'ovarian-type' patients including all three patients treated according to GM-CSF adaptation D. This regimen has been abandoned due to its toxicity. No chemotherapy or GM-CSF related deaths have occurred in our 57 patients.

From comparative observations between intravenous and subcutaneous GM-CSF regimens at the 225 mg/m^2 level of CBDCA, subcutaneous administration was chosen as apparently more active in promoting leucocytosis. At the 300 mg/m^2 level of CBDCA, we observed apparently better platelet support at the nadirs (mean 55×10^9/l) when daily doses of GM-CSF were begun the day following chemotherapy, and increased during the first week to 20 µg/kg (adaptation A). Further amelioration of platelet nadirs was suggested by our observations using the 20 µg/kg daily dose for both weeks of this same subcutaneous GM-CSF schedule (mean 84×10^9/l). With this regimen (adaptation B), leucocyte nadirs also appeared better (mean 2.7×10^9/l). When these same daily doses were split and given every 12 hours at 10 µg/kg (adaptation C), leucocyte nadirs were equally well supported (mean 2.8×10^9/l). Platelet support appeared even better (mean nadir 102×10^9/l with none below 50×10^9/l). At this same level of CBDCA the results of our basic subcutaneous GM-CSF regimen (10 µg/kg daily beginning 2 days after chemotherapy) produced mean leucocyte and platelet nadirs of 1.8×10^9/l and 19×10^9/l, respectively. Despite the sequential treatment format, variability of patients and quite small numbers, adaptations C, B, and probably A, seemed to be clinically superior to the basic 10 µg/kg daily subcutaneous regimen. When GM-CSF priming was added on days −6 to −3 prior to chemotherapy, using either 10 µg/kg or 5 µg/kg every 12 hours of GM-CSF, much greater amelioration of leucocyte and platelet nadirs was observed. Utilizing the 5 µg/kg priming dose regimen (adaptation E) mean leucocyte and platelet nadirs were 6.1×10^9/l and 143×10^9/l, respectively. Nevertheless, these patients received exactly the same amount of GM-CSF per day, as did those treated on the basic 10 µg/kg daily subcutaneous regimen, and they received it for 2 days less.

The mean total dose for CBDCA accompanied by GM-CSF according to adaptation C is approximately 700 mg/m² (with CYCLO 1000 mg/m²). Utilizing GM-CSF adaptation F (5 µg/kg every 12 hours subcutaneously on days 1–14) following the two-drug chemotherapy regimen at the 600 mg/m² level of CBDCA, has demonstrated generally satisfactory mean first cycle leucocyte (2.6×10^9/l) and platelet (62×10^9/l) nadirs with apparently less toxic effects than noted with adaptation C, which utilizes twice the 12-hourly dose of GM-CSF. We have not defined the mean total dose for carboplatin in this regimen when supported by GM-CSF according to adaptation E which includes pre-chemotherapy priming.

Fifty-three of our patients have finished participation in this study, and four (with ovarian, fallopian tube or peritoneal carcinomas) currently continue to receive protocol treatment. Among the 30 patients with these 'ovarian-type' cancers who have finished participating, 13 have died of progressive disease, five are alive with presumed active residual disease, and 12 are alive without known residual disease. One of our four patients with advanced lung cancer experienced partial tumour regression, and two of the six patients with metastatic carcinomas of undetermined origin also experienced objective tumour regression (one complete response). Partial regression also occurred in one of the three patients with pancreatic carcinoma, in our one patient with cervical carcinoma, and in the single patient with carcinoma of submandibular gland origin. None of our other patients derived any apparent therapeutic benefit from this treatment.

17.4 CONCLUSION

As a result of these studies, we recommend that GM-CSF be given subcutaneously 12-hourly on days 1–14 according to adaptation F as the initial support regimen for CYCLO 1000 mg/m² plus CBDCA 600 mg/m² which we plan to use in our next comparative randomized study in stage III–IV ovarian carcinoma. Because of its effectiveness in preventing the recurrence of serious GM-CSF toxicity (serositis, pulmonary infiltration, dermatitis) observed during our study, we also recommend the liberal use of prednisone 10 mg twice daily during post-chemotherapy GM-CSF treatment, in addition to non-prescription antihistamines and ibuprofen customarily given with GM-CSF. Following the initial cycle, or at any time when leucocyte and/or platelet nadirs are at dangerous levels, we plan to add 4 days of GM-CSF priming according to adaptation E. This may permit continued chemotherapy at the intended level of intensity. We also plan to continue GM-CSF after chemotherapy for the standard 14 days during later treatment cycles. Adjustments will be made to permit early cessation or longer continuation in individual cases. A phase III comparison between six cycles of this regimen and another comprising six cycles of ordinary doses of CYCLO and CBDCA only half as intensive without GM-CSF, should provide a reasonable test of the value of dose intensification in ovarian carcinoma.

REFERENCES

1. Edmonson, J.H., McCormack, G.W., Krook, J.E. *et al.* (1987) Pilot study of cyclophosphamide plus carboplatin in advanced ovarian carcinoma. *Cancer Treat. Rep.*, **71**, 199–200.
2. Edmonson, J.H., McCormack, G.M., Wieand, H.S. *et al.* (1989) Cyclophosphamide–cisplatin versus cyclophosphamide–carboplatin in stage III–IV ovarian carcinoma: A comparison of equally myelosuppressive regimens. *J. Natl Cancer Inst.*, **81**, 1500–4.
3. Laver, J. and Moore, M.A.S. (1989) Clinical use of recombinant human hematopoietic growth factors. *J. Natl Cancer Inst.*, **81**, 1370–82.

Chapter 18
Mitochondrial poisons and ovarian cancer

A. MANETTA, D. EMMA and G. GAMBOA

18.1 INTRODUCTION

Ovarian cancer is one of the four major causes of cancer-related deaths in women and, while treatable, still claims over 10 000 lives annually. Ovarian cancer is amenable to a variety of treatments which can result in a significant reduction in tumour burden. The effect of these varied treatment regimens is indicated by the five-year survival rates, which range from 12 to 60%, dependent upon the site and extent of the disease at the time of diagnosis. While these figures reflect an improvement in the treatment process, the adjusted death rate in women has remained constant for the past 35 years, underscoring the need for either novel or more aggressive forms of therapy. While newer surgical and radiological techniques are reflected in the improved five-year survivals, the greatest advances may be attributed to the effective/aggressive use of chemotherapy, either as a first line agent or in the adjuvant setting. Agents must therefore be identified that are capable of either augmenting the tumour cell killing by current chemotherapeutic agents or by reversing tumour cell resistance to those agents.

Lipophilic cationic compounds, such as rhodamine 123, have displayed cytotoxic activity against carcinoma cell lines both *in vitro* and *in vivo*[1–3]. Dequalinium chloride (DECA) is a lipophilic cationic compound with a structure similar to rhodamine. However, while rhodamine has a single positive charge, dequalinium has two[4]. DECA has been reported to inhibit protein kinase C[5] as well as to selectively accumulate and be retained by the mitochondria of neoplastic cells where it inhibits cellular energy production[6], and has been proven to be more cytotoxic than rhodamine in certain experimental systems[1,2,7,8]. Bodden and colleagues[8] have also shown a potent inhibitory effect on calmodulin by DECA, with an associated antiproliferative effect. The mechanism of effect of these compounds, although undetermined, seems to be related to inhibition of adenosine triphosphate synthesis in the mitochondria[9], specifically the inhibition of F_1-ATPase[10].

Bleday and colleagues[11] have demonstrated a significant inhibition of primary tumour growth of a rat colon tumour isograft following DECA treatment. Weiss and coworkers[12] have demonstrated prolongation of animal survival in mice implanted intraperitoneally with bladder carcinoma and treated with DECA. Furthermore, they found this to be more effective than 5-fluorouracil, cisplatin, vinblastine, bleomycin, methotrexate or cyclophosphamide in this animal model. Individually, DECA and tumour necrosis factor (TNF) display antitumour activity, and because both drugs seem to have a direct effect on mitochondrial function, *in vitro* pharmacological synergy against a panel of human ovarian cancer cell lines was observed following sequential dequalinium and TNF treatment[13].

Experiments were conducted to determine the acute and cumulative *in vivo* toxicities of DECA treatment. Its activity alone or in combination with cytotoxic agents was also examined *in vitro*, against a panel of human ovarian cancer cell lines.

18.2 MATERIALS AND METHODS

18.2.1 CELL LINES

The human ovarian carcinoma cell lines NIH:OVCAR-3; PA-I (American Type Culture Collection, Frederick, MD, USA); 222 (provided by B. Bonavida, University of California, Los Angeles, USA); A2780, A2780 doxorubicin (ADR) resistant and A2780 cisplatin (CDDP) resistant (provided by T. Hamilton, Fox Chase Cancer Center, Philadelphia, PA, USA); and UCI-101 were used. Cell lines were maintained in RPMI 1640 tissue culture medium supplemented with 10% fetal calf serum, glutamine (2 mM), insulin (0.2–0.3 U/ml) (required for NIH: OVCAR-3 and A2780), penicillin (100 U/ml) and streptomycin (100 µg/ml). Cell lines were incubated at 37 °C in an atmosphere of 5% CO_2 in air.

18.2.2 REAGENTS

Dequalinium chloride (Sigma, St Louis, MO, USA) stock solution was prepared by dissolving the reagent in distilled water using a waterbath sonicator. Further dilutions were prepared in complete tissue culture medium.

Doxorubicin (Adriamycin: Adria Laboratories, Columbus, OH, USA) and cisplatin (Bristol-Myers, Syracuse, NY, USA) were prepared and obtained from the pharmacy of the University of California, Irvine.

All reagents requiring sterilization were sterilized by passage through a 0.22-µm membrane filter.

18.2.3 *IN VITRO* ASSAY

The effect on cellular proliferation of DECA treatment either alone or in combination with other agents was assayed as follows.

Exponentially growing ovarian carcinoma cells were seeded into 96-well flat bottom microtitre plates (0.5–1.0 × 10^4 cells/well in 0.1 ml) and allowed to attach to the wells for 18–24 hours in the incubator prior to drug exposure. Test agents, alone or in combination, were then added to the individual wells in 0.1-ml final volume and the cells maintained in culture for two to three cell doublings. In combined agent experiments, DECA was added 18–24 hours prior to second agent exposure. Upon assay termination, supernatants were aspirated and viable cells stained with 50 µl of a 1% crystal violet solution as described by Yamamoto and colleagues[14]. Quantification of the staining was assessed by reading the absorbance at 570 nm on a Dynatech MR700 Microplate Reader (Dynatech, Inc., Chantilly, VA, USA). Results were expressed as a percentage of untreated controls. Data points represent the average of triplicate wells with experiments performed in triplicate. Variation between triplicate wells within a single data point was found to be less than 10%.

18.2.4 *IN VIVO* TOXICITY STUDIES

The acute and cumulative *in vivo* toxicity of DECA was evaluated using a female Balb/c mouse model. Briefly, DECA was prepared as described above with dilutions prepared in

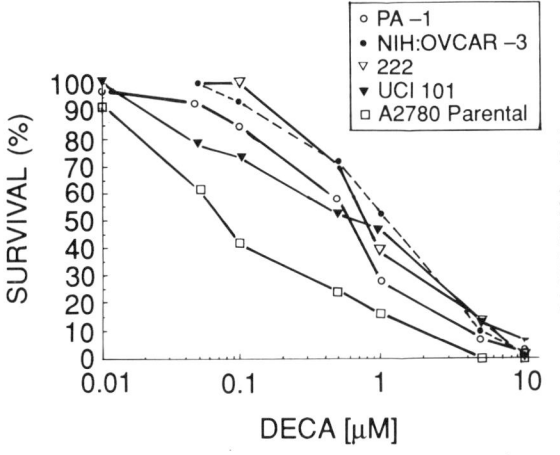

Figure 18.1 Effect of dequalinium chloride on the *in vitro* cell survival of a panel of human ovarian cancer cell lines.

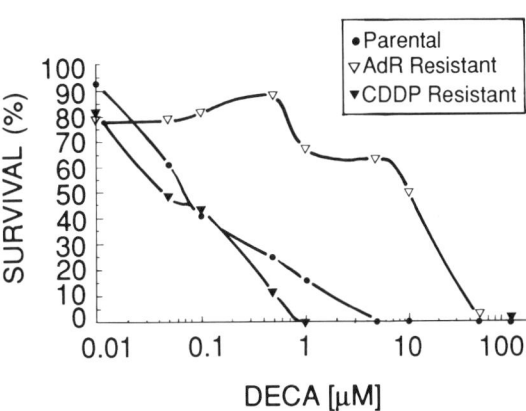

Figure 18.2 Effect of dequalinium chloride on the *in vitro* cell survival of parental, Adriamycin- and cisplatin-resistant A2780 ovarian cancer cells.

normal saline for injection. Drug concentrations were adjusted so that all animals received 0.02 ml/g body weight. For the acute intraperitoneal toxicity study, dose levels ranged from 10 to 25 mg/kg. Two drug dosing schedules were examined for the cumulative toxicity studies. DECA doses of 10–15 mg/kg were examined in a once per week treatment regimen, while doses from 5 to 9 mg/kg were tested on an every other day basis. The maximally tolerated dose of DECA (MTD; highest DECA dose without death) and LD_{50} dose (dose killing 50% of animals) were determined for each dose schedule. In all toxicity studies, animals were allowed to die with various stress parameters being evaluated. Post-mortem and histology were performed on all individual mice.

18.2.5 DATA ANALYSIS

Comparison of means was performed using a Student *t* test employing the Statgraphics computer statistical analysis system (STSC, Inc., Rockville, MD USA). Pharmacological synergism was confirmed by isobologram analysis[15–17].

18.3 RESULTS

18.3.1 DECA AS A SINGLE AGENT

As can be seen in Figure 18.1, comparable *in vitro* dose–response curves for the panel of cell lines tested were obtained following a single exposure to DECA. The concentration of DECA required to inhibit cell survival by 50% (IC_{50}) ranged from 0.08 μM for the A2780 cells, to approximately 1.2 μM for NIH:OVCAR3 cells. To date, resistance to DECA has only been observed in an ADR-resistant A2780 cell line derived from the parental cells. These cells were over 100 times more resistant than the parental cells to DECA, having an IC_{50} concentration of 10 μM. Similar cells, resistant to cisplatin, were not resistant to dequalinium and produced a dose–response curve comparable to the parental A2780 cell line (Figure 18.2).

18.3.2 DECA IN CONJUNCTION WITH CHEMOTHERAPEUTIC AGENTS

Pharmacological synergy was observed in a number of instances of combination chemotherapy using DECA pretreatment 18–24 hours prior to CDDP exposure. These drug

Table 18.1 Effect on human ovarian cancer cell survival of sequential dequalinium chloride and cisplatin exposure

Cell line	DECA (μm)	% Kill	CDDP (ng/ml)	% Kill	DECA + CDDP (% Kill)	Additive/ synergistic
UCI 101	0.10	34	0.1	3	47	S
			1.0	3	47	S
			10	29	52	A
			100	75	82	A
A2780 ADR resistant	0.01	21	100	19	54	S
			500	39	79	S
			1000	90	91	A
	0.10	18	100	19	39	S
			500	39	80	S
A2780 CDDP resistant	0.01	18	100	<1	28	S
			500	<1	46	S
			1000	8	52	S
	0.10	57	100	<1	61	A
			500	<1	75	S
			1000	8	78	S

A, additive; S, synergy; ADR, Adriamycin; CDDP, cisplatin.

combinations were not found to be protective, although for ADR exposure the resultant cell kill was slightly less than additive. As can be seen in Table 18.1, synergy was observed in the UCI-101 cell line for CDDP exposures between 0.1 to 1.0 ng/ml following a DECA pretreatment of 0.10 μM. For the A2780 ADR-resistant cells, synergy was observed with DECA pretreatment of 0.01–0.10 μM with CDDP from 100 to 500 ng/ml. The CDDP-resistant A2780 cells displayed similar synergy with the exception that CDDP exposure was between 500 and 1000 ng/ml. None of the DECA pretreatment doses was able to 'reverse' the resistance of the A2780 drug-resistant cells to a level comparable for the parental cell line.

18.3.3 TOXICITY STUDIES

Table 18.2 shows both the acute and cumulative toxicities for the intraperitoneal adminis-

Table 18.2 Intraperitoneal dequalinium toxicity

Schedule time	(days)	LD_{50}(mg/kg)	MTD (mg/kg)
Acute	8	20	15
Q7D	21	13	10–11
QOD	30	7–8	4–5

tration of DECA. The most overt symptom of DECA toxicity was a rapid and pronounced weight loss. Weight loss averaged 2–4 g/week either until animal weights stabilized or death occurred. Respiratory distress was also marked in many animals. Histological evaluation of biopsy materials could not account for the rapid loss of weight. The principal sites of drug damage occurred in the kidney and liver. Both renal and hepatic damage was observed in the acute dose studies, at doses of 15 mg/kg or higher. This damage consisted of microvascular destruction within the organ. The respiratory distress observed in animals was related to congestive lung problems secondary to the renal/hepatic damage.

18.4 CONCLUSION

As an antimicrobial agent, dequalinium has been in worldwide use for over 30 years in such non-prescription items as mouthwash, topical ointments, suppositories, vaginal/oral paints, or throat lozenges. Structurally similar to rhodamine 123 except for an additional positive charge, it has been postulated that the selective accumulation, retention and toxicity for carcinoma cells would be increased above that for the singly charged rhodamine analogue[12]. This selective accumulation and toxicity was observed to prolong animal survival in mice bearing the transplanted MB49 bladder carcinoma. Furthermore, its activity was greater than many of the standard chemotherapeutic agents employed. These observations, which need to be extended to human cells, would indicate that DECA could be effective in a number of treatment alternatives for human ovarian cancer.

Of the cell lines tested, only the ADR-resistant A2780 was resistant to DECA, requiring a 10-μM DECA exposure to suppress *in vitro* growth by 50%. While 100 times greater than for the other ovarian cell lines, the 10-μM dose was still within the clinically achievable intraperitoneal dose range in mice, and would be expected to be achievable in man under similar treatment conditions. Elevated levels of membrane p-glycoprotein or protein kinase C, as well as enhanced cellular drug efflux, have been associated with tumour cell resistance to ADR[18–20]. In theory, DECA should be able to overcome the resistance to ADR as it can inhibit protein kinase C[5] and cellular drug efflux by inhibiting calmodulin[21] or cellular energy production. The small molecular weight and lipophilic nature of dequalinium should allow it effectively to tie up the p-glycoprotein drug transport system, in a manner similar to verapamil and related compounds[22]. The reasons for the apparent cross-resistance between dequalinium and ADR are unknown and are under investigation, and studies are underway to determine whether DECA can be used as an agent for the reversal of drug resistance. The ability to synergize with cisplatin in drug-resistant cells may have clinical applications for patients refractory to that drug. Other drugs under investigation include 5-fluorouracil, etoposide, taxol and carboplatin. Studies will include both the determination of *in vitro* synergy, and the effects on prolongation of animal survival in a xenogeneic tumour system of parental and drug-resistant human tumour xenografts.

The mechanism of resistance to direct lysis by TNF is protein synthesis-dependent. Blockage of the synthetic pathway by pretreatment with actinomycin D or other agents allows for tumour cell kill and is the basis for measuring TNF units of activity[23]. In TNF-sensitive cells, addition of small molecular weight lipophilic compounds decreases the amount of actinomycin D necessary to synergize with TNF[22]. It is believed that the lipophilic compounds saturate the drug transporter efflux system, causing retention of the actinomycin D with the resultant synergy. However, in TNF-resistant cells, such synergy is not observed, indicating that the resistance mechanism to TNF may be, in part, different from that for chemotherapeutic agents [24,55]. Synergy between DECA and TNF has been observed in the panel of ovarian cancer cells[13], but a TNF-resistant cell line will be required to determine whether DECA can reverse resistance to this biological response modifier. TNF destruction of tumours *in situ* is not limited to direct cytotoxicity, and the observed synergy *in vitro* may translate into enhanced tumour control with prolongation of survival. This aspect is currently under study in a number of xenograft models.

Dequalinium presents itself as an agent with activity in a number of sites, all of which have direct application to the treatment of ovarian cancer. Previous reports of the activity of DECA have been in animal systems. This chapter extends those observations to human tumour cell lines *in vitro*. Agents are currently being sought that are active by themselves or are capable of reversing drug resistance. DECA is capable of both actions. As biological response modifier-based treatment protocols become active, the eventual observance of resistance can be anticipated. Thus, agents that can synergize with or reverse the resistance to the biological response modifiers will be needed. DECA requires further investigation to establish its suitability. The original toxicology studies for dequalinium are over 30 years old[4] and should be repeated in the light of newer technologies, so that DECA can be used in a phase 1 study. Due to the lipophilic nature of dequalinium chloride, the observed hepatic damage from the acute dosage schedule is not unexpected but may ultimately prove to be dose limiting for patient trials.

REFERENCES

1. Bernal, S.D., Lampidis, T.J., Summerhayes, I.C. and Chen, L.B. (1982) Rhodamine 123 selectively reduces clonal growth of carcinoma cells *in vitro*. *Science*, **218**, 1117–19.
2. Lampidis, T.J., Bernal, S.D., Summerhayes, I.C. and Chen, L.B. (1983) Selective toxicity of rhodamine 123 in carcinoma cells *in vitro*. *Cancer Res.*, **43**, 716–20.
3. Bernal, S.D., Lampidis, T.J., McIsaac, R.M. *et al.* (1983) Anticarcinoma activity *in vivo* of rhodamine 123, a mitochondrial specific dye. *Science*, **222**, 169–72.
4. Babbs, M., Collier, H.O.J., Austin, W.C. *et al.* (1956) Salts of decamethylene-*bis*-4-aminoquinaldinium (Dequadin), a new antimicrobial agent. *J. Pharmacol.*, **8**, 110–19.
5. Rotenberg, S., Smiley, S., Ueffling, M. *et al.* (1990) Inhibition of protein kinase C by the anticarcinoma agent dequalinium. *Cancer Res.*, **50**, 677–85.
6. Nadakavukaren, K.K., Nadakavukaren, J.J. and Chen, L.B. (1985) Increased rhodamine 123 uptake by carcinoma cells. *Cancer Res.*, **45**, 6093–9.
7. Davis, S., Weiss, M.J., Wong, J.R. *et al.* (1985) Mitochondrial and plasma membrane potentials cause unusual accumulation and retention of rhodamine 123 by human breast adenocarcinoma-derived MCF-7 cells. *J. Biol. Chem.*, **260**, 13844–50.
8. Bodden, W.L., Palayoor, S.T. and Hait, W.N. (1986) Selective antimitochondrial agents inhibit calmodulin. *Biochem. Biophys. Res. Commun.*, **135**, 574–82.
9. Modica-Napolitano, J.S., Weiss, M.J., Chen, L.B. *et al.* (1984) Rhodamine 123 inhibits bioenergetic function in isolated rat liver mitochondria. *Biochem. Biophys. Res. Commun.*, **118**, 717–23.
10. Zhou, S. and Allison, W.S. (1988) Inhibition and photoinactivation of the bovine heart mitochondrial F_1-ATPase by the cytotoxic agent, dequalinium. *Biochem. Biophys. Res. Commun.*, **152**, 968–72.
11. Bleday, R., Weiss, M.J., Salem, R.R. *et al.* (1986) Inhibition of rat colon tumor isograft growth with dequalinium chloride. *Arch. Surg.*, **121**, 1272–5.
12. Weiss, M.J., Wong, J.R., Ha, C.S. *et al.* (1987) Dequalinium, a topical antimicrobial agent displays anticarcinoma activity based on selective mitochondrial accumulation. *Proc. Natl. Acad. Sci. USA*, **84**, 5444–8.
13. Manetta, A., Emma, D., Lucci, J. and Granger, G. (1990) Combined effects of dequalinium and rhuTNF on human cell lines *in vitro*. *Proc. Am. Assoc. Cancer Res.*, **31**, 237.
14. Yamamoto, R.S., Kobayashi, M., Plunkett, J.M. *et al.* (1985) Production and detection of lymphotoxin *in vitro*: microassay for lymphotoxin, in *Investigation of Cell Mediated Immunity* (ed. T. Yoshida), Churchill Livingstone, London, pp. 126–34.
15. Steel, G.G. and Peckham, M.J. (1979) Exploitable mechanisms in combined radiotherapy–

chemotherapy: The concept of additivity. *Int. J. Radiat. Oncol. Biol. Phys.*, **5**, 85–91.
16. Tsai, C.M., Gazdar, A., Venzon, D.J. *et al.* (1989) Lack of *in vitro* synergy between etoposide and cis-diaminedichloroplatinium(II). *Cancer Res.*, **49**, 2390–7.
17. Deen, D.F. and Williams, M.E. (1979) Isobologram analysis of X-ray BCNU interactions *in vitro*. *Radiat. Res.*, **79**, 483–91.
18. Kato, S., Ideguchi, H., Muta, K. *et al.* (1990) Mechanisms involved in the development of adriamycin resistance in human leukemic cells. *Leuk. Res.*, **14**, 567–73.
19. Posada, J., Vichi, P. and Tritton, T.R. (1989) Protein kinase C in adriamycin action and resistance in mouse sarcoma 180 cells. *Cancer Res.*, **49**, 6634–9.
20. O'Brian, C.A., Fan, D., Ward, N.E. *et al.* (1989) Level of protein kinase C activity correlates directly with resistance to adriamycin in murine fibrosarcoma cells. *FEBS Lett.*, **246**, 78–82.
21. Hait, W.N. (1987) Targeting calmodulin for the development of novel cancer chemotherapeutic agents. *Anticancer Drug Des.*, **2**, 139–49.
22. Hofsli, E. and Nissen-Meyer, J. (1990) Reversal of multidrug resistance by lipophilic drugs. *Cancer Res.*, **50**, 3997–4002.
23. Espevik, T. and Nissen-Meyer, J. (1986) A highly sensitive cell line, WEHI 164 clone 13, for measuring cytotoxic factor/tumor necrosis factor from human monocytes. *J. Immunol. Methods*, **95**, 99–105.
24. Hofsli, E. and Nissen-Myer, J. (1989) Reversal of drug resistance by erythromycin: erythromycin increases the accumulation of actinomycin D and doxorubicin in multidrug-resistant cells. *Int. J. Cancer*, **44**, 149–54.
25. Hofsli, E. and Nissen-Myer, J. (1989) Effect of erythromycin and tumor necrosis factor on the drug resistance of multidrug resistant cells: reversal of drug resistance by erythromycin. *Int. J. Cancer*, **43**, 520–5.

Chapter 19

TNFα mediated lysis in gynaecological malignancies

J.L. COLLINS and D.G. MUTCH

This chapter focuses on our experiences over the past 13 years concerning the mechanism of action of tumour necrosis factor alpha (TNFα), its antitumour activity, and the resistance that cancer cells develop to avoid lysis by this cytokine.

TNFα was first described as a factor in the serum of animals treated with *Mycobacterium bovis*, strain BCG, that caused tumour necrosis[1]. TNFα is a small glycoprotein of approximately 157 amino acids secreted primarily by macrophages[2–4]. The gene encoding human TNFα has been isolated and the availability of recombinant TNFα has prompted considerable research *in vitro* and *in vivo*[2–7]. *In vitro*, TNFα has been shown to mediate a variety of biological responses including cytolysis of tumour cells[8–13]. TNFα has also been shown to mediate antitumour activity in mice[14–16]. The antitumour activity of TNFα would seem to make it a potential candidate for the immunotherapy of malignant disease, and a number of clinical trials of TNFα have been initiated [17–19]. Unfortunately, clinical trials of TNFα have not been very successful. The failure of TNFα as an anticancer agent could be related to the fact that most cancer cells are resistant to TNFα, despite the fact that they possess receptors for it[20].

Although many ovarian cancer cell lines are resistant to lysis by TNFα, they become sensitive to lysis when exposed to inhibitors of protein synthesis such as actinomycin D[21–25]. This indicates that the TNFα receptors expressed by these cells are functional, and that the maintenance of protein synthesis is required for their resistance to TNFα. Our recent efforts have focused on the identification of this protein synthesis-dependent resistance mechanism[21,26–28]. Once the mechanism of TNFα resistance is characterized, agents that would specifically decrease the resistance mechanism (and thereby increase sensitivity to TNFα-mediated lysis) might be identified. Here it is hoped that such agents would increase the therapeutic potential of TNFα for the treatment of malignant disease.

Our interest in TNFα began with the analysis of murine natural cytotoxic (NC) cells[29–31]. NC cells are a type of naturally occurring cytotoxic cell that use membrane-bound TNFα to effect lysis[30,32]. The other major type of naturally occurring cytotoxic cells are natural killer (NK) cells[33–35]. These, unlike NC cells, use cytolysin/per-

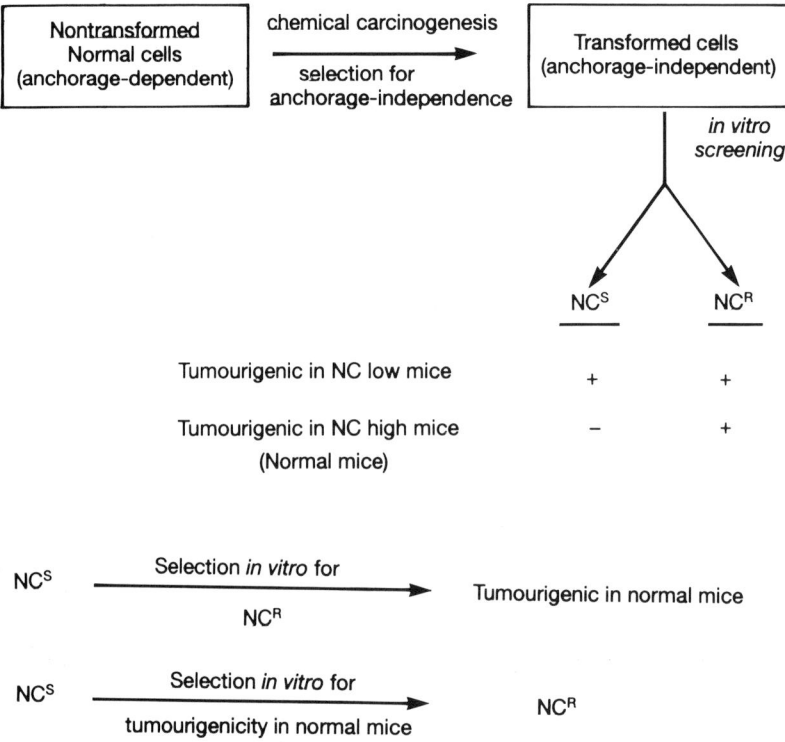

Figure 19.1 Involvement of natural cytotoxic cells in tumour surveillance. NC^R, resistant to lysis by natural cytotoxic cells; NC^S, sensitive to lysis by natural cytotoxic cells.

forins to mediate lysis rather than TNFα[36]. While there is good evidence that NC cells are involved with tumour surveillance, there is little evidence that NK cells, as originally defined[29,33,34] are capable of mediating tumour rejection *in vivo*.

Murine NC cells have been characterized by a number of investigators[37–40]. Humans also have NC activity which is indistinguishable from murine NC cell activity[41]. The evidence that NC cells can mediate antitumour activity is based on two types of experiments (Figure 19.1)[39,42–44]. In the first type, normal anchorage-dependent cells are exposed to chemical carcinogens, and transformed anchorage-independent cells are isolated. The transformed cells can be divided into two groups based on their *in vitro* sensitivity to NC-mediated lysis: those that are sensitive to natural cytotoxic cell lysis (NC^S), and those that are resistant to lysis by natural cytotoxic cells (NC^R). While the NC^R transformants are tumourigenic in normal, as well as in NC-low mice, the NC^S transformants are tumourigenic only in NC-low mice; that is, they are not tumourigenic in normal mice[42,43]. Thus, there is a correlation between sensitivity to NC-mediated lysis *in vitro* and tumour rejection in normal mice.

The second type of experiment begins with NC^S transformants (which are tumourigenic only in NC-low mice) and selects *in vitro* for NC resistance using NC effector cells. These NC^R cells grow as tumours in normal mice. The reverse selection starts with NC^S transformants (which are tumourigenic only in NC-deficient mice) and selects *in vivo* for cells that are tumourigenic in normal mice by injecting large numbers of cells into normal mice. Those cells that are tumourigenic in normal mice become resistant to NC-mediated lysis *in vitro*[42,43]. In the first selection procedure, the selected marker is NC^R and this turns out to be associated with the unselected marker, which is tumourigenicity in normal mice. In the second selection procedure, the selected marker is tumourigenicity which turns out to be associated with the unselected marker, NC^R. This reciprocal correlation between the selected and unselected markers strongly suggests that NC resistance and tumourigenicity are causally related.

Although NC activity could play a role in tumour surveillance, two experimental observations suggest that their actual effectiveness in eliminating tumour cells may be limited. First, most tumour cells are resistant to lysis by NC cells, in spite of the fact that they are recognized by NC effector cells[42,43]. Secondly, because NC activity is not regulated by antigen recognition, the level of NC cell activity remains constant as the number of tumour cells increases[44]. Clearly there would be an increase in the ability of the existing NC cells to eliminate tumour cells if the resistance to NC-mediated lysis by these effectors could be decreased.

There is evidence that cisplatin is capable of decreasing resistance to NC-mediated lysis[45,46]. Cisplatin has been shown to be a more effective anticancer agent in normal rather than irradiated animals[47]. Because irradiation has been shown to reduce NC activity[48], it is possible that some of the antitumour activity of cisplatin is mediated through NC activity. For this reason, we examined the effects of cisplatin on NC-mediated lysis. As a control we also examined the effect of cisplatin on NK-mediated lysis. 10ME cells have been shown to be resistant to both murine and human NK-mediated lysis, but sensitive to both murine and human NC-mediated lysis; YAC-1 cells are sensitive to murine NK-mediated lysis but resistant to murine NC-mediated lysis; and K-562 cells are sensitive to human NK-mediated lysis but resistant to human NC-mediated lysis.

These target cells were first exposed to cisplatin, then washed before they were exposed to NC and NK effector cells. This assured that the effect of cisplatin was on the target cells and not on the effector cells. As shown in Figure 19.2, both murine and human NC activity is increased by treatment of the target cells with non-toxic concentrations of cisplatin. The same treatment actually reduced the level of both murine spleen and human peripheral blood NK-mediated lysis. This indicates that cisplatin enhances the lytic mechanism initiated by NC effector cells, but has a negative effect on the lytic potential of NK cells. Extrapolating this to the human situation implies that cisplatin may have an added effect (in addition to its direct anticancer activity) on cancer cells through NC cell-mediated lysis.

It is now known that NC cells use a membrane-bound form of TNFα to effect lysis. This is based on the fact that anti-TNFα antibody blocks NC-mediated lysis, and more recently on the identification of a 26-kDa integral membrane form of TNFα on the surface of NC effector cells[41,49]. The observation that NC cells use TNFα to mediate lysis suggested that, at least from the point of anticancer activity, NC cells and TNFα might be similar. There may be an advantage of TNFα as compared to NC cells in mediating tumour rejection, in that the level of TNFα can be increased by giving patients TNFα exogenously. We know of no way to increase the number of endogenous NC cells.

To better understand how TNFα and cisplatin act synergistically, we evaluated their combined effect on two well-established ovarian cancer cell lines, SKOV3 and OVCAR3[50]. These cell lines differ in their

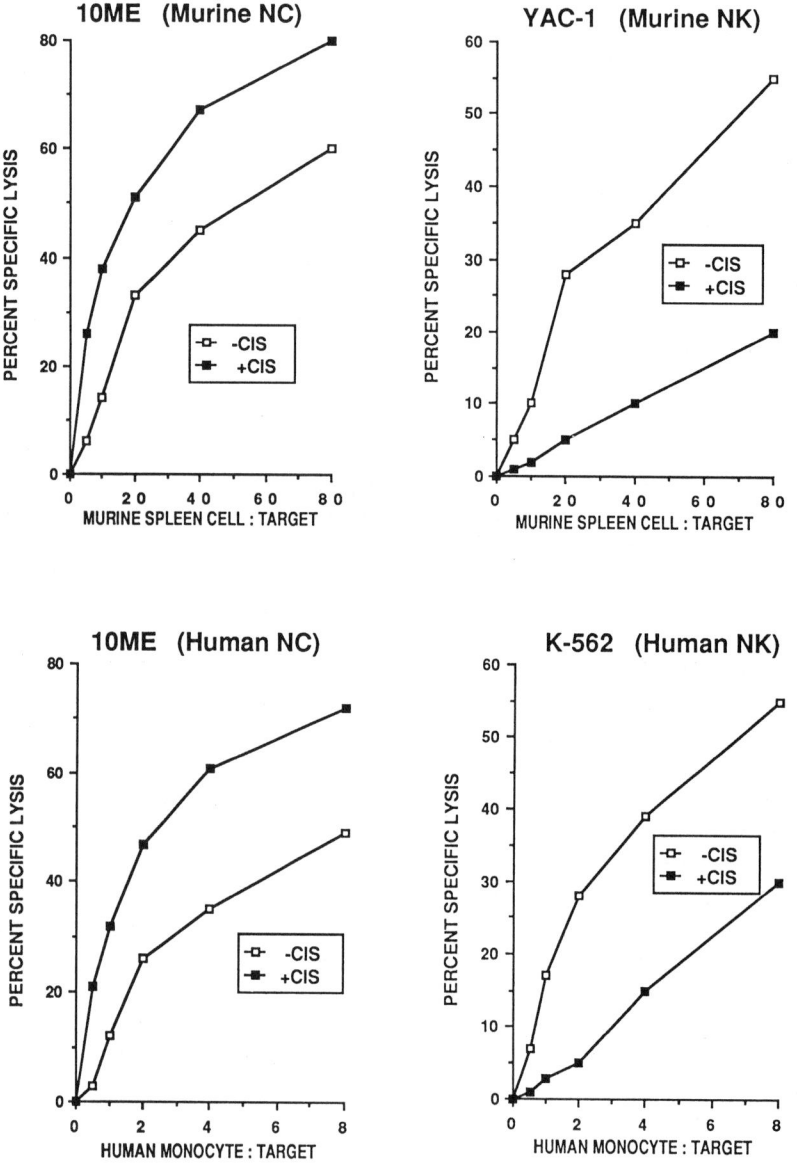

Figure 19.2 The effect of cisplatin on NC- and NK-mediated lysis. The percent specific lysis of target cells is plotted as a function of the ratio of effector cells to target cells. NC-mediated lysis was measured by the release of ^{51}Cr from target cells pretreated with cisplatin (-CIS) or at the indicated concentrations after 18 hours (+ CIS). NK-mediated lysis was measured by the release of ^{51}Cr from target cells pretreated with cisplatin (-CIS) or at the indicated concentrations after 6 hours (+CIS).

sensitivity to cisplatin and TNFα as single agents. SKOV3 cells are relatively sensitive to cisplatin but relatively resistant to TNFα, whereas OVCAR3 are relatively resistant to cisplatin but relatively sensitive to TNFα. As the specificity of all anticancer drugs resides in their ability to kill dividing but not non-dividing cells, lysis was evaluated using dividing and non-dividing populations of these two cell lines.

The rate of cell division is controlled by the density with which cells are plated[51]. In these experiments, the non-dividing populations of both cell lines are resistant to lysis at all concentrations of TNFα and cisplatin tested (Figure 19.3). In contrast, the dividing populations of SKOV3 cells are sensitive to lysis by cisplatin alone. When TNFα and cisplatin are combined, there is a synergistic increase in the sensitivity of dividing SKOV3 cells. Although dividing populations of OVCAR3 cells are relatively sensitive to TNFα, while resistant to cisplatin as single agents, these cells also show a synergistic increase in lysis when TNFα and cisplatin are combined (Figure 19.3). Thus, synergy which results from the combination of TNFα and cisplatin is seen in cell lines independent of their sensitivity to either TNFα or cisplatin as single agents[50]. This enhanced lysis observed with the combination of TNFα and cisplatin is very similar to that observed when the target cells were exposed to NC cells in the presence of cisplatin (as outlined above). Furthermore, the fact that non-dividing cells are resistant to the cytolytic effects of this combination suggests that while tumour cell lysis may be increased synergistically, toxicity may not.

Although we knew that there was synergy between TNFα and cisplatin to increase lysis, we did not know why they acted synergistically. A number of investigators, working with TNFα1-sensitive cell lines, have shown that the lytic potential of TNFα could be increased in these cells when they were exposed to actinomycin D (ACT-D), an inhibitor of mRNA synthesis[21–25]. It was possible that the ability of cisplatin to increase lysis in the presence of TNFα resides in its ability to inhibit protein synthesis at the level of mRNA. In support of this, we have previously shown that cells that are resistant to lysis by NC cells express a protein synthesis-dependent resistance mechanism, such that when protein synthesis is inhibited they become sensitive to NC-mediated lysis[41]. Because NC cells use TNFα to mediate lysis the resistance mechanism that prevents this NC-mediated lysis does so by preventing lysis by TNFα. Since most cells are resistant to TNFα-mediated lysis, as they are to NC-mediated lysis, we determined if ACT-D and other inhibitors of protein synthesis (cycloheximide (CHX) or emetine (EM)) could increase the sensitivity of resistant ovarian and cervical cells to lysis by TNFα[21]. These three agents were chosen because they all inhibit protein synthesis, but do so at different sites along the pathway of protein synthesis.

As shown in Figure 19.4, the three ovarian cancer cell lines, OVCAR3, SKOV3 and CaOV3, are all resistant to lysis by TNFα with variable sensitivity (dependent on the inhibitors and cell lines) to protein synthesis inhibitors alone. In all cell lines, lysis increases significantly when protein synthesis inhibitors and TNFα are used in combination. Isobolographic analysis demonstrates that TNFα and protein synthesis inhibitors act synergistically[21]. The same synergistic increase in lysis has also been shown in four of five cervical carcinoma cell lines exposed to TNFα and protein synthesis inhibitors[21].

We believe that the resistance of cancer cells to lysis by TNFα has presented a major block to the clinical success of TNFα as an anticancer agent. The results of the experiments above suggest a strategy for manipulating the TNFα resistance mechanism in cancer cells by inhibiting protein synthesis. If we extrapolate from these *in vitro* data to *in vivo* application it may be possible to increase the response rate of patients with ovarian

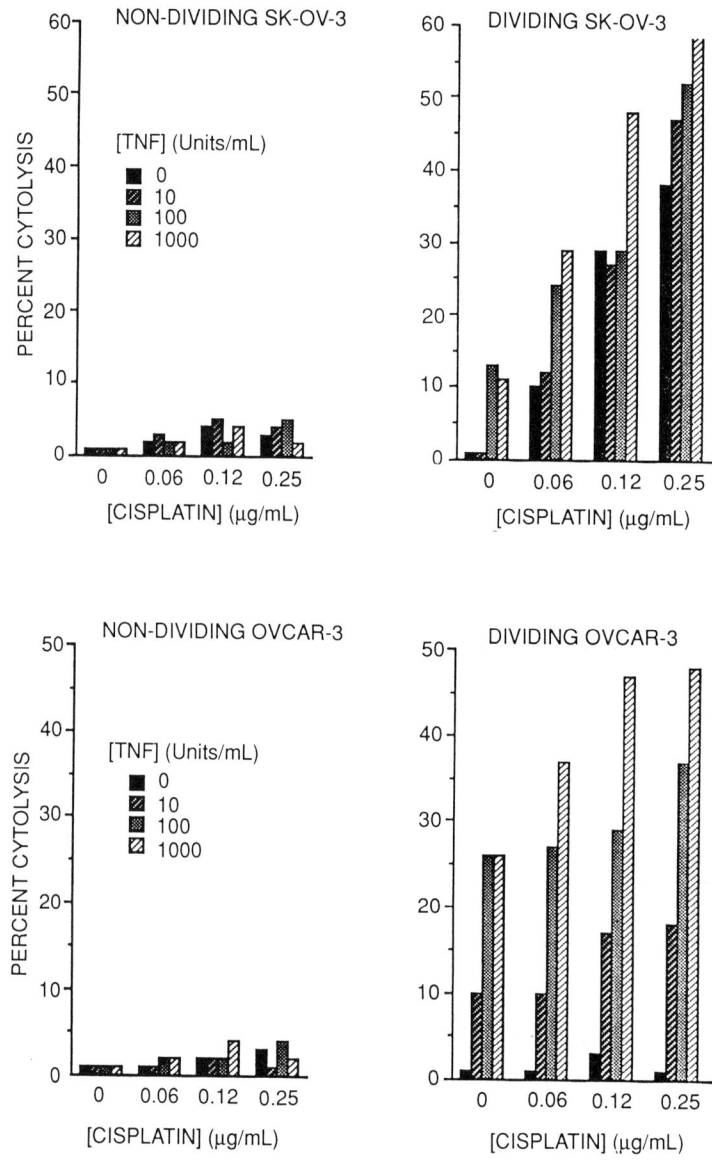

Figure 19.3 Effect of cisplatin and TNF-α on dividing and non-dividing populations of ovarian cancer cell lines. Percent cytolysis is plotted as a function of TNF-α and/or cisplatin at the indicated concentrations. Cells were exposed to cisplatin and TNF-α for 48 hours and the percent cytolysis measured by the incorporation of ^{51}Cr.

cancer by combining TNFα with chemotherapeutic drugs that inhibit protein synthesis.

At this point a number of questions remain. Foremost among these, do protein synthesis inhibitors increase lysis by TNFα or does TNFα increase the sensitivity of cells to protein synthesis inhibitors? Understanding how TNFα and protein synthesis inhibitors

TNFα mediated lysis in gynaecological malignancies

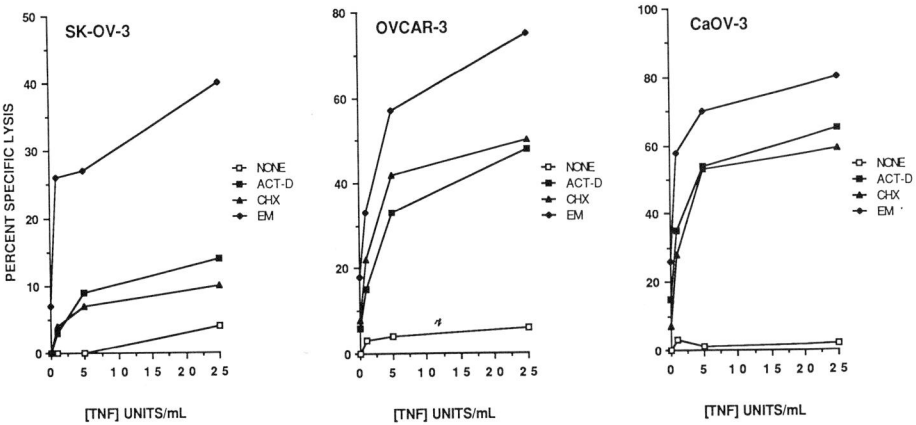

Figure 19.4 The effect of protein synthesis inhibitors in combination with TNF-α on ovarian carcinoma cell lines. Percent specific lysis of ovarian cancer cell lines is plotted as a function of the concentration of TNF-α in an 18-hour ^{51}Cr release assay with no protein synthesis inhibitor added (none), with actinomycin D added (ACT-D), with cycloheximide added (CHX), or with emetine added (EM). The concentrations of ACT-D were 10^{-5} M, 10^{-6} M, and 10^{-8} M, the concentrations of CHX were 10^{-4} M, 10^{-4} M, and 10^{-7} M, and the concentrations of EM were 10^{-4} M, 10^{-4} M, and 10^{-7} M for SK-OV-3, OVCAR-3, and CaOV-3.

increase the sensitivity of cells to lysis could be important to our understanding of the therapeutic potential of this combination.

Phospholipase A_2 (PLA$_2$) activity has been shown to be required for the TNFα-mediated lysis of cells that are sensitive to TNFα, even in the absence of protein synthesis inhibitors[28,52–54]. In an attempt to characterize the mechanism whereby the combination of TNFα and protein synthesis inhibitors increase lysis, we first determined if PLA$_2$ activity was required for the lysis of cells that are only sensitive in the presence of TNFα and protein synthesis inhibitors[28]. As shown in Figure 19.5, the PLA$_2$ inhibitors 4-bromophenacylbromide and quinacrine block the lysis of ovarian cancer cells exposed to TNFα and emetine, indicating that the activity of this enzyme is required for lysis of these cells in the presence of TNFα and inhibitors of protein synthesis.

Because PLA$_2$ mediates the enzymatic release of arachidonic acid from membrane phospholipids, the activity of PLA$_2$ can also be measured indirectly by the release of radiolabelled material from cells prelabelled with ^{3}H-arachidonic acid[54]. Although pre-labelled OVCAR3 cells and SKOV3 cells did not release radiolabelled material when exposed only to TNFα or only protein synthesis inhibitors, they released radiolabelled material when exposed to TNFα in the presence of protein synthesis inhibitors (Figure 19.6). The mechanical disruption (freeze/thaw) of these cells did not release ^{3}H-arachidonic acid, indicating that the release of radiolabelled material is enzymatically mediated. The fact that PLA$_2$ was not activated unless protein synthesis was inhibited indicates that these resistant cells synthesize a protein(s) that prevents the activation of PLA$_2$. In the presence of protein synthesis inhibitors this protein(s) is not made, and as a result PLA$_2$ is activated and the cells are lysed. Given that PLA$_2$ has been shown to be activated by TNFα in cells sensitive to TNFα, it seems reasonable to assume that the activation of PLA$_2$ in the presence of TNFα

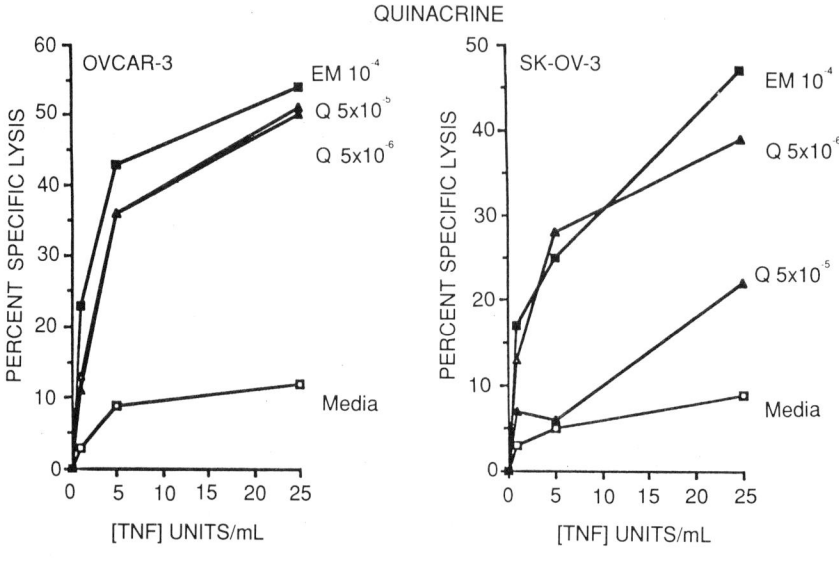
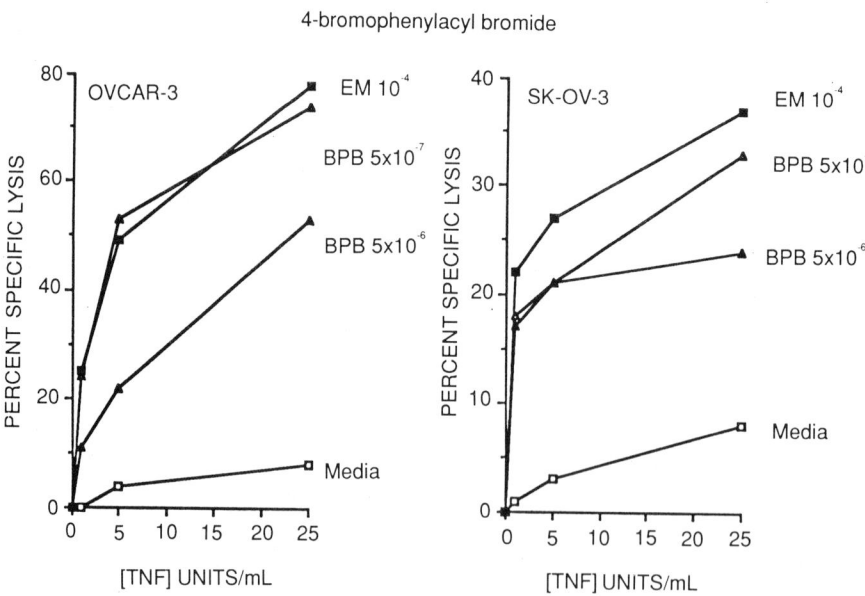

Figure 19.5 The effect of quinacrine and 4-bromophenylacyl bromide on the lysis of ovarian cancer cell lines in the presence of the protein synthesis inhibitor emetine and TNF-α. Percent specific lysis of ovarian cancer cell lines is plotted as a function of the concentration of TNF-α in an 18-hour ^{51}Cr release assay. Cells were exposed to only media and TNF-α, or TNF-α and emetine, or TNF-α, emetine and quinacrine (Q) or 4-bromophenylacyl bromide (BPB) at the indicated molar concentrations.

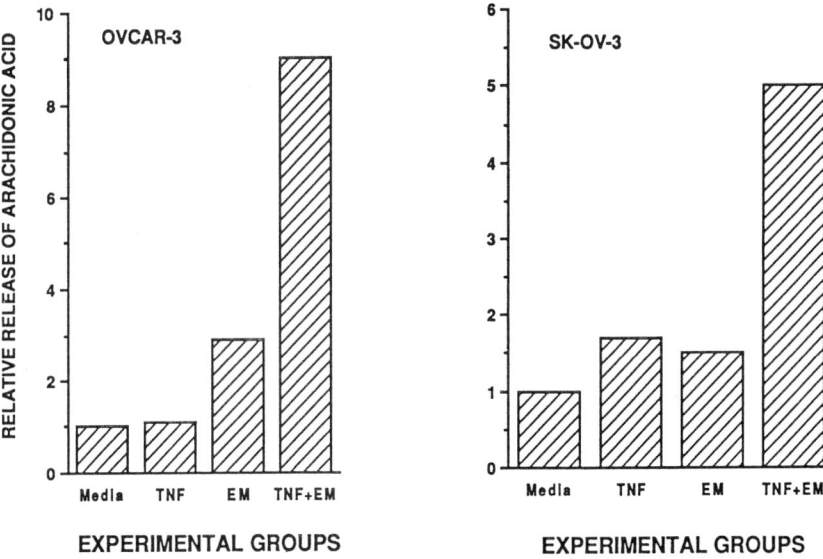

Figure 19.6 The effect of the protein synthesis inhibitor emetine and TNF-α on the release of arachidonic acid and/or its metabolites from ovarian cancer cell lines. Cells were prelabelled with [^{3}H]arachidonic acid then exposed to media, TNF-α (25 U/ml), emetine 10^{-4}M, or TNF-α and emetine (TNF+EM) at the same concentrations. The release of radiolabelled material was measured 16 hours later. The release of radioactive material from cells exposed only to media was normalized to 1 and all other experimental groups were compared to this.

and protein synthesis inhibitors is mediated by TNFα.

There is evidence to suggest that the mechanism by which TNFα mediates lysis involves the generation of toxic oxygen radicals [55,56]. These are thought to be generated during the metabolism of arachidonic acid[57–59] once it has been released from membrane phospholipids by PLA$_2$. In support of this, resistance to lysis by TNFα has been shown to be associated with the expression of superoxide dismutase, an enzyme capable of detoxifying oxygen radicals[10]. Clearly, resistance which results from the expression of superoxide dismutase would be sensitive to protein synthesis inhibitors. However, it would function at a step in the TNFα lytic mechanism that occurs after the release of arachidonic acid from membrane phospholipids (i.e. after the activation of PLA$_2$). The resistance mechanism described here is also sensitive to protein synthesis inhibitors. However, this resistance mechanism must function at a step in the TNFα lytic mechanism that occurs before the activation of PLA$_2$. The two resistance mechanisms are not mutually exclusive and both may be operative in some or all ovarian cancer cells (Figure 19.7).

The fact that inhibition of protein synthesis in many ovarian and cervical carcinoma cell lines permits them to be lysed by TNFα suggests that this mechanism of resistance to TNFα may be common among gynaecological malignancies. Therefore, the therapeutic potential of TNFα as an anticancer agent may be increased when combined with inhibitors of protein synthesis. This synergistic effect of TNFα and chemotherapeutic agents needs to be confirmed in an animal model. Then serious consideration should be given to human trials. As more information becomes

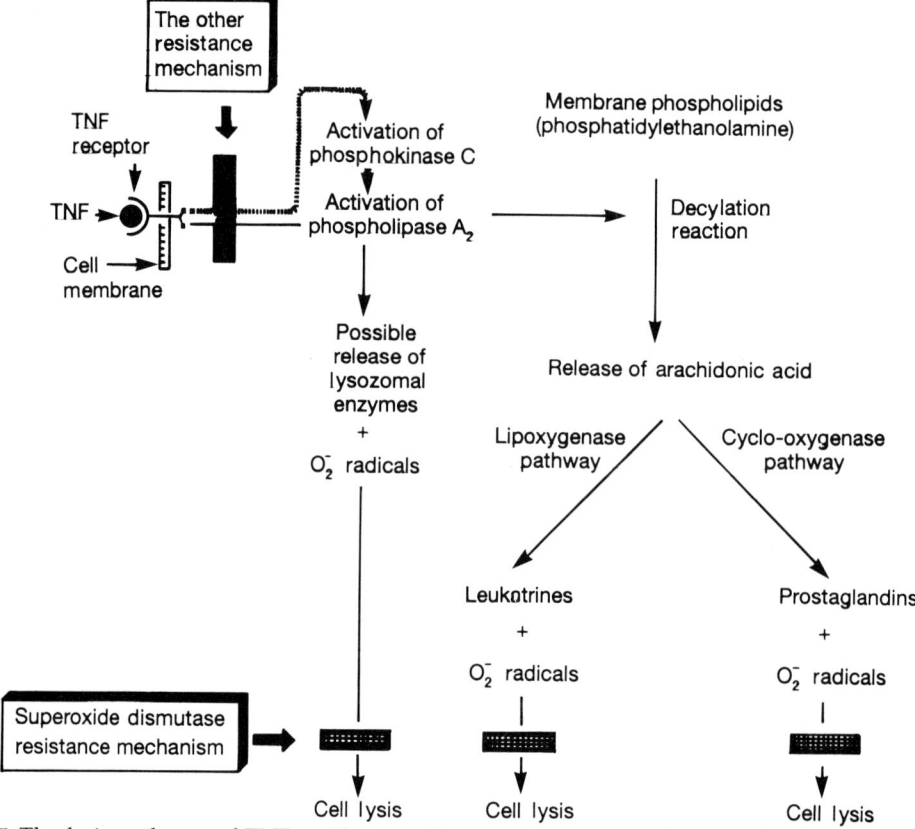

Figure 19.7 The lytic pathway of TNF-α. The possible resistance mechanisms are highlighted.

available on the mechanism of action of TNFα and the mechanism of resistance, it may be possible to identify agents that would specifically inhibit these resistance mechanisms.

REFERENCES

1. Carswell, E.A., Old, L.J., Kasse, R.L. et al. (1975) An endotoxin induced serum factor that causes necrosis of tumors *Proc. Natl Acad. Sci. USA*, **72**, 3666–70.
2. Pennica, D., Nedwin, G.E., Hayflick, J.S. et al. (1984) Human tumor necrosis factor: precursor structure, expression and homology to lymphotoxin. *Nature*, **312**, 721–9.
3. Aggarwal, B.B., Kohr, W.J., Hass, P.E. et al. (1985) Human tumor necrosis factor. *J. Biol. Chem.*, **260**, 2345–54.
4. Williamson, B.D., Carswell, E.A., Rubin, B.Y. et al. (1983) Human tumor necrosis factor produced by human B-cell lines; synergistic cytotoxic interaction with human interferon. *Proc. Natl Acad. Sci. USA*, **80**, 5397–401.
5. Urban, J.L., Shepard, H.M., Rothstein, J.L. et al. (1987) Tumor Necrosis Factor: A potent effector molecule for tumor cell killing by activated macrophages. *Proc. Natl Acad. Sci. USA*, **83**, 5233–7.
6. Shirai, T., Yamaguchi, H., Ito, H. et al. (1985) Cloning and expression in *Escherichia coli* of the gene for human tumor necrosis factor-α. *Nature*, **313**, 803–6.
7. Wang, A.M., Creasey, A.A., Ladner, M.B. et al. (1985) Molecular cloning of the complementary DNA for human tumor necrosis factor. *Science*, **228**, 149–51.
8. Fransen, L., Van der Heyden, J., Ruuyss-

chaert, R. and Fiers, W. (1986) Recombinant tumor necrosis factor: Its effects and its synergism with interferon-γ on a variety of normal and transformed human cell lines. *Eur. J. Clin. Oncol.*, **22**, 419–26.

9. Leibovich, S.J., Polverini, P.J., Shepard, H.M. *et al.* (1987) Macrophage induced angiogenesis is mediated by tumor necrosis factor (TNFα). *Nature*, **329**, 630–2.

10. Aggarwal, B.B., Traquina, P.R. and Eessalu, T.E. (1986) Modulation of receptors and cytotoxic response of tumor necrosis factor-α by various lectins. *J. Biol. Chem.*, **261**, 13652–6.

11. Haranaka, K. and Satomi, N. (1981) Cytotoxic activity of tumor necrosis factor (TNF) on human cancer cells *in vitro*. *Jpn. J. Exp. Med.*, **51**, 191–7.

12. Ruggerio, V., Latham, K. and Baglioni, C. (1987) Cytostatic and cytotoxic activity of tumor necrosis factor on human cancer cells. *J. Immunol.*, **138**, 2711–18.

13. Sugarman, B.J., Aggarwal, B.B., Hass, P.E. *et al.* (1985) Recombinant human tumor necrosis factor-α: effects on proliferation of normal and transformed cells *in vitro*. *Science*, **230**, 943–51.

14. Krosnick, J.A., McIntosh, J.K., Mule, J.J. and Rosenberg, S.A. (1989) Studies of the mechanisms of toxicity of the administration of recombinant tumor necrosis factor-α in normal and tumor bearing mice. *Cancer Immunol. Immunother.*, **30**, 133–8.

15. Asher, A.L., Mule, J.J., Reichert, C.M. *et al.* (1987) Studies of the antitumor efficacy of systematically administered tumor necrosis factor against several murine tumors *in vivo*. *J. Immunol.*, **138**, 963–70.

16. Alexander, R.B., Isaacs, J.D. and Coffey, D.S. (1987) Tumor nerosis factor enhances the *in vivo* and *in vitro* efficacy of chemotherapeutic drugs targeted at DNA topoisomerase II in the treatment of murine bladder cancer. *J. Urol.*, **138**, 427–38.

17. Blick, M., Sherwin, S.A., Rosenblum, M. and Gutterman, J. (1987) Phase I study of recombinant tumor necrosis factor in cancer patients. *Cancer Res.*, **47**, 2986–94.

18. Sherman, M.L., Spriggs, D.L., Arthur, K.A. *et al.* (1988) Recombinant tumor necrosis factor administered as a five-day continuous infusion in cancer patients: Phase I toxicity and effects on lipid metabolism. *J. Clin. Oncol.*, **6**, 344–50.

19. Creaven, P.J., Plager, J.E., Dupere, S. *et al.* (1987) Phase I clinical trial of recombinant tumor necrosis factor. *Cancer Chemother. Pharmacol.*, **20**, 137–45.

20. Shepard, M.R. and Lewis, G.D. (1988) Resistance of tumor cells to tumor necrosis factor. *J. Clin. Immunol.*, **8**, 333–41.

21. Powell, C.B., Mutch, D.G., Massad, L.S. *et al.* (1990) Common expression of a tumor necrosis factor resistance mechanism among gynecologic malignancies. *Cancer Immunol. Immunother.*, **32**, 131–6.

22. Darzynkiewicz, Z., Carter, S.P. and Old, L.J. (1986) Effects of recombinant tumor necrosis factor on HL-50 cells: Cell cycle specificity and synergism with actinomycin-D. *J. Cell. Physiol.*, **130**, 328–41.

23. DeFillippi, P., Poupart, P., Tavernier, J. *et al.* (1987) Induction and regulation of mRNA encoding 26-kDa protein in human cell lines treated with recombinant human tumor necrosis factor. *Proc. Natl Acad. Sci. USA*, **83**, 4557–69.

24. Kirstein, M. and Baglioni, C. (1986) Tumor necrosis factor induces synthesis of two proteins in human fibroblasts. *J. Biol. Chem.*, **261**, 9565–78.

25. Laster, S.M., Wood, J.G. and Gooding, L.R. (1988) Tumor necrosis factor can induce both apoptic and necrotic forms of cell lysis. *J. Immunol.*, **141**, 2629–40.

26. Mutch, D.G., Massad, L.S., Kao, M.S. and Collins, J.L. (1990) Proliferative and antiproliferative effects of interferon-γ and tumor necrosis factor-α on cell lines derived from cervical and ovarian malignancies. *Am. J. Obstet. Gynecol.*, **163**, 1920–4.

27. Massad, L.S., Mutch, D.G., Kao, M.S. *et al.* (1991) Inhibition of protein synthesis enhances the lytic effects of tumor necrosis factor α and interferon γ in cell lines derived from gynaecological malignancies. *Cancer Immunol. Immunother.* (in press).

28. Mutch, D.G., Powell, C.B., Kao, M.S. and Collins, J.L. Resistance to lysis by TNFα in malignant gynecologic cell lines is associated with the expression of protein(s) that prevent

the activation of phospholipase A_2 by TNFα. *Cancer Res.* (submitted).
29. Collins, J.L., Patek, P.Q. and Cohn, M. (1981) Tumorigenicity and lysis by natural killers. *J. Exp. Med.*, **153**, 89–99.
30. Patek, P.Q., Lin, Y. and Collins, J.L. (1987) Natural cytotoxic cells and tumor necrosis factor activate similar lytic mechanisms. *J. Immunol.*, **138**, 1641–6.
31. Brauer, S., Patek, P.Q., Collins, J.L. and Cohn, M. (1984) Different effectors and effector mechanisms are involved in the natural cell mediated lysis of lymphoid and fibroblast targets, in *UCLA Symposium of Molecular and Cellular Biology. B- and T-cell tumors*, Vol.24, UCLA, Los Angeles, pp.155–9.
32. Ortaldo, J.R., Mason, L.H., Mathieson, B.J. *et al.* (1986) Mediation of mouse natural cytotoxic activity by tumor necrosis factor. *Nature*, **321**, 700–3.
33. Herberman, R.B. and Holden, H.T. (1978) Natural cell mediated immunity. *Adv. Cancer Res.*, **27**, 305–77.
34. Haller, O., Hanson, M., Kiessling, R. and Wigzell, H. (1977) Role of nonconventional natural killer cells in resistance against syngeneic tumor cells *in vivo*. *Nature*, **270**, 609–11.
35. Kiessling, R.E., Klein, E., Pross, H. and Wigzell, H. (1975) 'Natural' killer cells in the mouse. II. Cytotoxic cells with specificity for mouse Moloney leukemia cells. Characteristics of the killer cell. *Eur. J. Immunol.*, **5**, 117–36.
36. Henkart, P.A. (1985) Lymphocyte mediated cytotoxicity. *Annu. Rev. Immunol.*, **3**, 31–58.
37. Stutman, O. and Lattime, E. (1981) Natural cell mediated cytotoxicity against tumors in mice: a heterogeneous system. *Transplant. Proc.*, **13**, 752–61.
38. Patek, P.Q., Collins, J.L. and Cohn, M. (1983) Evidence that cytotoxic T cells and natural cytotoxic cells use different lytic mechanisms to lyse the same targets. *Eur. J. Immunol.*, **13**, 433–6.
39. Patek, P.Q., Lin, Y., Collins, J.L. and Cohn, M. (1986) *In vivo* or *in vitro* selection for resistance to natural cytotoxic cell lysis selects for variants with increased tumorigenicity. *J. Immunol.*, **136**, 741–55.
40. Stutman, O. and Cuttito, M.J. (1981) Normal levels of natural cytotoxic cells against solid tumors in NK deficient beige mice. *Nature*, **290**, 254–5.
41. Collins, J.L., Kao, M.S. and Patek, P.Q. (1987) Humans express natural cytotoxic (NC) cell activity that is similar to murine NC cell activity. *J. Immunol.*, **138**, 4180–4.
42. Patek, P.Q., Collins, J.L. and Cohn, M. (1978) Transformed cell lines susceptible or resistant to *in vivo* surveillance against tumorigenesis. *Nature*, **276**, 510–11.
43. Collins, J.L., Patek, P.Q. and Cohn, M. (1982) *In vivo* surveillance of tumorigenic cells transformed *in vitro*. *Nature*, **299**, 169–71.
44. Lin, Y., Patek, P.Q., Collins, J.L. and Cohn, M. (1985) Analysis of immune surveillance of sequentially derived cell lines that differ in their tumorigenic potential. *J. Natl Cancer Inst.*, 1025–30.
45. Collins, J.L. and Kao, M.S. (1989) The anticancer drug cisplatin increases the naturally occurring cell-mediated lysis of tumor cells. *Cancer Immunol. Immunother.*, **29**, 17–22.
46. Rosenberg, B. (1980) Cisplatin: Its history and possible mechanism of action, in *Cisplatin current status and new developments*, Academic Press, New York, p.9.
47. Schlaefli, E., Ehrke, M.J. and Mihich, E. (1983) The effects of dichlorao-*trans*-hihydroxy-*bis* isopropyl-amine-platinum IV on the primary cell mediated response. *Immunopharmacology*, **6**, 107–22.
48. Stutman, O. and Cuttito, M.J. (1980) in *Natural Cell Mediated Immunity against Tumours*. (ed. R.B. Herberman), Academic Press, New York. pp.431–42.
49. Vanderslice, W.E. and Collins, J.L. (1991) Differences in the tumor necrosis factor-α-mediated lysis by fixed natural cytotoxic cells and fixed cytotoxic macrophages. *J. Immunol.*, **146**, 156–61.
50. Mutch, D.G., Powell, C.B., Kao, M.S. and Collins, J.L. (1989) *In vitro* analysis of the anticancer potential of tumor necrosis factor in combination with cisplatin. *Gynecol. Oncol.*, **34**, 328–33.
51. Kao, J.W. and Collins, J.L. (1989) A rapid *in vitro* screening system for the identification and evaluation of anticancer drugs. *Cancer Invest.*, **7**, 303–11.

52. Palombella, V. and Vilacek, J. (1989) Mitogenic and cytotoxic actions of tumor necrosis factor in BALB/c 3T3 cells: Role of phospholipase activation. *J. Biol. Chem.*, **264**, 18128–36.
53. Neale, M.L., Fiera, R.A. and Matthews, N. (1988) Involvement of phospholipase A_2 activation in tumour cell killing by tumour necrosis factor. *Immunology*, **64**, 81–5.
54. Knauer, M.F., Longmuir, K.J., Yamamoto, R.S. *et al.* (1990) Mechanism of human lymphotoxin and tumor necrosis factor induced destruction of cells *in vitro*: Phospholipase activation and deacylation of specific membrane phospholipids. *J. Cell. Physiol.*, **142**, 469–79.
55. Wong, G.H.W., Elwell, J.H., Oberly, L.W. and Goeddel, D.V. (1989) Manganous superoxide dismutase is essential for cellular resistance to cytotoxicity of tumour necrosis factor. *Cell*, **58**, 923–30.
56. Watanabe, N., Niitsu, Y., Neda, H. *et al.* (1988) Cytocidal mechanism of TNF: effects of lysosomal enzyme and hydroxyl radical inhibitors on cytotoxicity. *Immunopharmacol. Immunotoxicol.*, **10**, 109–16.
57. Ida, E.A., Sakata, M., Tominaga, H. *et al.* (1988) Arachidonic acid release is closely related to the Fcγ receptor-mediated superoxide generation in macrophages. *Microbiol. Immunol.*, **32**, 1127–43.
58. Schultz, R.M., Nanda, S.K.W. and Altom, M.G. (1985) Effects of various inhibitors of arachidonic acid oxygenation on macrophage superoxide release and tumoricidal activity. *J. Immunol.*, **135**, 2040–4.
59. Suga, K., Kawasaki, T., Blank, M. and Snyder, F. (1990) An arachidonyl (polyenoic)-specific phospholipase A_2 activity regulates the synthesis of platelet-activating factor in granulocytic HL-60 cells. *J. Biol. Chem.*, **265**, 12363–71.

Chapter 20

Chemo-immunotherapy with a combination of low-dose cisplatin and recombinant interleukin 2

M. BERNSEN, H.F.J. DULLENS, W. DEN OTTER
and S.P.M. HEINTZ

20.1 INTRODUCTION

Successful treatment of patients with metastatic ovarian cancer is a problem. Despite aggressive surgery and chemotherapy prognosis for these patients is still poor. The 5-year survival is less than 25%[1]. The combination of cytoreductive surgery with platinum-based combination chemotherapy has given the best results[2]. The main reason for this limited success is the high incidence of recurrent disease and the development of drug resistance by the tumour. New strategies for the treatment of ovarian cancer patients are therefore necessary. Efforts in this direction have involved studies and trials with combinations of cisplatin with other chemotherapeutic agents, use of pharmacological techniques for decreasing the toxicity of cisplatin, newly developed cisplatin analogues with supposed lower toxicities and alternative ways of administration (intraperitoneal)[2]. The aims of these studies and trials are mainly directed towards overcoming the problems of drug resistance and/or enabling dose escalations of cisplatin, in order to achieve therapeutic results. A different approach is the use of immunotherapy. Immunotherapy is a treatment modality that has been very much in focus for the past decade, with a dominant interest for recombinant interleukin 2 (rIL-2). IL-2 is a 15.5-kDa glycoprotein with many immune regulatory functions. Lymphocytes activated by specific antigen express IL-2 receptors which can interact with IL-2, resulting in proliferation of the cells and an immune response[3]. Recombinant IL-2 therapy, whether or not combined with immune lymphoid cells, was very successful in several animal studies[4]. Its application in humans however, has only been relatively successful, and less than hoped or anticipated. Except for renal cell carcinoma and malignant melanoma, immune therapy with rIL-2 has not resulted in better responses in other forms of cancer when compared to the responses obtained with conventional therapy[5]. In addition rIL-2 is very toxic at high doses. Dosages used are often close to the maximum tolerated dose. These are however not necessarily identical to the maximum therapeutic dose as is evidenced by studies in animals. In these studies locoregional application of low-dose rIL-2 resulted in effective antitumour responses[6,7]. Apparently 'high-dose' and 'low-dose' rIL-2 therapy differ in their mechanism. 'High-dose' rIL-2 mainly induces a non-specific reaction whereas 'low-dose' administration induces a specific immune reaction.

Like most chemotherapeutic agents, cisplatin has immunosuppressive side-effects. There are, however, indications that cisplatin has other immune-modulating effects too.

1. Despite the fact that there is no selective uptake of cisplatin by tumour cells and the fact that cisplatin shows a fast clearance shortly after injection, large tumour burdens are eradicated at dose levels of cisplatin that are considered too low to kill all tumour cells in a direct way[8].
2. When cisplatin is administered to immunosuppressed animals, the number of cures are reduced compared to immunocompetent animals in an identical system [9].
3. Mice 'cured' of large tumour burdens after treatment with cisplatin are immune to the tumour, as they resist a second challenge with tumour cells[10].

Studies on the immune-modulating properties of cisplatin show the following.

1. Cisplatin enhances natural killer cell cytotoxicity and spontaneous monocyte-mediated cytotoxicity[11].
2. Cisplatin is able to increase the antigenicity of tumour cells and/or increase their susceptibility to immune effector cells [12].
3. Cisplatin has a potentiating effect on T-cell effector functions as it increases the 24-hour delayed-type hypersensitivity reaction[13].

Based on these studies and our similar experience there is reason to believe that therapy with combinations of low doses of these agents can be a valuable option for improved therapeutic results, yet with less toxicity.

20.2 THE EFFECT OF LOW-DOSE rIL-2 AND LOW-DOSE CISPLATIN THERAPY IN THE SYNGENEIC MURINE TUMOUR MODEL DBA/2-SL2

SL2 is a DBA/2-derived lymphoma (H2d). It is sensitive to low-dose rIL-2 therapy. When mice are inoculated intraperitoneally with 2×10^4 SL2 cells on day 0, followed by daily intraperitoneal (i.p.) injections of 60 000 International Units (IU) rIL2 (Proleukin) from day 10 to 14, an average cure rate of 55% ($n = 55$) of the mice is obtained. Mice that receive only control solutions during treatment all die of disseminated tumour around day 17. SL2 is also sensitive to cisplatin, since a single i.p. injection of cisplatin (Platinol) at dose levels of ≥ 1.0 mg/kg given 2 days after i.p. inoculation of 2×10^4 SL2 cells results in significant extension of survival time of the mice and of complete cures (Figure 20.1).

When therapeutically suboptimal doses of cisplatin and rIL-2 are used in combination, an additive/synergistic effect on tumour eradication is obtained: mice were inoculated i.p. with 2×10^4 SL2 cells on day 0, and subsequently treated with a single injection of 1 mg/kg cisplatin on day 2, 'combined' with daily injections of rIL-2 at 5000, 20 000, or 60 000 IU from day 10 to 14. When mice were treated with 1 mg/kg cisplatin only, 6% of the mice were cured. The cure rates of mice treated with 5000, 20 000, 60 000 IU rIL-2 were 0%, 9% and 50% respectively. However, when these doses of rIL-2 were combined with a single injection of 1 mg/kg cisplatin the percentages of mice cured

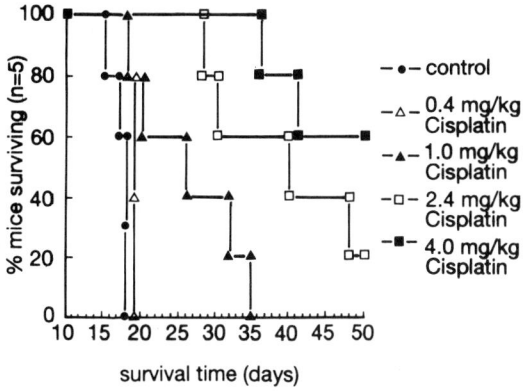

Figure 20.1 Antitumour effect of cisplatin in SL2-bearing mice. Groups of mice were inoculated with 2×10^4 SL2 cells i.p. on day 0, followed by injection on day 2 with cisplatin at various doses. Control mice were injected with PBS.

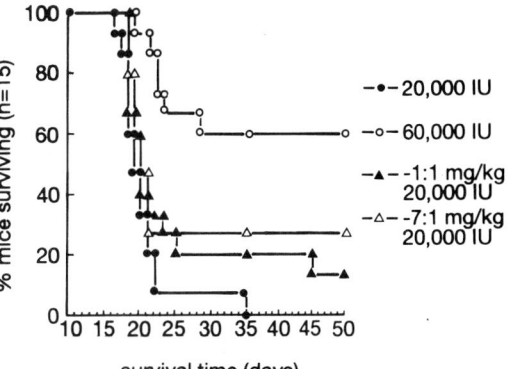

Figure 20.2 Potentiation of the immune system by cisplatin. Before tumour inoculation, groups of mice were injected with a 1 mg/kg cisplatin. This was done either 7 days (-7) or 1 day (-1) before tumour inoculation. Mice were also treated with 20 000 iu rIL-2 from day 10–14. Control groups consisted of mice treated with PBS or either 20 000 or 60 000 iu rIL-2 from day 10–14. Mice treated with control solutions only all died between days 17–19 (not shown).

increased to 25%, 34% and 65% respectively (Table 20.1). During treatment of the mice no adverse effects due to treatment were observed.

In order to be able to tell whether cisplatin does have direct effects on the immune system that contribute to its synergistic activity with rIL-2, we injected naive DBA/2 mice with 1 mg/kg cisplatin, seven or one day prior to challenge 2×10^4 SL2 cells. These mice then received treatment with 20 000 or 60 000 IU rIL-2 from day 10 to 14. Effects seen in these experiments could then not be ascribed to direct effects of cisplatin on tumour cells, as there were no tumour cells present at the time that cisplatin was injected. Therefore the effects seen had to be due to the direct influence of cisplatin on the immune system. The results of these experiments indicated a potentiating effect of cisplatin on rIL-2 therapy given from day 10 to 14 (Figure 20.2).

Mice that were cured of SL2 tumour were tested for their immunity against the tumour by rechallenging them intraperitoneally with 1×10^6 SL2 cells (Table 20.2). Cured mice that had been treated either with rIL-2 or with 1 mg/kg cisplatin (day 2) alone were immune, as were mice that had been treated with combination therapy. Mice that had received an injection of cisplatin seven or one day before tumour inoculation (respectively day -7 or -1), followed by treatment with rIL-2 from day 10 to 14 were also immune. An

Table 20.1 Synergistic antitumour effect of cisplatin and rIL-2 on disseminated SL2 tumour. On day 0 mice were inoculated with 2×10^4 SL2 cells i.p. Mice were treated with either 1 mg/kg cisplatin or with 5000, 20 000, or 60 000 iu rIL-2 on days 10–14, or with a combination of 1 mg/kg cisplatin and rIL-2. Control mice were treated with diluent

	% of cured mice	
Dose of rIL-2	No cisplatin	Cisplatin (1 mg/kg)
0	0 ($n = 34$)	6 ($n = 35$)
5 000	0 ($n = 20$)	25 ($n = 20$)
20 000	9 ($n = 35$)	34 ($n = 35$)
60 000	50 ($n = 20$)	65 ($n = 20$)

interesting observation was that only 2 out of the 8 mice that had rejected the tumour after treatment with 5 mg/kg cisplatin (day 2) were immune. Mice that rejected the tumour after treatment with 10 mg/kg (day 2) alone were not immune. So high doses of cisplatin probably damage the immune system.

From the results of the experiments in this tumour model, we can conclude that locoregional treatment of mice with a combination of therapeutically suboptimal and non-toxic doses of cisplatin and rIL-2 results in synergistic antitumour responsiveness, as there is an increase in the percentage of cured animals. Furthermore, cisplatin has a dose-dependent effect on the host's immune system. Whether mice develop an immunity against the tumour during the rejection process following treatment with cisplatin alone, depends on the dose of cisplatin used. There is an apparent shift in the effect at 5 mg/kg cisplatin. Cisplatin seems to have a direct stimulating effect on the immune system, as injection of mice with cisplatin before tumour inoculation has a potentiating effect on subsequently applied therapy with rIL-2.

20.3 THE EFFECTS OF CISPLATIN ON rIL-2 IN THE SYNGENEIC MURINE TUMOUR MODEL DBA/2-P815

P815 is a DBA/2-derived mastocytoma. This tumour line is also sensitive for low-dose rIL-2 therapy. A cure rate of up to 60% can be obtained when mice, after i.p. inoculation of 2×10^4 P815 cells, receive five injections of 60 000 IU rIL-2 from day 10 to 14. However, P815 is resistant to cisplatin as no cures are obtained, even at doses up to 10 mg/kg (Figure 20.3).

We studied the effect of the combination of cisplatin and rIL-2 in P815-bearing mice also. Mice were injected with 1 or 2.5 mg/kg cisplatin on day 2, and 20 000 or 60 000 IU rIL-2 injection from day 10 to 14. In contrast to the effect seen in the DBA/2-SL2 tumour model, we did not observe any synergistic effect. Moreover, the antitumour response obtained with the combination of cisplatin and rIL-2 was lower than the response obtained with rIL-2 alone, as is presented in Figure 20.4. This apparent detrimental effect of cisplatin on the antitumour effect of rIL-2 was most explicit at the dose of 2.5 mg/kg

Table 20.2 Immunity of mice cured of disseminated SL2 tumour. Mice that were cured of disseminated SL2 tumour after treatment with cisplatin or rIL-2 or both, and mice that had received an injection of cisplatin before tumour inoculation and were treated with rIL-2, were rechallenged on day 60 with 1×10^6 SL2 cells. Mice that were able to reject the rechallenge were considered immune

Day of cisplatin	Dose of cisplatin (mg/kg)	Dose of rIL-2	Number immune	Total number
AT	10	0	0	8
AT	5	0	2	7
AT	1	0	3	3
AT	1	5 000	2	2
AT	1	20 000	7	7
AT	1	60 000	5	5
−1	5	60 000	4	4
−1	1	60 000	1	1
−7	5	60 000	4	4
−7	1	20 000	1	1
−7	1	60 000	1	1

Combination therapy of cisplatin and rIL-2 in syngeneic murine tumour model C3HeB/FeJ-MOT

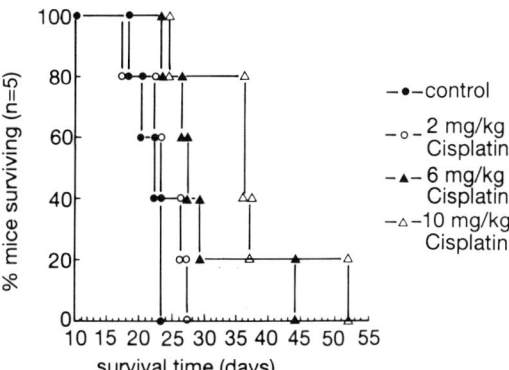

Figure 20.3 Antitumour effect of cisplatin on P815-bearing mice. Groups of mice were inoculated with 2×10^4 P815 cells i.p. on day 0, followed by injection on day 2 with cisplatin at various doses. Control mice were injected with PBS.

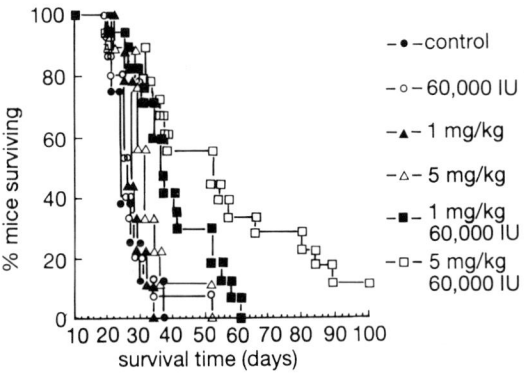

Figure 20.4 Effect of cisplatin on subsequent rIL-2 therapy in P815-bearing mice. Groups of mice were injected with: 2×10^4 P815 cells on day 0; PBS, 1 mg/kg, or 5 mg/kg cisplatin on day 2; and with 60 000 iu rIL-2 on days 10–14. Control mice were treated with diluent.

cisplatin, and thus seems to be dose dependent. It also seems to be related to tumour cell characteristics, as is evident from the contrasting results of the combination therapy in the two models (DBA/2-SL2, DBA/2-P815). The only apparent difference between these two models is the histological type of the tumour.

20.4 COMBINATION THERAPY OF CISPLATIN AND rIL-2 IN THE SYNGENEIC MURINE TUMOUR MODEL C3HeB/FeJ-MOT

MOT is a murine ovarian teratocarcinoma, which spontaneously arose in C3HeB/FeJ mice. In the international literature it is described as a representative model for ovarian cancer in humans. Its biological behaviour resembles the human situation. It was minimally sensitive to rIL-2 therapy as no cures could be obtained. Only occasional extended survival times were observed, of which representative results are shown in Figure 20.5. MOT is also very insensitive for cisplatin

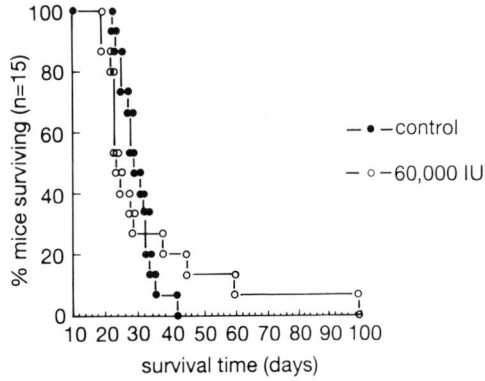

Figure 20.5 Recombinant IL-2 therapy in MOT-bearing mice. C3HeB/FeJ mice inoculated with 2×10^5 MOT cells i.p. on day 0 were treated on days 7–14 with 60 000 iu rIL-2. Control group were treated with diluent.

Mice were inoculated with 2×10^5 MOT cells, on day 4 they were injected i.p. either with 1 or 5 mg/kg cisplatin, followed by treatment with 60 000 IU rIL-2 injections from

193

day 10 to 14 and 17 to 21. Mice that received only cisplatin or rIL-2 did not show significant differences in survival time compared to control mice. However, mice that were treated with the combination showed a significant increase in survival time (Figure 20.6).

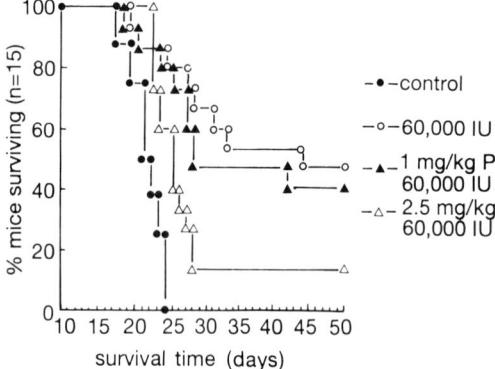

Figure 20.6 Synergistic antitumour effect of cisplatin and rIL-2 in MOT-bearing mice. On day 0, mice were injected with 2×10^5 MOT cells i.p. They were then treated with 1 or 5 mg/kg cisplatin on day 4 or with 60 000 iu rIL-2 on days 10–14, or with a combination of cisplatin and rIL-2. Control mice were treated with diluent.

The conclusion of this experiment is that the combination of cisplatin and rIL-2 exerts a relatively strong antitumour effect in a tumour system that is insensitive to the drugs used independently.

20.5 DISCUSSION

We studied the antitumour effects of low, non-toxic, doses of cisplatin and rIL-2 in three murine tumour models. Although the use of drug considered to be immunosuppressive in combination with a well-known immune stimulant might seem illogical, nonetheless we showed that this combination can be very effective in inducing antitumour responses. In the MOT tumour model, a system that is insensitive to either drug, the combination of cisplatin and rIL-2 resulted in increased survival time and/or long-term survivors.

An interesting observation was that the combination of cisplatin and rIL-2, although very effective in the MOT and SL2 tumour models, was not effective in the P815 model. Moreover, it decreased the responsiveness of P815 to rIL-2 in a dose-dependent way. As already stated the only apparent difference between this system and the SL2 tumour model is the histological type of the tumour. Apparently, the effects of the interaction of cisplatin, the P815 tumour and the immune system differ from the effects of such an interaction in the SL2 system.

We showed that cisplatin affects the immune system. The results of our experiments in which cisplatin was injected before inoculation of the tumour indicate that this can be a direct stimulatory effect. However, cisplatin can also damage the immune system, as can be concluded from the fact that mice treated with higher doses of cisplatin do not develop immunity against the tumour.

We suggest that cisplatin not only exerts direct tumouricidal effects, but also modulates the immune system. It is known that tumours affect the immune system too. Therefore, we hypothesize that the results of combination therapy with cisplatin and rIL-2 will depend on the extent of these separate effects, combined with the immune-modulating effects of IL-2 and their interactions. Detailed analysis of these effects, related to therapeutic efficacy, might form the basis for treatment modality for ovarian cancer patients with better therapeutic results with lower toxicity.

REFERENCES

1. Heintz, A.P.M. (1988) Surgery in advanced ovarian carcinoma: is there proof to show the benefit? *Eur. J. Surg. Oncol.*, **14**, 91–9.
2. Thigpen, J.T., Blessing, J.A., Vance, R.B. and

Lambuth, B.W. (1989) Chemotherapy in ovarian carcinoma: Present role and future prospects. *Semin. Oncol.*, **16**, 58–65.
3. Smith, K.A. (1988) Interleukin-2: Inception, impact and implications. *Science*, **240**, 1169–76.
4. Chang, A.E. and Rosenberg, S.A. (1989) Overview of interleukin-2 as an immunotherapeutic agent. *Semin. Surg. Oncol.*, **5**, 385–90.
5. Rosenberg, S.A., Lotze, M.T., Yang, J.C. *et al.* (1989) Experience with the use of high-dose interleukin-2 in the treatment of 652 cancer patients. *Ann. Surg.*, **210**, 474–85.
6. 6. Rutten, V.P.M.G., Klein, W.R., De Jong, W.A.C. *et al.* (1989) Local interleukin-2 therapy in bovine ocular squamous cell carcinoma. *Cancer Immunol. Immunother.*, **30**, 165–9.
7. Maas, R.A., Dullens, H.F.J., De Jong, W.H. and Den Otter, W. (1989) Immunotherapy of mice with a large burden of disseminated lymphoma with low-dose interleukin-2. *Cancer Res.*, **49**, 7037–40.
8. Hoeschele, J.D. and Van Camp, L. (1972) Whole-body counting and the distribution of cis [Pt(NH$_3$)$_2$-Cl$_2$] in the major organs of Swiss white mice, in *Advances in Antimicrobial and Antineoplastic Chemotherapy*, University Park Press 2, Baltimore, pp.291–2.
9. Conran, P.B. and Rosenberg, B. (1972) The role of the host defences in the regression of sarcoma-180 in mice treatment with cis-dichlorodiammineplatinum (II), in *Advances in Antimicrobial and Antineoplastic Chemotherapy*, University Park Press 2, Baltimore, pp.235–6.
10. Rosenberg, B. (1975) Possible mechanisms for the antitumour activity of platinum coordination complexes. *Cancer Chemother. Rep.*, **59**, 589–98.
11. Sodhi, A., Pai, K., Singh, R.K. and Singh, S.M. (1990) Activation of human NK-cells and monocytes with cisplatin *in vitro*. *Int. J. Immunopharmacol.*, **12**, 893–8.
12. Lichtenstein, A.K. and Pende, D. (1986) Enhancement of natural killer cytotoxicity by cis-diammine-dichloroplatinum (II) *in vivo* and *in vitro*. *Cancer Res.*, **46**, 639–44.
13. Scheper, R.J., Limpens, J., Tan, B.T.G. *et al.* (1987) Immunotherapeutic effects of local chemotherapy with an active metabolite of cyclophosphamide. *Methods Find. Exp. Clin. Pharmacol.*, **9**, 611–15.

Part Four
Pathology, Early Detection and Prognosis

Chapter 21
Early ovarian cancer

R.E. SCULLY

Early ovarian cancer has been defined in several ways: (i) any ovarian cancer that can be completely removed regardless of size or stage[1]; (ii) any stage IA ovarian cancer less than 5 cm in diameter; (iii) any ovarian cancer that is neither perceptible by the surgeon on examination of the ovary at operation nor visible to the pathologist on examination of the external and sectioned surfaces of the ovary[2]. This chapter will discuss the third group of early ovarian cancers, specifically these of surface epithelial lineage.

Early ovarian cancers of surface epithelial origin can be divided into two subtypes: (i) those that arise *de novo* from the surface epithelium or its inclusion cysts and (ii) those that are present within a pre-existing benign epithelial lesion either non-neoplastic or neoplastic.

Only a few examples of *de novo* epithelial cancers have been described in the literature[3–6]. The reasons for the paucity of reports include: (i) the usual destruction of the ovarian surface epithelium as a result of drying or rubbing by the surgeon, the pathologist, or both and (ii) inattention to the morphological details of the surface epithelium and its inclusion cysts by the pathologist. Dysplasia or carcinoma involving the surface epithelium, its inclusion cysts, or both have been reported after the detection of atypical or malignant epithelial cells in cul-de-sac puncture specimens[3,4], on examination of ovaries from women with a familial history of ovarian cancer, or on retrospective study of previously removed, apparently normal ovaries after a diagnosis of peritoneal carcinomatosis has been made[5,6]. These studies are biased toward the detection of surface or superficial lesions because of the indications for performing the investigations.

In an as yet incomplete review of my consultation files and the files of the Massachusetts General Hospital, I have identified six cases of early ovarian carcinoma, two of them in patients with a known family history of ovarian cancer. The largest of these tumours was in the interior of the ovary, was 0.8 cm in diameter, and was not appreciated by the pathologist on sectioning of the ovary. Four of the six carcinomas involved the surface of the ovary (Figures 21.1 and 21.2), and two of the patients with surface involvement had a known family history of ovarian cancer. Although the series is small the presence of surface involvement in four of the six cases raises the question whether an appreciable number of ovarian carcinomas involve the surface very early in their development and evolve rapidly into stage III tumours instead of passing gradually from stage I to stage III. Also, the retrospective

Early ovarian cancer

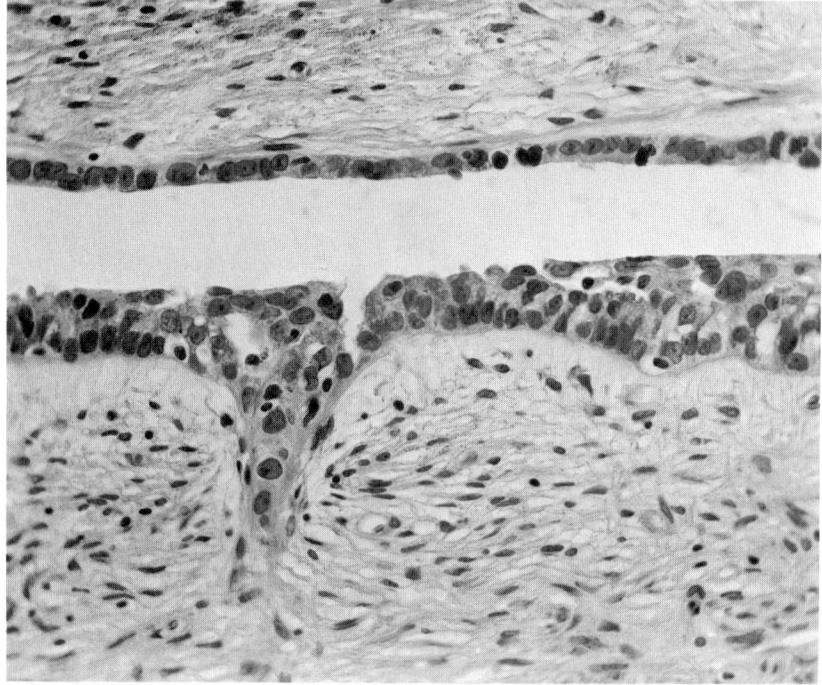

Figure 21.1 Crevice on surface of ovary. The epithelium lining one border of the crevice (above) appears normal; the epithelium lining the other border (below) is lined by carcinoma.

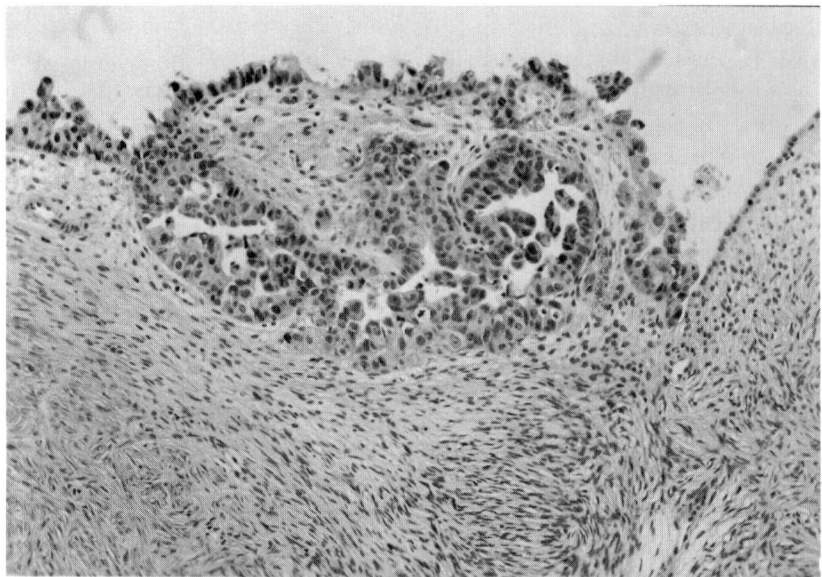

Figure 21.2 Another area from the surface of the ovary illustrated in Figure 21.1. The malignant surface epithelium is invading the underlying stroma.

Early ovarian cancer

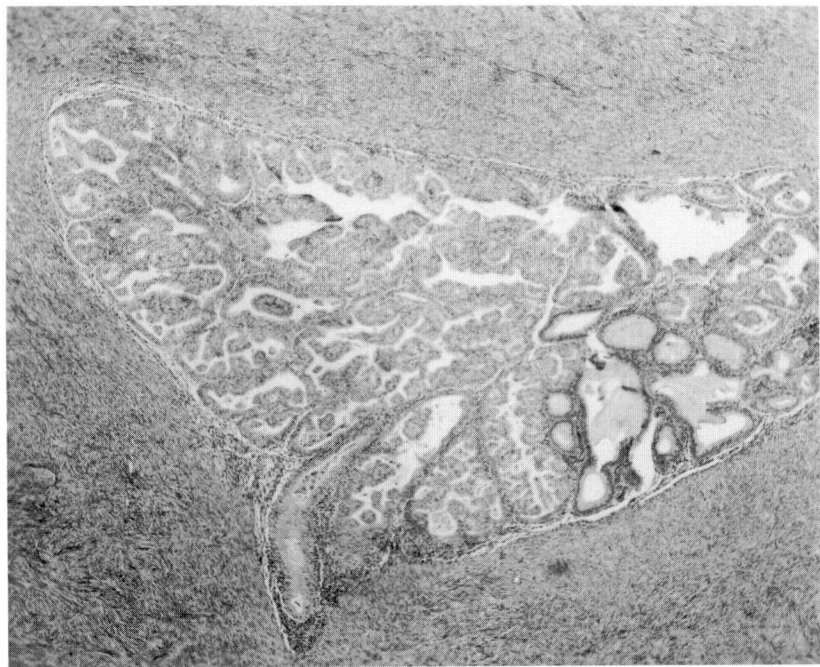

Figure 21.3 Island of endometriosis within ovarian stroma that has been largely replaced by grade 1 adenocarcinoma.

finding of microscopic carcinomas in prior oophorectomy specimens of patients presenting with presumed primary peritoneal of carcinomatosis[5,R.E. Scully, personal observation] raises the question of how many cases of the latter are correctly interpreted as such.

Obviously, much more investigation of the morphological genesis of ovarian cancer of surface epithelial type is warranted. Both surgeons and pathologists must be urged to attempt to preserve the surface epithelium for microscopic study in order to extend our limited knowledge of early ovarian cancer.

The most common non-neoplastic disorder of the ovary that is occasionally associated with the development of early carcinoma is endometriosis. It has been known for many years that a variety of malignant tumours can arise within ovarian endometriosis[7,8]. The most frequent of these are endometrioid carcinoma and clear cell carcinoma, but other types of epithelial cancer as well as endometrioid sarcomas and malignant müllerian mixed tumours have also been reported to originate rarely in endometriosis. Although such tumours are typically recognizable grossly, occasionally they are detected only on microscopic examination (Figures 21.3 and 21.4). It is well known that endometrioid carcinoma of the endometrium can evolve through a sequence of cystic hyperplasia and atypical hyperplasia, which can be demonstrated in successive biopsy specimens over an interval of time[9]. Although such a sequence is rarely observed in the ovary, the occasional observation of similar precancerous lesions in ovarian endometriosis and their occasional coexistence with carcinoma with transitions between the atypical hyperplasia and the carcinoma strongly suggest that some carcinomas arising in endometriosis evolve from hyperplasia[6,10–12].

Early ovarian cancer

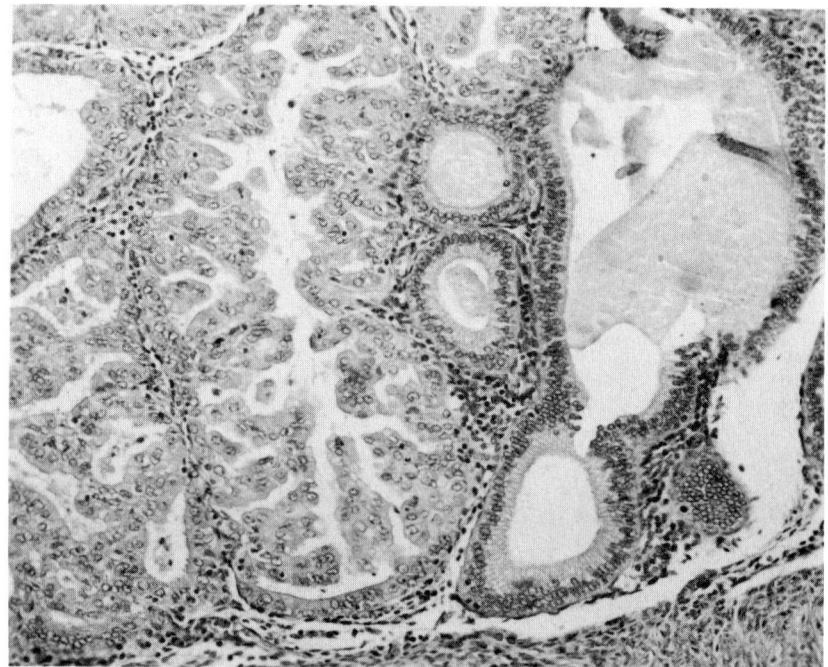

Figure 21.4 High power view of Figure 21.3 showing benign glands on the right and carcinomatous glands on the left.

Endometrioid carcinomas of the ovary may occur synchronously with carcinoma of the endometrium and may have a similar background of continuous, unopposed oestrogenic stimulation.

Ovarian cancers of epithelial type may also arise in association with benign epithelial tumours of the same cell type; similarly borderline tumours may be found in association with benign tumours, and carcinomas with borderline tumours of the same cell type (Figure 21.4)[6]. Unfortunately, as in cases of endometriosis and related carcinomas, it is impossible to follow sequentially the evolution of a benign epithelial tumour into a borderline tumour and thence into an invasive tumour. Two types of evidence, however, suggest such an evolution. A compilation of the results of five series of cases of serous and mucinous ovarian tumours totalling 2320 cases has shown that the benign forms of these tumours occurred at an average age of 44 years, the borderline forms at 48 years, and the carcinomas at 56 years[13–17]. The second form of evidence is the frequent coexistence of benign, borderline and invasive neoplasia in varying proportions in the same specimen (Figure 21.5). This finding is most common in cases of mucinous carcinoma. Most often in such cases gradual transitions are observed between the benign and borderline areas as well as between the borderline and frankly malignant components. The question arises whether the benign borderline and carcinomatous-appearing elements evolved simultaneously or whether the more malignant element arose from the less malignant component. The possibility of the former explanation is suggested by the observation that metastatic mucinous carcinomas from the gastrointestinal tract that appear uniformly carcinoma-

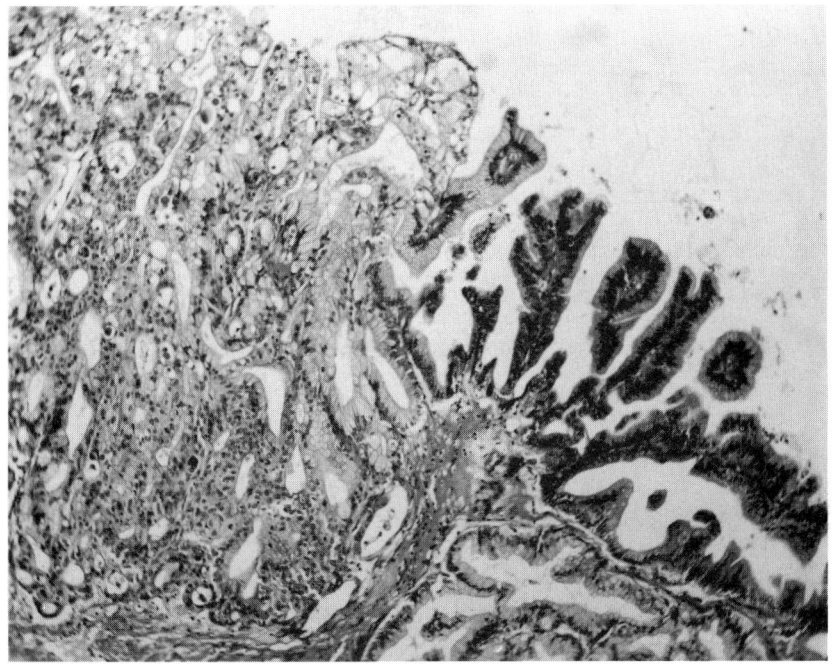

Figure 21.5 Borderline mucinous cystic tumour forming papillae on the right merging with adenocarcinoma on the left and below.

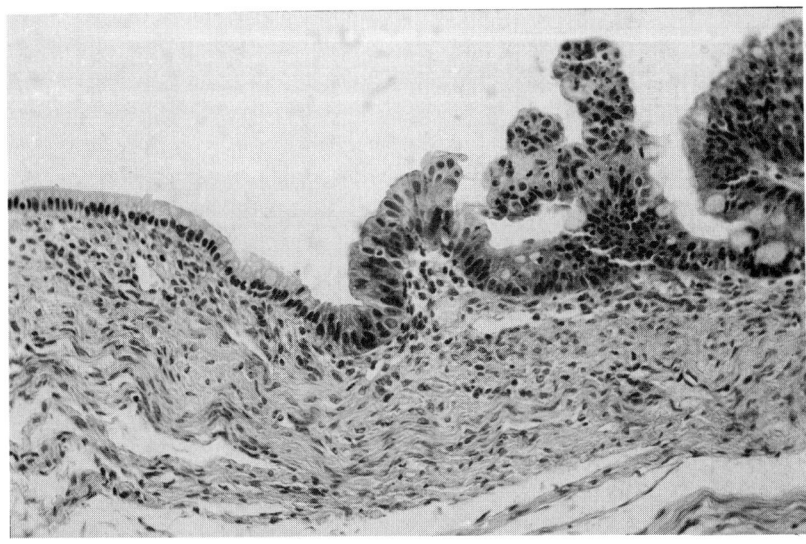

Figure 21.6 Mucinous carcinoma on the right arising abruptly from benign mucinous epithelium on the left.

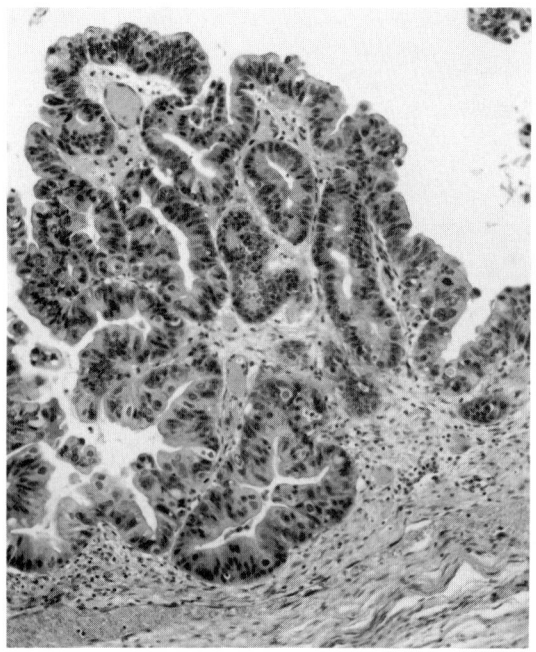

Figure 21.7 Closeup view of the carcinoma near its centre.

tous at the primary site often differentiate focally at the ovarian metastatic site into tumour with the morphological features of borderline or even benign neoplasia. The much less common situation in which an ovarian carcinoma arises as single or multiple discrete foci in an otherwise benign epithelial tumour is more suggestive of an origin of the carcinoma in a previously benign epithelial tumour (Figures 21.6 and 21.7). Unfortunately, there have been very few studies of the frequency of association of benign, borderline and invasive epithelial tumours. However, in one investigation[18], 24% of ovarian carcinomas were said to contain borderline components; almost all the carcinomas with borderline elements were grade 1.

A number of authors have reported the use of special techniques such as morphometry and immunohistochemistry to investigate dysplastic or early neoplastic lesions of the ovary as well as the surface epithelium and surface epithelial inclusion cysts. A variety of morphometric changes similar to those found in invasive carcinoma have been detected in adjacent dysplastic and *in situ* carcinomatous surface epithelium[18–21]. Several immunohistochemical reactions that are commonly positive in cases of epithelial cancer have been found to be more frequently positive in surface epithelial inclusion cysts, from which the majority of epithelial cancers are thought to arise, than in the surface epithelium[22–24]. This finding suggests the possibility that the cyst epithelium is closer biochemically to carcinoma than the surface epithelium. Bell and Scully [25], comparing uninvolved ovaries contralateral to ovaries containing epithelial cancers with ovaries of age-matched control patients without ovarian cancer found that both types of ovaries contained equal numbers of surface epithelial inclusion cysts, but that tubal metaplasia of the cyst linings was three times as common in the cancer patients' uninvolved ovaries as in the control ovaries and that staining for placental-like alkaline phosphatase was twice as common in the former group of specimens. More detailed comparative studies of these types of ovary with the use of newer investigative techniques may provide much needed information about the early stages of ovarian epithelial cancer.

REFERENCES

1. Guthrie, D., Davy, M.L.J. and Philips, P.R. (1984) A study of 656 patients with 'early' ovarian cancer. *Gynecol. Oncol.*, **17**, 363–9.
2. Scully, R.E (1982) Minimal cancer of the ovary. *Clin. Oncol.*, **1**, 379–87.
3. Graham, R.M., Schueller, E.F. and Graham, J.B. (1965) Detection of ovarian cancer at an early stage. *Obstet. Gynecol.*, **26**, 151–6.
4. Graham, J.B. and Graham, R.M. (1967) Cul-de-sac puncture in the diagnosis of early ovarian carcinoma. *J. Obstet. Gynaecol. Br. Cwlth*, **74**, 371–8.
5. Chen, K.T.K., Schooley, J.L. and Flam, M.S.

(1985) Peritoneal carcinomatosis after prophylactic oophorectomy in familial ovarian cancer syndrome. *Obstet. Gynecol.*, **66**, 93S–94S.
6. Scully, R.E. (1986) Ovary, in *Pathology of Incipient Neoplasia*, (eds D.E. Hemson and J. Albores-Saavedra), W.B. Saunders, Philadelphia, pp.279–93.
7. Mostoufizadeh, M. and Scully, R.E. (1980) Malignant tumors arising in endometriosis. *Clin. Obstet. Gynecol.*, **23**, 951–63.
8. Clement, P.B. (1987) Endometriosis, lesions of the secondary mullerian system, and pelvic mesothelial proliferations, in *Blaustein's Pathology of the Female Genital Tract*, 3rd edn (ed. R.J. Kurman), Springer-Verlag, New York, pp.516–59.
9. Welch, W.R. and Scully, R.E. (1977) Precancerous lesions of the endometrium. *Hum. Pathol.*, **8**, 503–12.
10. Czernobilsky, B. and Morris, W.J. (1979) A histologic study of ovarian endometriosis with emphasis on hyperplastic and atypical changes. *Obstet. Gynecol.*, **53**, 318–23.
11. LaGrenade, A. and Silverberg, S.G. (1988) Ovarian tumors associated with atypical endometriosis. *Hum. Pathol.*, **19**, 1080–4.
12. Moll, U.M., Chumas, J.C., Chala, E. and Mann, W.J. (1990) Ovarian carcinoma arising in atypical endometriosis. *Obstet. Gynecol*, **75**, 537–9.
13. Russell, P. (1979) The pathological assessment of ovarian neoplasms. I: Introduction to the common 'epithelial' tumours and analysis of benign 'epithelial' tumors. *Pathology*, **11**, 5–26.
14. Chenevart, P. and Gloor, E. (1980) Cystadenomes sereux et muqueux de l'ovaire a la limite de la malignité. *Schweiz. Med. Wochenschr.*, **110**, 531–9.
15. Stalsberg, H., Blom, P.G., Bostad, L.H. and Westgaard, G. (1983) Ovarian tumours and endometriosis in Norway General Hospital material, in *An International Survey of Distributions of Histologic Types of Tumours of the Testis and Ovary* (ed. H. Stalsberg), UICC, Geneva, p.307.
16. Salazar, H. (1983) Epidemiological observations on histologic types of ovarian tumours at Magee-Women's Hospital, a gynecological referral center in Pittsburgh, PA, USA, in *An International Survey of Distributions of Histologic types of Tumours of the Testis and Ovary* (ed. H. Stalsberg), UICC, Geneva, p.331.
17. Isarangkul, W. (1984) Ovarian epithelial tumors in Thai women: a histological analysis of 291 cases. *Gynecol. Oncol.*, **17**, 326–30.
18. Plaxe, S.C., Deligdisch, L., Dottino, P.R. and Cohen, C.J. (1990) Ovarian intraepithelial neoplasia demonstrated in patients with Stage I ovarian carcinoma. *Gynecol. Oncol.*, **38**, 367–72.
19. Deligdisch, L. and Gil, J. (1989) Characterization of ovarian dysplasia by interactive morphometry. *Cancer*, **63**, 748–55.
20. Gil, J. and Deligdisch, L. (1989) Interactive morphometric procedures and statistical analysis in the diagnosis of ovarian dysplasia and carcinoma. *Pathol. Res. Pract.*, **185**, 680–5.
21. Deligdisch, L., Heller, D. and Gil, J. (1990) Interactive morphometry of normal and hyperplastic peritoneal mesothelial cells and dysplastic and malignant ovarian cells. *Hum. Pathol.*, **21**, 218–22.
22. Nouwen, E.J., Hendrix, P.G., Dauwe, S. *et al.* (1987) Tumor markers in the human ovary and its neoplasms. A comparative immunohistochemical study. *Am. J. Pathol.*, **126**, 230–42.
23. Kabawat, S.E., Bast, R.C., Bhan, A.K. *et al.* (1983) Tissue distribution of a coelomic-epithelium-related antigen recognized by the monoclonal antibody OC125. *Int. J. Gynecol. Pathol.*, **2**, 275–85.
24. Cordon-Cardo, C., Mattes, M.J., Melamed, M.R. *et al.* (1985) Immunopathologic analysis of a panel of mouse monoclonal antibodies reacting with human ovarian carcinomas and other human tumors. *Int. J. Gynecol. Pathol.*, **4**, 121–30.
25. Bell, D.A. and Scully, R.E. In: *Progress in experimental tumor research. Risk factors of ovarian cancer. Histologic risk factors of epithelial ovarian cancer.* (eds F. Homburger and G.S. Richardson) Karger, Basel (in press).

Chapter 22

New techniques in the pathological assessment of ovarian cancer

M. WELLS

In the histopathological reporting of ovarian epithelial malignancy there is no substitute for the careful assessment of haematoxylin and eosin stained sections. However, the application of new tools may enhance and refine tissue diagnosis resulting in new information that may have an important influence on therapeutic decision making or may be of prognostic significance. The areas to be covered in this chapter include immunohistochemistry with special emphasis on the application of monoclonal antibodies, flow cytometry, morphometry and nucleolar organizer regions. The new molecular biological techniques of *in situ* hybridization and the polymerase chain reaction have not yet found a place in the histopathological assessment of ovarian cancer.

22.1 IMMUNOHISTOCHEMISTRY

Despite the intense activity in this field the value of immunohistochemical techniques in the routine diagnosis of ovarian epithelial tumours is limited. The antibodies that have been investigated most thoroughly are placental alkaline phosphatase, human milk fat globulin and OC125[1]. Immunohistochemistry is, arguably, of most value when one is confronted with a malignant small cell tumour of the ovary where the differential diagnosis may include true primary small cell carcinoma of the ovary, lymphoma, metastatic carcinoma, dysgerminoma or malignant melanoma.

A recent study investigated four monoclonal antibodies raised against distinct ovarian carcinoma-associated antigens: OC125, OV-TL3, MOV 18 and OV-TL23[2]. These antibodies did not facilitate discrimination between the various adenocarcinomas arising from the female genital tract. Furthermore the reactivity of OC125 and OV-TL3 with carcinomas of the breast, colon and lung limits the use of these antibodies in differential diagnosis, since these carcinomas comprise a major proportion of secondary ovarian tumours. MOV18 showed only very limited reactivity with non-gynaecological tumours (3%), whereas OV-TL23 was completely unreactive, making these antibodies suitable potential tools in gynaecological pathology practice.

It has been suggested that differences in

staining for carcinoembryonic antigen (CEA) may be useful in the histological distinction between primary ovarian cancer and metastatic ovarian adenocarcinoma of intestinal origin[1]. However, the positivity for CEA of mucinous tumours of enteric type probably militates against this. Studies have also shown large variations in the percentage of ovarian tumours positive for CEA. This may be due to the use of polyclonal antisera or a commercial CEA antibody not pre-absorbed with the closely related cross-reacting substance NCA (CEX). Negative weak or very focal immunoreactivity for CEA is more indicative of an ovarian than a colorectal or gastric tumour[3].

There is no antibody by itself that is specific enough for the purpose of distinguishing between ovarian and colorectal cancer though statistical analysis of antigen profiles may be of some value[4]. For this purpose an anti-keratin 7 antibody could be added to the panel, since application of a combination of anti-ovarian carcinoma antibodies and anti-keratin 7 antibodies has been shown to facilitate the discrimination between ovarian and colonic carcinomas[5,6]. However, it has been shown that cytokeratins as well as vimentin are expressed in a variety of sex cord stromal tumours including granulosa tumours as well as in some ovarian carcinomas, thus limiting their value in this aspect of differential diagnosis[7].

Hyperamylasaemia and alterations of serum isoamylase profiles have been recorded in a variety of ovarian tumours and have been suggested as potentially useful tumour markers. In a recent study the presence of large amounts of immunoreactive amylase in 6 of 8 cases of mucinous cystadenocarcinoma contrasted sharply with the low level of expression seen in mucinous cystadenomas. Mucinous cystadenomas of borderline malignancy showed an intermediate degree of amylase immunoreactivity [8].

The literature on the histochemical demonstration of oestrogen receptors (ER) in the normal and neoplastic human ovary has been clouded by the assumption that the presence of oestrogen binding sites (particularly those in the cytoplasm) is equivalent to the presence of ER. The pitfalls inherent in such histochemical techniques have now been appreciated with regard to ovarian cancer. Only with the advent of monoclonal antibodies to the oestrogen receptor has it been shown that authentic immunohistochemical reactivity for high affinity oestrogen receptor is confined to the nuclei of cells[9]. Immunohistochemical determination of ER appears to be a significant prognostic indicator in ovarian cancer[10].

Growth factor receptor[11] and oncogene expression (by the immunohistochemical detection of the oncogene product) have also been studied in ovarian neoplasia[12–14], though their practical application appears, at present, to be limited. In a recent study, amplification of the c-*myc* proto-oncogene was not detected in any cases of normal ovary, benign ovarian tumour or ovarian tumour of borderline malignancy[15]. Proto-oncogene amplification was frequently associated with morphological nuclear anaplasia and high mitotic count. It was suggested that proto-oncogene amplification may be involved in the pathogenesis of aggressive common epithelial tumours of the ovary. The prognostic significance of c-*erb*B2 overexpression in ovarian cancer is currently controversial[16,17]

The monoclonal antibody Ki-67 reacts with a nucleolar antigen in normal and neoplastic cells in G1, S, G2 or M phases of the cell cycle. In a study of ovarian cancer the growth fraction assessed by the percentage of Ki-67 positive cells correlated well with S phase values determined by DNA flow cytometry[10]. High Ki-67 ($\geq 15\%$) levels correlated with advanced stage disease and patient survival. Ki-67 immunohistochemistry is simple and easily applicable in routine pathology laboratories.

22.2 FLOW CYTOMETRY

The analysis of archival material by flow cytometry has expanded greatly in recent years following the publication of a technical method describing the extraction of individual cell nuclei from formalin-fixed, paraffin-processed tissue blocks. The application of flow cytometry to ovarian tumour pathology may be considered principally in terms of measuring the ploidy status of a tumour. The fraction of cells with S-phase DNA content as an estimate of proliferative activity may also be measured. There are now more than 50 published studies on the application of flow cytometry to the study of ovarian neoplasia with a remarkable consensus in their findings (for a review see reference 1). The most important points may be summarized as follows: Firstly, most borderline tumours are diploid and flow cytometric detection of aneuploid tumours may be of value in identifying those borderline ovarian epithelial neoplasms which will show progression. Secondly, well-differentiated malignant tumours are more likely to be diploid, whereas poorly differentiated tumours are more likely to be aneuploid. Thirdly, tumour ploidy is an important prognostic indicator in ovarian cancer: aneuploid tumours have a worse prognosis than diploid tumours at all stages, and in some studies ploidy has proved to be an independent prognostic variable. The combination of flow cytometry and morphometry may provide particularly powerful prognostic indices.

In a recent study of some importance, flow cytometry was performed on tumour samples from 34 patients with epithelial tumours of borderline malignancy who had relapsed or died of disease[18]. There were 15 diploid and 19 aneuploid tumours; four of the latter were serous tumours and 15 were mucinous. In addition ploidy was measured in 30 patients without recurrence; 16 patients had serous tumours and 14 had mucinous tumours. There were no aneuploid serous tumours and three mucinous tumours were aneuploid. Patients with an aneuploid tumour regardless of histology and stage had a less than 20% chance of long-term survival. However there were also patients with diploid tumours who relapsed and succumbed to disease.

22.3 MORPHOMETRY

Morphometry is defined as the quantitation of structure and its great advantages are objectivity and reproducibility. The use of morphometry is aimed at identifying cytological features that predict the prognosis of an ovarian tumour with greater accuracy than is possible by unaided light microscopy.

Morphometry has been used to correctly predict a poor prognosis in two out of three mucinous tumours of borderline malignancy that behaved badly from a total group of 18 tumours. All patients whose tumours were morphometrically graded as having a good prognosis were alive and tumour-free at intervals ranging from 4 to 14 years[19].

In a large series of ovarian cancers an analysis of morphometric features was the second best indicator of prognosis after FIGO staging and was especially useful for patients with stage I disease[20]. Variables indicating a relatively poor outcome in stepwise linear regression analysis were the mitotic index (mitoses/25 h.p.f.), the volume percentage of epithelium and the standard deviation of the shortest nuclear axis.

The volume-corrected mitotic index (M/V index) expresses the mitotic activity as the number of mitotic figures per square millimetre of neoplastic epithelium in the microscope field. In a recent study Cox's multivariate regression model showed that the clinical stage was the best predictor of prognosis in ovarian cancer followed by the M/V index. For stage I tumours it was the only parameter selected by the Cox's model as a significant and independent prognostic predictor[21]. Reassuringly, mitotic estimation by the conventional method (mitoses/10 h.p.f.) and the M/V index appear to be reproducible with good intra- and inter-observer correlation[22].

Nuclear density (the fraction of the number of nuclei per area of parenchyma) has also been assessed morphometrically in a series of patients with advanced ovarian cancer (stage III–IV)[23]. The combination of nuclear density and age allowed the identification of a subgroup of patients with an excellent prognosis.

22.4 NUCLEOLAR ORGANIZER REGIONS

Cell nuclei contain large loops of DNA whose rRNA genes are transcribed by RNA polymerase I; such a loop is known as a nucleolar organizer region (NOR). In humans, NORs are located on the secondary constrictions of acrocentric chromosomes on the short arms of D (13,14,15) and G (21,22). Thus in the diploid 10 cell it is possible to see up to 10 NORs.

NOR-associated proteins can be stained with silver (AgNORs); the proteins are acidic and rich in sulphydryl and disulphide groups. The silver staining method can be applied to paraffin sections and many studies have linked AgNOR counts to malignancy.

Recently the potential of AgNORs as a diagnostic aid in serous and mucinous ovarian tumours was examined[24]. Mucinous cystadenomas had significantly lower counts than borderline tumours and cystadenocarcinomas ($P \leq 0.01$) but no significant difference was found between mucinous borderline tumours and cystadenocarcinomas. Serous tumours showed no significant difference between cystadenomas and borderline tumours, but cystadenocarcinomas had significantly higher AgNOR counts than borderline tumours. AgNOR counts may therefore be useful in distinguishing borderline serous tumours from serous cystadenocarcinomas, but do not seem to be useful when applied to mucinous tumours. These findings confirm the limited value of AgNOR counts in the routine differentiation between borderline and malignant mucinous ovarian tumours [25,26]. No correlation has been demonstrated between the AgNOR count and survival and there are clearly problems of inter-observer variation in counting AgNORs.

REFERENCES

1. Wells, M. (1991) Application of new techniques in gynaecological histopathology in *Female Reproductive System* (ed. M.C. Anderson), Churchill Livingstone, Edinburgh, pp. 447–65.
2. Boerman, O.C., van Niekerk, C.C., Makkink, K. *et al.* (1991) Comparative immunohistochemical study of four monoclonal antibodies directed against ovarian carcinoma-associated antigens. *Int. J. Gynecol. Pathol.*, **10**, 15–25.
3. Hammond, R.H., Bates, T.D., Clarke, D.G. *et al.* (1991) The immunoperoxidase localization of tumour markers in ovarian cancer: the value of CEA, EMA, cytokeratin and DD9. *Br. J. Obstet. Gynaecol.*, **98**, 73–83.
4. Henzen-Logmans, S.C., Schipper, N.W., Poels, L.G. *et al.* (1988) Use of statistical evaluation of antigen profiles in differential diagnosis between colonic and ovarian adenocarcinomas. *J. Clin. Pathol.*, **41**, 644–9.
5. Van Niekerk, C.C., Jap, P.H.K., Thomas, C.M.G. *et al.* Marker profile of mesothelial cells versus ovarian carcinoma cells. *Int. J. Cancer*, **43**, 1065–71.
6. Ramaekers, F., van Niekerk, C., Poels, L. *et al.* (1990) Use of monoclonal antibodies to keratin 7 in the differential diagnosis of adenocarcinomas. *Am. J. Pathol.*, **136**, 641–55.
7. Benjamin, E., Law, S. and Bobrow, L.G. (1987) Intermediate filaments cytokeratin and

vimentin in ovarian sex cord stromal tumours with correlative studies in adult and fetal ovaries. *J. Pathol.*, **152**, 253–63.

8. Griffin, N.R. and Wells, M. (1990) Immunolocalisation of α-amylase in ovarian mucinous tumours. *Int. J. Gynecol. Pathol.*, **9**, 41–6.

9. Press, M.F., Holt, J.A., Herbst, A.L. and Greene, G.L. (1985) Immunocytochemical identification of estrogen receptor in ovarian carcinomas. Localization with monoclonal estrophilin antibodies compared with biochemical assays. *Lab. Invest.*, **53**, 349–61.

10. Isola, J., Kallioniemi, O.-P., Korte, J.-M. et al. (1990) Steroid receptors and Ki-67 reactivity in ovarian cancer and in normal ovary: correlation with DNA flow cytometry, biochemical receptor assay and patient survival. *J. Pathol.*, **162**, 295–301.

11. Bauknecht, T., Janz, I., Kohler, M. and Pfleiderer, A. (1989) Human ovarian carcinoma: correlation of malignancy and survival with the expression of epidermal growth factor receptors (EGF-R) and EGF-like factors (EGF-F). *Med. Oncol. Tumor Pharmacother.*, **6**, 121–7.

12. Watson, J.V., Curling, O.M., Munn, C.F. and Hudson, C.N. (1987) Oncogene expression in ovarian cancer: a pilot study of c-*myc* oncoprotein in serous papillary ovarian cancer. *Gynecol. Oncol.*, **28**, 137–50.

13. Rodenburg, C.J., Koelma, I.A., Nap, M. and Fleuren, G.J. (1988) Immunohistochemical detection of the *ras* oncogene product p21 in advanced ovarian cancer. Lack of correlation with clinical outcome. *Arch. Pathol. Lab. Med.*, **112**, 151–4.

14. Polacarz, S.V., Hey, N.A., Stephenson, T.J. and Hill, A.S. (1989) c-*myc* Oncogene product P62$^{c\text{-}myc}$ in ovarian mucinous neoplasms: immunohistochemical study correlated with malignancy. *J. Clin. Pathol.*, **42**, 148–52.

15. Sasano, H., Garrett, C.T., Wilkinson, D.S. et al. (1990) Protooncogene amplification and tumor ploidy in human ovarian neoplasms. *Hum. Pathol.*, **21**, 383–91.

16. Slamon, D.J., Godolphin, W., Jones, L.A. et al. (1989) Studies of the HER-2/*neu* protooncogene in human breast and ovarian cancer. *Science*, **244**, 707–12.

17. Haldane, J.S., Hird, V., Hughes, C.M. and Gullick, W.J. (1990) c-*erb*B-2 oncogene expression in ovarian cancer. *J. Pathol.*, **162**, 231–7.

18. Kaern, J., Trope, C., Kjorstad, K.E. et al. (1990) Cellular DNA content as a new prognostic tool in patients with borderline tumours of the ovary. *Gynecol. Oncol.*, **38**, 452–7.

19. Baak, J.P.A., Fox, H., Langley, F.A. and Buckley, C.H. (1985) The prognostic value of morphometry in ovarian epithelial tumours of borderline malignancy. *Int. J. Gynecol. Pathol.*, **4**, 186–91.

20. Baak, J.P.A., Wisse-Brekelmans, E.C.M., Langley, F.A. et al. (1986) Morphometric data to FIGO stage and histological type and grade for prognosis of ovarian tumours. *J. Clin. Pathol.*, **39**, 1340–6.

21. Haapasulo, H., Collan, Y., Atkin, N.B. et al. (1989) Prognosis of ovarian carcinomas: prediction by histoquantitative methods. *Histopathology*, **15**, 167–78.

22. Haapasalo, H., Collan, Y., Montironi, R. et al. (1990) Consistency of quantitative methods in ovarian tumour histopathology. *Int. J. Gynecol. Pathol.*, **9**, 208–16.

23. Ludescher, C., Weger, A.-R., Lindholm, J. et al. (1990) Prognostic significance of tumour cell morphometry, histopathology and clinical parameters in advanced ovarian carcinoma. *Int. J. Gynecol. Pathol.*, **9**, 343–51.

24. Griffiths, A.P., Pickles, A. and Wells, M. (1989) AgNORs in diagnosis of serous and mucinous ovarian tumours. *J. Clin. Pathol.*, **42**, 1311.

25. Kinsey, W., Randall, B. and Brown, L.J.R. (1989) AgNOR counts in borderline ovarian tumours, in *Ovarian Cancer, Biological and Therapeutic Challenges* (eds F. Sharp, W.P. Mason and R.E. Leake), Chapman & Hall, London, pp.113–24.

26. Mauri, F.A., Barbareschi, M., Scampini, S. et al. (1990) Nucleolar organiser regions in mucinous tumours of the ovary. *Histopathology*, **16**, 396–8.

Chapter 23
Flow cytometry in ovarian cancer

P.S. BRALY

23.1 INTRODUCTION

Despite the widespread use of aggressive surgical debulking and multi-agent chemotherapy, the long-term prognosis for patients with epithelial ovarian cancer has not been greatly affected and appears to be related to individual tumour characteristics rather than the treatment administered[1–7]. An accurate method of determining the prognostic factors for a tumour would greatly aid the clinician in individualizing postoperative management [8]. The traditionally used prognostic factors, including tumour grade, stage and size of residual disease after tumour debulking, have often been subjectively and inaccurately assigned[7,9,10] and only partially successful in determining which patients will be most likely to benefit from aggressive combination therapy. Some of the determinants of tumour response (e.g. DNA content, growth kinetics and hormone receptor expression) are expressed at the cellular level and, therefore, are not only not readily recognized by standard histopathological techniques but also seem to vary between tumour cells in the same individual. Newer techniques of quantitative cytology are greatly facilitating the objective evaluation of tumour cell heterogeneity. Flow cytometry is a technique that can rapidly and quantitatively measure a variety of individual cell characteristics, thus adding valuable prognostic information to the management of patients with cancer. The prognostic significance of the cellular DNA content in ovarian carcinoma was recognized as early as 1971[11] but because the technique of absorption cytometry was tedious and time consuming it found little application in routine clinical practice. With the introduction of flow cytometry[12–14], the measurement of cellular DNA content has become simpler, rapid and more accurate. Because of the original requirement for fresh or cryopreserved tissue for flow cytometric analysis, however, the early studies utilizing this technique were performed prospectively on small numbers of patients. With the description of a flow cytometric method to determine the DNA in content in archival paraffin-embedded tumour tissue from patients whose clinical outcome was already known [15], application of this technique has become widespread. Numerous early studies demonstrated the importance of DNA content (ploidy) on the survival of patients with lung[16–18], ovarian[19–24], breast[25,26], colon[27] and genitourinary malignancies[28–

30] as well as other solid and haematological cancers. More recent studies have confirmed the importance of tumour cell DNA content as an independent prognostic tumour factor for both early and late stage ovarian cancer and have combined this analysis with the measurement of other cellular characteristics such as cell cycle information[31–35], CA-125 levels and major histocompatibility complex antigens[36] and *in vitro* resistance predictive tests[22].

23.2 REVIEW OF TECHNIQUE

Flow cytometry provides a high speed (>1000 cells/s), precise method for analysing and sorting cells according to morphological, molecular, biophysical and functional cellular characteristics. This information has significantly contributed to the current knowledge about tumour heterogeneity, including the occurrence of multiclonality and proliferative patterns which may account for the diverse clinical course of comparably staged and treated neoplasms.

The major requirement for the use of flow cytometry is the preparation of single cell suspensions when solid tumour samples are studied. Even though preparation procedures are presently available using either mechanical or enzymatic techniques, concern persists regarding the degree of representation of biopsy and surgical specimens and multiple samples from the same site should be analysed whenever possible[37]. In one study[38], cell suspensions obtained from 23 tumours by enzymatic and mechanical techniques were compared for the percentage of aneuploid cells and demonstrated that the mechanical method of disaggregation provided better cell suspensions and a higher percentage of aneuploid cells than the enzymatic procedure. When aneuploid cells were identified by either method, however, there were no significant differences between the DNA indices and coefficient of variation (CV) for the aneuploid cell populations.

Since the introduction of DNA-specific fluorescent dyes to the technique of flow cytometry, studies of the cellular DNA content of human tumours has become the most common flow cytometric method of characterizing these malignancies. The DNA content abnormalities reflect not only variations in chromosome number but also the presence of chromosomal abnormalities and have been consistently observed in a wide variety of human malignancies[12,13,39]. The detection of very small changes in chromosomal material, however, is difficult in DNA flow cytometry, and in most instances two cell populations can be discriminated as distinct peaks in the DNA histogram only if their DNA content varies by >10%. The DNA index (DI), defined as the ratio of the modal DNA fluorescence of abnormal to normal G1/0 cells, is the currently accepted term to quantify cytometric ploidy[39]. The incidence of cytometric aneuploidy in human solid tumours of different primary sites and histology varies from 19% for basal cell cancer of the skin to 100% for squamous cell cancers of the oesophagus and head and neck[37].

Another important aspect of flow cytometric analyses is the availability of specific, quantitatively binding fluorochromatic dyes to identify and adequately describe the cytogenetic, phenotypic and proliferative heterogeneity of cancer cells[37]. A promising new technique developed to measure both the percentage of cells in the S phase of the cell cycle and the growth fraction or percentage of proliferating cells is based on the incorporation of 5-bromodeoxyuridine (BUdR), a halogenated analogue of thymidine, into tumour cell DNA[40,41]. The detection of cells with BUdR-substituted DNA can be achieved in two ways. The first method involves incorporation of BUdR into DNA strands, which results in a modification of DNA-specific dye fluorescence. With this

Table 23.1 Ovarian cancer ploidy vs. survival

Reference	No. of patients	% Aneuploid	Independent prognostic factor	Median survival (months)	
				Diploid	Aneuploid
Barnabei [31]	115	76	No	36	23
Blumenfeld [19]	84	52	Yes	48	19
Brescia [32]	99	48	Yes	60	24
Erba [33]	101	NG	No	25	18
Friedlander [48]	128	73	Yes	60	13
Iverson [47]	51	52	Yes	18	8
Khoo [46]	53	68	Yes	33	13
Murray [46]	40	60	Yes	60	18
Rodenburg [50]	74	79	Yes	60	24
Volm [34]	37	60	Yes	41	14

technique, BUdR enhances the fluorescence of DNA stained with chromomycin and mithramycin and quenches the fluorescence of other dyes such as acridine orange, ethidium bromide, propidium iodide and Hoechst 32258[42,43]. The second way to detect BUdR-labelled cells is by utilizing a fluorescein-isothiocyanate labelled goat anti-mouse antibody directed against the thymidine analogue[44]. At present, this technique is based on a pulse incorporation of BUdR, followed by exposure of the cells to the antibody after fixation and DNA denaturation. The fraction of labelled cells can then be determined by flow cytometry and has been shown to approximate the percentage of S phase cells.

23.3 PLOIDY STUDIES IN OVARIAN CANCER

Tumour ploidy is considered to be a major prognostic indicator in ovarian and other malignancies and several authors have described an association between aneuploidy and high tumour grade or stage and poor prognosis. Most of the recent published studies of ovarian cancer (Table 23.1) have documented that approximately 50–80% of the tumours analysed are aneuploid. Eight of the 10 studies presented in Table 23.1 have shown that ploidy is an independent prognostic factor with an aneuploid tumour predicting a significantly shorter survival. Several previous studies have compared DNA ploidy to histological tumour grading alone, with somewhat varying and contradictory results. Even though histological grade does seem to give some prognostic information, comparison of results from different authors is difficult and reflects the difficulties arising due to variation of grade within the tumour, the diversity of grading criteria used and the poor reproducibility of histopathological grading between observers [45]. As compared to grading, DNA ploidy is a more objective parameter and is probably more stable throughout the tumour[51], and thus may be a more reliable prognostic factor that could be used as a routine adjunct to the clinicopathological classification of ovarian malignancies.

Even though there is fairly widespread consensus regarding the ability of DNA ploidy to discriminate patients with a poor prognosis, there is much disagreement in the literature as to what definition should be used to differentiate diploid from aneuploid DNA content. In the past, the resolution of static cytometric analysis of cellular DNA content was limited and tumours could be classified only as 'near diploid' or as hyperdiploid and a number of investigators using flow cytometric techniques to study ovarian cancer have continued to use a DNA index of less than or equal to 1.3–1.5 to define a near diploid group of tumours with a better prognosis than those tumours with an aneuploid tumour content. A study by Friedlander and co-workers[48] demonstrated that patients with near diploid tumours had a better prognosis than those with more aneuploid tumours, but that the statistical significance of this survival difference increased the closer the cut-off DNA index was to 1.0.

A study by Klemi and colleagues[23] demonstrated that a DNA index of >1.3 was the most important prognostic factor in a multivariate analysis and was more accurate than a diploid vs. aneuploid designation in determining survival. These authors also reported that 3 of the 27 histologically benign (11%) and 7 of the 43 borderline malignant (16%) ovarian neoplasms were aneuploid but in all of these cases the DNA index was low (<1.3), and in the follow-up period none of these patients with histologically benign and borderline malignancies died of their disease. Two studies have suggested that ploidy can select a subgroup of patients with borderline malignant ovarian cancers who will demonstrate a more aggressive disease course than is typical for this diagnosis. In the report by Friedlander and co-workers[51], 2 of 44 borderline ovarian tumours were found to have an aneuploid DNA content and one of these two patients died of progressive disease 7 months after diagnosis. Kaern and colleagues[52] reported a difference in aneuploidy related to histology in borderline tumours, with 47% of the mucinous tumours and 15% of the serous tumours being aneuploid. In this preliminary report, it appeared that there was a significant difference in survival with diploid and aneuploid tumours but that those patients who had a recurrence with a diploid tumour all died of their disease within the first 5 years after diagnosis. Those patients with aneuploid tumours, on the other hand, were much more likely to relapse and die from their disease (3 of 4 serous tumours, 18 of 18 mucinous tumours), but this occurred over the entire 15 years of follow-up.

As described in Table 23.2, several investigators have evaluated the relationship between tumour ploidy and residual disease following surgical debulking. In the study by Erba and co-workers[33], even though the DNA index and percentage of cells in S phase were not independent prognostic variables to predict survival, ploidy distribution in the tumour samples was significantly related to residual disease in the peritoneal cavity following primary tumour debulking. Hamaguchi and co-workers[53] described a 'mosaic' subtype in 11 of 19 aneuploid tumours that contained both diploid and aneuploid cell populations and described a significant association of this mosaic pattern in tumours that could not be adequately debulked. Iversen[47] noted in his study that the proportion of patients considered to be without grossly apparent residual disease after the initial operation was significantly higher among those patients with diploid tumours. Kallioniemi and colleagues[54] described a FCM determined classification that combined the simultaneous evaluation of DNA index, the number of aneuploid cell clones and S phase fraction (SPF) to identify three distinct prognostic groups for both early and late stage ovarian cancer and found that all patients in the high risk group, even with early stage ovarian cancer, died of disease within 3 years of diagnosis. They did not note a relationship between ploidy and the amount of residual disease after tumour debulking. A study by Khoo and co-workers[46] described that in patients with stage III ovarian cancer treated by a standard-

Table 23.2 Ploidy vs. residual disease

Reference and size of residual disease	Diploid		Aneuploid		P value
	No.	%	No.	%	
Erba [33]					
< 2 cm	11	58	16	22	< 0.01
> 2 cm	8	42	58	78	
Hamaguchi [53]					
< 2 cm	18	78	7	37	< 0.05
> 2 cm	5	22	12	63	
Iversen [47]					
0	13	54	4	15	
< 2 cm	5	21	9	35	< 0.01
> 2 cm	6	25	13	50	
Kallioniemi [54]					
0	57	86	71	78	NS
< 1.5 cm	5	8	12	13	
> 1.5 cm	4	6	8	9	
Khoo [46]					
0	8	47	5	14	NG
< 2 cm	5	29	8	22	
> 2 cm	4	24	23	64	
Rodenburg [24]					
< 1.5 cm	11	73	12	20	NG
> 1.5 cm	4	27	47	80	

ized protocol (maximum tumour debulking and intravenous cisplatin, cyclophosphamide and adriamycin chemotherapy) those with aneuploid tumours were more likely to have residual disease greater than 2 cm (64% of aneuploid tumours vs. 23% of diploid tumours), whereas nearly 50% of patients with diploid tumours had no residual disease. They also noted that the amount of residual disease after primary surgery influenced survival only in patients with diploid tumours. Rodenburg and co-workers [24] noted that tumour ploidy was strongly associated not only with the size of residual disease but also with the bulk of disease at the time of presentation. Although most investigators agree that the proliferative activity or SPF of ovarian cancers are of prognostic significance [20,22,31,34,54], there is some disagreement regarding its contribution[32,33]. Other authors have suggested that a low SPF is associated with benign or well-differentiated histological subtypes and with diploid malignancies[55–57].

23.4 FCM IN THE EVALUATION OF EFFUSIONS

Despite the fact that the documentation of malignant cells in effusions or washings is often of critical clinical significance, the standard cytological interpretation of these specimens is often subjective. False negative cytology results, even in the presence of disseminated carcinoma, occurs frequently [58] and may be attributed to the scarcity of recognizable exfoliated cancer cells or because the cancer cells cannot be distinguished from the normal cellular components that are invariably present. The occasional false positive result is most likely due to the abnormalities of mesothelial cells and macrophages in the cytological preparation. Therefore, there is a need for more objective and complementary methods of evaluation of body cavity fluids for the presence or absence of malignant cells. Numerous studies evaluating the potential contribution of flow cytometry in the evaluation of effusions have been published recently[59–68] but there is little consensus regarding the value of this technique compared to that of conventional cytological evaluation. Some studies[59–65] suggest that the addition of FCM evaluation can be a very useful adjunct to standard cytological evaluation of body fluids. Those investigators[66–68] who feel that use of flow cytometry to evaluate effusions adds little to routine cytological evaluation note the high false negative rate of FCM (19–43%). Since the definition of malignancy as determined by FCM is the presence of aneuploid tumour cells, it is obvious that this technique cannot add to the detection rate of diploid tumours. In a study by Sinton and colleagues[60], because of the ability of FCM to detect aneuploidy in cases where conventional cytological examination could not detect malignant cells, the number of patients with documented malignant effusions was increased by 39% over those detected by cytological examination only.

23.5 TREATMENT MONITORING AND MODIFICATION

Even though many *in vitro* and animal studies on the effects of chemotherapeutic agents, partial surgical debulking and radiation therapy on tumour cells have been described over the last decade, surprisingly little is known about the effects of these treatments on residual *in vivo* human malignancies. The increasing availability of flow cytometry in the clinical setting promises to facilitate the elucidation of this type of information and, to date, several preliminary studies have been published. Because of the obvious ethical limitations to repeated biopsies of human *in vivo* malignancies, one of the first reports on the effect of a cell cycle phase-specific agent in an *in vivo* gynaecological cancer was reported by O'Quinn and co-workers[69] using flow cytometric techniques to evaluate the cell cycle changes after treatment in advanced cervical cancer. These authors speculated that using this information, a second chemotherapeutic agent could be administered at a time in the cell cycle when it would be most effective. Rutgers and colleagues[70] have monitored the radiation-induced G2 arrest of *in vivo* human cervix cancers following low dose-rate irradiation and have suggested that this synchronization might be exploited in fractionation schemes in which treatment intervals are adjusted to these cell cycle changes. In a study of patients with refractory or relapsed ovarian cancer, Russo and co-workers[71] described that ifosfamide could be used to enhance the tumour cell proliferative activity (and thereby potentially increase the tumour susceptibility to follow-up chemotherapy administration). Studies from our laboratory [72–75] have been designed to evaluate the effect of tumour debulking on the residual tumour cell cycle kinetic changes in *in vivo* advanced ovarian cancer. Using flow cytometric evaluation of serial intraperitoneal tumour samples obtained via repeated peritoneal cavity irrigations in the postoperative period, we have described a cyclical pattern of tumour cell in proliferation nearly 12 hours out of phase with non-tumour cell proliferation[72]. In addition, in patients whose tumour was significantly surgically debulked (but not

Summary

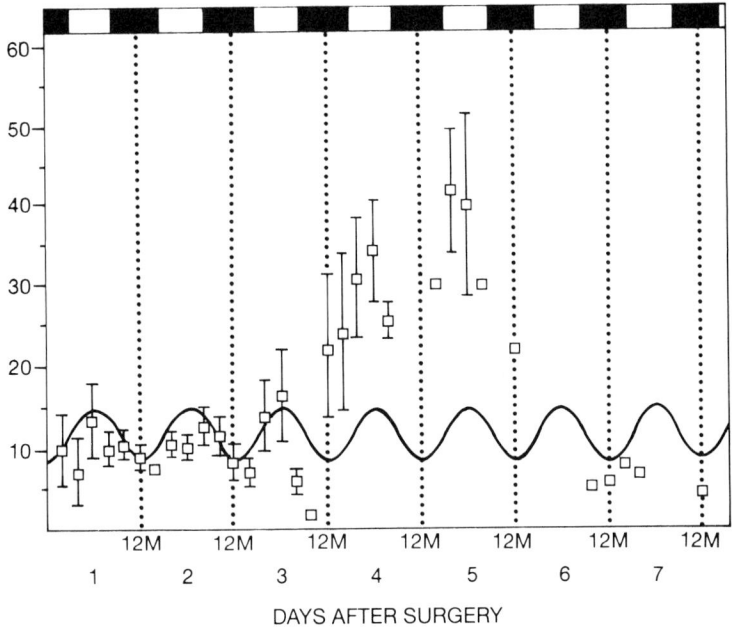

Figure 23.1 S + G$_2$ fraction vs. time in cells from peritoneal cavity irrigations in pooled data from patients sampled through 5 days. Dotted lines mark midnight for the 24-hour period. Dark and light bars (top) indicate 6 a.m. and 6 p.m. In addition to the daily rhythmicity there appears superimposed on the daily peak an increase in the value of the S + G$_2$ fraction on day 4 and 5 following surgery.

with biopsy only) there is an apparent increase in the proliferative pool of tumour cells that occurs between postoperative days 5 and 7, followed by a decline to baseline levels (Figure 23.1). These studies are now being repeated following *in vivo* preoperative BUdR administration in an attempt to further elucidate the postoperative changes in tumour cell proliferation.

23.6 SUMMARY

In the last decade, the flow cytometer has evolved from a research tool to a valuable addition to the important field of quantitative pathology. With the increasing awareness of the need to predict tumour response to therapy and to be able to stratify individual patients to various treatment options that are currently available, it is expected that over the next 5–10 years, many more diagnostic pathology laboratories will begin to use flow cytometric analyses to report DNA ploidy and cell cycle kinetic information for ovarian cancer and other gynaecological malignancies. Flow cytometry will also aid in the delineation of the likely tumour response to treatment protocols and allow for individualization of therapy for both early and advanced malignancies as well as to facilitate the prediction of premalignant conditions that will progress to invasive cancer.

REFERENCES

1. Dyson, J.L., Beilby, J.O. and Steele, E.J. (1971) Factors influencing survival in carcinoma of the ovary. *Br. J. Cancer*, **25**, 237–49.
2. Day, T.G., Gallagher, H.S. and Ruthledge, E.N. (1975) Epithelial carcinoma of the ovary: Prognostic importance of histologic grade. *Natl Cancer Inst. Monogr.*, **42**, 15–21.
3. Ozols, R.F., Garvin, J., Costa, J. et al. (1980) Advanced ovarian cancer: Correlation of histologic grade with response to therapy and survival. *Cancer*, **45**, 572–81.
4. Sorbe, B., Frankendal, B. and Veress, B. (1982) Importance of histologic grading in the prognosis of epithelial ovarian cancer. *Obstet. Gynecol.*, **59**, 576–82.
5. Sigurdsson, K., Alm, P. and Gullberg, B. (1983) Prognostic factors in malignant epithelial ovarian tumors. *Gynecol. Oncol.*, **15**, 370–80.
6. Swenerton, K.D., Hislop, J.G., Spinelli, J. et al. (1985) Ovarian carcinoma: A multivariate analysis of prognostic factors. *Obstet. Gynecol.*, **65**, 264–70.
7. Baak, J.P.A., Langley, F.A., Talerman, A. and Delemarre, J.F.M. (1986) Inter-pathologist and intrapathologist disagreement in ovarian tumor grading and typing. *Anal. Quant. Cytol.*, **8**, 354–7.
8. Dembo, A.J. and Bush, R.S. (1982) Choice of postoperative therapy based on prognostic factors. *Int. J. Radiat. Oncol. Biol. Phys.*, **8**, 893–7.
9. McGowan, L., Lesher, L.P., Norris, H.J., and Barnett, M. (1985) Misstaging of ovarian cancer. *Obstet. Gynecol.*, **65**, 568–72.
10. Hernandez, E., Bhagavan, B.S., Parmley, T.H. and Rosensheim, N.B. (1984) Interobserver variability in the interpretation of epithelial ovarian cancer. *Gynecol. Oncol.*, **17**, 117–23.
11. Atkin, N.B. (1971) Modal DNA value and chromosome number in ovarian neoplasia: A clinical and histopathological assessment. *Cancer*, **27**, 1064–73.
12. Barlogie, B., Drewinko, B., Schumann, J. et al. (1980) Cellular DNA content as a marker of neoplasia in man. *Am. J. Med.*, **69**, 195–203.
13. Barlogie, B., Gohde, W., Johnston, D.A. et al. (1978) Determination of ploidy and proliferative characteristics of human solid tumors by pulse cytophotometry. *Cancer Res.*, **38**, 3333–9.
14. Laerum, O.D. and Farsund, T. (1981) Clinical application of flow cytometry: A review. *Cytometry*, **2**, 1–13.
15. Hedley, D.W., Friedlander, M.L., Taylor, S.W. et al. (1983) Method for analysis of cellular DNA content of paraffin-embedded pathological material using flow cytometry. *J. Histochem. Cytochem.*, **21**, 1333–5.
16. Bunn, P., Schlam, M. and Gazdar, A. (1980) A comparison of cytology and DNA content analysis by flow cytometrics (FCM) in specimens from lung cancer patients. *Proc. Am. Assoc. Cancer Res.*, **21**, 160.
17. Johnson, T.S., Barlogie, B., Valdivieso, M. and Bedrossian, C. (1982) Ploidy and proliferative activity in human lung cancer, in *Cytometry in the Clinical Laboratory*, Eng. Foundation Conference, Santa Barbara, California.
18. Olzewski, W., Darzynkiewicz, Z., Claps, M.L. and Melamed, M.R. (1982) Flow cytometry of lung carcinoma: a comparison of DNA stemline and cell cycle distribution with histology. *Anal. Quant. Cytol.*, **4**, 90–4.
19. Blumenfeld, D., Braly, P.S., Ben-Ezra, J. and Klevecz, R.R. (1987) Tumor DNA content as a prognostic feature in advanced epithelial ovarian carcinoma. *Gynecol. Oncol.*, **27**, 389–98.
20. Friedlander, N.L., Taylor, I.W., Russell, P. et al. (1983) Ploidy as a prognostic factor in ovarian cancer. *Int. J. Gynecol. Pathol.*, **2**, 55–63.
21. Iverson, O.E. and Laerum, O.D. (1985) Ploidy disturbances in endometrial and ovarian carcinomas: A review. *Anal. Quant. Cytol. Histol.*, **7**, 327–36.
22. Volm, M., Bruggermann, A., Gunther, M. et al. (1985) Prognostic relevance of ploidy, proliferation, and resistance-predictive tests in ovarian carcinoma. *Cancer Res.*, **45**, 5180–5.
23. Klemi, P.J., Joensuu, H., Maenpaa, J. et al. (1989) Influence of cellular DNA content on survival in ovarian carcinoma. *Obstet. Gynecol.*, **74**, 200–4.
24. Rodenburg, C.J., Cornelisse, C.J., Heintz, P.A.M. et al. (1987) Tumor ploidy as a major prognostic factor in advanced ovarian cancer. *Cancer*, **59**, 317–23.

25. Olzewski, W., Darzynkiewicz, Z., Rosen, P.P. et al. (1981) Flow cytometry of breast cancer: I. Relation of DNA ploidy level to histology and estrogen receptor. *Cancer*, **48**, 980–4.
26. Raber, M.N., Barlogie, B., Latreille, J. et al. (1982) Ploidy, proliferative activity and estrogen receptor content in human breast cancer. *Cytometry*, **3**, 36–41.
27. McGuire, W.L. and Dressler, L.G. (1985) Emerging impact of flow cytometry in predicting recurrence and survival in breast cancer patients. *J. Natl Cancer Inst.*, **75**, 405–10.
27. Wolley, R.C., Schreiber, K., Koss, L.G. et al. (1982) DNA distribution in human colon carcinoma and its relationship to clinical behavior. *J. Natl. Cancer Inst.*, **69**, 15–22.
28. Bichel, P., Frederiksen, P., Kjaer, T. et al. (1977) Flow microfluorometry and transrectal fine-needle biopsy in the classification of human prostatic carcinoma. *Cancer*, **40**, 1206–11.
29. Tribukait, B. and Esposti, P.L. (1978) Quantitative flow-microfluoro-metric analysis of the DNA in cells from neoplasms of the urinary bladder: correlation of aneuploidy with histological grading and cytological findings. *Urol. Res.*, **6**, 201–5.
30. Tribukait, B., Esposti, P.L. and Ronstrom, L. (1980) Tumor ploidy for characterization of prostatic carcinoma: flow-cytofluorometric DNA studies using aspiration biopsy material. *Scand. J. Urol. Nephrol.*, **55** (Suppl.), 59–64.
31. Barnabei, V.M., Miller, D.S., Bauer, K.D. et al. (1990) Flow cytometric evaluation of epithelial ovarian cancer. *Am. J. Obstet. Gynecol.*, **162**, 1584–92.
32. Brescia, R.J., Barakat, R.A., Beller, U. et al. (1990) The prognostic significance of nuclear DNA content in malignant epithelial tumors of the ovary. *Cancer*, **65**, 141–7.
33. Erba, E., Ubezio, P., Pepe, S. et al. (1989) Flow cytometric analysis of DNA content in human ovarian cancers. *Br. J. Cancer*, **60**, 45–50.
34. Volm, M., Klein, W. and Pfleiderer, A. (1989) Flow-cytometric prognostic factors for the survival of patients with ovarian carcinoma: A 5-year follow-up study. *Gynecol. Oncol.*, **35**, 84–9.
35. Kallioniemi, O.P., Mattila, J., Punnonen, R. and Koivula, T. (1988) DNA ploidy and cell cycle distribution in ovarian cancer: Relation to histopathological features of the tumor. *Int. J. Gynecol. Pathol.*, **7**, 1–11.
36. Fowler, W.C., Maddock, M.B., Moore, D.H. and Haskill, S. (1988) Significance of multi-parameter flow cytometric analysis of ovarian cancer. *Am. J. Obstet. Gynecol.*, 838–45.
37. Mauro, F., Teodori, L., Schumann, J. and Gohde, W. (1986) Flow cytometry as a tool for the prognostic assessment of human neoplasia. *Int. J. Radiat. Oncol. Biol. Phys.*, **12**, 625–36.
38. Frankfurt, O.S., Slocum, H.K., Rustum, Y.M. et al. (1984) Flow cytometric analysis of DNA aneuploidy in primary and metastatic human solid tumors. *Cytometry*, **5**, 71–80.
39. Barlogie, B., Raber, M.R., Schumann, J. et al. (1983) Flow cytometry in clinical cancer research. *Cancer Res.*, **43**, 3982–97.
40. Rabinovitch, P.S., Kubbies, M., Chen, Y.C. et al. (1988) BrdU-Hoechst flow cytometry: A unique tool for quantitative cell cycle analysis. *Exp. Cell Res.*, **174**, 309–18.
41. Riccardi, A., Danova, M. and Ascari, E. (1988) Bromodeoxyuridine for cell kinetic investigations in humans. *Haematologica*, **73**, 423–30.
42. Swartzendruber, D.E. (1976) Microfluorometric analysis of cellular DNA following incorporation of BUdR. *J. Cell. Physiol.*, **90**, 445–54.
43. Bohmer, R.M. (1979) Flow cytometric cell cycle analysis using the quenching of 33258 Hoechst fluorescence by bromodeoxyuridine incorporation. *Cell Tissue Kinet.*, **12**, 101–10.
44. Dolbeare, F., Gratzner, H., Pallavicini, M.G. and Gray, J.W. (1983) Flow cytometric measurement of total DNA content and incorporated bromodeoxyuridine. *Proc. Natl Acad. Sci. USA*, **80**, 5573–7.
45. Hiddemann, W., Schumann, J., Andreef, M. et al. (1984) Convention on nomenclature for DNA cytometry. *Cytometry*, **5**, 445–6.
46. Khoo, S.K., Hurst, T., Kearsley, J. et al. (1990) Prognostic significance of tumor ploidy in patients with advanced ovarian carcinoma. *Gynecol. Oncol.*, **39**, 284–8.
47. Iversen, O.E. (1988) Prognostic value of the flow cytometric DNA index in human ovarian carcinoma. *Cancer*, **61**, 971–5.
48. Friedlander, M.L., Hedley, D.W., Swanson,

C. and Russell, P. (1988) Prediction of long-term survival by flow cytometric analysis of cellular DNA content in patients with advanced ovarian cancer. *J. Clin. Oncol.*, **6**, 282–90.
49. Murray, K., Hopwood, L., Volk, D. and Wilson, F. (1989) Cytofluorometric analysis of the DNA content in ovarian carcinoma and its relationship to patient survival. *Cancer*, **63**, 2456–60.
50. Rodenburg, C.J., Cornelisse, C.J., Hermans, J. and Fleuren, G.J. (1988) DNA flow cytometry and morphometry as prognostic indicators in advanced ovarian cancer: A step forward in predicting the clinical outcome. *Gynecol. Oncol.*, **29**, 176–87.
51. Friedlander, M.L., Russell, P., Taylor, I.W. *et al.* (1984) Flow cytometric analysis of cellular DNA content as an adjuvant to the diagnosis of ovarian tumors of borderline malignancy. *Pathology*, **16**, 301–6.
52. Kaern, J., Trope, C., Kjorstad, K.E. *et al.* (1990) Cellular DNA content as a new prognostic tool in patients with borderline tumors of the ovary. *Gynecol. Oncol.*, **38**, 452–7.
53. Hamaguchi, K., Nishimura, H., Miyoshi, T. *et al.* (1990) Flow cytometric analysis of cellular DNA content in ovarian cancer. *Gynecol. Oncol.*, **37**, 219–23.
54. Kallioniemi, O.-P., Punnonen, R., Mattila, J. *et al.* (1988) Prognostic significance of DNA index, multiploidy and S-phase fraction in ovarian cancer. *Cancer*, **61**, 334–9.
55. Iversen, O.-E. and Skaarland, E. (1987) Ploidy assessment of benign and malignant ovarian tumors by flow cytometry. A clinicopathologic study. *Cancer*, **60**, 82–7.
56. Feichter, G.E., Kuhn, W., Czernobilsky, B. *et al.* (1985) DNA flow cytometry of ovarian tumors with correlation to histopathology. *Int. J. Gynecol. Pathol.*, **4**, 336–45.
57. Christov, K. and Vassilev, N. (1987) Flow cytometric analysis of DNA and cell proliferation in ovarian tumors. *Cancer*, **60**, 121–5.
58. Pretorius, R.G., Lee, K.R., Papillo, J. *et al.* (1986) False-negative peritoneal cytology in metastatic ovarian carcinoma. *Obstet. Gynecol.*, **68**, 619–23.
59. Lovecchio, J.L., Budman, D.R., Susin, M. *et al.* (1986) Flow cytometry of peritoneal washings in gynecologic neoplasia. *Obstet. Gynecol.*, **67**, 675–9.
60. Sinton, E.B., Carver, R.K., Morgan, D.L. *et al.* (1990) Prospective study of concurrent ploidy analysis and routine cytopathology in body cavity fluids. *Arch. Pathol. Lab. Med.*, **114**, 188–94.
61. Unger, K.M., Raber, M., Bedrossian, C.W.M. *et al.* (1983) Analysis of pleural effusions using automated flow cytometry. *Cancer*, **52**, 873–7.
62. Rijken, A., Dekker, A., Taylor, S. *et al.* (1991) Diagnostic value of DNA analysis in effusions by flow cytometry and image analysis. *Am. J. Clin. Pathol.*, **95**, 6–12.
63. Cibas, E.S., Malkin, M.G., Posner, J.B. and Melamed, M.R. (1987) Detection of DNA abnormalities by flow cytometry in cells from cerebrospinal fluid. *Am. J. Clin. Pathol.*, 570–7.
64. Stonesifer, K.J., Xiang, J., Wilkinson, E.J. *et al.* (1987) Flow cytometric analysis and cytopathology of body cavity fluids. *Acta Cytol.*, **31**, 125–30.
65. Hostmark, J., Vigander, T. and Skaarland, E. (1985) Characterization of pleural effusions by flow cytometric DNA analysis. *Eur. J. Respir. Dis.*, **66**, 315–19.
66. Zarbo, R.J. (1991) Flow cytometric DNA analysis of effusions. A new test seeking validation. *Am. J. Clin. Pathol.*, 2–4.
67. Hedley, D.W., Philips, J., Rugg, C.A. and Taylor, I.W. (1984) Measurement of cellular DNA content as an adjunct to diagnostic cytology in malignant effusions. *Eur. J. Clin. Oncol.*, **20**, 749–52.
68. Schneller, J., Eppich, E., Greenebaum, E. *et al.* Flow cytometry and Feulgen cytophotometry in evaluation of effusions. *Cancer*, **59**, 1307–13.
69. O'Quinn, A.G., Barranco, S.C. and Costanzi, J.J. Tumor cell kinetics-directed chemotherapy for advanced squamous carcinoma of the cervix. *Gynecol. Oncol.*, **18**, 135–44.
70. Rutgers, D.H., van Oostrum, I.E., Noorman van der Dussen, M.F. and Wils, I.S. (1989) Relationship between cell kinetics and radiation-induced arrest of proliferating cells in G2: relevance to efficacy of radiotherapy. *Anal. Cell Pathol.*, **1**, 53–62.
71. Rosso, R., Alama, A., Repetto, L. and Conte, P.F. (1990) Timed sequential chemotherapy

following ifosfamide-induced kinetic recruitment in refractory ovarian cancer. *Cancer Chemother. Pharmacol.*, **26** (Suppl.), S43–S44.
72. Klevecz, R.R., Shymko, R.M., Blumenfeld, D., and Braly, P.S. (1987) Circadian gating of S phase in human ovarian cancer. *Cancer Res.*, **47**, 6267–71.
73. Klevecz, R.R. and Braly, P.S. (1991) Circadian and ultradian rhythms of proliferation in human ovarian cancer. *Chronobiol. Int.*, **4**, 313–23.
74. Klevecz, R.R. and Braly, P.S. (1991) Circadian and ultradian cytokinetic rhythms of spontaneous human cancer. *Ann. N.Y. Acad. Sci.*, **618**, 257–76.
75. Braly, P.S. and Klevecz, R.R. (1991) Proliferative response of human tumors to surgical debulking (submitted).

Chapter 24
Ultrasound for early cancer screening

W.P. COLLINS, T.H. BOURNE, K. REYNOLDS,
V. BHAN, J. HAMPSON, P. ROYSTON,
M.I. WHITEHEAD and S. CAMPBELL

24.1 INTRODUCTION

Screening has been defined as 'the identification, among apparently healthy individuals, of those who are sufficiently at risk of a specific disorder to justify a subsequent diagnostic test or procedure, or in certain circumstances, direct preventive action'[1]. The implications of screening for ovarian cancer have been discussed, and criteria for the evaluation of alternative and complementary procedures have been described[2]. In this context, early ovarian cancer is usefully defined as the stage of the disease that can be removed completely by surgery. The presence of the tumour may not be apparent from an examination of the ovarian surface, and the maximum diameter of the lesion is usually <5 cm.

To date, first stage screening procedures for early cancer have either involved the immunoassay of tumour-associated antigens in peripheral serum, or the detection of persistent abnormal ovarian morphology, growth and blood flow by pelvic ultrasonography. We believe that a high detection rate is the most important characteristic of a screening procedure for this lethal disease. The false positive rate, however, should be low and ideally restricted to those women with ovarian masses which have the potential to become malignant. Second stage tests are being used to help achieve this objective. This chapter is concerned with the use of pelvic ultrasonography as the first stage of a screening procedure for early ovarian cancer.

24.2 ULTRASONOGRAPHY

This technique involves the use of high frequency sound waves to produce an image of the internal morphology of organs and obtain information about localized blood flow [3]. Ultrasound passes relatively freely through body fluids, but is reflected in varying amounts by solid tissues. A probe is used to emit the sound waves and collect the reflected signals. Traditionally the beam (usually 3.5 MHz) has been directed transabdominally through a full bladder. More recently transvaginal probes have been developed. This approach enables the use of higher frequency ultrasound (around 7.0 MHz), which improves the resolving power of the technique. A major practical advantage is that the woman does not need to have a full bladder. The same type of probe can be used in gynaecological practice for colour flow imaging and blood flow analysis.

24.3 TRANSABDOMINAL SCREENING

Transabdominal ultrasonography was introduced into gynaecological practice by Donald and used in the diagnosis of overt abdominal swellings[4]. Subsequently, progressively improved equipment was used to study intra-ovarian morphology[5], monitor follicular development[6] and record changes that are associated with presumed ovulation[7]. These applications led to the suggestion that the same technique might be used to screen for early ovarian cancer[8]. Initially, the maximum ovarian diameters in three planes were determined by ultrasonography and shown to correlate well with the same measurements with a ruler after laparotomy[8]. Subsequently, the persistence of areas of hypo- or hyper-echogenicity at two or more scans was shown to reflect the presence of abnormal morphology.

24.3.1 PROSPECTIVE STUDY

The technique has been used to screen self-referred, asymptomatic women on three separate successive occasions. The end-point was the presence of persistent abnormal intra-ovarian morphology over a minimum of two scans, 3–6 weeks apart. Details of ovarian volumes, histopathology and epidemiological data were recorded for retrospective analysis.

Screening outcomes

The study has been reported in detail[9,10]. The mean age (years) of the women was 52 (range 18–78). At the time of the first screening 45% were premenopausal, 43% were

Table 24.1 Initial outcomes of screening self-referred, asymptomatic women for sporadic ovarian cancer by transabdominal ultrasonography

		Result (first scan/final scan)		
Screening	No. of women	Negative/ negative (%)	Positive/ negative* (%)	Positive/ positive† (%)
1	5479	93.9	2.5	3.6
2	4914	93.0	5.2	1.9
3	4201	92.6	6.2	1.2

* Result initially positive but abnormality disappeared on rescanning.
† One or more rescans showed abnormality and woman referred for surgical investigation.

naturally postmenopausal and 12% had undergone a hysterectomy. A summary of the initial outcomes at each screening is shown in Table 24.1. About 90% of women attended for the second screening and 77% for the third. The proportion of women with a positive result at screening 1 that became negative at a subsequent scan (indicating the presence of a transient mass) increased over successive screenings, whereas the proportion with a consistent positive result decreased markedly. The outcome after surgical investigations of women with an overall positive result is shown in Table 24.2. Five women had primary ovarian cancer (four had stage Ia and one stage Ib disease). Two cases were detected at the first screening and three at the second. The detection rate (sensitivity) of the procedure was 100%, based on the expected prevalence of the disease in the population, the outcome of successive screenings and the results of a 12–18-month follow-up study. The false positive rate decreased by two-thirds over three screenings. The predictive value of a positive result was 1.0% at screening 1 and 3.3% at screening 2 (1.5% overall). The overall odds in favour or against finding different types of ovarian pathology in women referred for surgical investigation are shown in Table 24.3.

Table 24.2 Outcome after surgical investigation of women with a positive/positive screening result

Screening	No. of positive/positive results	No. of primary cancers	Prevalence/ 1000 women	Detection rate (%)	False positive results No.	False positive results %	PPV (odds)
1	195	2	0.365	100	193	3.5	1 : 97*
2	92	3	0.548	100	89	1.8	1 : 30
3	51	0	–	–	51	1.2	–
All	338	5	0.913	100	333	2.3	1.67

* odds of 97 : 1 against finding primary ovarian cancer.
PPV, positive predictive value.
odds, PPV (1−PPV).

24.3.2 RETROSPECTIVE ANALYSIS

The data on ovarian volumes were analysed retrospectively[11]. The use of abnormal morphology, or the maximum ovarian volume (MOV) >96th centile as alternative criteria for a positive result, together with a defined volume change (VC) at rescan (>0.63 of value at scan 1) would have given a false positive rate of 3.1% at screening 1 and 2.0% overall. The presence of abnormal morphology alone at scan 1, followed the persistence of the abnormality at rescan together with the defined VC, would have given an overall false positive rate of 1.6%. These criteria for a positive screening result would change the overall odds to 2 : 1 in favour of finding a tumour at surgery and 1 : 50 against the presence of primary ovarian cancer (Table 24.3).

The implications of these findings are debatable[12]. The removal of 50 benign ovarian masses for each malignant tumour is only acceptable if a proportion of the former have the potential to become malignant, or cause other problems. Related studies have shown areas of malignant transformation in some serous tumours[13], and in cases of atypical ovarian endometriosis[14]. Accordingly, the removal of apparently benign lesions may lead to a reduction in the incidence of ovarian cancer. The development of a test for premalignancy is required urgently. Currently, it is of particular interest that nine malignant and 91 benign epithelial tumours were detected at the first two screenings[10], whereas no cancers and only four benign epithelial tumours were detected at the third. A policy of performing a prevalence examination, followed by a second screening 18 months later, might be a practical approach to the start of population screening for early ovarian cancer. The recent advances in pelviscopic surgery also have the potential to alter the implications of a false positive screening

Table 24.3 Odds of finding different types of ovarian pathology in women referred for surgical investigation after a positive/positive screening result

Ovarian pathology	Odds Prospective study	Odds Retrospective analysis
Any persistent mass	1 : 0.25*	1 : 0.25
Any persistent tumour	1 : 2	1 : 5
All cancers†	1 : 37	1 : 26
Primary ovarian cancer	1 : 67	1 : 50

* 4 : 1 in favour.
† Four cases of secondary (metastatic) ovarian cancer were detected.
odds, positive value/(1 − positive predictive value).

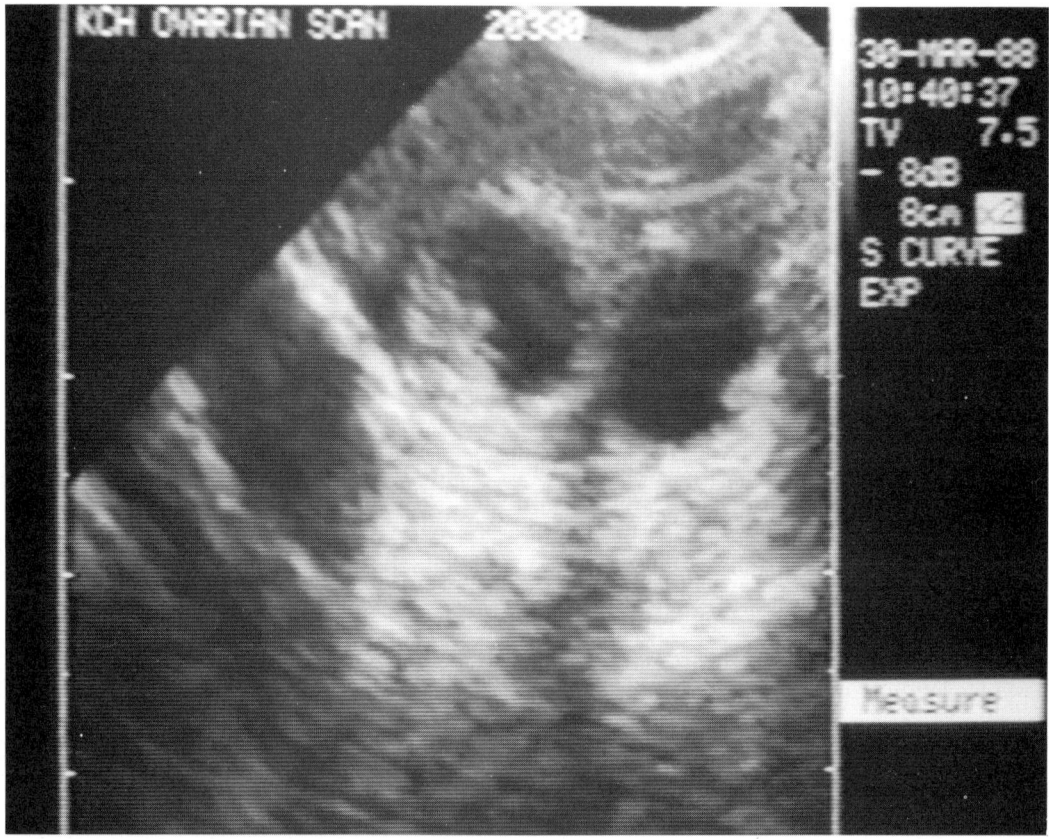

Figure 24.1 Transvaginal sonograms of an ovary containing a serous cystadenocarcinoma of borderline malignancy (identification no. 0330, Table 24.5). (A) Transverse section. (B) Transverse section, in artificial colour based on signal intensity.

result[15]. However, other workers have screened 801 women attending a gynaecological outpatients clinic and concluded that ultrasonography does not have a role in screening for ovarian cancer[16].

24.3.3 MORPHOLOGICAL CHARACTERIZATION

The limited resolution of transabdominal ultrasonography is a major factor which negates attempts to characterize tissues on the basis of their morphological appearance by this technique. Nevertheless, a premise has arisen that the appearance of an ovarian mass as a simple unilocular cyst signifies a low risk of malignancy. For example, Andolf and colleagues found no evidence of cancer in 58 anechoic simple cystic lesions that had a maximum diameter of <5.0 cm[17]. However, Meire and co-workers found two malignant tumours amongst 42 unilocular cystic lesions[18]. All five primary cancers that we detected in our screening study for early ovarian cancer were described as unilocular[9]. We now have unpublished data which

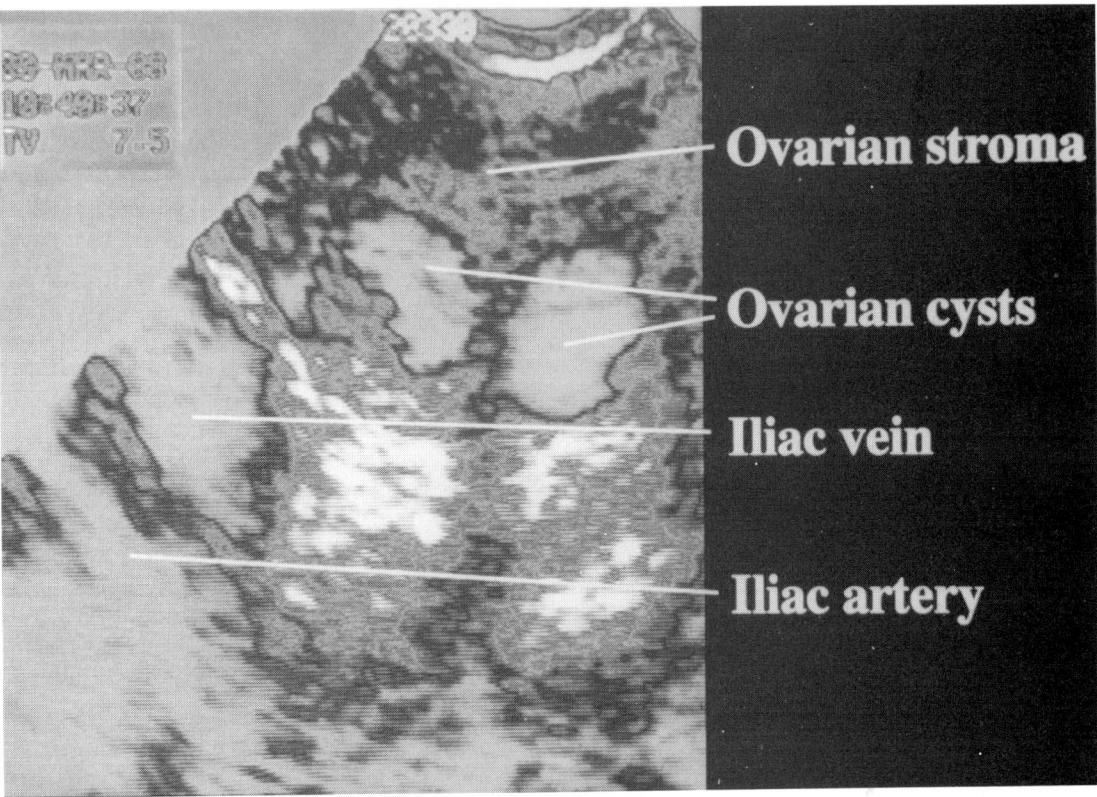

Figure 24.1(B)

show that small multilocular cysts may be described as unilocular on the basis of transabdominal ultrasonography. Accordingly we do not recommend the use of transabdominal probes for a detailed morphological assessment of small ovarian masses. We are optimistic, however, that transvaginal ultrasonography can be used more successfully for this purpose.

24.4 TRANSVAGINAL SCREENING

The practical aspects of transvaginal ultrasonography have been described in detail[19]. Preliminary data have also been published on the use of transvaginal probes (with and without colour Doppler) in screening programmes for primary ovarian cancer.

24.4.1 POPULATION-BASED STUDY

The findings from a transvaginal-based screening programme of 506 asymptomatic women have been reported[20]. There were 12 screenings with a positive result and 10 women consented to exploratory surgery. One metastatic ovarian cancer from a primary in the colon was detected. There were four cases, two with benign serous cystadenomas, three with endometriomas and two with cystic teratomas. Data on the first 1000 women who have undergone a first and second screening (12 months later) have now been reported[21]. The criterion for a positive screening result was the presence of persistent abnormal morphology in an enlarged ovary. The volume for a normal premenopausal ovary was defined as <18.0 cm^3 and for a postmenopausal ovary <8.0 cm^3. Although 9.2% of the premenopausal women had an ovarian abnormality at their initial scan, 41% of the masses had resolved when the examination was repeated one week after the next menses. A total of 24 women (2.4%) underwent surgery, but no primary ovarian cancer was found. The authors concluded that transvaginal ultrasonography is an appropriate technique for detecting subtle changes in ovarian morphology and volume. More data are required to determine the detection rate of the procedure. The potential use of transabdominal and transvaginal ultrasonography as first stage tests in screening procedures for sporadic ovarian cancer has been reviewed[22].

24.4.2 MORPHOLOGICAL SCORES

We believe that transvaginal ultrasonography will enable ovarian lesions to be detected and classified at an earlier stage than has been possible with transabdominal probes. Normal ovarian structures and abnormal lesions in 200 patients have been classified according to the sonographic presence and appearance of septae, papillae, loculations, daughter cysts, solid areas and fluid[23]. There is a greater probability that with this approach an ovarian mass <5.0 cm in diameter with the appearance of a simple unilocular cyst is not malignant. For example, a study of 180 ovarian lesions scanned prior to laparotomy revealed that no unilocular simple cyst was malignant, whereas the presence of papillary formations were associated with an increased risk of cancer[24]. Criteria for the assessment of preoperative masses, however, are not necessarily applicable to the detection of early ovarian cancer in a screening programme. To date, we have detected two masses of borderline malignancy in normal sized ovaries and a detailed assessment of both has indicated that the morphological evidence as observed by transvaginal ultrasonography can occur at a very early stage in the disease process. An example of a very early pre-invasive tumour of borderline malignancy as observed under transvaginal ultrasonography is shown in Figure 24.1A after computerized image enhancement and Figure 24.1B in artificial colour based on signal intensity. Both approaches gave a positive screening result, but it is apparent that much more detail can be observed with a transvaginal probe. Various morphology scoring systems have been developed. The system that we are assessing retrospectively is weighted so that a unilocular cystic lesion with a regular outline has a low score, whereas a multilocular cyst with a solid area and an irregular outline receives a high score. Other workers using a different scoring system (based on data from 128 ovarian masses) found that it was possible to discriminate between benign and malignant lesions with a specificity and sensitivity of 83% and 100% respectively[25]. These scoring systems should be applied retrospectively to other databases before being used prospectively in screening programmes.

24.4.3 CANCER FAMILIES STUDY

The overall performance of a screening procedure may be improved if it is applied to a population who are at an increased risk of developing the disease. A family history of ovarian cancer has been recognized as a significant risk factor for unaffected relatives[26]. We are currently screening a cohort of self-referred asymptomatic women, who were probably at an increased risk of ovarian cancer because they had at least one close relative develop the disease[27]. A summary

Transvaginal screening

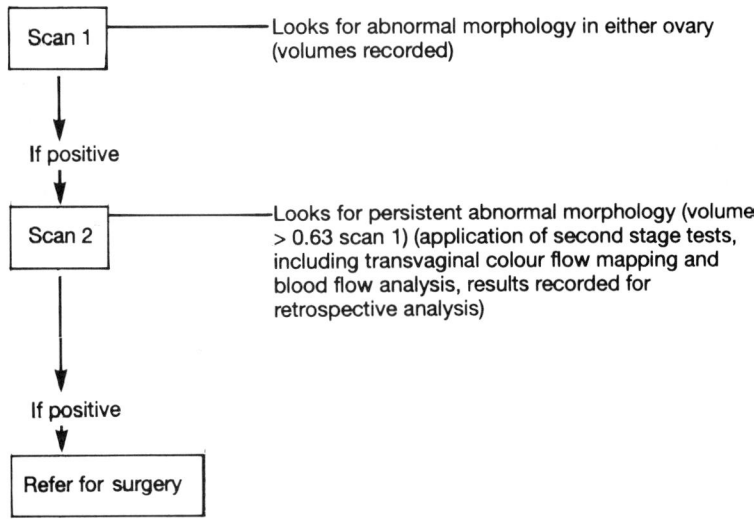

Figure 24.2 Flow diagram of ultrasound-based screening procedure for familial ovarian cancer.

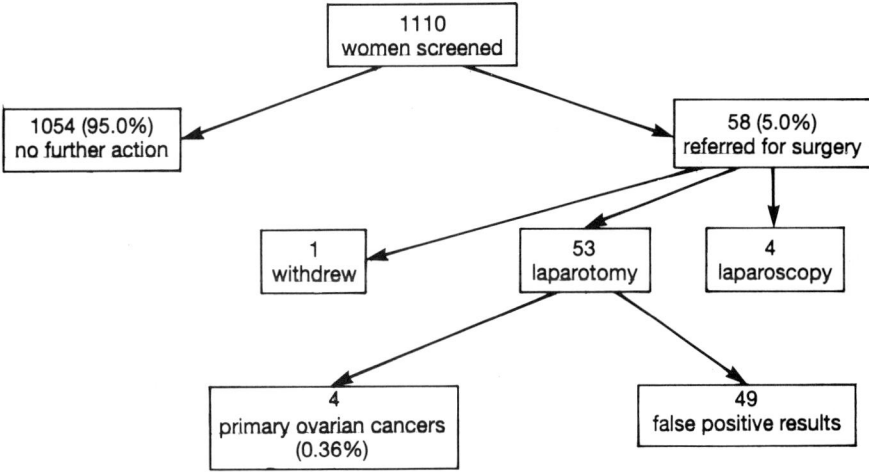

Figure 24.3 Initial outcomes of screening self-referred, asymptomatic women by transvaginal ultrasonography for familial ovarian cancer.

of the protocol and the criteria for a positive screening result is shown in Figure 24.2. All the women have been scanned transabdominally and transvaginally and had peripheral blood taken for the measurement of tumour-associated antigens. Detailed results from the first 776 screenings have been published[28]. Their mean age was 51 years (range 24–78); 52% were premenopausal, 43% were naturally postmenopausal, and 12% had undergone a hysterectomy. To date, 1110 women have been screened. The initial outcome of screening is summarized in Figure 24.3. Four primary ovarian cancers have been detected, all at FIGO stage Ia (giving a prevalence of 3.6/1000 women). Of the false positive results 48% were found in premenopausal women, 16% in women who had undergone a hyster-

231

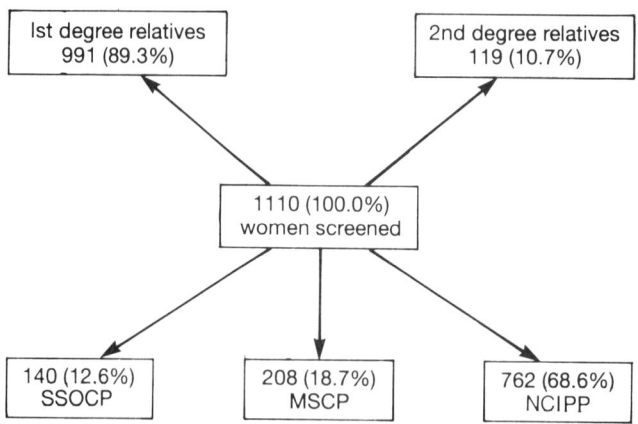

Figure 24.4 Classification of women attending the screening clinic according to index patient (the relative who had ovarian cancer) and family pedigrees (SSOCP, site-specific ovarian cancer pedigree; MSCP, multiple site cancer pedigree; NCIPP, no clear inheritance pattern pedigree).

ectomy and 14% in women who were naturally postmenopausal. The prevalence of the disease and the predictive value of positive screening results were significantly higher than the corresponding values from our population-based study[28].

Pedigree analysis

Detailed family pedigrees were taken by a trained research nurse from every woman attending the screening clinic. A record was made whether a first degree relative (mother, sister or daughter) or second degree relative had developed ovarian cancer. The family pedigrees were divided retrospectively into three groups: (i) site-specific ovarian cancer (SSOC) in which only ovarian cancer was present in two or more relatives in a pattern suggestive of an autosomal dominant inheritance; (ii) multiple site cancer syndrome (MSCS) in which there was evidence for the dominant inheritance of a variety of cancers (excluding lung and cervix), often called the Lynch Type II cancer family syndrome; and (iii) no clear inheritance pattern (NCIP) in which there was no evidence for the dominant inheritance of either ovarian or other cancers. The classification of women according to the relationship with the index patient (the relative who had ovarian cancer) and family pedigree is shown in Figure 24.4. Approximately 30% showed evidence for the inheritance of a predisposing gene for cancer.

The outcome from the referral of women with a positive screening result for surgical investigation is shown in Table 24.4. The data are shown by family pedigree and overall. At this stage of the project the apparent difference in the predictive value of a positive screening result between the subgroups is not statistically significant. Consequently, we are continuing to recruit more women into the study. Some details of the primary cancers detected at screening and those reported at follow-up are listed in Table 24.5. Three women were premenopausal and the fourth had undergone hysterectomy. The two cancers from women with a family history of cancer were serous cystadenocarcinomas. All of the cancers were classified as FIGO stage Ia.

To date, two interval cancers have been reported at follow-up (see Table 24.5). The first (Id. no. 0220) was classified as a peritoneal adenocarcinoma. The woman had not

Table 24.4 Outcome after the referral of women with a positive/positive screening result for surgical investigation (the four women who underwent laparoscopy are classified as false positive results for this analysis)

Pedigree	No. of women	Referrals for surgery No.	Referrals for surgery (%)	No. of primary cancers	False positive rate (%)	PPV (odds)	95% CI
NCIP	762	40	5.4	2	5.1	1 : 20	1 : 168–1 : 5
MSC	208	13	6.3	1	5.8	1 : 13	1 : 528–1 : 2
SSOC	140	5	3.6	1	2.9	1 : 5	1 : 196–1 : 0.4
All	1110	58	5.2	4	4.9	1 : 15	1 : 51–1 : 5

PPV, positive predictive value.
odds, PPV/(1−PPV).
NCIP, no clear inheritance pattern pedigree.
MSC, multiple site cancer.
SSOC, site-specific ovarian cancer pedigree.

Table 24.5 Primary cancers detected by screening for familial ovarian cancer and those reported at follow-up

Identification no.	Age (years) at screening or diagnosis	Menopausal state	Pedigree	Histological classification	FIGO stage
Detected by screening					
0115	38	pre	SSOC	Papillary serous cystadenocarcinoma	Ia
0330	63	art*	MSC	Borderline serous cystadenocarcinoma	Ia
0464	54	pre	NCIP	Endometrioid cystadenocarcinoma	Ia
1058	46	pre	NCIP	Borderline endometrioid cystadenocarcinoma	Ia
Reported at follow-up					
0220	55	post	MSC	Peritoneal adenocarcinoma	–
0452	53	pre	MSC	Papillary serous cystadenocarcinoma	III

pre, premenopausal; post, postmenopausal; art, post-hysterectomy.
NCIP, no clear inheritance pattern pedigree.
MSC, multiple site cancer pedigree.
SSOC, site-specific ovarian cancer pedigree.

Table 24.6 Relative risks of ovarian cancer in first or second degree relatives of index patients with ovarian cancer, subdivided according to type of pedigree

Group	No. of women	No. of primary cancers	Rate/ 1000	Relative risk
General population	5479	2	0.365	–
Cancer families	1110	4	3.60	9.9
NCIP	762	2	2.62	7.2
MSCS	208	1	4.81	13.2
SSOC	140	1	7.14	19.6

NCIP, no clear inheritance pattern pedigree.
MSCS, multiple site cancer pedigree.
SSOC, site-specific ovarian cancer pedigree.

menstruated for 4 years. She contacted her family doctor about 11 months after the last scan and described a 3-month history of abdominal pain and distension. Laparoscopy revealed the presence of disseminated nodules in the peritoneum. A CT scan and repeat ultrasonograms (at the referral hospital and subsequently in our clinic) showed that the ovaries were apparently normal. The second case (Id. no. 0452) was premenopausal. She contacted her family doctor about 24 months after her last scan following 3 months of abdominal discomfort and pain. Subsequently surgery revealed the presence of bilateral papillary serous cystadenocarcinoma (FIGO stage III). The timing of this incident cancer is consistent with our previous, preliminary findings that a second screening should be performed about 18 months after the first[9].

The relative risk of ovarian cancer in first or second degree relatives of women who have developed the disease, compared with the general population, is shown in Table 24.6.

Table 24.7 A summary of advances in first stage ultrasound screening for early ovarian cancer

Technique	Population	Screening	End-points	Study	Odds against finding primary ovarian cancer at surgery
TAU	General	1	M	pro	1 : 97
		1	M, VC	ret	1 : 75
		1–3	M	pro	1 : 67
		1–3	M, VC	ret	1 : 50
TVU	Cancer families	1	M, VC	pro	1 : 15
	NCIPP	1	M, VC	ret	1 : 20
	MSCP	1	M, VC	ret	1 : 13
	SSOCP	1	M, VC	ret	1 : 5

TAU, transabdominal ultrasonography.
TVU, transvaginal ultrasonography.
NCIPP, no clear inheritance pattern pedigree.
MSCSP, multiple site cancer cyndrome pedigree.
SSOCP, site-specific ovarian cancer pedigree.
M, persistent abnormal morphology.
VC, defined volume change at rescan.
pro, prospective study.
ret, retrospective analysis.

The overall risk of 9.9 is over twice the value (4.5) calculated from a detailed analysis of the first 518 family pedigrees[29], and may indicate the detection of some slow growing tumours, i.e. length time bias[2]. Advances in the development of a first stage ultrasound-based screening test for early ovarian cancer are summarized in Table 24.7. These data show that the odds against finding cancer at surgery in a woman with a positive screening result can be reduced from 1 : 97 (95% CI 1 : 26 to 1 : 780) to about 1 : 4 (95% CI 1 : 0.4 to 1 : 196) by the use of appropriate end-points and the study of high risk groups. The technique has the added advantage that endometrial cancer can be detected at the same examination[30].

24.5 CONCLUSION

Pelvic ultrasonography, using the presence of persistent abnormal morphology as an end-point, can be used effectively as a first stage screening test for early ovarian cancer. The false positive rate can be reduced by the use of a defined change in ovarian volume at rescan as an additional end-point. The false positive rate is lower for postmenopausal than for premenopausal women. The use of a morphological score may reduce the false positive rate even further. The predictive value of a positive screening result is increased significantly by screening first degree relatives of women who have developed ovarian cancer. There should be at least two screenings less than 18 months apart. Family pedigrees can be used to identify women who are at an increased risk of developing the disease.

ACKNOWLEDGEMENTS

The authors are grateful to the Cancer Research Campaign and the Imperial Cancer Research Fund for financial support; Aloka Ltd, Tokyo, Japan for the use of their ultrasound equipment; previous colleagues for their contributions to the database; and J. Slack and R. Houlston from the Royal Free Hospital for the analysis of family pedigrees. The manuscript was typed by Mrs Jill Monk.

REFERENCES

1. Cuckle, H.S. and Wald, N.J. (1984) Principles of screening, in *Antenatal and Neonatal Screening* (ed N.J. Wald), Oxford University Press, Oxford.
2. Cuckle, H.S. and Wald, N.J. (1990) The evaluation of screening tests for ovarian cancer, in *Ovarian Cancer, Biological and Therapeutic Challenges* (eds. F. Sharp, W.P. Mason, R.E. Leake), Chapman & Hall, London, pp.229–39.
3. Wells, P.N.T. (1989) Doppler ultrasound in medical diagnosis. *Br. J. Radiol.*, **62**, 399–420.
4. Donald, I. (1963) Use of ultrasonics in the diagnosis of abdominal swellings. *Br. Med. J.*, ii, 1154–5.
5. Kratochwill, A., Jenstsch, K. and Bresina, K. (1973) Ultraschall Anatomie des weiblichen Beckens und ihre klinische Bedeutung. *Arch. Gynekol.*, **214**, 273–7.
6. Hackeloër, B.J., Fleming, R., Robinson, H.P. et al. (1979) Correlation of ultrasonic and endocrinologic assessment of human follicular development. *Am. J. Obstet. Gynecol.*, **135**, 122–8.
7. Queenan, J.T., O'Brien, G.D., Bains, L.M. et al. (1980) Ultrasound scanning of ovaries to detect ovulation in women. *Fertil. Steril.*, **34**, 99–105.
8. Campbell, S., Goessons, L., Goswamy, R. and Whitehead, M.I. (1982) Real-time ultrasonography for the determination of ovarian morphology and volume. A possible early screening test for ovarian cancer. *Lancet*, **i**, 425–6.
9. Campbell, S., Bhan, V., Royston, P. et al. (1989) Transabdominal ultrasound screening for early ovarian cancer. *Br. Med. J.*, **229**, 1363–7.
10. Bhan, V., Amso, N., Whitehead, M.I. et al. (1989) Characteristics of persistent ovarian masses in asymptomatic women. *Br. J. Obstet. Gynaecol.*, **96**, 1384–91.

11. Campbell, S., Royston, P., Bhan, V. et al. (1990) Novel screening strategies for early ovarian cancer by transabdominal ultrasonography. *Br. J. Obstet. Gynaecol.*, **97**, 304–11.
12. Bourne, T.H., Reynolds, K. and Campbell, S. (1991) Ovarian cancer screening. *Eur. J. Cancer*, **27**, 655–9.
13. Stenback, F. (1981) Benign, borderline and malignant serous cystadenomas of the ovary. *Pathol. Res. Pract.*, **172**, 58–72.
14. Moll, U.M., Chumas, J.C., Chalas, E. and Mann, W.J. (1990) Ovarian carcinoma arising in atypical endometriosis. *Obstet. Gynecol.*, **75**, 537–9.
15. Parker, W.H. and Berek, J.S. (1990) Management of selected cystic adnexal masses in postmenopausal women by operative laparoscopy: a pilot study. *Am. J. Obstet. Gynecol.*, **163**, 1574–7.
16. Andolf, E., Jorgensen, C. and Astedt, B. (1990) Ultrasound examination for detection of ovarian cancer in risk groups. *Obstet. Gynecol.*, **75**, 106–9.
17. Andolf, E. and Jorgensen, C. (1989) Cystic lesions in elderly women, diagnosed by ultrasound. *Br. J. Obstet. Gynaecol.*, **96**, 1076–9.
18. Meire, H.B., Farrant, P. and Guha, T. (1978) Distinction of benign from malignant ovarian cysts by ultrasound. *Br. J. Obstet. Gynaecol.*, **85**, 893–9.
19. Timor-Tritsch, I.E., Bar-Yam, U., Elgali, S. and Rottem, S. (1988) The technique of transvaginal sonography with the use of the 6.5 MHz probe. *Am. J. Obstet. Gynecol.*, **158**, 1019–24.
20. Higgins, R.V., Van Nagell, J.R., Donaldson, E.S. et al. (1989) Transvaginal ultrasonography as a screening method for ovarian cancer. *Gynecol. Oncol.*, **34**, 402–6.
21. Van Nagell, J.R., Higgins, R.V., Donaldson, E.S. et al. (1990) Transvaginal sonography as a screening method for ovarian cancer. A report of the first 1000 cases screened. *Cancer*, **65**, 573–7.
22. Sparks, J.M. and Varner, R.E. (1991) Ovarian cancer screening. *Obstet. Gynecol.*, **77**, 787–92.
23. Rottem, S., Levit, N., Thaler, I. et al. (1990) Classification of ovarian lesions by high-frequency transvaginal sonography. *J. Clin. Ultrasound*, **18**, 359–63.
24. Granberg, S., Norstrom, A. and Wikland, M. (1990) Tumors in the lower pelvis as imaged by transvaginal sonography. *Gynecol. Oncol.*, **37**, 224–9.
25. Timor-Tritsch, I.E., Sassone, M., Artner, A. et al. (1991) Transvaginal sonographic characterisation of ovarian pathology: evaluation of a new scoring system to predict ovarian malignancies. *Ultrasound Obstet. Gynecol.* Suppl. 1, 49.
26. Lynch, H.T., Watson, P., Bewtra, C. et al. (1991) Hereditary ovarian cancer. *Cancer*, **67**, 1460–6.
27. Campbell, S., Bourne, T.H., Whitehead, M.I. et al. (1990) New developments in ultrasound screening for ovarian cancer. *J. Ultrasound Med.*, **9** (Suppl. 1), 93.
28. Bourne, T.H., Whitehead, M.I., Campbell, S. et al. Ultrasound screening for familial ovarian cancer. *Gynecol. Oncol.*, **43**, 92–7.
29. Houlston, R., Bourne, T.H., Davies, A. et al. (1991) Use of family history to identify relatives at high risk of ovarian and other cancer. *Gynecol. Oncol.* (submitted).
30. Bourne, T.H., Campbell, S., Steer, C.V. et l. (1991) Detection of endometrial cancer by transvaginal ultrasonography with color flow imaging and blood flow analysis: a preliminary report. *Gynecol. Oncol.*, **40**; 253–9.

Chapter 25

Role of colour Doppler in an ultrasound-based screening programme

S. CAMPBELL, T.H. BOURNE, K. REYNOLDS,
J. HAMPSON, P. ROYSTON, M.I. WHITEHEAD
and W.P. COLLINS

25.1 INTRODUCTION

Transvaginal ultrasonography is being used to screen women from the general population[1,2], or high risk families[3,4], for early ovarian cancer. We are optimistic that the detection rate of the procedure is high, but the false positive rate will vary from around 1.2 to 5.2% depending on the population being screened and the number of previous examinations. A false positive result invariably arises from the presence of a persistent, benign ovarian mass. Consequently, various second stage tests, based on the immunometric assay of serum antigens, radio-immunoscintigraphy, or intra-ovarian colour flow mapping are being evaluated for their ability to distinguish between early malignant and benign masses. The aim is to identify a test (or combination of tests) which will maintain the detection rate of the screening procedure, but reduce the false positive rate to a more acceptable level. Our aim is to ensure that only women with ovarian cancer, or the type of mass that has a known malignant potential, are referred for surgery.

The relationship between the development of tumours and angiogenesis has been reviewed[5]. In particular, there is evidence to show that neo-angiogenesis may be induced during the transition from hyperplasia to neoplasia[6]. Furthermore it has been suggested that local tumour invasion and angiogenesis may involve many of the same enzymes and processes[7]. Substances that suppress tumour growth and inhibit angiogenesis have been discovered[8]. Transvaginal ultrasonography with colour flow mapping (colour Doppler) can be used to identify ovarian intratumoural neo-angiogenesis. This chapter contains a progress report on the use of this technique, together with derived indices of intratumoural blood flow, as a second stage test in ultrasound-based screening procedures for early ovarian cancer.

25.2 PRINCIPLES OF COLOUR DOPPLER

Traditional pelvic ultrasonography (B-mode imaging) has been developed to produce on-line pictures of reproductive organs and tissues. Furthermore, the change in frequency of reflected ultrasound waves from moving red cells has been used to derive indices of blood flow impedance and velocity[9]. This technique is called pulsed or spectral Doppler (after the man who first described the changes in frequency of reflected light or sound from moving objects). The information is obtained by a systematic evaluation of the organ under investigation and sampling errors may occur. In particular small blood vessels and minor vascular changes are difficult to detect and study. These problems are being overcome by the advent of colour Doppler.

The B-mode image (which is produced from the excitation of many pixels on the display screen) can be frozen and the Doppler flow signals superimposed in real-time. The information is colour coded, i.e. an electronic convertor produces colour according to the direction and variance of the frequency shifts. Red usually indicates flow towards the transducer and blue away. The brightness of the colour reflects the velocity of flow within the vessels, whilst turbulence may appear as various shades of green[10]. Accordingly, regions of intratumoural vascularization (with a blood velocity between 3 and 7 cm/s depending on the equipment) can be located as areas of colour. A range gate can be placed over a vessel of interest to obtain flow velocity waveforms (FVWs) for analysis. A transvaginal probe enables the transducer to be placed nearer to each ovary, a higher pulsed repetition frequency is used, and there are fewer limitations on the maximum shifted frequencies that can be measured.

25.3 END-POINTS

FVWs which illustrate the distribution and intensity of the shifted Doppler frequencies with time are displayed on-line. In practice, the angle of the transducer is moved to obtain the maximum waveform amplitude and clarity. Blood flow impedance can be expressed as the resistance index (RI) or the pulsatility index (PI). These values are usually calculated electronically from smooth curves fitted manually to good quality waveforms over three cardiac cycles according to the formulae:

$$RI = S/D \text{ and } PI = (S-D)/\text{mean}$$

where S is the peak Doppler shifted frequency at systole, D is the minimum Doppler shifted frequency at diastole, and mean is the mean maximum shifted frequency over the cardiac cycle.

A reduction in the value for the RI or PI reflects a decrease in the impedance to blood flow in the distal vasculature[11,12]. The peak systolic blood velocity can be measured reproducibly after the transducer has been positioned so that the angle between the pulsed Doppler beam and the vessel is probably close to 0°.

25.4 DIAGNOSIS OF OVERT OVARIAN MASSES

Initial results on the use of colour Doppler to distinguish between overt benign and malignant tumours were encouraging. Bourne and colleagues[3] studied blood flow within the ovaries of 30 women with no apparent pathology and 18 women with ovarian tumours (eight of which were subsequently shown to be malignant). All cases of invasive cancer showed evidence of neo-angiogenesis with low impedance blood flow (i.e. a PI <1.0). One serous cystadenoma of borderline malignancy did not show abnormal flow and there was one false positive test result (a dermoid cyst). Two of the invasive cancers were at FIGO stage Ia demonstrating that the technique can detect ovarian cancer before the capsule has ruptured and therefore might be used to screen for early disease. These results were supported by the work of Kurjak

Table 25.1 An analysis of intratumoural blood flow as a diagnostic test for primary ovarian cancer

	Colour Doppler, PI <1.0		
	Positive	Negative	Total
Histopathology			
Primary cancers			
Positive	20	1	21
Negative	4	65	69
Total	24	66	90

PI, pulsatility index.

and co-workers[14], who studied infertile women with normal pelvic morphology and patients with known pelvic masses. Low impedance intratumoural flow (a RI <0.4) was seen in four cases of unstaged primary ovarian cancer. There was one false positive result (a granulosa cell tumour) amongst the 15 benign cystic lesions that were studied. Hata and colleagues[15] found colour and low impedance blood flow (mean RI 0.5) in eight cases of ovarian cancer and four benign tumours (two haemorrhagic corpus luteum cysts, two endometrial ovarian cysts). We have now accumulated data on 90 women attending the Gynaecological Ultrasound Clinic with known ovarian masses. Data from an analysis of intratumoural blood flow (i.e. the presence of colour and a PI <1.0) as an index of ovarian cancer are shown in Table 25.1. The prevalence of ovarian cancer was 30.4%. The detection rate (sensitivity) was 95.2% and the false positive rate 5.8%. The false positive results were obtained from two dermoid cysts, an endometrial ovarian cyst and a corpus luteum cyst. The positive predictive value was 83.3% (i.e. the odds in favour of finding ovarian cancer in women with a positive test result were about 5 : 1). These data suggest that the use of colour Doppler as a second stage test in a population-based screening programme might reduce the odds against finding ovarian cancer at surgery from about 1 : 50[4] to as low as 1 : 3.

25.5 NORMAL OVARIAN FUNCTION

We have used transvaginal ultrasonography with colour flow imaging to study intrafollicular blood flow during the peri-ovulatory period[16] and at the exact time of presumed ovulation[17]. The reproducibility of intra-ovarian blood flow indices have been studied by replicate analysis (at least triplicate) of seven pre-ovulatory follicles and six corpora lutea. The results from a one-way analysis of variance are shown in Table 25.2. The lowest within-woman and highest between-women coefficients of variation were obtained from the measurement of blood velocity.

Intra-ovarian blood FVWs are observed clearly around the time of the luteinizing hormone (LH) rise in the peripheral circulation and intrafollicular coloured vessels are visible (as shown in Figure 25.1) around the time that the hormone level reaches a peak. The RI and PI tend to decrease slightly during the peri-ovulatory period and the peak systolic blood velocity is highest immediately after follicular rupture (i.e the presumed time of ovulation). The mean and range of values for the RI, PI and velocity in pre-ovulatory follicles, corpora lutea and early invasive cancers are listed in Table 25.3. None of the indices can be used to distinguish between the physiological and pathological tissues. Consequently, colour Doppler can only be used to screen postmenopausal or premenopausal women after physiological growths have been identified by at least two scans at different times during the menstrual cycle.

25.6 EARLY FAMILIAL CANCER SCREENING

We are evaluating the use of transvaginal ultrasonography with colour flow imaging as a second stage test in our screening pro-

Role of colour Doppler in an ultrasound-based screening programme

Table 25.2 Reproducibility of blood flow analysis within the ovaries (7 preovulatory follicles, 6 corpora lutea) by transvaginal colour flow imaging

				Coefficient of variation (%)*	
Variable	Mean	Min.	Max.	Between women	Within woman
Resistance index (RI)	0.48	0.40	0.58	8.8	5.6
Pulsatility index (PI)	0.66	0.48	0.94	13.3	7.6
Velocity (cm/s)	30.5	10.2	73.5	61.3	4.4

* From one-way analysis of variance.

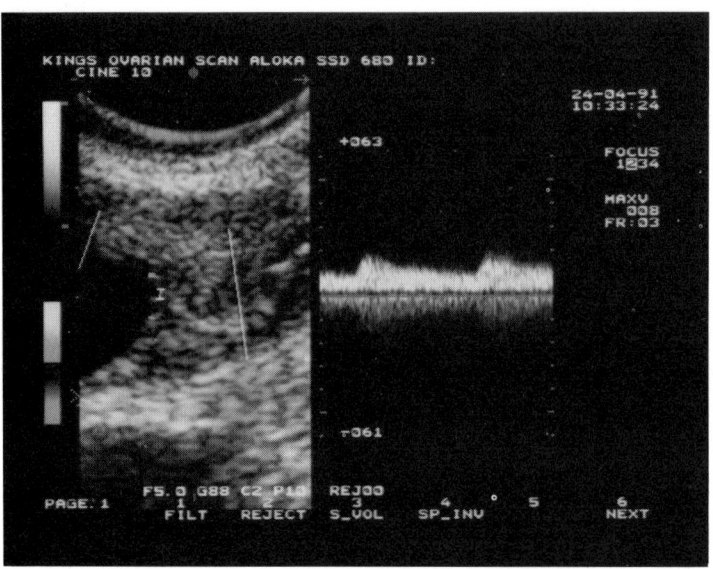

Figure 25.1 A pre-ovulatory follicle (post LH surge) showing neo-angiogenesis (red and blue areas) and distinct flow velocity waveforms (RI 0.53, PI 1.10, velocity 19.1 cm/s).

gramme for early familial ovarian cancer[3,4]. The criterion for a positive screening result (leading to a referral for surgery) is the presence of persistent abnormal morphology over a minimum of two scans. In addition, the volume of the affected ovary at the repeat scan must be >67% of the value recorded at the first examination.

The equipment that we have used (Aloka SSD 680 Colour Doppler) was not available for the first three referrals, which included one case of primary ovarian cancer. For an interim analysis we have used the presence of intratumoural colour with a vessel giving a PI >.0 as the criterion for a positive second stage test result. The outcomes, to date, are

General population screening

Table 25.3 Range of values for indices of intra-ovarian blood flow

Variable	Follicle (post LH) (n = 12)		Corpora lutea (n = 30)		Early invasive cancer (n = 7)	
	Mean	Range	Mean	Range	Mean	Range
Resistance index (RI)	0.48	0.36–0.58	0.43	0.28–0.54	0.46	0.33–0.78
Pulsatility index (PI)	0.62	0.39–0.99	0.56	0.30–0.73	0.61	0.40–0.96
Velocity (cm/s)	26.1	14.3–45.2	43.2	16.1–73.4	44.0	35.6–57.5*

* $n = 3$.

Table 25.4 An analysis of transvaginal ultrasonography with colour flow imaging as a second stage test in a screening programme for familial ovarian cancer

	Colour Doppler, PI <1.0		
	Positive	Negative	Total
Primary cancer			
Positive	2	1	3
Negative	9	43	52
Total	11	44	55

PI, pulsatility index.

summarized in Table 25.4. The prevalence of ovarian cancer, in women with a positive first stage screening result, is 5.5%. The detection rate is 66.6%, the false positive rate 17.3%, and the positive predictive value 18.2%. The odds against finding ovarian cancer at surgery would have been reduced from 1 : 15[4] to about 1 : 5 if colour Doppler had been used prospectively.

25.6.1 DETECTION RATE

The false negative result was obtained from a very early borderline serous cystadenocarcinoma. The woman only had one ultrasound examination and referred herself for surgery elsewhere. A further four FIGO stage I cancers have been detected in women referred from the Gynaecological Outpatients Clinic. An example of a complex cyst showing signs of neo-angiogenesis and low impedance blood flow is shown in Figure 25.2. Some details of all cases are shown in Table 25.5. The overall detection rate is 85.7%.

25.6.2 FALSE POSITIVE RATE

Some details of the cases giving a false positive result in the screening programme for familial ovarian cancer are shown in Table 25.6. The most common are endometriosis (49) and teratomas (2/9). Both of these conditions have been shown to be potentially malignant [18,19]. An example of a persistent corpus luteum cyst with a coloured area is shown in Figure 25.3.

25.7 GENERAL POPULATION SCREENING

We have estimated the potential value of transvaginal ultrasonography with colour flow imaging as a second stage test in a screening procedure for women from the general population. The prevalence of persistent ovarian masses at the first screening in two populations of self-referred asymptomatic women who have participated in our prospective studies is shown in Table

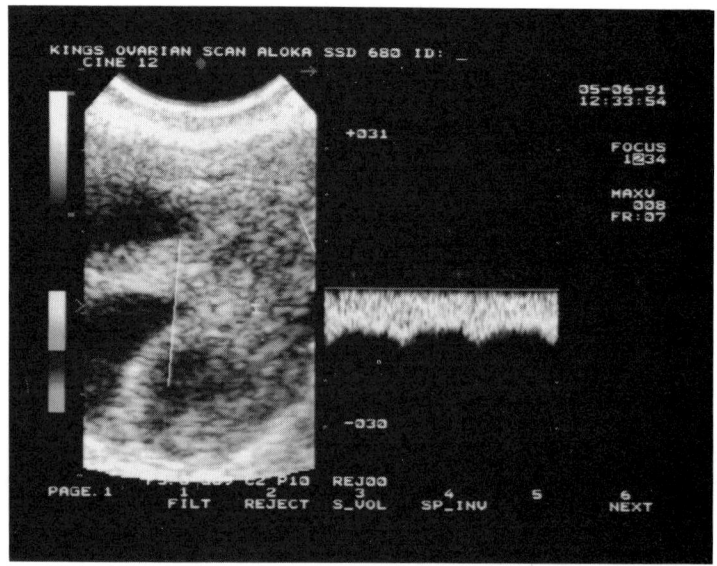

Figure 25.2 A complex cyst showing areas of colour and flow velocity waveforms (RI 0.35, PI 0.52, velocity 12.2 cm/s).

Table 25.5 The detection of early ovarian cancer by transvaginal ultrasonography with colour flow imaging and blood flow analysis

Case no.	Age (years)	Menopausal status	Histological classification	FIGO stage	Test result
1	63	art	Borderline serous cystadenocarcinoma	Ia	Negative*
2	54	pre	Endometrioid cystadenocarcinoma	Ia	Positive
3	46	pre	Borderline endometrioid cystadenocarcinoma	Ia	Positive
4	52	post	Serous cystadenocarcinoma	Ia	Positive
5	52	post	Serous cystadenocarcinoma	Ib	Positive
6	53	post	Endometrioid cystadenocarcinoma	Ic	Positive
7	37	pre	Serous cystadenocarcinoma	Ic	Positive

* Woman had one scan only.
pre, premenopausal; post, postmenopausal, art, post-hysterectomy.

25.7. From the number of each type of mass detected, and the proportion giving a false positive result in the cancer families study, we have calculated that transvaginal colour

Positive predictive values

Table 25.6 False positive results by colour flow imaging and blood flow analysis in a screening programme for familial ovarian cancer

Identification no.	Age (years)	Menopausal status	Pedigree type	Histological diagnosis
0341	45	art	NCIP	Endometriosis
0385	36	pre	MSC	Bilateral teratoma
0401	43	art	MSC	Endometriosis
0507	45	art	SSOC	Endometriosis
0554	53	pre	NCIP	Teratoma
0578	45	pre	NCIP	No abnormality detected
0583	51	pre	NCIP	Follicular cyst
0993	50	pre	NCIP	Corpus luteum cyst
1020	43	pre	NCIP	Endometriosis

MSC, multiple site cancer.
SSOC, site-specific ovarian cancer.
NCIP, no clear inheritance pattern.
pre, premenopausal; post, postmenopausal; post-hysterectomy.

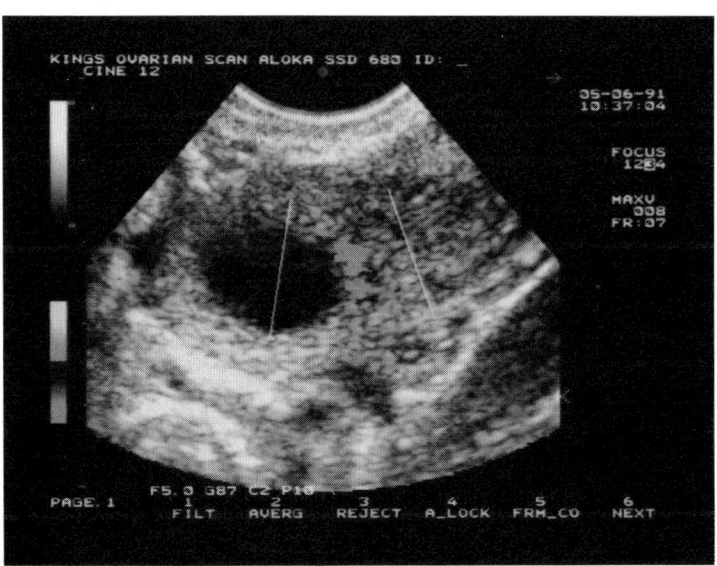

Figure 25.3 A persistent corpus luteum cyst showing retained vascularity.

flow imaging might have reduced the odds against finding ovarian cancer at surgery from 1 : 50 overall to about 1 : 8. The true value, however, must be determined by prospective studies.

25.8 POSITIVE PREDICTIVE VALUES

A summary of advances in the development of an ultrasound screening procedure for early ovarian cancer (based on the presence

Role of colour Doppler in an ultrasound-based screening programme

Table 25.7 The prevalence of persistent ovarian masses in two populations of self-referred, asymptomatic women attending ultrasound-based screening programmes

Ovarian mass	Study group		Rate ratio
	General population (rate/1000 women)	Cancer families (rate/1000 women)	
Malignant epithelial	0.3	1.9	7.0
Benign epithelial	6.6	10.3	1.6
Sex cord stromal/germ cell	1.3	2.6	2.0
Tumour-like	6.3	20.8	3.3
Simple cysts	2.7	6.5	2.4
Follicular cysts	1.9	3.2	1.7
Corpus luteum cysts	0.7	7.1	10.1

Table 25.8 A summary of advances in ultrasound screening for early ovarian cancer

Technique	Population	Screening	End-points	Study	PPV (odds)
TAU	General	1	M	pro	1 : 97
TAU	General	1–3	M	pro	1 : 97
TAU	General	1–3	M, VC	ret	1 : 50
TVU	High risk	1	M, VC	pro	1 : 15
TVUCD	High risk	1	M, VC, BF	pro	1 : 5
TVUCD	High risk	1	MD, VC, BF	pro	1 : 1

TAU, transabdominal ultrasonography.
TVU, transvaginal ultrasonography.
CD, colour Doppler.
M, persistent abnormal morphology.
VC, defined volume change.
BF, defined change in blood flow (pulsatility index).
MS, morphological score.
pro, prospective study.
ret, retrospective analysis.
PPV, positive predictive value.
odds, PPV (1−PPV).

of persistent abnormal morphology) is shown in Table 25.8. The positive predictive value, expressed as the odds of finding ovarian cancer at surgery, can be increased by successive screenings, the introduction of a defined volume change at rescan as a second variable for a positive screening result, the study of high risk (cancer) families, the introduction of a morphological score[4] and the use of colour Doppler. Under optimum conditions, the procedure, to date, would have detected ovarian cancer and benign masses which may have a malignant potential (i.e. teratomas and endometriotic cysts).

25.9 OTHER GYNAECOLOGICAL CANCERS

We have used transvaginal ultrasonography with colour flow imaging to detect endo-

Technological developments

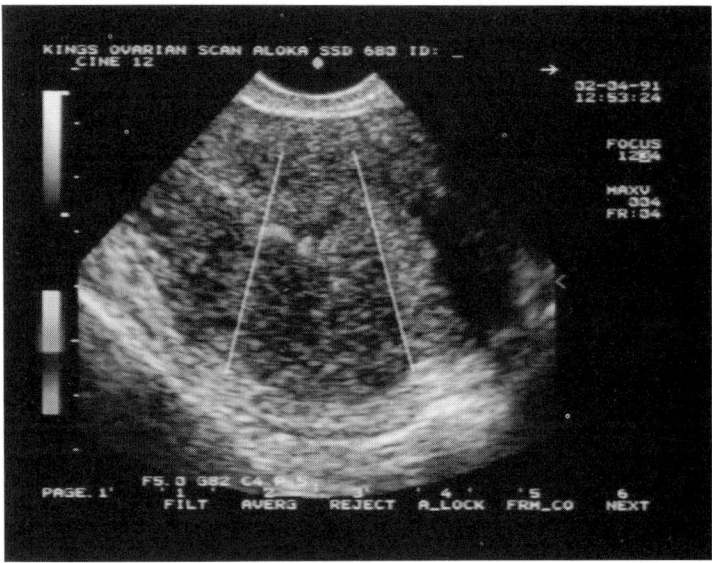

Figure 25.4 Longitudinal section of uterus. Note thickened endometrium and extensive vascular changes associated with stage Ic endometrial cancer.

metrial cancer in women with and without postmenopausal bleeding[20,21]. The data suggest that it should be possible to screen for ovarian and endometrial cancers at the same examination. An example of neo-angiogenesis in early endometrial cancer is shown in Figure 25.4.

25.10 CONCLUSIONS

Transvaginal ultrasonography with colour flow imaging can be used as a second stage test to reduce the false positive rate of an ultrasound-based screening procedure. The criteria for a positive screening result are the presence of persistent abnormal morphology, a defined volume change in the ovary at rescan (3 weeks later), the presence of intra-tumoral colour and a PI <1.0. To date, no false positive results have been obtained from screening postmenopausal women. The use of this algorithm for screening women from cancer families would have reduced the odds against finding cancer at surgery from 1 : 15 to 1 : 5. More data are required to determine precise cut-off values for the blood flow indices of early cancer. A prospective, randomized clinical trial is required to assess if screening, and hence early treatment, will reduce the high mortality rate from this disease.

25.11 TECHNOLOGICAL DEVELOPMENTS

New approaches to the study of pelvic blood flow are being investigated. The measurement of volume flow is one possibility. This index is the product of the mean blood velocity and vessel area. However, it is difficult to obtain reproducible data from small pulsating vessels. Attempts to quantify colour have been disappointing, but it may be possible to determine the number of pixels which have been excited and hence the area of neo-angiogenesis. The new velocity flow scanners may have a useful role to play. Colour velocity imaging (CVI) involves the use of data contained in the grey scale B-mode scan lines to determine blood flow velocity and tissue motion[22]. Both indices are obtained with lower power outputs than are used for pulsed Doppler. The relative importance of this new technique for the detection of early ovarian cancer awaits evaluation in clinical trials.

ACKNOWLEDGEMENTS

The authors are grateful to the Cancer Research Campaign and the Imperial Cancer Research Fund for financial support, and Aloka Ltd, Tokyo, Japan, for the use of their ultrasound equipment.

REFERENCES

1. Higgins, R.V., Van Nagell, J.R., Donaldson, E.S. et al. (1989) Transvaginal ultrasonography as a screening method for ovarian cancer. *Gynecol. Oncol.*, **34**, 402–6.
2. Van Nagell, J.R., Higgins, R.V., Donaldson, E.S. et al. (1990) Transvaginal sonography as a screening method for ovarian cancer. A report of the first 1000 cases screened. *Cancer*, **65**, 573–7.
3. Bourne, T.H., Whitehead, M.I., Campbell, S. et al. (1991) Ultrasound screening for familial ovarian cancer. *Gynecol. Oncol.* (in press).
4. Collins, W.P., Bourne, T.H., Reynolds, K. et al. (1992) Ultrasound for early cancer screening, in *Ovarian Cancer. Biology, diagnosis and management* (eds F. Sharp, W.P. Mason and W. Creasman), Chapman & Hall, London, pp.225–36.
5. Paweletz, N. and Knierim, M. (1989) Tumor-related angiogenesis. *Crit. Rev. Oncol. Hematol.*, **9**, 197–242.
6. Folkman, J., Watson, K., Ingber, D. and Hanahan, D. (1989) Induction of angiogenesis during the transition from hyperplasia to neoplasia. *Nature*, **339**, 58–61.
7. Liotta, L.A., Steeg, P.S. and Stetler-Stevenson, W.G. (1991) Cancer metastasis and angiogenesis: an imbalance of positive and negative regulation. *Cell*, **64**, 327–36.
8. Ingber, D., Fujita, T., Kishimoto, S. et al. (1990) Synthetic analogues of fumagillin that inhibit angiogenesis and suppress tumour growth. *Nature*, **348**, 555–7.
9. Wells, P.N.T. (1989) Doppler ultrasound in medical diagnosis. *Br. J. Radiol.*, **62**, 399–420.
10. Omoto, R. and Kasai, C. (1987) Physics and instrumentation of Doppler color flow mapping. *Echocardiography. A review of cardovascular ultrasound*, **4**, 467–83.
11. Thompson, R.S., Trudinger, B.J. and Cook, C.M. (1988) Doppler ultrasound waveform indices: A/B ratio, pulsatility index and Pourcelot ratio. *Br. J. Obstet. Gynaecol.*, **95**, 51–8.
12. Trudinger, B.J. and Thompson, R.S. (1991) Do velocity indices measure resistance? *Ultrasound Obstet. Gynecol.* **3**, 10–1.
13. Bourne, T.H., Campbell, S., Steer, C.V. et al. (1989) Transvaginal colour flow imaging: a possible new screening technique for ovarian cancer. *Br. Med. J.*, **299**, 1367–70.
14. Kurjak, A., Zalud, I., Jurkovic, D. et al. (1989) Transvaginal color Doppler for assessment of pelvic circulation. *Acta Obstet. Gynecol. Scand.*, **68**, 131–5.
15. Hata, T., Hata, K., Senoh, D. et al. (1989) Doppler ultrasound assessment of tumor vascularity in gynecologic disorders. *J. Ultrasound Med.*, **8**, 309–14.
16. Collins, W.P., Jurkovic, D., Bourne, T. et al. (1991) Ovarian morphology, endocrine function and intra-follicular blood flow during the peri-ovulatory period. *Hum. Reprod.*, **6**, 319–24.
17. Bourne, T.H., Jurkovic, D., Waterstone, J. et al. (1991) Intrafollicular blood flow during human ovulation. *Ultrasound Obstet. Gynecol.*, **1**, 63–9.
18. Moll, U. M., Chumas, J.C., Chalas, E. and Mann, W.J. (1990) Ovarian carcinoma arising in atypical endometriosis. *Obstet. Gynecol.*, **75**, 537–9.
19. Stenback, F. (1981) Benign, borderline and malignant serous cystadenomas of the ovary. *Pathol. Res. Pract.*, **172**, 58–72.

20. Bourne, T.H., Campbell, S., Steer, C.V. *et al.* (1990) Detection of endometrial cancer in postmenopausal women by transvaginal ultrasonography and colour flow mapping. *Br. Med. J.*, **301**, 369.
21. Bourne, T.H., Campbell, S., Steer, C.V. *et al.* (1991) Detection of endometrial cancer by transvaginal ultrasonography with color flow imaging and blood flow analysis: A preliminary report. *Gynecol. Oncol.*, **40**, 253–9.
22. Tegeler, C.H., Kremkau, F.W. and Hitchings, L.P. (1991) Color velocity imaging: Introduction to a new ultrasound technology. *J. Neuroimaging*, **1**, 85–90.

Chapter 26

Transvaginal colour Doppler in the differentiation between benign and malignant ovarian masses

A. KURJAK and I. ZALUD

26.1 INTRODUCTION

Many malignant tumours have bizarre vascular morphology with abnormal blood flow that can be detected by Doppler ultrasound. In general, the tumour vasculature consists of: (i) vessels recruited from the pre-existing vascular anatomy; and (ii) vessels recruited from the angiogenic response of host vessels to cancer cells[1–13]. Although the tumour vasculature originates from the host vasculature, its organization may be completely different depending upon the tumour type, its growth rate and its location. The architecture is different not only among various tumour types, but also between a primary tumour and its metastases[3]. The tumour vasculature can be studied in terms of two idealized categories: peripheral and central. In tumours with peripheral vascularization, the centres are usually poorly perfused. In those with central vascularization, one would expect the opposite. In reality, a tumour may consist of many territories, each exhibiting one or other of these two types of idealized vascular patterns.

26.2 TUMOUR NEOVASCULARIZATION

Solid tumour growth in animals and in man is accompanied by neovascularization. New capillary growth is stimulated by a diffusible factor generated by malignant tumour cells. In the absence of neovascularization, most tumours might become dormant at a tiny diameter, perhaps 2–3 mm[4,5].

Folkman and collaborators have demonstrated that the development of an adequate vascular supply is critical for the growth and development of a cancer as well as its metastases[6]. When they injected cancer cells into the non-vascularized anterior chamber of a rabbit eye, nodules greater than 1–2 mm in diameter did not develop. However, the same cells introduced into a vascularized area of the eye developed a stromal blood supply and rapidly grew over the eye[7].

Often, primary tumours and their metastases outgrow their blood supply and undergo central ischaemic necrosis.

The vascular morphology of one tumour differs from another and is determined to some extent by the growth pattern of cancer cells. Quantitative morphometric studies in induced animal tumours show that vascular volume, length and surface area increase during the early stages of tumour growth, and then decrease after the onset of necrosis. Frequency of large diameter vessels increases in the later stages of growth[8]. New blood vessels and vascular channels in a tumour arise from older, pre-existing vessels. Tumour vessels have a relative paucity of smooth muscle in their walls in comparison to their calibre. Since most of the resistance to flow resides at the level of the muscular arterioles, vessels deficient in these muscular elements present diminished resistance to flow and thereby receive a larger volume flow than vessels with a high impedance.

Obviously, microcirculation plays an important role in the growth, metastasis, detection and treatment of tumours. Transabdominal pulsed Doppler imaging offers a view of the surrounding anatomy and evidence of blood flow in major pelvic vessels, but not in the microcirculation[9,10]. By contrast, transvaginal colour Doppler offers a qualititative picture of blood flow in the vascular system in relation to surrounding anatomy[11–14].

26.3 TISSUE CHARACTERIZATION BY PULSED AND COLOUR DOPPLER

Wells and colleagues[15] and Burns and co-workers[16] documented the presence of abnormal flow spectra around the periphery of malignant tumours. These results were confirmed by several groups[17–19]. The abnormal flow signals consisted of an increase in signal amplitude when compared with the contralateral normal side (corresponding to a great number of moving cells within the beam), an increase in peak systolic flow; a characteristic distribution of the Doppler spectrum, showing a predominance of high power, low frequency element; high diastolic shift (in some tumours the systolic/diastolic variation is absent).

Kujipers and colleagues[20] and Hata and co-workers[21] were the first to apply pulsed Doppler and transabdominal colour Doppler in the study of the female pelvic circulation. In all cases of endometrial carcinoma, ovarian carcinoma and trophoblastic disease, typically abnormal flows were observed by the Japanese team. However, all cases of cervical carcinoma with abnormal flows were stage II-B and above.

Kurjak and colleagues have reported that transvaginal colour flow imaging can be used in the assessment of the pelvic circulation, and to differentiate between benign and malignant pelvic tumours[11–14,22,23]. Using transvaginal colour Doppler, Bourne and colleagues showed that the absence of intratumoral neovascularization and a normal (high) pulsatility index can be used to exclude the presence of invasive primary ovarian cancer[24]. Transvaginal colour flow mapping may be used to identify potentially malignant ovarian masses and help elucidate the early stages of tumorigenesis.

The presence of arteriovenous communications is thought to be an important factor producing sufficient velocity above the minimal threshold on colour Doppler imaging. When tumoral blood flow is not visualized on transvaginal colour Doppler examination, the following factors should be considered.

1. There is no blood flow due to a lack of the newly formed vessels characteristic of malignant tumours.
2. The velocity of flow may be too slow to exceed the minimal threshold for measurement with current colour Doppler systems.
3. Intratumoral blood flow is non-uniform

and turbulent, and detectable blood flow on colour Doppler imaging is distributed in certain regions of a tumour. It sometimes requires considerable effort on the part of the ultrasonographer to obtain a good angle of incidence to the target for colour flow imaging.

26.4 CLINICAL METHODS FOR DIFFERENTIATION OF OVARIAN MASSES

The standard evaluation of adnexal masses includes history, physical examination, CA-125 measurements and ultrasound. Although few data are available regarding the accuracy of physical examination in the differentiation of benign from malignant ovarian masses, it is generally felt that clinical impression has few predictive value. The findings of McFarlane and colleagues[25] are often quoted to support this view. They discovered only six ovarian cancers during a total of 18 753 pelvic examinations performed in 1319 women over a 15-year period (1938–1952).

Some evidence to support the view that vaginal examination lacks sensitivity for the detection of an adnexal mass (either benign or malignant) was provided by the study of Andolf and co-workers[26]. Each patient recruited underwent a pelvic examination prior to ultrasonography. Pelvic examination was reported as normal in 18 of 24 benign ovarian cysts, two borderline malignancies and one ovarian carcinoma.

Encouraging levels of specificity were achieved by the combination of CA-125 measurement with either pelvic examination (100%) or ultrasound (99%). This multimodal approach does appear to provide acceptable levels of specificity[27].

Ultrasound examination of the pelvis and abdomen has become the standard diagnostic test for evaluating adnexal masses. Its principal value in this setting involves confirming the mass, differentiating ovarian from uterine or tubal origin, delineating the internal appearance of the mass, and defining any associated abdominal findings. Whether ultrasound can differentiate between benign and malignant pelvic masses has been the subject of many studies. Meire and colleagues[28] first examined the accuracy of ultrasound in delineating a malignant ovarian mass based on size and appearance. In this study, fixed septa, tumour size exceeding 5 cm and multiloculation were considered prognostic for ovarian malignancy. Only 16 of 27 patients with such findings were found to have ovarian cancer. Furthermore, ultrasound has proved disappointing as a means of delineating disease outside the pelvis. Thus, to determine whether the mass is benign or malignant, operative intervention remains necessary.

The first evaluation of the use of pelvic sonography as a screening procedure for early ovarian neoplasia has been done by Campbell and his group[29]. Recently, the results of a prospective study of 5479 self-referred women without symptoms have been described in terms of the ovarian masses detected, the value of the screening procedure over time and the development of new screening strategies entailing the use of defined changes in ovarian volume[30]. Five primary ovarian cancers were detected at stage Ia or Ib, and evidence from a follow-up study at least one year after the last screening showed that the detection rate was 100%, within the limitation of the study design. A screening procedure based on the presence of abnormal ovarian morphology at the first scan and a defined volume change on rescanning would have given a false positive rate of 1.6% and a positive predictive value of 2.0%, i.e. the odds against a positive screen result indicating the presence of primary ovarian cancer are 1 : 50. This odds ratio is mainly due to the difficulty of distinguishing malignant tumours from benign masses, tumour-like conditions or hydrosalpinges.

The value of routine ultrasound has been examined by Andolf and colleagues[26]. In their study, 805 women attending gynaecological outpatient clinics in Sweden were examined. As a result of positive ultrasound findings 39 women had surgery, with one woman with malignant and two with borderline ovarian tumours as well as one with cancer of the caecum. Because the results were obtained from symptomatic women, it is difficult to know how to apply the findings to population screening. None of the four tumours detected by ultrasound could be found by digital pelvic examination.

Bourne and co-workers[24] used transvaginal colour Doppler in the evaluation of women selected on the basis of their medical history and the result of a previous transvaginal ultrasound scan. Thirty women (10 premenopausal (scan taken on days 1 to 8 of the menstrual cycle) and 20 postmenopausal) had normal ovaries, and 20 had at least one ovary with an abnormal morphology or volume or both. Two women with a positive result on screening had hydrosalpinges, 10 a benign tumour or a tumour-like condition, and eight primary ovarian cancers. No areas of neovascularization were seen in the 30 women with morphologically normal ovaries and the two patients with hydrosalpinges; the pulsatility index ranged from 3.1 to 9.4. Similarly, nine patients (10 affected ovaries with a non-malignant mass) had no signs of neovascularization and the pulsatility index varied from 3.2 to 7.0. One patient with bilateral dermoid cysts containing nests of thyroid-like cells had vascular changes and pulsatility index values of 0.4 and 0.8. Seven patients (eight ovaries) with primary ovarian cancer (one stage IV, four stage III, and two stage Ia) showed clear evidence of neovascularization and pulsatility index values from 0.3 to 1.0. One patient with a serous cystadenocarcinoma in a small ovary (5 ml volume) had no signs of any vascular change and the pulsatility index was 5.5. The authors concluded that transvaginal colour flow imaging may be used to identify potentially malignant ovarian masses and help elucidate the early stages of tumourigenesis.

26.5 EXPERIENCE IN ZAGREB

Diagnosis of ovarian carcinoma at an early stage was very unusual before colour Doppler ultrasound was introduced by us in 1987 for examinations of the pelvic organs (Figures 26.1–26.8)[11]. Pulsed Doppler is used to quantify such colour-coded flow using Pourcelot's resistance index (RI). In this way ovarian malignancy is differentiated from other lesions which produce similar but not identical patterns of flow (Figures 26.9–26.16).

In the first study we presented 14 317 asymptomatic or minimally symptomatic women who had been evaluated for ovarian carcinoma with colour Doppler ultrasound and who had been followed up with this technique; 8620 asymptomatic women were referred by 'Well Woman' or similar clinics for colour Doppler examination, and 5697 were referred by gynaecologists because of a suspected adnexal mass. Of the women 7495 (87%) were premenopausal and 1125 were postmenopausal. Every examination included abdominal B-mode and vaginal colour Doppler scanning with pulsed Doppler examination.

The equipment used was SSD-680 with 5-MHz transvaginal probe (Aloka Co., Japan). The RI was calculated from at least five separate cardiac cycles and the mean of these was calculated[31]. In premenopausal women the examinations were carried out at day 3–10 of the menstrual cycle to avoid false positive results as the consequence of increased ovarian flow during the luteal phase[32]. Those who had a diagnosed adnexal mass had laparotomy within 6 weeks: all were carried out by the same

Experience in Zagreb

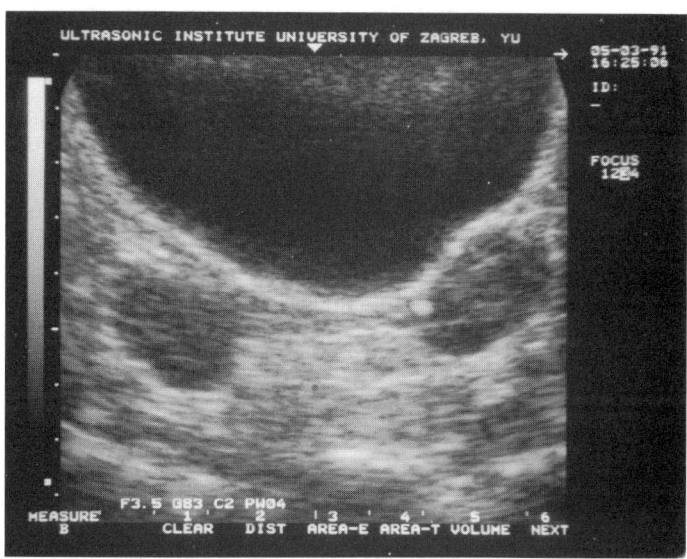

Figure 26.1 Enlarged ovaries diagnosed after hysterectomy.

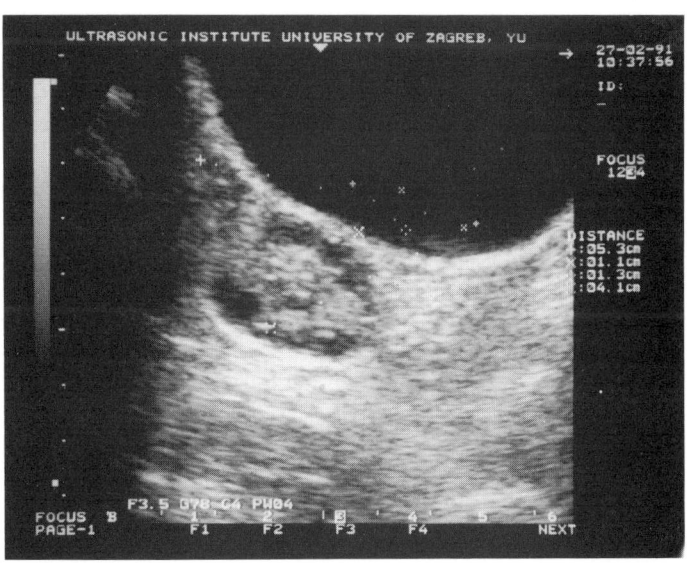

Figure 26.2 Slightly enlarged right ovary visualized by B-mode transabdominal ultrasound.

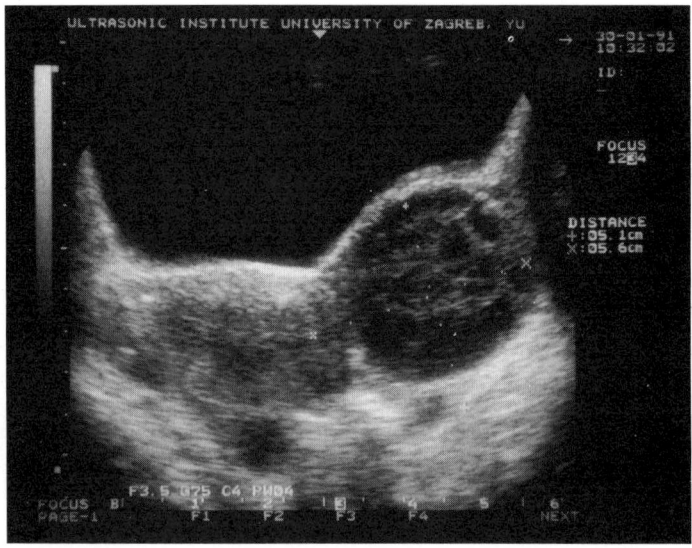

Figure 26.3 Enlarged left ovary.

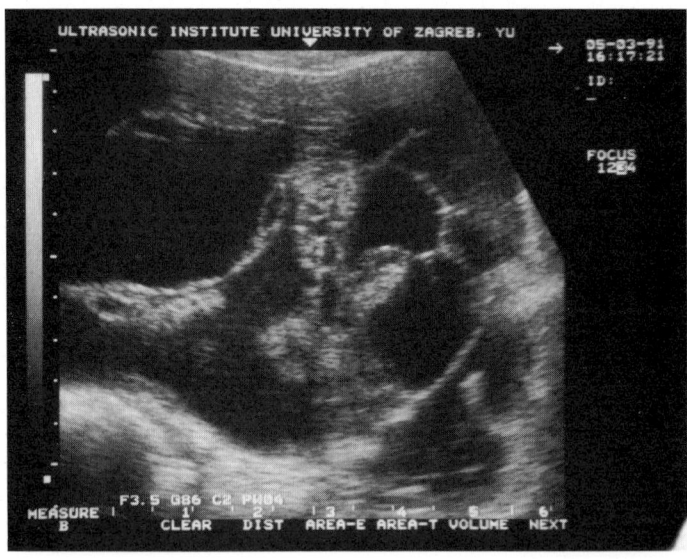

Figure 26.4 Complex, dominantly cystic adnexal mass. Malignancy was confirmed after surgery.

Experience in Zagreb

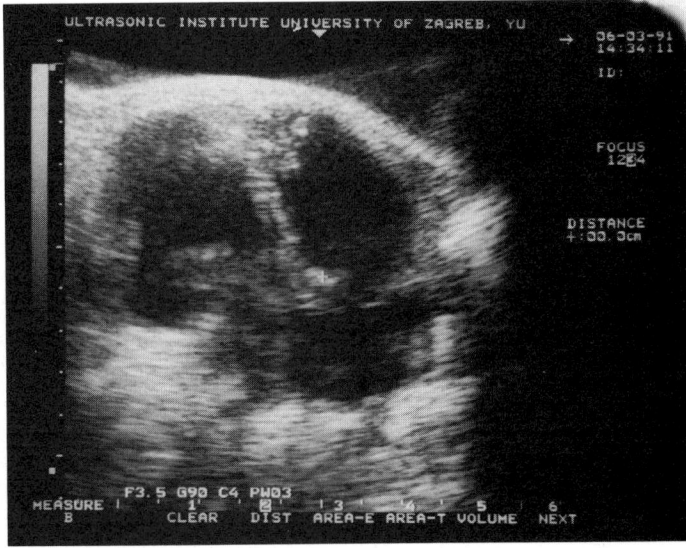

Figure 26.5 Bizarre ultrasound finding in one case of ovarian cancer.

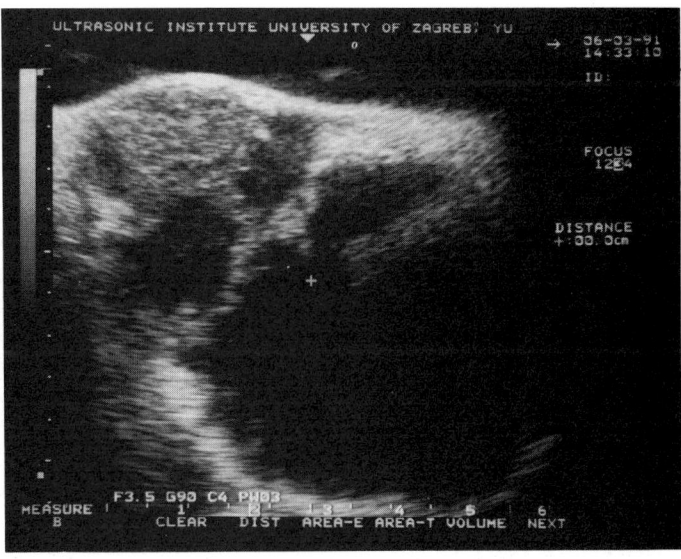

Figure 26.6 Another example of bizarre adnexal mass. Pelvic inflammatory disease was diagnosed at surgery.

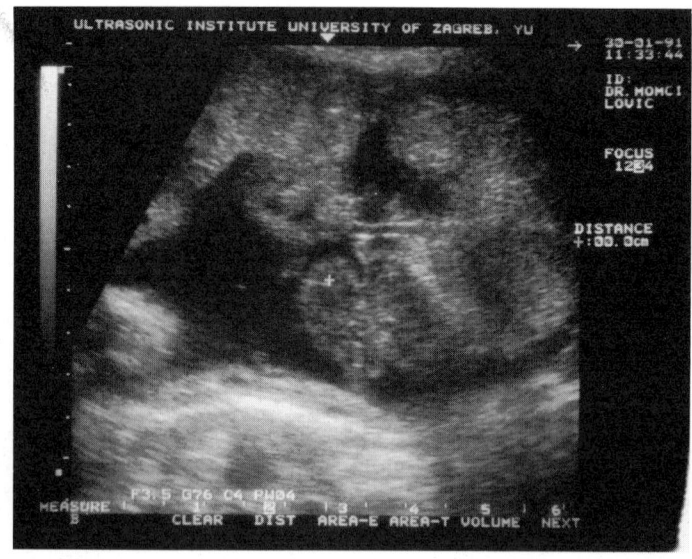

Figure 26.7 Advanced stage of ovarian cancer.

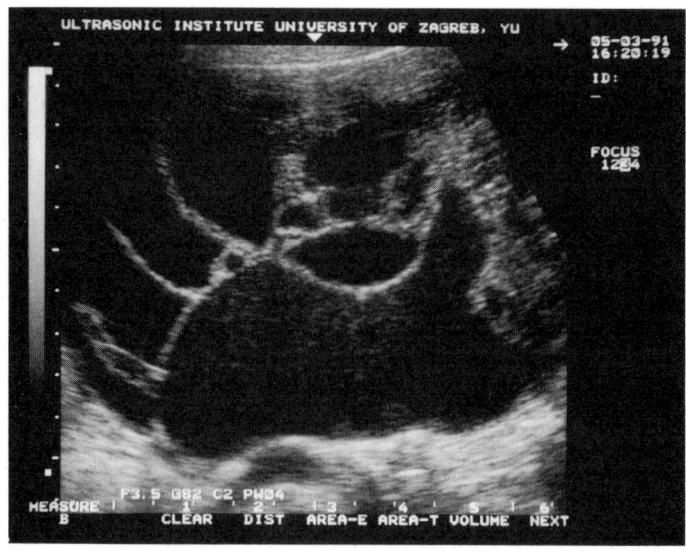

Figure 26.8 Huge cystic, septated ovarian mass. Benign nature (cystadenoma serosum) was diagnosed on histopathology.

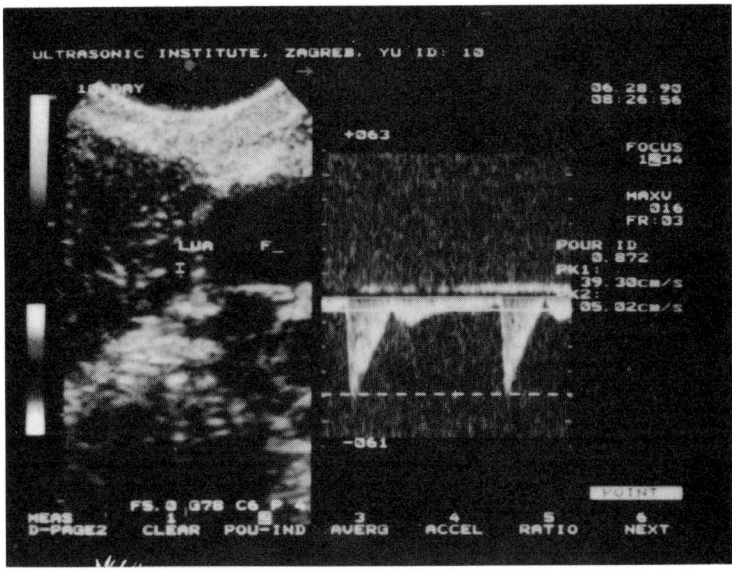

Figure 26.9 Transvaginal colour Doppler shows blood flow on the periphery of ovarian follicle. Pulsed Doppler waveform (right) shows high resistance and very low diastolic blood flow. Normal finding.

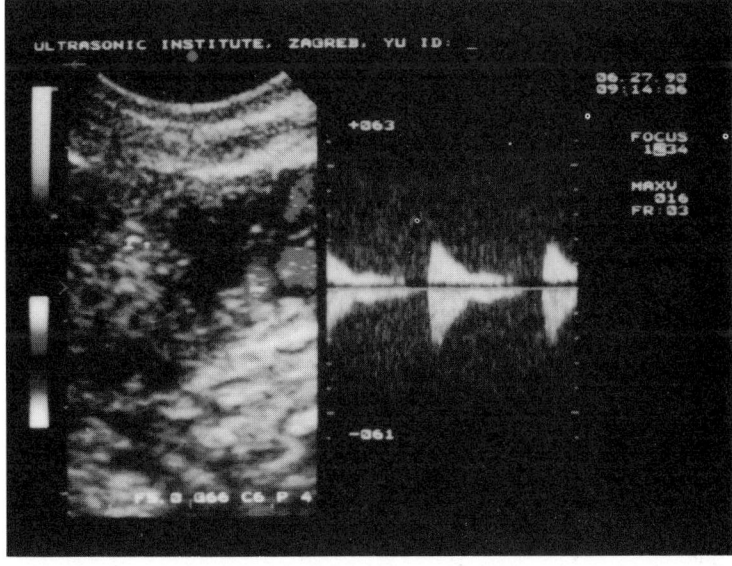

Figure 26.10 Enlarged, benign solid right ovary. Colour Doppler presents tumour blood flow. Pulsed Doppler (right) shows very high resistance blood flow.

Transvaginal colour Doppler in differentiation between benign and malignant masses

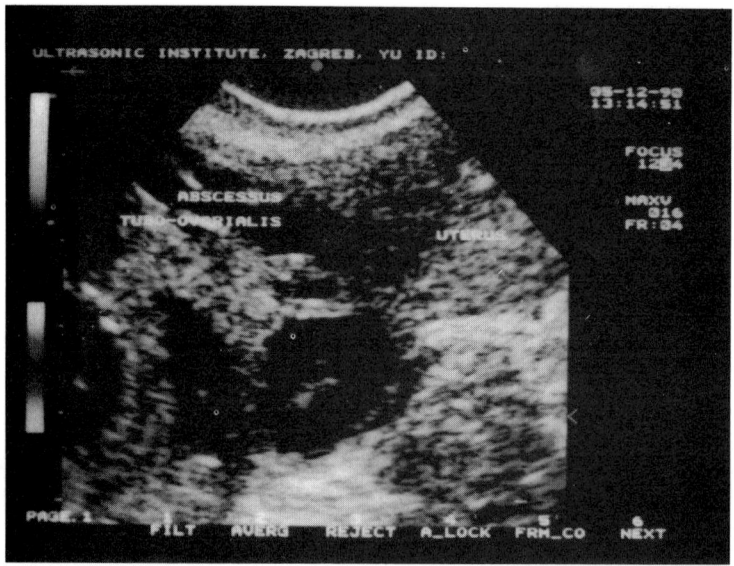

Figure 26.11 Colour Doppler finding in the case of tubo-ovarian abscess. Such a finding may be possible source of error in the diagnosis of ovarian malignancy.

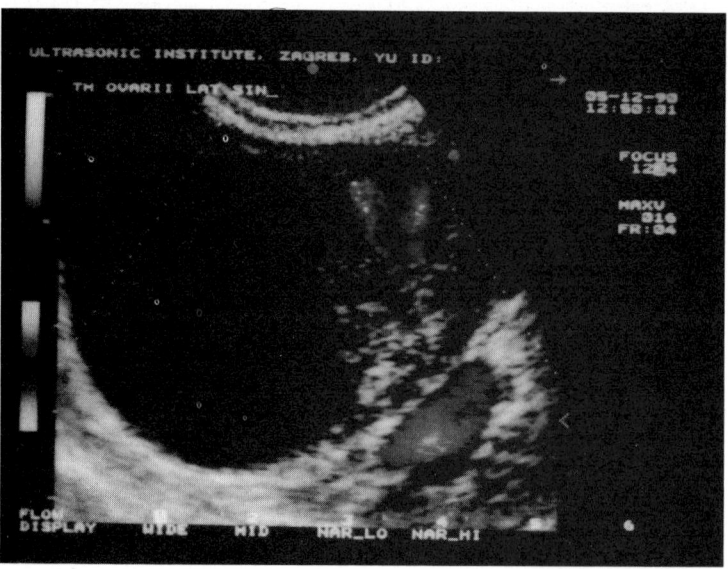

Figure 26.12 Large ovarian cyst. Blood flow was detected on the tumour periphery.

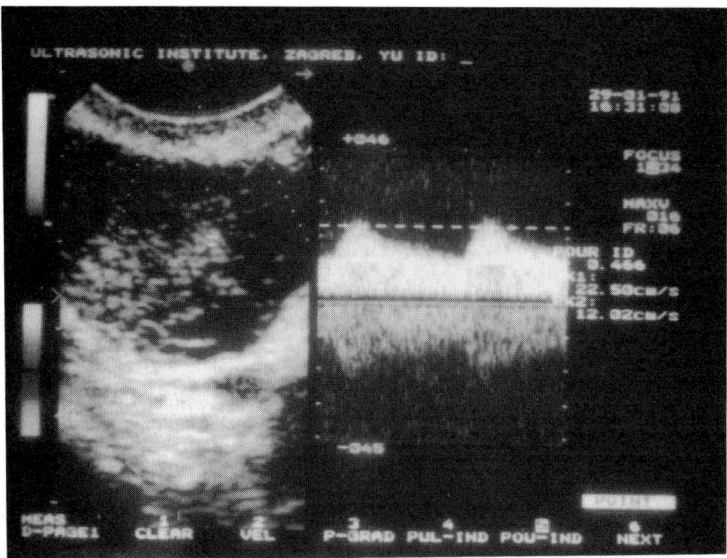

Figure 26.13 Solid adnexal mass. Blood flow was detected in the central part of the tumour. Pulsed Doppler (right) shows increased diastolic blood flow. Resistance index was borderline.

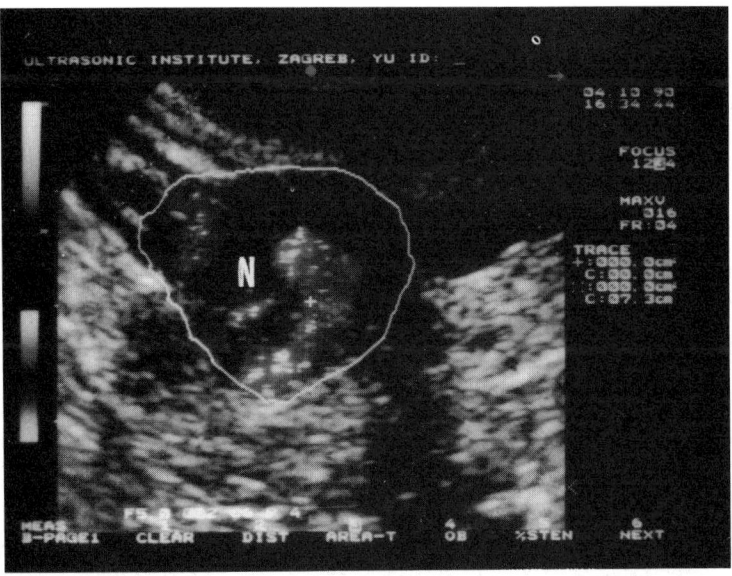

Figure 26.14 Very rich malignant tumour vascularization (N) was visualized by transvaginal colour Doppler.

Transvaginal colour Doppler in differentiation between benign and malignant masses

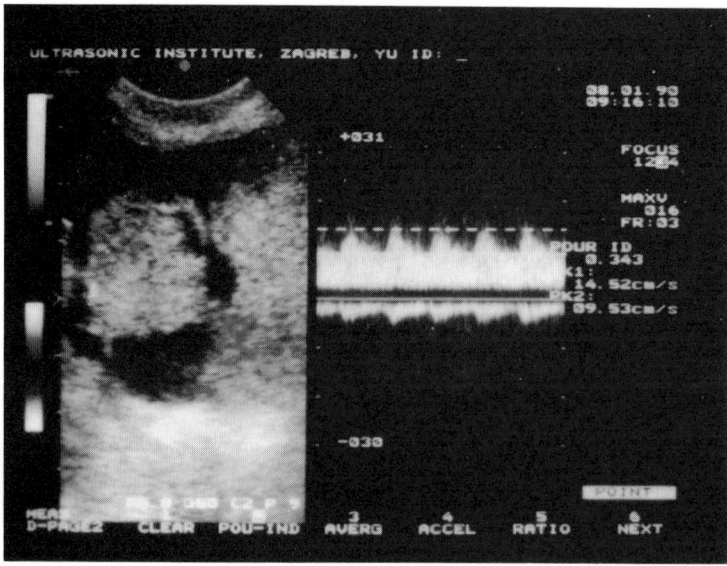

Figure 26.15 Complex malignant ovarian mass. Blood flow was detected in the solid part of the tumour. Pulsed Doppler (right) shows very low resistance of blood flow.

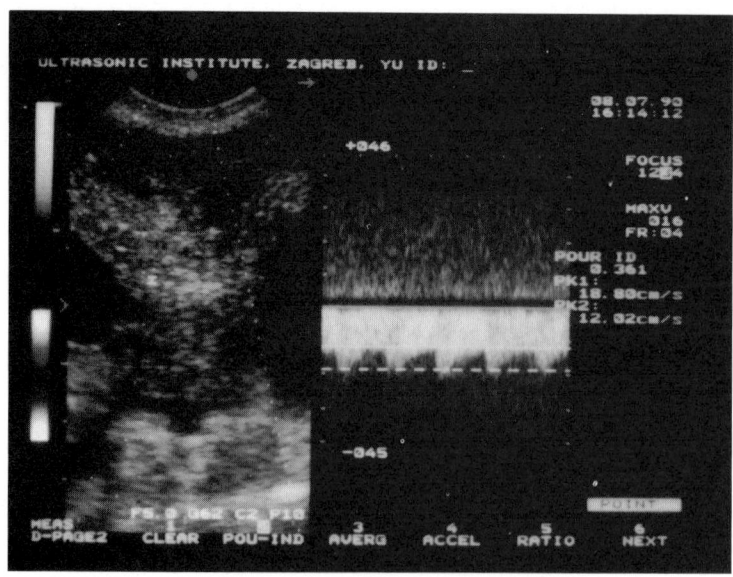

Figure 26.16 Another example of malignant ovarian mass diagnosed by transvaginal colour Doppler.

surgeon. The same pathologist examined every specimen.

There were 624 benign adnexal tumours: in each of these the RI was greater than 0.40, with one exception of a chronic appendicitis with hydrosalpinx in which the RI was 0.39. Eleven primary and nine secondary ovarian neoplasms were discovered. Five of the primary ovarian neoplasms were found in women known to have an adnexal mass, and six were in the asymptomatic group. Each of these tumours was stage I, and their diameters were between 3.5 and 5 cm. Neovascularization was found in 9 of 11 primary neoplasms: in these nine the RI was 0.32–0.40 (mean 0.36). Nine secondary ovarian neoplasms were found; the primary growths were in the breast, thyroid and rectum. Their mean diameter was 4.6 cm (range 3.8–5.6 cm). In each the RI was 0.40 or less (mean 0.38: range 0.28–0.40) and the colour was not so prominent as in primary neoplasms. In five of these nine cases clinical examination, transabdominal or transvaginal ultrasound did not reveal any pelvic abnormality.

The remaining 40 malignant neoplasms were ovarian cancers stage III and IV with characteristic abnormal flow patterns in 39 cases. Three of them were found in the asymptomatic group. A secondary tumour from an endometrial cancer was not detected by means of colour Doppler. However, sensitivity, specificity and accuracy of this diagnostic method are high (Table 26.1).

The second study included 1623 women studied by transvaginal colour Doppler, B-mode and pulsed Doppler. The observations were described according to our own scoring system (Figure 26.17). Among them, 38 women were operated on as the result of adnexal masses detected on ultrasound. Every patient was examined by transvaginal colour Doppler one day before operation. Five ovarian cancers were diagnosed: two were at stage I, two at stage II and one at

Table 26.1 Transvaginal colour Doppler in the detection of adnexal tumour malignancy

Colour flow present and RI = 0.40	Histopathology		Total
	Malignant	Benign	
Yes	54	1	55
No	2	623	625
Total	56	624	680

RI, resistance index.
Positive predictive value = 98.2%.
Negative predictive value = 99.7%
Sensitivity = 96.4%.
Specificity = 99.8%.
Accuracy = 99.5%.

stage III. When only the cut-off resistance index 0.40 was used to differentiate between benign and malignant ovarian masses, true positive diagnosis was made in two cases, false negative diagnosis in three cases and false positive diagnosis in three cases. When the scoring system (limit c.4) was used, true positive diagnosis was made in all five patients with ovarian cancer, false positive diagnosis in four patients and no false negative diagnoses were made. When both criteria were used (RI and score limit), false positive results were reduced to a minimum (7%) and no false negative results were obtained. In the case of false positive results, pelvic inflammatory disease or corpus luteum neovascularization were diagnosed.

26.6 CONCLUSION

Abdominal ultrasound can be used to identify and delineate the characteristics of ovarian masses. Features of a mass which are of importance are unilocular or multilocular apearance, opaque versus clear fluid, presence of solid components, presence of papillary projections, roughness or smoothness of the internal lining, and the presence of ascites. Many of these features can help delineate benign from suspected malignant

Transvaginal colour Doppler in differentiation between benign and malignant masses

(Circle characteristics seen and add for score)

PATIENT NAME _____ DATE _____ INSTITUTION _____

	FLUID		INTERNAL BORDERS		SIZE
UNILOCULAR	clear	(0)	smooth	(0)	
	internal echoes	(1)	irregular	(2)	
MULTILOCULAR	clear	(1)	smooth	(1)	
	internal echoes	(1)	irregular	(2)	
CYSTIC-SOLID	clear	(1)	smooth	(1)	
	internal echoes	(2)	irregular	(2)	
PAPILLARY PROJECTIONS	suspicious	(1)	definite	(2)	
SOLID	homogeneous	(1)	echogenic	(2)	
PERITONEAL FLUID	absent	(0)	present	(2)	

COLOUR DOPPLER		RI (index)		VELOCITY
no vessels seen	(0)	(0)		
regular separate vessels	(1)	>0.40	(1)	
randomly dispersed vessels	(2)	<0.41	(2)	

If suspected corpus luteum, repeat in next menstrual cycle in proliferative phase.

SCORE	ULTRASOUND	COLOUR
< 2	Benign	Benign
3 – 4	Questionable	Questionable
> 4	Suspicious	

OUTCOME:

Spontaneous resolution.

Surgery, Type _____

Pathologic Diagnosis _____

Complications _____

Follow up, date and status _____

Figure 26.17 Ovarian tumour ultrasound–Doppler classification.

growths, but once again, sensitivities beyond 75% have not been achieved. This figure is too low for a screening test.

Doppler ultrasound identifies the movement of erythrocytes in blood vessels. We have learned from Doppler ultrasound that vessels with high resistance and low diastolic blood flow can be differentiated from those with low resistance and high diastolic blood flow. Normal pelvic vessels are usually of high resistance and low diastolic flow. Malignant neoplasms may have a low resistance and very high diastolic blood flow. When colour is added to the Doppler system, the examiner is able to see many more vessels clearly, particularly small vessels. In most cancers, there is a proliferation of vessels which were formerly identified in radiographic angiology as a tumour 'blush'. Doppler measurements can then be made on these previously inaccessible vessels. We believe that transvaginal colour Doppler is a technology which offers the promise of high sensitivity, specificity and positive predictive value for the differentiation of ovarian masses and consequently is promising with regard to the early detection of ovarian cancer. Transvaginal ultrasound accurately identifies significant morphological components of ovarian enlargement. Colour flow Doppler identifies the presence of abnormal blood vessels, and measurements of the systolic–diastolic components of this flow differentiates vessels into high and low velocity states. Our experience also stresses the importance of careful examination of the ovaries with colour Doppler regardless of their B-mode ultrasound appearance and their size.

REFERENCES

1. Gulino, P.M. (1975) Extracellular compartments of solid tumors, in *Cancer* (ed. H. Becker), Plenum Press, New York, pp. 327–35.
2. Folkman, J. (1985) Tumor angiogenesis. *Adv. Cancer Res.*, **43**, 175–82.
3. Jain, R.K. (1988) Determination of tumor blood flow: a review. *Cancer Res.*, **48**, 2641–5.
4. Jain, R.K. (1987) Transport of molecules across tumor vasculature. *Cancer Metastasis Rev.*, **6**, 559–63.
5. Jain, R.K. and Baxter, L.T. (1988) Mechanisms of heterogeneous distribution of monoclonal antibodies and other macromolecules in tumors: Significance of elevated interstitial pressure. *Cancer Res.*, **48**, 7022–6.
6. Folkman, J. (1963) Growth and metastasis of tumor in organ culture. *Cancer*, **16**, 453–60.
7. Gimbrone, M.A. (1972) Tumor dormancy *in vivo* by prevention of neovascularisation. *J. Exp. Med.*, **136**, 261–4.
8. Jain, R.K. and Ward-Hartley, K.A. (1987) Dynamics of cancer cell interaction with microvasculature and interstitium. *Biorheology*, **24**, 117–20.
9. Taylor, K.J.W., Burns, P.N., Wells, P.N.I. and Conway, D.I. (1985) Ultrasound Doppler flow studies of the ovarian and uterine arteries. *Br. J. Obstet. Gynaecol.*, **92**, 240–3.
10. Long, M.G., Boulbee, J.E., Hanson, M.E. and Begent, R.H.J. (1989) Doppler time velocity waveform studies of the uterine artery and uterus. *Br. J. Obstet. Gynaecol.*, **96**, 588–91.
11. Kurjak, A., Zalud, I., Jurkovic, D. *et al.* (1989) Transvaginal color Doppler for the assessment of pelvic circulation. *Acta Obstet. Gynecol. Scand.*, **68**, 131–5.
12. Kurjak, A., Zalud, I., Alfirevic, Z. and Jurkovic, D. (1991) The assessment of abnormal pelvic blood flow by transvaginal color Doppler. *Ultrasound Med. Biol.*, **16**, 437–42.
13. Kurjak, A., Jurkovic, D., Alfirevic, Z. and Zalud, I. (1990) Transvaginal color Doppler imaging. *J. Clin. Ultrasound*, **18**, 227–34.
14. Kurjak, A. (1990) *Transvaginal Color Doppler*. Parthenon Publishing, Carnforth, NJ.
15. Wells, P.N.T., Halliwell, M., Skidmore, R. *et al.* (1977) Tumor detection by ultrasonic Doppler blood flow signals. *Ultrasonics*, **15**, 231–5.
16. Burns, P.N., Halliwell, M., Wells, P.N.T. and Webb, A.J. (1987) Ultrasonic Doppler studies of the breast. *Ultrasound Med. Biol.*, **8**, 127–30.
17. Srivastava, A., Huges, L.E., Woodcock, J.P. and Laider, P. (1987) Vascularity in cutanous melanoma detected by Doppler sonography

and histology: correlation with tumor behavior. *Br. J. Cancer*, **59**, 89–93.
18. Taylor, K.J.W., Ramos, I., Morse, S.S. *et al.* (1987) Focal liver masses: Differential diagnosis with pulsed Doppler ultrasound. *Radiology*, **164**, 643–6.
19. Taylor, K.J.W. and Morse, S.S. (1988) Doppler detects vascularity of some malignant tumors. *Diagn. Imaging*, **10**, 132–6.
20. Kujipers, D. and Jaspers, R. (1989) Renal masses: differential diagnosis with pulsed Doppler ultrasound. *Radiology*, **170**, 59–64.
21. Hata, H., Hata, K., Senoh, D. *et al.* (1989) Doppler ultrasound assessment of tumor vascularity in gynecological disorders. *J. Ultrasound Med.*, **8**, 309–12.
22. Kurjak, A. and Zalud, I. (1991) Transvaginal color Doppler sonography, in *Transvaginal Sonography*, 2nd edn (eds I.E. Timor-Tritsch and S. Rottem), Elsevier, Amsterdam, pp. 451–62.
23. Kurjak, A. and Zalud, I. (1990) Transvaginal colour Doppler, in *Handbook of Ultrasound in Obstetrics and Gynecology* (ed. A. Kurjak), CRC Press, Boca Raton, pp. 447–52.
24. Bourne, T., Campbell, S., Steer, C. *et al.* (1989) Transvaginal colour flow imaging: a possible new screening technique for ovarian cancer. *Br. Med. J.*, **299**, 1367–9.
25. McFarlane, C., Strugis, M.C. and Fetterman, F.S. (1985) Results of an experiment in the control of cancer of the female pelvis organ and report of a fifteen year research. *Am. J. Obstet. Gynecol.*, **69**, 294–7.
26. Andolf, E., Svalenius, E. and Astedt, B. (1986) Ultrasonography for early detection of ovarian carcinoma. *Br. J. Obstet. Gynaecol.*, **93**, 1286–9.
27. Jacobs, I.J., Stabile, I., Bridges, J. *et al.* (1988) Multimodal approach to screening for ovarian cancer. *Lancet*, **i**, 268–70.
28. Meire, H.B., Farrant, P. and Guha, T. (1978) Distinction of benign from malignant ovarian cysts by ultrasound. *Br. J. Obstet. Gynaecol.*, **85**, 893–5.
29. Campbell, S., Goessens, L., Goswamy, R. and Whitehead, M.I. (1982) Real-time ultrasonography for the determination of ovarian morphology and volume. A possible early screening test for ovarian cancer. *Lancet*, **i**, 425–8.
30. Bhan, V., Amso, N., Whitehead, M.I. *et al.* (1989) Characteristics of persistent ovarian masses in asymptomatic women. *Br. J. Obstet. Gynaecol.*, **96**, 1384–7.
31. Thompson, R.S., Trudinger, B.J. and Cook, C.M. (1988) Doppler ultrasound waveform indices: A/B ratio, pulsatility index and Pourcelot ratio. *Br. J. Obstet. Gynaecol.*, **95**, 581–604.
32. Zalud, I. and Kurjak, A. (1990) The assessment of luteal blood flow in pregnant and non-pregnant women by transvaginal color Doppler. *J. Perinat. Med.*, **18**, 215–21.

Chapter 27

Role of CA-125 in screening for ovarian cancer

I. JACOBS, A. PRYS DAVIES and D. ORAM

27.1 THE CA-125 ANTIGEN

The CA-125 is an antigenic determinant on a high molecular weight glycoprotein expressed by epithelial ovarian tumours and by other tissues of müllerian origin. It is recognized by the monoclonal antibody OC 125, which was raised using an ovarian cancer cell line as an immunogen[1]. Serum levels of CA-125 measured by radioimmunoassay were initially reported to be elevated in 83% of patients with epithelial ovarian cancer and to correlate with the clinical course of the disease[2]. Subsequent studies have confirmed that serum CA-125 levels are elevated preoperatively in 80–85% of women with epithelial ovarian cancer[3]. Although CA-125 was not originally thought to be expressed by mucinous carcinoma of the ovary, subsequent reports indicate elevation of serum CA-125 in up to 66% of this histological type. Serum levels of CA-125 are, however, less frequently elevated in mucinous than other histological types of epithelial ovarian cancer[4]. In addition to ovarian cancer serum CA-125 levels are elevated in a variety of physiological states (pregnancy, menstruation), benign disorders (endometriosis, hepatic failure, pelvic infection) and malignant conditions (endometrial, breast, colon cancers). The proportion of patients in these groups with an elevated serum CA-125 level (> 35U/ml) is summarized in Figure 27.1. It should be noted that although some of these conditions are associated with elevation of CA-125 above 35 U/ml in a high percentage of cases (e.g. early pregnancy and advanced stage endometriosis) the peak serum CA-125 levels are several fold less than those observed in ovarian cancer. Numerous studies have established that serum CA-125 measurement is of value in the management of clinically apparent ovarian cancer. In preoperative differential diagnosis of the benign or malignant nature of a pelvic mass, CA-125 measurement has a diagnostic accuracy of approximately 80%[5–7]. Diagnostic accuracy can be improved by combined consideration of CA-125, ultrasound and menopausal status[7]. In patients receiving treatment for ovarian cancer a rising or persistently elevated serum CA-125 level is a reliable indicator of poor response to therapy and may provide a basis for discontinuing or altering the therapeutic regimen. A falling serum CA-125 indicates response to treatment. However, a fall to < 35U/ml is not a reliable indicator of complete pathological

Role of CA-125 in screening for ovarian cancer

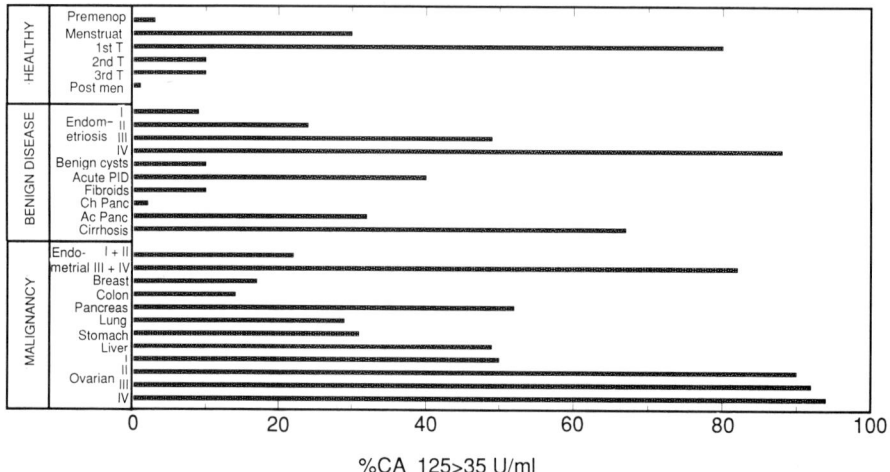

Figure 27.1 The proportion of individuals with serum CA-125 levels >35 U/ml in various physiological states, benign disorders and malignancy (adapted from reference 3).

response. The rate of fall of CA-125 during treatment (CA-125 half-life) is a strong prognostic indicator for survival[8].

27.2 ELEVATION OF SERUM CA-125 LEVELS IS RELATED TO DISRUPTION OF TISSUE BARRIERS

Epithelial ovarian cancer disseminated outside the ovary (stages II, III and IV) is associated with elevation of serum CA-125 in over 90% of cases. In stage I disease high tissue expression of the CA-125 antigen is observed in over 90% of cases but serum CA-125 levels are elevated preoperatively in only 50% (Figure 27.1). The phenomenon of high tissue CA-125 levels in association with normal serum levels is not confined to early stage malignancy. Benign ovarian tumours, normal endometrium and second trimester amniotic fluid express high levels of CA-125 activity but are in the majority of cases associated with serum CA-125 levels less than 35 U/ml[9,10]. It is clear that elevation of serum CA-125 is not only related to production of the CA-125 antigen. The specificity of CA-125 activity for ovarian cancer at tissue level is poor compared to serum measurement. The clinical value of serum CA-125 measurement appears to be related more closely to factors influencing release of the antigen into the circulation than to its synthesis.

A common factor in all physiological and pathological events associated with elevation of serum CA-125 levels is the occurrence of a process which alters normal tissue barriers in relation to tissues known to produce CA-125. Such events may occur in malignancy, pregnancy, menstruation, endometriosis, pelvic inflammatory disease and benign tumours. It is not surprising therefore that elevation of serum CA-125 is most consistently associated with invasive, widely spread malignancy and less commonly with early stage, localized disease. When increased CA-125 synthesis is not usually associated with alteration in tissue barriers, an increase in serum levels is uncommon (e.g. benign ovarian tumours). It is likely that the intact capsule of most benign ovarian tumours is an effective barrier to the high molecular weight moiety expressing the CA-125 antigen. When increased tissue production of CA-125 coincides with changes in surface barriers (e.g. decidua in the first trimester of pregnancy) a rise in serum CA-125 is usual. As there is no evidence of increased endometrial production of CA-125 at the time of menstruation it would seem that serum CA-125 levels may also rise due to alteration in tissue barriers alone without any increase in synthesis, although this phenomenon may be attributable to retrograde menstruation. The one clear exception to this pattern is elevation of serum CA-125 in liver disease. Normal liver does not express CA-125 (Eerderkens 1985) and peritoneal irritation cannot explain serum CA-125 elevation in cirrhotic patients without ascites and jaundice. A plausible explanation is that CA-125 is metabolized in the liver and that elevated CA-125 levels in cirrhotic patients are a consequence of liver failure (Ruibal 1986).

The hypothesis that elevation of serum levels of CA-125 is dependent upon an alteration in tissue barriers has important implications for the use of CA-125 measurement as a screening test for early stage ovarian cancer. If this hypothesis is valid it follows that the malignant process will only be associated with an elevation of serum CA-125 levels when it has caused a degree of disruption of tissue barriers. This process may occur either by the process of invasion itself or by precipitating an inflammatory response. Such events probably occur relatively late in the process of carcinogenesis but may occur relatively early on the time scale of clinical staging. As screening must be performed at intervals, which could not in practice be less than 1 year, a crucial point is the time lapse from the initial disruption of tissue barriers to the initiation of metastatic spread. If metastatic spread follows closely on disruption of tissue barriers, screening with CA-125 is unlikely to reduce mortality. This limitation will apply even if serum CA-125 measurement has high sensitivity for early stage disease, since the opportunity to 'catch' the disease at an early stage will arise infrequently. If the time lapse to metastatic spread is months or years a reduction in overall mortality may be achieved by interval screening. Unfortunately, it may be the case that the tumours with a poor prognosis progress rapidly from tissue disruption to metastasis whilst only those with a relatively good prognosis have a longer time span and are amenable to screening.

27.3 SERUM CA-125 LEVELS ARE ELEVATED IN SOME CASES OF PRECLINICAL ASYMPTOMATIC OVARIAN CANCER

Although the precise temporal relationship of serum CA-125 levels to the process of ovarian carcinogenesis and metastatic spread is unclear there is now good evidence for elevation of serum CA-125 prior to clinical presentation. Bast and colleagues[11] reported the case of a patient with ovarian cancer whose serum had been stored prior to diagnosis for investigation of another condition. Retrospective analysis of the stored samples revealed an elevated serum CA-125 level 12 months prior to clinical presentation of the disease. The first case of ovarian cancer diagnosed by prospective serum CA-125 measurement was described in the preliminary report of the London Hospital Ovarian Cancer Screening Study[12]. Further evidence that serum CA-125 elevation may precede clinical presentation of ovarian cancer was provided by retrospective analysis of the JANUS serum bank[13]. Sera obtained from 39 300 healthy female volunteers was stored and tested for CA-125 after ovarian cancer had presented clinically in 105 women. CA-125 levels in the serum samples from 105 women who subsequently developed ovarian malignancy (interval 1–143 months) were significantly greater than 323 matched controls. Of 12 samples collected within 18 months of diagnosis 6 had a serum

CA-125 greater than 30 U/ml and 4 were greater than 65 U/ml. Of 59 samples collected more than 60 months before diagnosis 14 had a serum CA-125 greater than 30 U/ml.

27.4 THE LONDON HOSPITAL OVARIAN CANCER SCREENING STUDY

The data outlined above confirm that serum CA-125 levels are elevated in a significant proportion of cases of asymptomatic preclinical ovarian cancer but also highlight limitations of both specificity and sensitivity. The results of the first phase of the London Hospital Study [12,14], presented at the Second Helene Harris Memorial Trust conference, indicated that the limitations of specificity encountered on screening with CA-125 alone could be overcome by screening with a sequential combination of CA-125 and ultrasound. Subsequently a phase 2 study was commenced to further evaluate this screening protocol. The following is a summary of an interim report of the results of the phase 2 study in the first 20 000 women recruited[15].

27.4.1 STUDY DESIGN

Women over 45 years of age and with greater than 12 months amenorrhoea were invited to volunteer for the study by the following methods: (i) articles describing the project in the national press; (ii) leaflets distributed by the occupational health departments of companies collaborating with the project; and (iii) postal invitations from the age–sex register of general practitioners in England, Scotland and Wales collaborating with the project. The eligibility criteria were the same as previously described for the phase 1 study[12,15]. After giving informed consent volunteers completed a questionnaire and underwent venepuncture for CA-125 radioimmunoassay. A serum CA-125 level of 30 U/ml or more was defined as abnormal and followed up by recall for a real-time ultrasound scan. Real-time ultrasonography was performed via the abdominal route and ovarian diameter was measured in three dimensions. Ovarian volume was calculated using the formula for an ovoid as described by Campbell and colleagues[16] and a volume greater than 8.8 ml was defined as abnormal. If the result of the ultrasound scan was abnormal the patient was referred to a gynaecologist for further management after consultation with her general practitioner. Women with a normal ultrasound scan were followed up at 3-monthly intervals with repeat serum CA-125 measurements. In this group, repeat ultrasound scans were performed if serum CA-125 levels doubled between consecutive samples or remained greater than 100 U/ml in two consecutive samples.

27.4.2 STUDY RESULTS

The first 20 000 volunteers were recruited over a 2-year period. The age distribution of the study population is shown in Table 27.1. The median age was 56 years (range 45–85). The number of years of amenorrhoea reported by these women ranged from 1 to 43 years with a median of 6.2 years. The mean number of pregnancies beyond 28 weeks in the past obstetric history of this population was 2.2 (range 0–14). Of the study population 4450 (22.8%) had undergone hysterectomy with conservation of one or both ovaries; 892 women had a past history of malignancy (of which 403 had been treated for cancer of the breast), but at the date of recruitment had no evidence of recurrent disease.

The distribution of CA-125 results in the study population is summarized in Figure 27.2. On the basis of the results of their initial screening visit the population has been divided into three groups (Figure 27.3).

Group I. Screen negative: CA-125 <30 U/ml ($n = 19\ 719$)

Of the 20 000 volunteers 19 719 had serum CA-125 levels less than 30 U/ml. At the time

Table 27.1 Age distribution, expected incidence of ovarian cancer[17] and actual incidence of ovarian cancer amongst the 20 000 volunteers in phase 2 of the London Hospital Study[15]

Age group (years)	No. of volunteers	Incidence of ovarian cancer (rate/100 000)	Expected cases/year	Actual cases
45–49	2 308	20.5	0.47	1
50–54	5 818	33.0	1.92	2
55–59	6 459	36.4	2.35	4
60–64	3 224	46.7	1.51	2
65–69	1 536	47.7	0.73	1
70–74	467	46.7	0.22	1
75–79	115	47.9	0.06	0
80–84	81	48.7	0.04	0
>84	2	43.0	0.00	0
Total	20 000	—	7.30	11

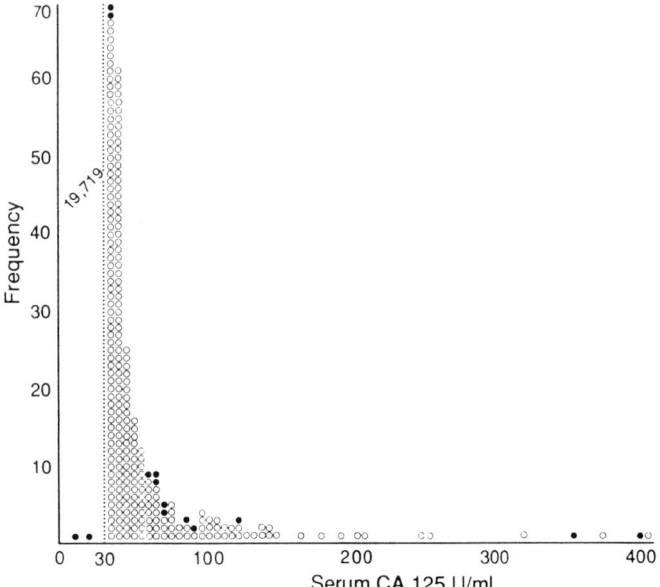

Figure 27.2 Distribution of serum CA-125 levels in 20 000 apparently healthy postmenopausal volunteers [15]. Volunteers with a serum CA-125 level <30 U/ml not diagnosed as having ovarian cancer at the time of this report are not represented by individual symbols ($n = 19\,717$). Solid symbols represent volunteers with a diagnosis of ovarian cancer ($n = 14$). Open symbols represent volunteers with serum CA-125 >30 U/ml not diagnosed as having ovarian cancer ($n = 269$).

of this report a subsequent diagnosis of ovarian cancer had been made in two of these patients. A diagnosis of stage III germ cell tumour was made 8 months after the registration visit of a volunteer with an initial serum CA-125 level of 10 U/ml. A diagnosis

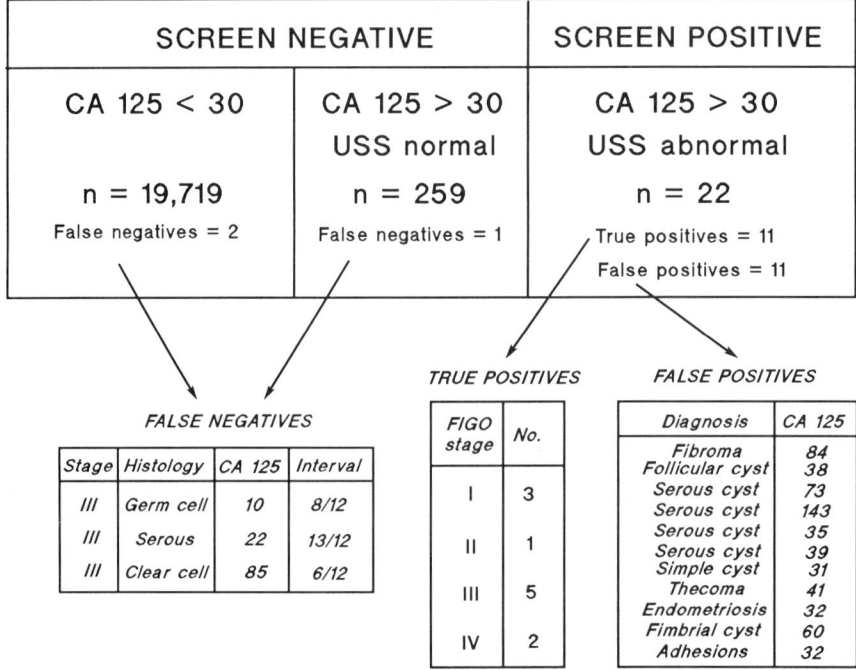

Figure 27.3 Summary of interim findings from phase 2 of the London Hospital study[15]. Volunteers have been divided into screen-negative and screen-positive groups on the basis of their CA-125 level and ultrasound findings. The details of patients with false negative and true positive results are shown.

of stage III serous cystadenocarcinoma was made 13 months after the registration visit of a volunteer with an initial serum CA-125 level of 22 U/ml.

Group II. Screen negative: CA-125 >30 U/ml ($n = 259$)

The ultrasound findings were normal in 257 volunteers with a CA-125 level >30 U/ml. Two other volunteers with a serum CA-125 level >30 U/ml declined to attend for an ultrasound scan and have therefore been classified as screen negative. One of these volunteers with an initial serum CA-125 level of 85 U/ml presented with stage III clear cell carcinoma 6 months after her registration visit.

Group III. Screen positive ($n = 22$)

Laparoscopy or laparotomy was performed on 22 volunteers with an elevated serum CA-125 level and either abnormal ultrasound findings or equivocal findings which became abnormal on follow-up at 3, 6 or 9 months. Eleven of these volunteers had benign pelvic pathology. The remaining 11 screen-positive volunteers were found to have invasive epithelial ovarian cancer at laparotomy (Figure 27.3).

27.4.3 YEARS OF CANCER DETECTED

The expected incidence of ovarian cancer for the study population can be estimated from Office of Population Censuses and Surveys (OPCS) cancer incidence data[17], (Table 27.1). Table 27.1 also lists the number of cases

Table 27.2 Summary of the performance of CA-125 alone and CA-125 combined sequentially with ultrasound in screening for ovarian cancer amongst 20 000 apparently healthy postmenopausal women in phase 2 of the London Hospital ovarian cancer screening study [15]

	CA-125 and ultrasound		CA-125 alone	
	Positive result	Negative result	Positive result	Negative result
Ovarian cancer present	11 (a)	3 (b)	12 (a)	2 (b)
Ovarian cancer absent	11 (c)	19 975 (d)	270 (c)	19 716 (d)
Specificity [d/(d + c)]	99.95%		98.65%	
Detection rate [a/(a + b)]	78.6%		85.7%	
Positive predictive value [a/(a + c)]	50.0%		14.6%	

of ovarian cancer actually detected in each age group. A total of 11 cases were detected by the screening programme in a population with an expected incidence of 7.3 cases per year.

27.4.4 TEST PERFORMANCE

Table 27.2 summarizes the performance of the screening programme at the time of this report. The sequential combination of CA-125 and ultrasound achieved a specificity of 99.95%, a positive predictive value of 50.0% and an apparent sensitivity (with follow-up still in progress) of 78.6%. The comparable figures for CA-125 alone were 98.65%, 14.6% and 85.7%.

27.5 ESTIMATING THE LEAD TIME ACHIEVED BY SCREENING WITH CA-125

The results above represent a preliminary report of the initial prevalence screen for phase 2 of the London Hospital Study[15]. Follow-up and attendance for subsequent screens are currently underway and will provide further information about the screening programme on completion. These preliminary results have confirmed the high specificity and positive predictive value of the screening protocol as well as providing a basis for estimating the lead time achieved compared to clinical presentation.

Figure 27.4 is a model of the progress of ovarian cancer from a pre-invasive phase (A) to the development of symptoms of invasive disease (E). Between these extremes are phases B and C during which the cancer is invasive but screening test negative and positive respectively. The model assumes that the screening test is negative in the pre-invasive phase and incorporates a separate pathway (D) for cases in which the screening test remains negative at the time of clinical presentation. For CA-125 pathway D would account for approximately 15% of patients[3]. The expected incidence of symptomatic invasive ovarian cancer in phase E for the study population is approximately 7.3 per year (Table 27.1). If the proposed model is valid the incidence of symptomatic invasive ovar-

Role of CA-125 in screening for ovarian cancer

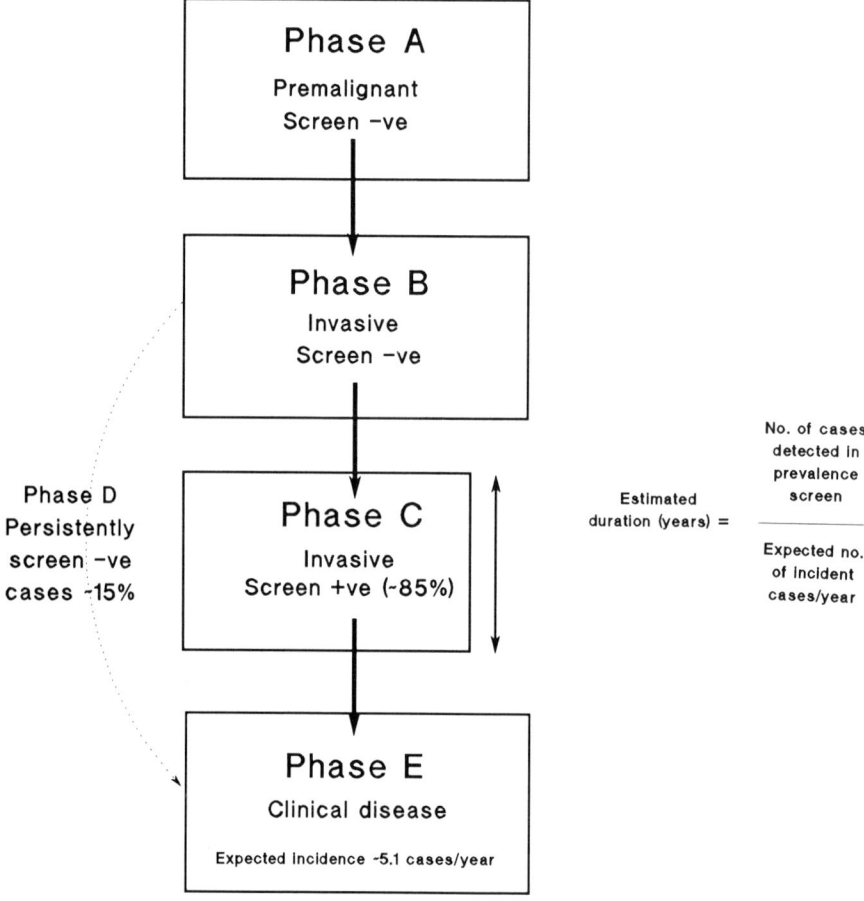

Figure 27.4 A model of the progress of ovarian cancer from a pre-invasive phase to the development of symptoms of invasive disease.

ian cancer in phase E must approximate to the incidence of disease in phase A, in phase B and in phases C and D combined. Furthermore, all ovarian cancers detected by the screening programme must be in phase C at the time of detection as symptomatic volunteers (phase E) were excluded at registration and volunteers in other phases (A, B and D) would be screen negative. The length of time spent in phase C can therefore be estimated from the results of this study. The expected number of incident cases entering phase C is estimated by the expected incidence for phase E (7.3 cases/year, Table 27.1) minus the expected incidence for phase D (0.15 × 7.3 cases/year) which equals 6.2 cases/year. The length of time spent in phase C is therefore estimated by the actual number of cases detected (11.0) divided by the number of cases expected each year (6.2). The results of this study suggest that the length of time spent in phase C is approximately 1.8 years (11/6.2).

The validity of this estimate of the length of time in phase C is dependent upon the assumption that the age-related incidence of ovarian cancer in the study population is similar to the national incidence for each age group provided by OPCS data. It is likely that women with a family history of ovarian cancer were over-represented in the study population compared to the general population as they have a particularly strong motivation to volunteer. As family history is a well-established risk factor for ovarian cancer this factor would tend towards an overestimate of the duration of phase C. Conversely, the proportion of women with a previous hysterectomy was higher than would be expected in the general population. This factor would tend toward an underestimate of the duration of phase C as available evidence [18,19] suggests that there is a reduction in risk of ovarian cancer amongst women who have had a previous hysterectomy with ovarian conservation. There is an additional cause for underestimation of the CA-125-positive preclinical phase; volunteers with an elevated CA-125 result but normal ultrasound scan did not undergo further investigation and some of them may develop ovarian cancer on long-term follow-up.

The estimate obtained from this study is slightly less than that suggested by the only other source of data for estimating the duration of phase C. In the JANUS study [13] sera taken from 39 300 healthy volunteers was stored and tested for CA-125 after ovarian cancer had presented clinically in 105 women. The incidence of ovarian cancer in this population over a 12-year follow-up period was 8.8 cases/year and 25 of the 105 volunteers who subsequently developed ovarian cancer had an initial CA-125 >30 U/ml. The estimated length of time spent in phase C calculated as described above was therefore 3.3 years $(25/8.8 - (8.8 \times 0.15)]$. The directly comparable figure from our own study would be 1.9 years (12/6.2) (as for direct comparability the volunteer with a CA-125 of 85 U/ml who declined to return for an ultrasound scan would be classified as screen positive).

The ultrasound-positive preclinical phase can be estimated from the Kings College Study[20] in which 5479 apparently healthy women were scanned annually for 3 years (90% and 77% compliance with second and third scans respectively). The expected number of cases of ovarian cancer per year estimated from OPCS age–sex cancer incidence data was 1.64 and the actual number of cases detected at the initial prevalence screen was two (Table 27.1). No interval screen-negative cases were reported in this study. With an apparent detection rate of 100% the estimated length of the ultrasound-positive preclinical phase was therefore $2/1.64 = 1.2$ years. (Three further cases of ovarian cancer were detected at the second screen but these cases cannot provide information about the length of the ultrasound-positive preclinical phase. The expected number of screen-detected cases on interval screening after 1 year would be equal to the expected incidence of ovarian cancer, i.e. approximately 1.64 cases.)

Only four of the 11 cases detected to date in phase 2 of the London Hospital Study had stage I or II ovarian cancer. This finding may in part reflect the fact that the available results are the findings of a prevalence screen. The remaining seven cases with advanced stage disease may represent cases which have spent a relatively long period of time in phase C prior to the prevalence screen. If so, follow-up screening of the same population should detect a larger proportion of early stage disease than the prevalence screen. Cases detected at follow-up screens will have been in phase C for a period of time limited by the screening interval.

27.6 SUMMARY

Evidence for the effect of screening to detect early stage ovarian cancer on mortality from the disease must await the performance of randomized controlled trials. The data available at present provide a basis for conclusions concerning the performance characteristics of potential screening tests (Table 27.3).

Table 27.3 Summary of test performance data for CA-125 and ultrasound alone and in combination in screening for ovarian cancer[12–15, 20]

	CA-125	Ultrasound	CA-125 + ultrasound
Sensitivity	~50%	~100%	~50%
Specificity	98.5%	97.7%	99.9%
Positive predictive value	3.5%	1.5%	26.8%
Lead time (years)	1.9–3.3	~1.2	1.9–3.3

1. Specificity. Our previous conclusion that the specificity of CA-125 is not sufficient to achieve an acceptable positive predictive value on screening a population of postmenopausal women remains unchanged [14]. Acceptable levels of specificity (>99.6%) can be achieved by the sequential combination of CA-125 and ultrasound scanning[12,15]. The specificity of abdominal ultrasound alone is less than that of CA-125 alone[15,20] but may be higher when performed via the vaginal route.
2. Sensitivity. The apparent sensitivity of CA-125 measurement in the London Hospital Study was 78.6%. This figure is likely to be an overestimate as follow-up is still in progress. It is unlikely that the sensitivity of serum CA-125 measurement for preclinical disease will exceed the value of 50% established for clinically apparent preoperative stage I disease. The apparent sensitivity of ultrasound scanning in the King's College Study[20] was 100%. However, only two cases of invasive ovarian cancer were detected (plus three borderline) and follow-up was incomplete. In addition it is difficult to explain the detection of stage I disease but no cases of more advanced stage disease at the initial prevalence screen. The sensitivity of the sequential combination of CA-125 and ultrasound would be that of the least sensitive test.
3. Lead time over clinical presentation. The available evidence suggests that although the sensitivity of serum CA-125 measurement is limited, the mean length of time from CA-125 elevation to clinical presentation in cases which do become CA-125-positive is 1.9–3.3 years. In contrast, available evidence suggests that ultrasound has high sensitivity but a relatively short interval from screen-positive test to clinical presentation.

Due to limitations of both specificity and sensitivity the role of CA-125 measurement in screening for ovarian cancer is likely to be as part of a multimodal approach. There is currently no single test or combination of tests available which can achieve both the combinations of specificity and sensitivity required to screen for early stage ovarian cancer. The choice at present lies between a test with apparently high sensitivity but low specificity (ultrasound) and a combination of tests with relatively low sensitivity and high specificity (CA-125 and ultrasound). Ultimately it may be possible to improve both specificity and sensitivity by using a panel of complementary tumour markers as an initial screen and transvaginal ultrasonography as a secondary test.

REFERENCES

1. Bast, R.C., Feeney, M., Lazarus, H. *et al.* (1981) Reactivity of a monoclonal antibody with human ovarian carcinoma. *J. Clin. Invest.*, **68**, 1331–7.
2. Bast, R.C., Klug, T.L., St John, E. *et al.* (1983) A radioimmunoassay using a monoclonal

antibody to monitor the course of epithelial ovarian cancer. *N. Engl. J. Med.*, **309**, 169–71.
3. Jacobs, I.J. and Bast, R.C. (1989) The CA 125 tumour associated antigen; a review of the literature. *Hum. Reprod.*, **4**, 1–12.
4. Zurawski, V.R., Knapp, R.C., Einhorn, N. *et al.* (1988) An initial analysis of preoperative serum CA 125 levels in patients with early stage ovarian carcinoma. *Gynecol. Oncol.*, **30**, 7–14.
5. Einhorn, N., Bast, R.C, Jr, Knapp, R.C. *et al.* (1986) Preoperative evaluation of serum CA 125 levels in patients with primary epithelial ovarian cancer. *Obstet. Gynecol.*, **67**, 414–16.
6. Vasilev, S.A., Schaerth, J.B., Campeau, J. and Morrow, C.P. (1988) Serum CA 125 levels in preoperative evaluation of pelvic masses. *Obstet. Gynecol.*, **71**, 751–6.
7. Jacobs, I.J., Oram, D.H., Fairbanks, J. *et al.* (1990) A risk of malignancy index incorporating CA 125, ultrasound and menopausal status for the accurate preoperative diagnosis of ovarian cancer. *Br. J. Obstet. Gynaecol.*, **97**, 922–9.
8. Van der Burg, M.E.L., Lammes, F.B., van Putten, W.L.J. and Stoter, G. (1988) Ovarian cancer: The prognostic value of the serum half-life of CA125 during induction chemotherapy. *Gynecol. Oncol.*, **30**, 307–12.
9. Fleuren, G.J., Nap, M., Aalders, J.G. *et al.* (1987) Explanation of the limited correlation between tumour CA 125 content and serum CA 125 antigen levels in patients with ovarian tumours. *Cancer*, **60**, 2437–42.
10. Jacobs, I.J., Fay, T.N., Stabile, I. *et al.* (1988) The distribution of CA 125 in the reproductive tract of pregnant and non pregnant women. *Br. J. Obstet. Gynaecol.*, **95**, 1190–4.
11. Bast, R.C., Siegal, F.P., Runowicz, C. *et al.* (1985) Elevation of serum CA 125 prior to diagnosis of an epithelial ovarian carcinoma. *Gynecol. Oncol.*, **22**, 115.
12. Jacobs, I.J., Stabile, I., Bridges, J. *et al.* (1988) Multimodal approach to screening for ovarian cancer. *Lancet*, **i**, 268–71.
13. Zurawski, V.R., Orjaseter, H., Andersen, A. and Jellum, E. (1988) Elevated serum CA 125 levels prior to diagnosis of ovarian neoplasia; relevance for early detection of ovarian cancer. *Int. J. Cancer*, **42**, 677–80.
14. Jacobs, I.J. and Oram, D.H. (1990) Potential screening tests for ovarian cancers, in *Ovarian Cancer. Biological and Therapeutic Challenges* (eds F. Sharp, W.P. Mason, and R.E. Leake), Chapman & Hall, London, pp.197–205.
15. Jacobs, I.J. (1991) CA 125 in the physiology and pathology of the female reproductive tract; with particular reference to the diagnosis of ovarian cancer. MD thesis, London University.
16. Campbell, S., Goessens, L., Goswamy, R. and Whitehead, M.I. (1982) Real time ultrasonography for determination of ovarian morphology and volume. *Lancet*, **i**, 425–6.
17. OPCS Cancer Statistics, Registrations (1983) *Cases of Diagnosed Cancer Registered in England and Wales*. HMSO, London, pp.26–31.
18. Booth, M. (1986) Aspects of the epidemiology of ovarian cancer. PhD thesis, London University.
19. Hildreth, N.G., Kelsey, J.L., LiVolsi, V.A. *et al.* (1981) An epidemiologic study of epithelial carcinoma of the ovary. *Am. J. Epidemiol.*, **114**, 398–405.
20. Campbell, S., Bhan, V., Royston, P. *et al.* (1989) Transabdominal ultrasound screening for early ovarian cancer. *Br. Med. J.*, **299**, 1363–7.

Chapter 28

Advantages and disadvantages of randomized controlled trials of ovarian cancer screening

IAN V. SCOTT

28.1 INTRODUCTION

The purpose of this chapter is to develop concepts discussed in a previous forum[1] and to examine the justification for trials of screening for epithelial ovarian cancer (EOC) and why they need to be randomized. Further issues requiring clarification are the size of trial required; which groups in the population would be most appropriate to target for such screening; which tests would be the most effective for the purpose and whether there is sufficient justification in the light of present knowledge to justify the investment in time and effort that such an exercise would involve if launched now, or whether this should be postponed until better tests become available.

28.2 JUSTIFICATION FOR SCREENING

The objective of health care is firstly to avoid premature mortality and secondly to reduce morbidity in diseases of a chronic nature. The criteria which should be satisfied in order to justify screening an asymptomatic population for a disease have been defined by Wilson and Jungner[2]. The requirement that the natural history of the disease should be well understood is not strictly necessary since there are many examples of conditions for which an empirical treatment has been found to work before detailed understanding is acquired of its mode of action in curing or interrupting the course of the disease. Ovarian cancer is the commonest cause of death from gynaecological malignancy in the UK, causing approximately 4000 deaths per year in England and Wales alone, as opposed to 2000 deaths per year from cancer of the cervix[3].

The mortality from ovarian cancer worldwide has increased four-fold since 1911 and shows considerable variation between different countries, being commonest in Sweden and least common in Japan[4]. The rate of increase in mortality appears greater than that for the increase in cancer of the breast. The high mortality is mainly due to more than 70% of cases presenting at stage III or

IV. The overall 5-year survival for stage I disease is 75%, varying from 95 to 50% according to substage and histological grade. In properly confirmed stage I cases the 5-year survival is 90% although the relapse-free rate at 5 years is only 79%. For stage II, 5-year survival is 46%, for stage III, 20–25% and for stage IV, 5%. Notwithstanding, it is necessary to weigh the potential benefits of screening for this disease against costs both in financial and quality of life terms.

The concept of future years of life lost (FYLL) allows comparison with other causes of death as a cause of premature mortality. A life expectancy of 75 years is assumed and the FYLL calculated by substracting the woman's age at death from this figure. Ovarian cancer causes an average loss of 15.5 future life years per victim and ranks ahead of cervical cancer (10.7 years) and breast cancer (10.3 years) in this respect[5] (Table 28.1). These figures are based on one year's mortality experience in a population of approximately 600 000 and hence may be subject to fluctuation from year to year. Nevertheless, local experience suggests that these rankings are maintained. Moreover, by comparison with the commoner causes of death such as coronary heart disease and strokes, ovarian cancer is a more significant cause of premature death (Figure 28.1).

Table 28.1 Female deaths and (FYLL), Southern Derbyshire Health Authority, 1988

Cancer	Deaths	FYLL	Average
Ovarian	37	575	15.5
Cervical	22	235	10.7
Breast	149	1535	10.3
Lung	106	1075	10.
Rectal	29	200	7
Stomach	41	195	5
Colon	60	280	4.7

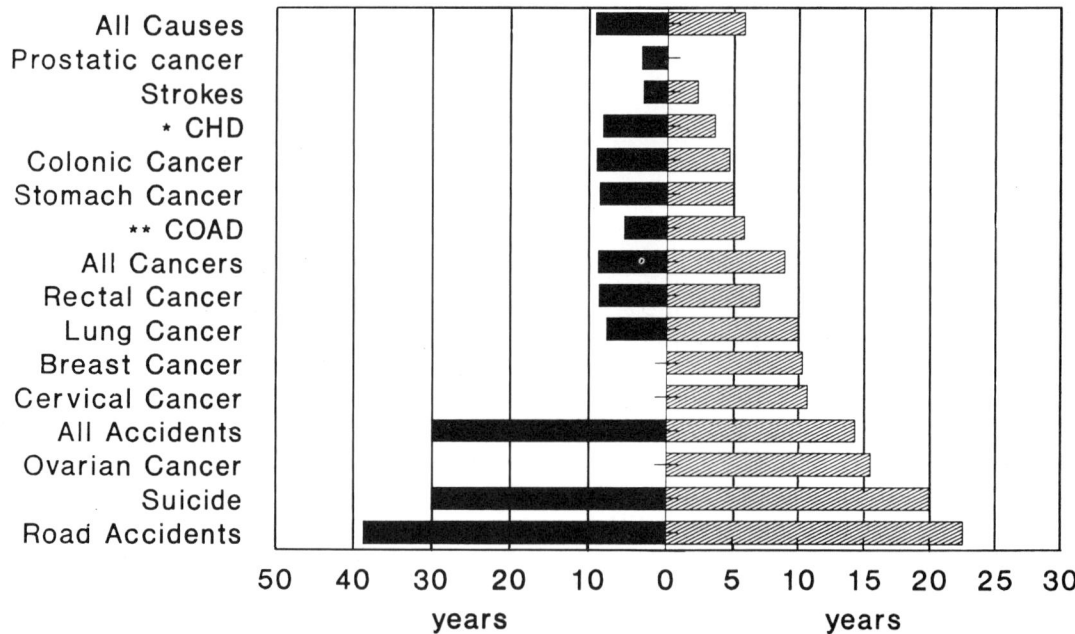

Figure 28.1. Average future years of life lost per victim by selected causes, Southern Derbyshire Health Authority, 1988. ■, Males; □, females. Source: OPCS, vol. 3, 1988.

28.3 WHY RANDOMIZED TRIALS?

The need for randomized trials is determined by the possibility of two differing types of bias confounding the attempt to reduce mortality by early detection. These are identified below.

28.3.1 LEAD TIME BIAS

Simply detecting an increased number of early stage cases may not lead to any alteration in ultimate mortality from the disease. If a standard growth pattern from one cell typical of solid tumours is assumed, the time from screening diagnosis to death may be longer in the screened cases (i.e. apparent increase in 5-year old survival) but these patients still die of their disease and are simply burdened by the knowledge of its presence for longer than would otherwise have been the case. This is known as lead time bias.

28.3.2 LENGTH BIAS

Additional bias may arise as a consequence of the heterogeneity of the tumours in respect of their growth rates. The most rapidly growing aggressive tumours will lead to rapid progression and death. Hence there will be a disproportionate number of slowly growing tumours prevalent in the community. These of course will be the ones most likely to be picked up by screening and will bias the results. This is known as length bias.

Neither improvement in crude 5-year survival data alone nor simply demonstrating an increased detection rate of early stage tumours is adequate proof of benefit to the screened population. The only objective proof lies in a clear and unequivocal reduction in registrations of ovarian cancer deaths in the screened as opposed to unscreened populations. Hence the need for randomized population-based trials.

28.4 WHAT SIZE?

What size should such studies be in order to determine whether screening produces a reduction in mortality? Much will depend on the magnitude of effect one hopes to find. Ovarian cancer has a low prevalence but a high mortality. A reasonable target to seek would be at least a 30% reduction in mortality similar to that sought by established breast cancer screening programmes and actually achieved by pilot studies of that disease[6]. Calculations suggest that given the low prevalence of ovarian cancer it would require in the order of 75 000 to 100 000 women to be screened in each arm of a trial to give an 80% probability of determining that such a reduction in mortality in the screened arm was statistically significant at the 5% level[7,8].

28.5 TARGET POPULATION

Cost-benefit considerations define the need to target screening on high risk groups. The means of identifying these groups must be simple and readily available.

The highest risk group consists of women with two or more first-degree relatives who have suffered from ovarian cancer[9]. Even individuals with one first-degree relative affected have a roughly doubled lifetime risk of developing the disease. Although screening studies are already underway in this group in the UK under the auspices of the Family Cancer Study Group and the United Kingdom Committee for the Coordination of Cancer Research (UKCCCR), the percentage of all ovarian cancer registrations associated with such family histories is unlikely to exceed 5–10% at most. It is also possible that the biology of the disease is different in this group, with an increased chance of epithelial ovarian cancer-like disease of multifocal origin throughout the peritoneal cavity. This impression is based largely on anecdote and remains highly speculative, although there are some pointers to this possibility[10–12]. Whether screening and/or prophylactic oophorectomy have a role to play in reducing mortality in this group remains to be proven.

Even covering the whole of England and Wales it is estimated that this very high risk group accounts only for 40–60 cases of ovarian cancer per year and that possibly 3% of all cases under 50 years occur in carriers of a putative susceptibility gene. Although this proportion is small, it is calculated that there are about 2000 such women aged 25–40 years. This group will suffer about 500 deaths from ovarian cancer by the age of 60 years. The much higher incidence and mortality in this group means that screening trials in this situation would have much greater statistical power for smaller numbers screened. Nevertheless, ethical considerations would make it extremely difficult to argue in favour of an unscreened arm in trials in such a high risk population. Notwithstanding uncertainty as to whether early detection will avert mortality, there is a strong case for identifying such families and offering them appropriate counselling, together with such reassurance as presently available screening methods can offer. These individuals can at least benefit from reassurance as regards their current status at the time of each screening. Valuable blood and tissue specimens can be obtained to help in the process of identification of specific genetic defects and understanding of the biology of the disease. The criteria for screening such families have been discussed elsewhere in this book.

Two other high risk groups deserve mention. One consists of women whose ovaries have been exposed to the effects of pelvic radiotherapy or major radiation exposure. Ovarian cancer may develop in this group many years after cure of the original condition for which pelvic radiotherapy was given[13,14]. Although numerically only a very small number of women fall into this risk category, it would be sensible to consider this group for screening trials. Another group deserving special attention are certain subfertile women, particularly those who have been subjected to treatment by superovulation techniques[15,16]. Much of the concern expressed regarding this is at present highly speculative and based at least in part on the negative correlation between anovulation and ovarian cancer in women using combined oral contraception[17]. Nevertheless there are some sporadic reports of women treated by superovulation therapy who subsequently present with ovarian cancer in their early forties and two such cases have been seen by the author in recent months. It would be inappropriate to assume a cause and effect relationship. It could well be that patients at risk of subfertility are inherently at greater risk of developing ovarian cancer for some quite separate reason. This issue could best be resolved by establishing national registers of all women receiving such treatments from specialist clinics and infertility centres, together with regular cross-referenced checks with cancer registries. It may also be relevant to compare the prevalence of 'involuntary infertility' in the female relatives of ovarian 'cancer family' cases with that of the normal population. The overwhelming majority of ovarian cancer cases occur unrelated to these factors. The only consistent risk determining factor for the majority of cases is age.

Cost–benefit considerations based on the hypothetical costs of screening and the size of the relevant age cohort determine the most appropriate target population. The cost is assumed for the purpose of these calculations to lie in the region of £15 per person screened, (although current developments support the view that a considerably lesser cost may be attainable). In terms of cost per cancer detected, the age range of 45 and over would be the least costly to screen. The objective of screening, however, must be to gain good quality future life years saved for each potential victim. If this is taken as the basis for the benefit sought, the 45- to 60-year age group becomes the most cost-effective target group (although it only encompasses 54% of annual ovarian cancer registrations). Nevertheless, slightly different issues need to be taken into account when considering the ideal age range to screen in a screening trial, as opposed to a proven and practical screening programme. In the case of a trial, the target population must be large enough to produce sufficient numbers of cases to achieve the necessary statistical power. In addition, in the UK there would be considerable logistical advantage in using a minimum of three large health districts with stable and compliant populations (such as Derby, Nottingham and Sheffield) as the basis for such a trial, encompassing 264 093 women aged 45–74 years, rather than using 15–20 centres of similar size, as would be necessary to generate the same numbers if a narrower age range were screened (e.g. 50–64 years).

Although the latter approach would be relatively easily attached to the existing breast screening programme in the UK, it would be much less easy to achieve the same degree of consistency in follow-up of screened and control cases in terms of investigations, management and histopathology, as would be possible between large geographically linked districts with staff already motivated to pursue such an approach. This would be particularly true of the motivation needed on the part of the operating gynaecologist to build up carefully documented banks of tissue and fluids from epithelial ovarian cancer cases occurring in the trial population.

A further potentially confounding feature is the evidence that, stage for stage, ovarian cancer patients diagnosed at less than 50 years of age have a better prognosis than older patients[18]. This implies that screening and early detection in a younger age cohort may be relatively more effective in reducing mortality than it would be for older subjects. Within the UK context, it may still be possible to integrate ovarian with breast screening in a younger series of women, since a multicentre trial of annual breast screening is about to be launched in women, starting at age 40 and continuing for 7 years. This would be a particularly suitable group in which to evaluate annual serum CA-125 assays and pelvic ultrasound. Nevertheless, at least 70 000 women would be required in each arm of a trial for adequate statistical power to be achieved.

These considerations throw doubt on the value of current proposals by the Board of Scientific Counsellors of the Division of Cancer Prevention and Control of the National Cancer Institute in the USA, to combine an ovarian screening trial with a larger study to evaluate screening for prostate, lung and colon cancer (PLCO trial)[19]. The objective is to screen some 37 000 women over the age of 60 with annual transvaginal ultrasound and serum CA-125 estimation. It is intended that there should be an equal number of controls. Such small numbers indicate that even with a 30% mortality difference for ovarian cancer between the arms, the power of the trial would be only 77%. Also, mortality reduction may be more difficult to achieve in the older age group and may confound the trial. The health economics of screening this age cohort alone are highly questionable in terms of the cost of future life years saved and quality of life costs. Furthermore, the intent to screen annually for colon cancer by flexible sigmoidoscopy may actively discourage participation of women whose risk of ovarian cancer may be more strongly linked to breast cancer than colon cancer epidemiologically.

28.6 WHICH TESTS?

Currently only abdominal or pelvic ultrasound examination and serum CA-125 measurement, separately or in combination, offer any practical hope of screening for this disease.

28.6.1 ULTRASOUND

Transabdominal ultrasound, as used in the original King's College Hospital studies, clearly has high sensitivity for detecting ovarian lesions, but nevertheless lacked specificity and gave rise to concern that the high false positive rate could lead to large numbers of unnecessary invasive surgical procedures, let alone the associated anxiety[20]. More recent developments using transvaginal ultrasound probes, and particularly the use of colour Doppler flow techniques[21], offer refinements in specificity which appear promising in skilled hands as reported elsewhere in this book. The size of the age cohorts eligible for screening in any given centre (and the need for rescreening on at least a 3 yearly basis), suggests that although likely to be important for screening cancer families, it is not likely to be easily applicable as a first-line screening technique for the general population, unless a high risk target group has been defined by some other preliminary test. Consideration of the size of the relevant age cohorts in a typical large district, such as Southern Derbyshire Health Authority (SDHA), reveals that there are 83 000 females aged 45–74 who even if screened only every 3 years would generate over 20 000 scans per annum. Restriction to an age range of 50–64 years would on the same basis required between 10 000 and 14 000 scans per annum, which would just be feasible and would be ideally suited to integration with the breast screening programme.

Nevertheless, there is some discrepancy in the stage distribution of ovarian cancer, as shown by the King's group. In particular, their cases have been almost exclusively borderline lesions or stage Ia tumours, which would be a most unusual finding in a population-based prevalence screen. In addition, in the screening of high risk cancer families this group found an increased prevalence of benign ovarian disease.

No easy explanation can be produced so far to explain these findings. The numbers screened in these studies are small in relation to the prevalence of the condition, and certainly by comparison with the numbers who would be eligible for screening in a single large health district in the UK. These results may simply be due to chance or selection bias since the subjects are self-selecting in response to media publicity and not called on the basis of an unselected population register. If, however, there are two types of ovarian cancer (one that starts in the ovary, growing slowly and spreading late, and another that is of epithelial ovarian cancer type but starts as part of a multifocal peritoneal transformation process and grows more rapidly), then one would expect to detect the former more commonly in an asymptomatic population due to length time bias. This might be one possible explanation for the difference in stage distribution found by the ultrasound-only screening studies as against the more standard stage distribution found by the London Hospital study based on preliminary serum CA-125 measurement with secondary ultrasound examination of positives.

28.6.2 SERUM MARKER ESTIMATION

In a retrospective case-controlled study on a population-based serum bank, blood tests for CA-125 have been shown to be positive in up to 25% of cases 5 years prior to clinical diagnosis and in up to 50% 18 months prior to clinical diagnosis[22] (Table 28.2).

A self-referral study based on preliminary screening with CA-125 and subsequent screening of positive cases by ultrasound shows promise[23]. A total of 20 000 women have now been screened[24] and produced

Table 28.2 Elevated serum CA-125 prior to diagnosis of ovarian cancer[22]

Months before clinical diagnosis	Percentage with CA-125>30U/ml
>60	25
60≤36	11
36≤18	19
≤18	50*

* 30%>65 U/ml 18 months prior to diagnosis.

281 positive blood tests. Only two clinical cases have appeared subsequently in those with negative blood tests (i.e. two false negatives). In 259 with positive blood tests but negative subsequent ultrasound examination, only one false negative has occurred. In 22 cases with positive blood tests and an abnormal ultrasound scan there were 11 cancers (true positives) and 11 benign lesions (false positives). These figures compare well with the 10 cases one might have predicted in a prevalence screen of 20 000 women based on the blood bank case control study mentioned previously. The 11 cases of cancer detected in the 20 000 women screened in the study by Jacobs and colleagues must be set against eight cases expected in one year in a population that size based on an incidence of 1 in 2500.

Using the concept of 'years of cancer detected' (YCD)[25], it would have taken only 11/8 = 1.4 years or approximately 17 months to accumulate the same number of cases at clinical presentation, as was detected by screening. The YCD for breast cancer screening is between 2 and 3 years. It is suggested that a YCD of one year or less is unlikely to be of much value. Nevertheless, we know little about the natural history of ovarian cancer, and it is still possible that even as short a lead time as 18 months might be of some advantage to patient survival. It must also be stated that these studies give only a very crude idea of the potential lead time for screened individuals over unscreened, since they are uncontrolled and based on relatively small self-selected populations.

An alternative strategy for the use of serum markers in the detection of occult disease is to evaluate serial measurements of CA-125 over time[26]. Between January 1986 and February 1988 serum CA-125 levels were measured annually in 5550 women at least 40 years of age. Where a CA-125 level > 30–35 U/ml was found, the test was repeated every 3 months and a pelvic examination and abdominal ultrasound examination were undertaken every 6 months. The same was done for an age-matched cohort of women with CA-125 < 30–35 U/ml. Among 175 women with elevated CA-125, six ovarian cancers were detected. In each of these cases, CA-125 doubled or reached 95 U/ml during follow-up (median 32 months). No ovarian cancers were detected in the 175 matched controls who did not have CA-125 elevation. However, although these studies may point to an effective screening programme, the true value of CA-125 and ultrasound screening for early stage disease can only be judged by a populated-based randomized trial of screening versus nonscreening, as concluded by the UKCCCR workshop on screening for early stage ovarian cancer[27]. Only by such a trial can differences in mortality be compared and the potential lead time bias and length time bias of the more limited studies be assessed. It is also the only way in which the sensitivity, specificity and positive predictive value of these tests can be determined. The most logical and cost-effective approach would be to use serum CA-125 estimation as the first tier of screening followed by additional marker tests[28,29], and ultrasound as a second tier in those found to be positive. Blood would be taken annually for 5 years from the whole target population. Test and control groups would be randomized and aliquots of serum from both groups stored in a serum bank.

Table 28.3 Randomized trial of ovarian cancer screening in 210 000 women, 105 000 in screened arm (Derby, Nottingham and Sheffield)

Laboratory (recurrent and capital)	£1 868 345 (£998 058 in house)
Ultrasound (recurrent and capital) (3 scanners and equipment)	£508 000 (£169 333 for one)
Radiographer (x 3)	£180 000 (£60 000 for one)
Data manager	£100 000
Secretary	£40 000
Computer and office equipment	£10 000
Nurse counsellor (x 3)	£210 000 (£70 000 for one)
Total costs over 5 years	£2 916 345 (£1 447 391 minimum)

Assuming 500 deaths expected over 5 years, the population of, for example, Derby is sufficiently large to enable a trial to determine whether mortality could be improved by early detection at an adequate statistical power[30,31].

Study centres should also have access to imaging techniques in addition to good quality ultrasound, such as CT scans, immunoscintigraphy and MRI[32]. Evaluation of the ability of these techniques to reduce the need for unnecessary invasive procedures would be crucial to such an exercise. It would also be of interest to assess the possible benefits to presymptomatic patients with non-ovarian benign and malignant conditions coincidentally discovered by the screening.

On the basis of the data from Jacobs' studies we would predict a need for about 500 ultrasound scans per centre in the first year of screening (i.e. prevalence screen). As a result of these, each centre would need to undertake a maximum of 35–40 invasive surgical investigations. These figures would decrease in subsequent years of rescreening the same population. Screened positive cases in each centre should be managed in the first instance by one named gynaecologist from that centre to ensure adequacy of data collection and uniformity of investigative approach.

Effective screening could give a considerable quality adjusted life year (QALY)[33] gain to patients detected at an early (stage Ia) rather than late stage (III–IV). There would also be a predictable direct cost saving in the form of surgical treatment alone, without the need for costly additional anticancer drugs if cases were detected sufficiently early. In a trial of this nature it would be possible to compare the treatment and quality of life costs and benefits of screened vs. unscreened patients within single health districts with stable populations in a way which is quite impossible for the pre-existing self-referral studies. The cost of the programme per centre is estimated to be in the region of £200 000 per year (or £100 000 if a new in-house assay is used; Table 28.3). This compares favourably with the current annual cost of cervical screening services. The cost per ovarian cancer detected should not exceed £11 000–£15 000.

Taken that there is a 30% reduction in mortality and given that 500 expected mortalities will occur over 5 years in the entire trial population, at most 75 deaths might be avoided in the screened half. This approach would avoid the loss of an average of 16 future life years per victim, and gain approximately 1200 life years in all, at a cost in the region of £1200–£2430 per life year saved [34,35]. These calculations are very crude and do not take account of a number of complicating factors, such as those patients who will already have clinically advanced disease at first screening and the effects of ageing and mortality on the numbers left at risk in each successive year of screening.

28.7 TIMING OF TRIALS

Concern exists about the costs and benefits of screening in financial, emotional, and physical terms for the population being screened[1,27]. If a screening project were launched with a serum marker test of inadequate sensitivity, then future credibility for any similar marker-based strategy would be destroyed. It might be best to await the development of new and more sensitive markers or multiple marker strategies. Paradoxically, however, the public is more likely to accept the necessity of randomized trials of screening vs. non-screening while reasonable doubt remains about the sensitivity of the screening test. If the test were known to be 100% sensitive it would be difficult to convince lay opinion of the justification for a non-screened arm even though the fundamental scientific arguments for this remain the same. There seems to be a proliferation of private screening clinics offering various combinations of existing tests, even though there is as yet no evidence to support the view that early detection will necessarily avert mortality from the disease. Unless this issue is tackled properly it will be impossible to determine if screening has any value.

28.8 THE PURPOSE OF A PROPOSED RANDOMIZED TRIAL

The primary objective of the trial is to determine a difference in mortality between the screened and unscreened groups. The secondary objectives are to determine: the sensitivity, specificity and predictive value of the tests in a defined population; the costs of secondary screening required; patient acceptability of the procedures; the morbidity of secondary screening tests; the ideal screening interval and natural history of the disease; the QALY benefits to patients detected and treated at an early rather than late stage of the disease; and an estimate of the cost saving in the form of simple surgery without the need for costly anticancer drug treatments.

28.9 CONCLUSIONS

Screening of ovarian cancer families is justified for reassurance of patients and facilitating research into screening methods. Ethical considerations rule out randomized trials of screening vs. no screening in this group.

A combination of pre-screening by serum CA-125 estimation and selective ultrasound examination seems the most practical way of screening large populations, preferably using vaginal probes with colour Doppler facility.

Large randomized population-based trials are needed to determine whether or not mortality can be reduced by screening and to collect serum and tissue for further research.

Trials need to have at least 100 000 subjects in each arm to have a 90% power of showing that a 30% improvement in mortality in the screened arm is significant at the 5% level. To show that smaller improvements in mortality are significant proportionately larger numbers need to be included in the trial.

No trial to date has been both population-based, randomized or sufficiently large to produce statistically valid results. The proposed PLCO trial in the USA with only 37 000 subjects in each arm is far too small to produce any meaningful result unless a greater than 35% improvement in mortality is shown in the screened arm.

A trial based on three or four large centres and covering the age range 45–74 would provide the most effective basis for a randomized study of screening. A reasonable compromise in order to spread the costs would be to integrate the study with existing breast cancer screening in the age range 50–64 and to exploit the proposed annual breast screening trial in a cohort of women aged 40 to study ovarian screening in the younger group. To produce the necessary numbers 15–20 centres would need to collaborate.

REFERENCES

1. Scott, I.V., Milford Ward, A., Selby, C., Whitehead, S. and Wilcox, M. (1990) Development of population based studies of ovarian cancer screening, in *Ovarian Cancer: biological and therapeutic challenges* (eds F. Sharp, W.P. Mason and R.E. Leake), Chapman & Hall, London, pp. 245–58.
2. Wilson, J.M.G. and Jungner, G. (1968) *Principles and Practice of Screening for Disease*, Public Health Papers 34, World Health Organization.
3. Beral, V. (1987) The epidemiology of ovarian cancer, in *Ovarian Cancer — The Way Ahead* (eds F. Sharp and W.P. Soutter), RCOG, London, pp. 21–31.
4. Greene, M.H., Clark, J.W. and Blayney, D.W. (1984) The epidemiology of ovarian cancer. *Seminars Oncol.*, **11(3)**, 209–26.
5. 'Going for Health' (1989) The Annual Report of the Director of Public Health, Southern Derbyshire Health Authority, pp. 53–4.
6. Shapiro, S., Venet, W., Strax, P., Venet, L. and Roeser, R. (1982) Ten- to fourteeen-year effect of breast cancer screening on mortality. *J. Natl Cancer Inst.*, **69**, 349–55.
7. Miller, D.K. and Homan, S.M. (1988) Graphical aid for determining power of clinical trials involving two groups. *Br. Med. J.*, **297**, 672–6.
8. Freedman, L.S. (1989) The size of clinical trials in cancer research — what are the current needs? *Br. J. Cancer*, **59**, 396–400.
9. Ponder, B.A.J., Eaton, D.F. and Peto, J. (1990) The risk of ovarian cancer associated with a family history. Preliminary report of the OPCS study, in *Ovarian Cancer: biological and therapeutic challenges* (eds F. Sharp, W.P. Mason and R.E. Leake), Chapman & Hall, London, pp. 3–6.
10. Parmley, T.H. and Woodruff, J.D. (1974) The ovarian mesothelioma. *Am. J. Obstet. Gynecol*, **120(2)**, 234–41.
11. Tobacman, J.K., Tucker, M.A., Kase, R., Greene, M.H., Costa, J. and Fraumeni, J.F. (1982) Intra-abdominal carcinomatosis after prophylactic oophorectomy in ovarian-cancer-prone families. *Lancet*, **ii**, 795–8.
12. Lynch, H.T., Bewtra, C. and Lynch, J.F. (1986) Familial peritoneal ovarian carcinomatosis: a new clinical entity? *Med Hypotheses*, **21**, 171–7.
13. Tokuoka, S., Kawai, K., Shimizu, Y., et al. (1987) Malignant and benign ovarian neoplasms among atomic bomb survivors, Hiroshima and Nagasaki, 1950–80. *JNCI*, **79(1)**, 47–57.
14. Kleinerman, R.A., Curtis, R.E., Boice, J.D. et al. (1982) Second cancers following radiotherapy for cervical cancer. *J. Natl Cancer Inst.*, **69**, 1027–33.
15. Fishel, S. and Jackson, P. (1989) Follicular stimulation for high tech pregnancies: are we playing it safe? *Br. Med. J.*, **299**, 309–11.
16. Smith, B.H. and Cooke, I.D. (1991) Ovarian hyperstimulation: actual and theoretical risks. *BMJ*, **302**, 127–8.
17. The Cancer and Steroid Hormone Study of the Centers for Disease Control and the National Institute of Child Health and Human Development (1987) The reduction in risk of ovarian cancer associated with oral-contraceptive use. *N. Engl. J. Med.*, **316**, 650–5.
18. Kolstad, P. (1980) Prognostic indicators and staging, in *Ovarian Cancer* (eds C.E. Newman, C.H.J. Ford and J.A. Jordan), Pergamon Press, Oxford, Advances in the Biosciences, **26**, 67–8.
19. Trial of Cancer Screening Methods (1991) *The Cancer Letter*, **17(7)**, 4–6.
20. Campbell, S., Collins, W.P., Royston, P., Bourne, T.H., Bhan, V. and Whitehead, M.I. (1990) Developments in ultrasound screening for early ovarian cancer, in *Ovarian Cancer: Biological and therapeutic challenges* (eds F.

21. Bourne, T., Campbell, S., Steer, C., Whitehead, M.I. and Collins, W.P. (1989) Transvaginal colour flow imaging: a possible new screening technique for ovarian cancer. *Br. Med. J.*, **299**, 1367–70.
22. Zurawski, V.R., Orjeseter, H., Anderson, A. and Jellum, E. (1988) Elevated CA 125 levels prior to diagnosis of ovarian neoplasia: relevance for early detection of ovarian cancer. *Int. J. Cancer*, **42**, 677–80.
23. Jacobs, I.J., Stabile, I., Bridges, J. *et al* (1988) Multimodal approach to screening for ovarian cancer. *Lancet*, **i**, 268–71.
24. Jacobs, I.J., Prys Davies, A., Grudzinkas, J.G. and Oram, D.H. (1989) Interim report on London Hospital Ovarian Cancer Screening Project reported to November 1989 meeting of the British Gynaecological Cancer Society.
25. Cuckle, H.S. and Wald, N.J. (1990) The evaluation of screening tests for ovarian cancer, in *Ovarian Cancer: biological and therapeutic challenges* (eds F. Sharp, W.P. Mason and R,.E. Leake), Chapman & Hall, London, pp. 229–39.
26. Einhorn N., Svojall, K., Schoenfeld, D.A. *et al.* (1990) Early detection of ovarian cancer using the CA 125 radioimmunoassay (RIA). *Proc. ASCO*, **9**, 157.
27. Screening for Early Ovarian Cancer (1989) Report of a Workshop to the UKCCCR Subcommittee for Coordination of Research into Gynaecological Cancers (Chairman F. Sharp), in *Ovarian Cancer: biological and therapeutic challenges* (eds F. Sharp, W.P. Mason and R.E. Leake), Chapman & Hall, London, pp. 461–5.
28. Dhokia, B., Canney, P.A., Pectasides, D. *et al.* (1988) A new immunoassay using monoclonal antibodies HMFG1 and HMFG2 together with an existing marker CA 125 for the serological detection and management of epithelial ovarian cancer. *Br. J. Cancer*, **54**, 891–5.
29. Bast, R.C., Boyer, C.M., Olt, G.J. *et al.* (1990) in *Ovarian Cancer: biological and therapeutic challenges* (eds F. Sharp, W.P. Mason and R.E. Leake), Chapman & Hall, London, pp. 265–75.
30. Miller, D.K. and Homan, S.M. (1988) Graphical aid for determining power of clinical trials involving two groups. *Br. Med. J.*, **297**, 672–6.
31. Freedman, L.S. (1989) The size of clinical trials in cancer research—what are the current needs? *Br. J. Cancer*, **59**, 396–400.
32. Perkins, A.C., Powell, M.C., Wastie, M.L., *et al.* (1990) A prospective evaluation of OC125 and magnetic resonance imaging in patients with ovarian carcinoma. *Eur. J. Nucl. Med.*, **16**, 311–16.
33. Gudex, C. and Kind, P. (1988) The QALY toolkit, Discussion Paper 38, Centre for Health Economics, Health Economics Consortium, University of York.
34. Working Group (1986) Breast cancer screening. Report to the Health Ministers of England, Wales, Scotland, and Northern Ireland. HMSO, London.
35. Rees, G.J.G. (1991) Cancer treatment: deciding what we can afford. *Br. Med. J.*, **302**, 799–800.

Part Five
Management

Chapter 29

Intraperitoneal therapy in the management of ovarian cancer

M. MARKMAN

29.1 INTRODUCTION

Since the late 1970s, investigators at a number of centres around the world have actively investigated the intraperitoneal administration of cytotoxic agents as therapy for ovarian cancer[1,2]. Both the potential benefits and major limitations of this approach have been clearly defined. A number of drugs with known activity in patients with ovarian cancer have been administered by the intraperitoneal route with a major pharmacokinetic advantage for cavity exposure, compared to the systemic compartment, being demonstrated (Table 29.1)[2]. In addition, phase 2 trials have confirmed the fact that the pharmacokinetic advantage associated with this approach can be translated into surgically defined responses, including complete responses, in patients previously treated with cisplatin or carboplatin-based systemic therapy[3–9].

In this chapter, the major conclusions reached from a decade of clinical and laboratory research in the area of intraperitoneal therapy will be summarized. This will be followed by a brief discussion of the role intraperitoneal therapy may be considered to play in the standard management of patients with ovarian cancer, and potential avenues for future clinical research employing this route of drug delivery.

29.2 IMPORTANT LESSONS LEARNED FROM A DECADE OF RESEARCH WITH INTRAPERITONEAL THERAPY OF OVARIAN CANCER

29.2.1 DRUG DISTRIBUTION

If there is to be any advantage associated with the intraperitoneal administration of

Table 29.1 Pharmacokinetic advantage associated with intraperitoneal administration of selected antineoplastic agents with known activity in ovarian cancer

Agent	Mean peak peritoneal cavity/ plasma concentration ratio
Cisplatin	20
Carboplatin	18
Mitoxantrone	640
Doxorubicin	474
Melphalan	93
Mitomycin-C	71
5-Fluorouracil	298

cytotoxic agents, it is crucial that the drugs actually reach the site of tumour. Despite the fact that a patient may be an 'excellent candidate' to receive intraperitoneal therapy (e.g. microscopic residual disease following initial systemic chemotherapy), it may be impossible to consider this treatment strategy due to the presence of extensive adhesions resulting from the tumour itself or the prior surgeries. Currently, there are no methods available to prevent adhesion formation in patients considered for this regional treatment approach. Individual patients with extensive adhesions observed at surgery, even if lysed at the time of surgery, should probably not be considered for an intraperitoneal treatment strategy. In patients who experience major difficulty with intraperitoneal drug instillation (e.g. excessively slow flow, pain associated with the administration of the treatment volume), either the intraperitoneal administration of a radiolabelled tracer or contrast material in a large volume may help to further define the extent of the problem[10,11]. Unfortunately, such patients will usually demonstrate inadequate distribution of the instilled volume and further therapy should probably be administered by the systemic route.

The importance of the adequacy of drug distribution throughout the abdominal cavity has been highlighted by the recent report of a phase 2 trial of intraperitoneal mitoxantrone as salvage therapy for patients with advanced ovarian cancer[7]. The 'blue colour' of mitoxantrone persisted within the peritoneal cavity following drug administration and at the time of surgery it was possible to estimate the adequacy of drug distribution during treatment. Not only did many patients fail to achieve complete distribution of the drug-containing fluid throughout the abdominal cavity, but several patients demonstrated major surgically documented antitumour responses within the area of drug distribution (blue colour of mitoxantrone present), while in areas where there was no blue stain tumour progression was observed.

29.2.2 DRUG PENETRATION

A number of preclinical models have been utilized to examine the depth of penetration of cytotoxic agents directly into tumour or normal tissue[12–15]. These models are quite relevant to the question of the relative advantage of intraperitoneal therapy, compared to systemic drug delivery in ovarian cancer, as the benefits to be derived from regional drug delivery come from the delivery of the drug to tumour by free-surface diffusion, rather than by capillary flow. Unfortunately, regardless of the agent employed, the penetration of drug directly into tissue is quite limited and ranges from several cell layers to 1–3 mm[12–15].

Of particular relevance to the issue of the potential benefits of intraperitoneal therapy in patients with ovarian cancer is data from Los and colleagues[15] who examined the penetration of cisplatin into the peritoneal lining following either intraperitoneal or intravenous delivery of the agent. While cisplatin was found to a depth of >2 mm from the peritoneal surface after either treatment, the only significant concentration advantage for the regional approach over systemic administration was found at a depth of 0.1–1 mm from the surface of the cavity.

This experimental analysis is fully supported by considerable clinical data, which have demonstrated that the patients with ovarian cancer who are most likely to benefit from intraperitoneal cytotoxic drug administration, compared to systemic delivery of the agent, will be those individuals with very small volume residual (minimal) disease when treatment is initiated. For example, in several phase 2 clinical trials of cisplatin-based intraperitoneal therapy in patients with persistent or recurrent disease following front-line cisplatin or carboplatin-based systemic therapy, surgically defined responses were observed principally in patients whose largest tumour mass was <0.5–1 cm in diameter (including patients with microscopic disease only)[4–6].

Overall, approximately 30–40% of patients with very small volume residual ovarian cancer (largest tumour mass <1 cm) treated on these trials experienced a surgically defined response, compared to a response rate of <10–15% in patients whose largest residual disease was >1 cm. In addition, surgically defined complete responses are rarely observed with second-line intraperitoneal cisplatin-based therapy in ovarian cancer patients with any tumour mass >0.5–1 cm in diameter, unless the patient has experienced a long treatment-free interval between the completion of the initial systemic treatment regimen and the initiation of the salvage regional treatment programme (see next section for further discussion of the treatment of persistent/refractory versus recurrent ovarian cancer).

29.2.3 RECURRENT VERSUS PERSISTENT OVARIAN CANCER

The importance of distinguishing recurrent from persistent or refractory ovarian cancer when attempting to define a role for the intraperitoneal route of drug delivery has recently been highlighted by several reports, which have documented that patients with ovarian cancer who have previously responded to a cisplatin or carboplatin-based regimen may demonstrate a second response to the same or similar regimens[16,17]. In addition, the likelihood that a secondary response will be observed increases as the time between the two treatment programmes lengthens.

It is known that a major portion of cisplatin or carboplatin administered intraperitoneally reaches the systemic compartment[18–21].

Thus, in a patient demonstrating a response to intraperitoneal cisplatin (or carboplatin) where the individual has previously responded to systemic cisplatin (or carboplatin) with a long treatment-free interval (>1–2 years) before the initiation of the salvage intraperitoneal treatment programme, it is not possible to know if the route of drug delivery contributed significantly to the clinical activity observed.

In our experience, major antitumour responses to intraperitoneal cisplatin in individuals with bulky residual disease (>1 cm) will rarely occur if the patient has a short treatment-free interval (<6 months to 1 year) [5–6]. However, in patients with a long treatment-free interval (>2 years), objective antitumour responses to intraperitoneal cisplatin are common, even in patients with large volume disease[16]. At the present time, we cannot state that the intraperitoneal route of drug administration contributed anything to the responses observed in this particular patient population, except to serve as an indirect method to deliver drug to the systemic circulation.

In sharp contrast, patients with small volume persistent residual disease after front-line chemotherapy who start a second-line platinum-based intraperitoneal chemotherapy approach shortly after the completion of the intravenous programme, may be anticipated to have a 20–40% chance of achieving a surgically documented complete response to this treatment strategy[4–6]. While a randomized controlled trial comparing this approach to continued intravenous cisplatin has not been conducted, several small trials which have continued patients on an intravenous cisplatin-based second-line programme after documenting persistent disease following five to six courses of therapy, have demonstrated that <10% of individuals treated in this manner will ultimately achieve a surgically documented complete response [22].

29.2.4 INFLUENCE OF A PRIOR RESPONSE TO SYSTEMIC PLATINUM THERAPY

We have recently retrospectively evaluated patients treated at the Memorial Sloan-Kettering Cancer Center with a second-line cisplatin-based intraperitoneal programme to analyse what influence, if any, a prior response to platinum therapy has on a secondary response to the agent delivered by the intraperitoneal route. Not surprisingly, we found that this factor strongly influenced the outcome of our second-line programme. Of 50 patients with minimal residual disease (largest tumour mass <1 cm in diameter) who were evaluable for a response to both their initial intravenous and second-line intraperitoneal cisplatin programme, 42% (15/36) who had previously responded to cisplatin achieved a surgically documented complete response to intraperitoneal cisplatin, compared to 7% (1/14) of patients who had failed to respond to the initial chemotherapy programme ($P <0.025$).

Thus, in addition to the bulk of disease present at the time of initiation of intraperitoneal therapy, a second important prognostic factor in selecting patients appropriate for this therapeutic approach (at least for a cisplatin-based regimen) is the individual's prior response to systemic cisplatin or carboplatin. This conclusion is consistent with the hypothesis that any therapeutic advantage of intraperitoneal therapy, over that achieved by administering the drug systemically, will be relative rather than absolute. Thus, if a tumour is clinically resistant to the cytotoxic agent at the concentrations achievable in the systemic compartment following intravenous delivery, it is not surprising that the 10–20 fold increased exposure to cisplatin achieved with intraperitoneal therapy would not significantly impact on tumour growth. In contrast, if the tumour demonstrated sensitivity to the drug one might anticipate that the relatively higher concentrations of the drug in the peritoneal cavity could produce additional tumour cell kill.

29.3 THE FUTURE OF INTRAPERITONEAL THERAPY IN THE MANAGEMENT OF OVARIAN CANCER

While it is clear that patients with minimal residual ovarian cancer following front-line intravenous cisplatin-based treatment can achieve a high surgically documented complete response rate, the ultimate impact of this response on survival remains uncertain[23]. Several reports from non-randomized series have suggested a survival benefit associated with this treatment strategy[24,25], but it is only through the performance of randomized clinical trials that this important question can be appropriately addressed.

Unfortunately, a major problem in the design of such a trial (at least for second-line therapy) is the selection of an appropriate 'control' regimen with which to compare the intraperitoneal treatment strategy. As suggested earlier, there are currently no established standard intravenous regimens which have demonstrated a major impact on surgically documented complete responses or survival when used as second-line treatment in ovarian cancer. Thus, for the present, the selection of a second-line intraperitoneal cisplatin (or carboplatin) regimen for patients with minimal persistent residual ovarian cancer must be regarded as a reasonable treatment strategy for individuals who have demonstrated a response to their initial therapeutic programme.

A number of institutions have begun to explore the use of intraperitoneal chemotherapy as part of an upfront programme in ovarian cancer[26,27]. These programmes have ranged from the administration of all drugs by the intraperitoneal route (e.g. intraperitoneal cisplatin and etoposide)[26], to the use of a limited number of cycles of high dose intravenous chemotherapy, followed by intraperitoneal drug delivery[27]. The major limitation of this strategy is the fact that few patients with advanced disease start chemotherapy with their largest tumour mass measuring <0.5 cm. Thus, it will be the minority of patients with advanced disease who can potentially benefit from the extremely high local concentrations achievable following regional drug delivery. Ultimately, randomized trials will be required to determine if any upfront chemotherapy programme which includes the intraperitoneal route of drug delivery will produce a higher response rate and longer survival than that achieved with the systemic administration of the agents.

Finally, while most of the efforts involving the intraperitoneal administration of antineoplastic agents as therapy of ovarian cancer have focused on cisplatin or carboplatin, there is considerable interest in other agents, including mitoxantrone[7], taxol[28], and a variety of biologicals (recombinant interferon-α, recombinant interferon-γ, recombinant interleukin-2, tumour necrosis factor)[29], for which there may be an even greater relative advantage associated with the regional route of drug administration. However, as with intraperitoneal cisplatin-based programmes, it remains to be determined if the responses observed following the regional administration of these agents, which include surgically documented complete remissions, can be translated into long-term survival for patients with advanced ovarian cancer.

REFERENCES

1. Dedrick, R.L., Myers, C.E., Bungay, P.M. and Devita, V.T. Jr (1978) Pharmacokinetic rationale for peritoneal drug administration in the treatment of ovarian cancer. *Cancer Treat. Rep.*, **62**, 1–9.
2. Markman, M., Hakes, T., Reichman, B. *et al.* (1989) Intraperitoneal therapy in the management of ovarian carcinoma. *Yale J. Biol. Med.*, **62**, 393–403.
3. Hacker, N.F., Berek, J.S., Pretorius, R.G. *et al.* (1987) Intraperitoneal cis-platinum as salvage therapy for refractory epithelial ovarian cancer. *Obstet. Gynecol.*, **70**, 759–64.
4. Piver, M.S., Lele, S.B., Marchetti, D.L. *et al.* (1988) Surgically documented responses to intraperitoneal cisplatin, cytarabine and bleomycin after intravenous cisplatin-based chemotherapy in advanced ovarian adenocarcinoma. *J. Clin. Oncol.*, **6**, 1679–84.
5. Reichman, B., Markman, M., Hakes, T. *et al.* (1989) Intraperitoneal cisplatin and etoposide in the treatment of refractory/recurrent ovarian carcinoma. *J. Clin. Oncol.*, **7**, 1327–32.
6. Markman, M., Hakes, T., Reichman, B. *et al.* (1991) Intraperitoneal cisplatin and cytarabine in the treatment of refractory or recurrent ovarian carcinoma. *J. Clin. Oncol.*, **9**, 204–10.
7. Markman, M., George, M., Hakes, T. *et al.* (1990) Phase 2 trial of intraperitoneal mitoxantrone in the management of refractory ovarian carcinoma. *J. Clin. Oncol.*, **8**, 146–50.
8. Kirmani, S., Lucas, W.E., Kim, S. *et al.* (1991) A phase II trial of intraperitoneal cisplatin and etoposide as salvage treatment for minimal residual ovarian carcinoma. *J. Clin. Oncol.*, **9**, 649–57.
9. Speyer, J.L., Beller, U., Colombo, N. *et al.* (1990) Intraperitoneal carboplatin: favorable results in women with minimal residual ovarian cancer after cisplatin therapy. *J. Clin. Oncol.*, **8**, 1335–41.
10. Dunnick, N.R., Jones, R.B., Doppmen, J.L. *et al.* (1979) Intraperitoneal contrast infusion for assessment of intraperitoneal fluid dynamics. *Am. J. Roentgenol.*, **133**, 221–3.
11. Van Weelde, B.J., Pauwels, E.K., Jones, B. and Van Oosterom, A.T. (1984) Scintigraphic peritoneography in advanced ovarian malignancies: its value for chemotherapeutic distribution studies. *Clin. Radiol.*, **35**, 465–8.
12. Ozols, R.F., Locker, G.Y., Doroshow, J.H. *et al.* (1979) Pharmacokinetics of adriamycin and tissue penetration in murine ovarian cancer. *Cancer Res.*, **39**, 3209–14.
13. West, G.W., Weichselbau, R. and Little, J.B. (1980) Limited penetration of methotrexate

into human osteosarcoma spheroids as a proposed model for solid tumor resistance to adjuvant chemotherapy. *Cancer Res.*, **40**, 3665–8.
14. Nederman, T. and Carlsson, J. (1984) Penetration and binding of vinblastine and 5-fluorouracil in cellular spheroids. *Cancer Chemother. Pharmacol.*, **13**, 131–5.
15. Los, G., Mutsaers, P.H.A., van der Vijgh, W.J.F. *et al.* (1989) Direct diffusion of cis-diamminedichloroplatinum(II) in intraperitoneal rat tumors after intraperitoneal chemotherapy: a comparison with systemic chemotherapy. *Cancer Res.*, **43**, 3380–4.
16. Markman, M., Rothman, R., Hakes, T. *et al.* (1991) Second-line platinum therapy in patients with ovarian cancer previously treated with cisplatin. *J. Clin. Oncol.*, **9**, 389–93.
17. Gore, M.E., Fryatt, I., Wiltshaw, E. and Dawson, T. (1990) Treatment of relapsed carcinoma of the ovary with cisplatin or carboplatin following initial treatment with these compounds. *Gynecol. Oncol.*, **36**, 207–11.
18. Howell, S.B., Pfeifle, C.E., Wung, W.E. *et al.* (1982) Intraperitoneal cisplatin with systemic thiosulfate protection. *Ann. Intern. Med.*, **97**, 845–51.
19. Lopez, J.A., Krikorian, J.G., Reich, S. *et al.* (1985) Clinical pharmacology of intraperitoneal cisplatin. *Gynecol. Oncol.*, **20**, 1–9.
20. Pretorius, R.G., Hacker, N.F., Berek, J.S. *et al.* (1983) Pharmaockinetics of Ip cisplatin in refractory ovarian carcinoma. *Cancer Treat. Rep.*, **67**, 1085–92.
21. Casper, E.S., Kelsen, D.P., Alcock, N.W. and Lewis, J.L. (1983) Ip cisplatin in patients with malignant ascites: pharmacokinetic evaluation and comparison with the iv route. *Cancer Treat. Rep.*, **67**, 235–8.
22. Markman, M., Hakes T., Reichman, B. *et al.* (1990) Intraperitoneal versus intravenous cisplatin-based therapy in small-volume residual refractory ovarian cancer: evidence supporting an advantage for local drug delivery. *Reg. Cancer Treat.*, **3**, 10–12.
23. Ozols, R.F. (1991) Intraperitoneal therapy in ovarian cancer: time's up. *J. Clin. Oncol.*, **9**, 197–9.
24. Howell, S.B., Zimm, S., Markman, M. *et al.* (1987) Long term survival of advanced refractory ovarian carcinoma patients with small-volume disease treated with intraperitoneal chemotherapy. *J. Clin. Oncol.*, **5**, 1607–12.
25. Menczer, J., Ben-Baruch, G., Modan, M. and Brenner, H. (1989) Intraperitoneal cisplatin chemotherapy versus abdominopelvic irradiation in ovarian carcinoma patients after second-look laparotomy. *Cancer*, **63**, 1509–13.
26. Howell, S.B., Kirmani, S., Lucas, W.E. *et al.* (1990) A phase II trial of intraperitoneal cisplatin and etoposide for primary treatment of ovarian epithelial cancer. *J. Clin. Oncol.*, **8**, 137–45.
27. Hakes, T.B., Markman, M., Reichman, B. *et al.* (1989) High intensity intravenous cyclophosphamide/cisplatin and intraperitoneal cisplatin for advanced ovarian cancer. *Proc. Am. Soc. Clin. Oncol.*, **8**, 152.
28. McGuire, W.P., Rowinsky, E.K., Rosenshein, N.B. *et al.* (1989) Taxol: a unique antineoplastic agent with significant activity in advanced ovarian epithelial neoplasms. *Ann. Intern. Med.*, **111**, 273–9.
29. Markman, M. (1987) Intracavitary administration of biological agents. *J. Biol. Response Mod.*, **6**, 404–11.

Chapter 30

Intraperitoneal cisplatin and interferon-α in persistent epithelial ovarian cancer: summary of phase I–II trials

J.S. BEREK and M. MARKMAN

30.1 INTRODUCTION

Epithelial ovarian carcinoma is responsive to a number of chemotherapeutic agents, but translating these rates of response into prolonged disease-free survival has proved difficult. Despite continuing improvements in therapy, the five-year survival remains only 10–20% for advanced disease [1–3]. Because ovarian cancer tends to remain confined to the peritoneal cavity, the intraperitoneal administration of a variety of cytotoxic agents and biological response modifiers has been used in an attempt to increase the total drug exposure to the tumour[4–12].

Over the past decade investigators at a number of institutions have explored a possible role for intraperitoneal (i.p.) therapy in the management of patients with persistent or recurrent ovarian cancer following intravenous chemotherapy[3–12]. Justification for these efforts includes mathematical modelling studies, preclinical experimental systems, and retrospective analysis of clinical data which suggest that the i.p. administration of certain antineoplastic agents may result in increased exposure of tumour within the peritoneal cavity to drugs whose cytotoxic activity appears to be concentration dependent[5,13–15].

Cisplatin has been the agent most extensively investigated for i.p. administration in patients with ovarian cancer. Cisplatin is one of the most active cytotoxic agents against ovarian carcinoma[16], and the i.p. administration of cisplatin has been shown to have a significant pharmacokinetic advantage, to be well tolerated, and produce a significant response rate as a salvage treatment[7–12]. Overall, surgically documented responses have been observed in 20–30% of patients with 'small-volume' persistent disease at treatment initiation[12]. Patients whose largest tumour mass measures <5 mm in diameter have been reported to experience response rates of between 40–50%, while responses occur in <10–15% of patients with larger tumour bulk.

The use of interferon-α has received considerable attention in the treatment of malignancies. Ovarian cancer cells have been shown to be sensitive *in vitro* to the antiproliferative effect of human leucocyte interferon[17]. In phase II trials, systemically administered interferon-α has produced responses in up to 18% of patients with advanced ovarian cancer[18,19]. Recombinant human (rh) interferon-α administered by the i.p. route has also demonstrated activity in patients with very small volume, i.e. minimal residual disease (MRD) (microscopic disease or tumour nodules <5 mm) that is persistent ovarian cancer following front-line chemotherapy[20,21].

There has been substantial evidence to suggest synergy between various interferons and standard cytotoxic agents[22–27]. Synergy has been noted *in vitro* with anthracyclines [22,23], actinomycin-D[24], vinca alkaloids [25], 5-fluorouracil (5-FU)[26], mitomycin-C[27], and cisplatin[24]. In several reports, the synergy is seen only when the cells are exposed to the interferon prior to the cytotoxic agent[22–24,26], presumably because the cytokine stimulates the proliferation of the cells making them potentially more susceptible to the cytotoxic effects of the drug. *In vivo* studies also have shown interferon to potentiate the cytotoxicity of cyclophosphamide and cisplatin in non-small cell lung cancer xenografts[28]. The preclinical data suggests that enhancement of the cytotoxic effect can be achieved when the cancer cells are exposed to the antiproliferative effects of interferon prior to the administration of the cytotoxic agent[24–26,28].

Recently, investigators at the UCLA and Bowman Gray Medical Centers have reported that cisplatin and rh interferon-α can be combined in an i.p. treatment programme with an acceptable toxicity profile and with efficacy being demonstrated[29]. Justification for this therapeutic strategy includes the individual activities of the two agents[12,20,21], as well as synergy between the two drugs demonstrated in experimental systems[24,28]. Another trial[30] used the same combination of agents in a different treatment schedule.

In an effort to expand upon this early experience with i.p. cisplatin plus rh interferon-α, the Gynecologic Oncology Group initiated a phase II trial of the two-drug i.p. combination. As will be discussed, while the therapeutic results were disappointing, the experience gained through the conduct of this trial is helpful in more critically defining the ovarian cancer patient population with 'small volume residual disease' who will not benefit from further treatment with a cisplatin-based i.p. programme and who should be considered for alternative treatment options.

30.2 INTRAPERITONEAL PHASE I–II TRIALS

30.2.1 INTRAPERITONEAL INTERFERON-α

A Gynecologic Oncology Group pilot study[20] administered intraperitoneal recombinant interferon-α (r-IFN 2b, Schering Corporation, Kenilworth, NJ). This was performed as a phase I trial in patients with persistent epithelial carcinoma after cisplatin-based combination chemotherapy (Table 30.1). Of 14 patients enrolled, 11 underwent

Table 30.1 Surgically documented responses to intraperitoneal interferon-α

Study	Patients	Complete	Partial	Total
Berek *et al.* [20]	11	4 (36%)	1 (9%)	5 (45%)
Willemse *et al.* [21]	17	5 (29%)	4 (24%)	9 (53%)
Total	28	9 (32%)	5 (18%)	14 (50%)*

* All responses in patients with MRD (microscopic or macroscopic <5 mm).

surgical re-evaluation and four complete responses (36%) were achieved. There was one partial response. Thus, the overall response rate in this trial was 45%. If one confined the analysis to patients with microscopic or <5 mm maximum residual disease, 4 of 7 patients (57%) had a complete response.

The toxicity of i.p. interferon-α as a single agent was also defined in this study[20]. Administration of the drug in the range of $25–50 \times 10^6$ iu three times a week was not tolerated because of persistent general malaise, fever and gastrointestinal toxicity. However, in those patients treated with the same dose once a week, the treatment was tolerated for 8–16 consecutive weeks. Of the 146 courses monitored throughout the study in all patients, fatigue was noted in 65%, chills in 53%, temporary anorexia in 39%, headaches in 38%, nausea and/or vomiting in 37%, diarrhoea in 27%, abdominal pain in 22% and total body pruritus in 12%. Other common complaints included 'abnormal' taste, 'dry' mouth, back pain and muscle aches, all of which were mild to moderate. Fever of 38°C or greater was noted in 58% of treatment cycles. Grade 2 and 3 anaemias were seen in 28% of the cycles by the seventh day, and leucopenias in 10% of cycles by one day after treatment. Notably, there was an absence of significant neurotoxicity, and renal toxicity. Although gastrointestinal toxicity was never severe, grade 1 and 2 toxicity was noted in 37% of patients. Therefore, while most of the side-effects of single agent interferon-α appear to be complementary with cisplatin, the general malaise and gastrointestinal toxicity produced by each one could potentially be additive when the agents are combined.

Willemse and colleagues[21] reported similar results in another trial of i.p. interferon-α in 20 patients with ovarian cancer, and of 17 who had a re-laparotomy, five (29%) had complete responses and four (24%) had partial responses. Responses in both studies were confined to patients whose disease was minimal residual (<5 mm). The toxicity encountered in this trial was similar.

Thus, overall, 28 surgically evaluated patients have been treated on these two trials, and 14 (50%) responded, with nine (32%) complete responses (Table 30.1). All of the responding patients had microscopic or small (<5 mm) residual disease, which we define as having minimal residual disease. These data suggest that the use of high-dose i.p. interferon given frequently can produce the regional control of very small volume disease confined to the peritoneal cavity. However, survival data are not available on these patients, and as such it is unclear if this approach can produce prolonged disease progression-free intervals.

30.2.2 COMBINED INTRAPERITONEAL INTERFERON-α AND CISPLATIN

Based on the laboratory data that demonstrate potential synergy between interferon and cisplatin against ovarian cancer and on the phase I–II data from the trials using each as a single i.p. agent, a phase I clinical study was conducted to evaluate the tolerance, safety and the maximum tolerable dose (MTD) of the combination of i.p. cisplatin and interferon-α in patients with residual carcinoma after conventional chemotherapy [29]. Response to therapy was assessed when feasible. Twenty-four patients were enrolled in this study, with 12 patients being enrolled at each of the two study centres, the UCLA Medical Center and the Bowman Gray School of Medicine. All patients had histologically documented epithelial carcinoma of the ovary confined to the peritoneal cavity and all had prior systemic chemotherapy. The status of disease must have been determined by a second-look laparotomy. Therapy started within 12 weeks after the second-look operation, and at least 3 or more weeks from the last treatment with chemotherapy.

Additional eligibility criteria included:

1. age of 18 years or older;
2. leucocyte count >3000/µl;
3. platelet count >100 000/µl;
4. bilirubin <1.5 mg%;
5. aspartate transaminase (SGOT) within normal limits;
6. absence of significant cardiac disease or arrhythmias;
7. creatinine ≤1.5 mg%, creatinine clearance >35 ml/min;
8. Gynecologic Oncology Group performance status of 0, 1 or 2.

Reassessment surgery was performed on patients within the month prior to being enrolled. Patients with disease outside of the peritoneal cavity were ineligible. If possible, cytoreductive surgery was performed to minimize the extent of tumour prior to the initiation of therapy. The extent of disease at study entry was classified as follows: (1) microscopic disease; (2) minimal residual disease, with (2a) largest residual mass <5 mm, or (2b) largest residual mass 5–20 mm; (3) bulky residual disease, with (3a) largest residual mass 20–60 mm or (3b) tumour >60 mm.

An i.p. treatment catheter was placed in each patient according to our previously reported technique[31]. The catheter was a Hickman or Tenckhoff type synthetic catheter, and in most patients the catheter was attached to a subcutaneous implantable reservoir (Port-a-Cath, Pharmacia, Piscataway, NJ) that allowed the administration of drug with an intravenous needle placed transdermally.

After admission to the hospital, routine intravenous fluids were administered at 200–250 ml/hour. The intraperitoneal infusion of 1.75 l of 1.5% dextrose dialysate was performed to adequately distend the peritoneal cavity to maximize the distribution of intraperitoneal drugs. Interferon-α was infused in the evening in an aliquot of 250 ml normal saline. The cisplatin was administered 12 hours later. Again, the peritoneal cavity was re-infused with dialysate (up to 1 more litre), and the cisplatin (45–90 mg/m^2) was then infused in an aliquot of 250 ml of normal saline. No attempt was made to drain the treatment volume from the peritoneal cavity, and thus an 'infinite dwell time' was used for both drugs in the study. Prior to the administration of the cisplatin, a routine hydration regimen was used[10]. Antiemetics were used as needed, and careful monitoring and replacement of serum electrolytes, including magnesium, were performed.

The starting dose (I) was cisplatin 45 mg/m^2 and an interferon-α dose of 10×10^6 iu. Three patients were to be entered at this dose level unless toxicity precluded administration. Dose modifications were allowed for patients experiencing toxicity as per Gynecologic Oncology Group criteria. The next dose level (II) would be an of the escalation of the interferon-α dose to 10×10^6 iu total dose combined with cisplatin 60 mg/m^2 for three patients, and then (III) 10×10^6 iu interferon and cisplatin 75 mg/m^2 for three patients. The dose would be increased to (IV) interferon 25×10^6 iu total dose and cisplatin 60 mg/m^2, (V) with cisplatin 75 mg/m^2, and (VI) with cisplatin 90 mg/m^2. As tolerated, the dose would be increased to (VII) interferon 50×10^6 iu with 75 mg/m^2 cisplatin, and (VIII) with cisplatin 90 mg/m^2. The planned treatment was for a maximum of 12 courses given as tolerated every 2 weeks. The actual dose level reached, the number of patients, and the number of courses per dose level are presented in Table 30.2.

In order to be evaluable for response, patients must have had an adequate trial of therapy, and they had to be alive at least 4 weeks after the last treatment. An adequate trial was defined as the patient having received at least two treatments given within 4 weeks. For patients with clinically measur-

Table 30.2 Planned dose schedule and dosages administered

Dose level	Interferon-α (× 10^6 iu)	Cisplatin (mg/m^2)	Courses	Patients
I	10	45	22	3
II	10	60	26	3
III	10	75	19	3
IV	25	45	20	4
V	25	60	26	4
VI	25	75	15	3
VII	50	75	13	4
VIII	50	90	0*	0
Total			141	24

* Not escalated to this dose level secondary to toxicity seen at prior level.
Reproduced, with permission, from reference 29.

Table 30.3 Residual tumour at entry on protocol

Residual tumour at entry	Patients
Microscopic	4 (16%)
MRD <5 mm	5 (21%)
MRD 5–20 mm	7 (30%)
Bulky 20–60 mm	5 (21%)
Bulky >60 mm	3 (12%)
Total	24 (100%)

MRD, minimal residual disease.
Reproduced, with permission, from reference 29.

able disease, a complete response was defined as disappearance of all measurable disease determined by two observations separated by at least 4 weeks. Partial response required a 50% decrease in all measurable lesions. Stable disease consisted of a decrease by <50% or an increase of >25% provided that there were no new lesions over 8 weeks. Progressive disease was defined as a 25% or more increase in the size of one or more measurable lesions or the appearance of any new lesions.

For patients with subclinical disease, responses could be documented only by a reassessment operation. A surgically documented complete response means the absence of all cytological and histological evidence of disease. A surgically documented partial response means at least a 50% regression of macroscopic disease, or the complete regression of all visible disease with the persistence of microscopic disease. No further therapy was given to patients who had a pathological complete response, as long as they remained disease free. Treatment would be discontinued if there was evidence of tumour progression or if unacceptable toxicity occurred. Toxicity was graded according to the criteria of the Gynecologic Oncology Group toxicity scale[32]. When grade 3 toxicity was encountered, the patient's treatment was held until recovery, and then restarted with one dose level reduction. Grade 3 renal toxicity would require a one dose level reduction of the cisplatin dose only.

The median age of the patients was 57 years (range 34–71). Of the 24 patients 23 (96%) had received prior systemic chemotherapy with cisplatin-containing regimens. The median prior total dose of cisplatin was 627 mg. The duration of their disease from diagnosis to the initiation of this protocol was 16 months range (7–21). Of the 18 evaluable patients, 15 had responded to cisplatin chemotherapy initially (cisplatin sensitive), and three had developed progressive disease while being treated with cisplatin (cisplatin resistant). Four patients had failed prior intraperitoneal therapy; two patients with cisplatin alone, one with cisplatin and 5-FU, and one with ^{32}P. Only nine patients (37%) began the study with MRD <5 mm, a status that has been identified with a favourable prognosis[33]. The maximum residual tumour at entry is presented in Table 30.3. Only one of these patients had their disease resected at their second-look to this level — she had a single pelvic mass of 5 cm that when removed left her with 1 mm

Table 30.4 Haematological toxicity versus dose level

Dose level*	Haematological toxicity by grade†							
	WBC		Granulocytes		Platelets		Anaemia	
	Grades							
	1–2	3	1–2	3	1–2	3	1–2	3
I (10/45)	28	0	22	0	0	0	42	0
II (10/60)	32	0	24	0	0	0	50	0
III (10/75)	40	0	28	0	0	0	61	0
IV (25/60)	41	0	31	0	3	0	59	0
V (25/75)	57	15	43	14	3	0	71	0
VI (25/90)	73	17	50	14	5	0	78	0
VII (50/75)	75	0	47	0	0	0	52	0
Total	49	8	39	6	2	0	58	<1

* The numbers in parentheses represent the doses of interferon-α and cisplatin (interferon dose × 10^6 iu/cisplatin dose mg/m²).
† Each number indicates the percentage of courses associated with the individual grade of toxicity.
Reproduced, with permission, from reference 29.

maximum residual tumour. Three patients had ascites that required frequent paracenteses prior to being enrolled in the study.

One hundred and forty-one courses of intraperitoneal therapy were administered (Table 30.2). The majority of the tolerated treatments used 25 × 10^6 iu of interferon-α and cisplatin doses of 45–90 mg/m². Cisplatin 60 mg/m² was the highest dose of the drug requiring the fewest schedule modifications. Reasons for discontinuation of therapy included mechanical (4) or infectious catheter complications (3); disease progression (3); toxicity (8); and patient preference (5). One patient completed 12 courses of therapy. The dose-limiting toxicity was either gastrointestinal, consisting primarily of nausea and vomiting in three patients; or general malaise in three patients; or neutropenia in two patients.

Twenty-four patients were evaluable for toxicity, and 18 of these were evaluable for response. The other six patients were not evaluable for response because the treatment was never initiated after enrollment in four patients (three had catheter failure or infection and declined placement of a new catheter, and one decided not to proceed and withdrew), and only one treatment was administered in two patients (one refused further treatment, and one had catheter failure).

Haemopoietic toxicity is summarized in Table 30.4. Life-threatening (grade 4) toxicity was not seen. Grade 3 toxicity was uncommon with 8% of treatment courses associated with decrements of the total white blood count (WBC), and 6% had grade 3 neutropenia. Grade 1–2 toxicity was common, however, with 39% of treatment courses associated with neutropenia of these grades. The majority of treatments (58%) were associated with grades 1–2 anaemia, while thrombocytopenia was rare.

Other toxicities encountered in this study are summarized in Table 30.5. No grade 4 toxicities were observed. Flu-like symptoms, including myalgias, chills and sweats, were universal; and at a severity of grade 3 in 11% of courses when the interferon dose was

Table 30.5 Non-haematological toxicity versus dose level

Dose level*		Toxicities and grades†								
		Renal		Gastrointestinal				General malaise		
				Nausea		Diarrhoea		Fever‡		Fatigue
		Grades								
		1–2	3	1–2	3	1–2	3	1–2	3	1–2	3
I	(10/45)	0	0	91	0	36	0	86	0	5	0
II	(10/60)	8	0	96	0	22	0	100	0	12	0
III	(10/75)	23	0	89	5	33	0	89	6	17	3
IV	(25/60)§	12	0	92	0	37	0	97	0	27	0¶
V	(25/75)	26	0	69	31¶	32	0	84	11¶	44	13¶
VI	(25/90)	31	0	56	44¶	37	0	85	15¶	36	11¶
VII	(50/75)	22	0	72	28¶	38	0	88	9¶	29	14¶
Total		27	0	82	17	35	0	89	5	27	7

* Numbers in parentheses represent the doses of interferon-α and cisplatin (interferon dose × 10^6 iu/cisplatin mg/m²).
† Each number indicates the percentage of courses associated with the individual grades of toxicity.
‡ Fever with 'flu-like' symptoms, including myalgias, chills, sweats.
§ MTD, maximum tolerated dose.
¶ Dose-limiting toxicity.
Nausea associated with vomiting in most courses.
Reproduced, with permission, from reference 29.

$25–50 \times 10^6$ iu. The most severe of the flu-like symptoms were seen in the initial courses. Fatigue and lethargy were common and seen at all dose levels, but were grade 3 (requiring bedrest for >72 hours) in 11–14% of courses at the highest dose levels of interferon ($25–50 \times 10^6$ iu). Gastrointestinal toxicity was common, with nausea and vomiting present in most patients. However, this toxicity was grade 3 in 28–44%, in treatment courses with 75–90 mg/m² cisplatin. The toxicity was most often seen in the initial course. Diarrhoea was grade 1 in most cases where this problem occurred, and there was no grade 3 toxicity. Renal toxicity was mild, and of grade 2 in only 18% of courses. Neuropathy and auditory toxicity were not clinically evident.

The dose level IV was tolerated well by the majority of the patients so treated (Tables 30.4 and 30.5). However, as the doses were escalated to the level V–VI, the probability of significant toxicity increased appreciably. Furthermore, at the dose levels IV or greater, almost half of the treatment cycles were delayed by 1–2 weeks to allow for resolution of the toxicity encountered in the previous cycle. Therefore, based on this experience and toxicity data, the MTD was 25×10^6 iu interferon and 60 mg/m² cisplatin.

Eighteen patients were evaluable for response. Eight patients underwent reassessment (third-look) laparotomy, while 10 additional patients were clinically evaluable. The results are summarized in Table 30.6. Of those 10 patients with measurable disease evaluable for a clinical response, one patient achieved a complete clinical response (CCR) and one had a partial clinical response (PCR).

Table 30.6 Responses in patients who were clinically or surgically evaluable

Residual	No.	Clinical				Surgical				Cycles
		CR	PR	SD	PD	CR	PR	SD	PD	
Microscopic	3	–	–	–	–	1	–	2	–	6,6,4
MRD <5 mm	3	–	–	–	–	1	1	–	1	5,4,4
MRD 5–20 mm	5	1	–	2	–	–	2	–	–	8,8,8,9,12
20–60 mm	5	–	1	1	3	–	–	–	–	7,8,8,9,9
>60 mm	2	–	–	–	2	–	–	–	–	4,5
Total	18	1	1	3	5	2	3	2	1	

CR, complete response; PR, partial response; SD, stable disease; PD, progressive disease; MRD, minimal residual disease.
Reproduced, with permission, from reference 29.

The patient who responded completely started her treatment with a 2-cm pelvic mass. The patient with a 5-cm mass who partially responded had already responded partially to intravenous cisplatin-combination chemotherapy. The patient with a CCR had a disease progression-free interval of 11 months, and died of disease at 19 months, while the one with a PCR was free of disease at 6 months and died of disease at 11 months. The mean survival of the non-responding patients was 8 months.

Eight patients who had no clinical evidence of disease at the completion of their chemotherapy were re-explored, and two of these patients were found to have had a complete pathological response (CPR). One of these patients had microscopic disease when her therapy was begun, and one had minimal residual macroscopic disease <5 mm. Neither patient required resection of tumour at the second-look in order to achieve this favourable tumour status. Three additional patients had a partial pathological response (PPR), one who started with MRD <5 mm, and one with residual disease 5–20 mm, neither one undergoing resection at second-look. The disease progression-free intervals of the two patients who had a CPR were 15 and 19 months, respectively. The first of these patients died of disease at 24 months, and the other is alive with persistent disease at 38 months. The mean progression-free interval for those patients who had a PPR was 9 months with mean survival of 18 months. Non-responding patients had a mean survival of 12 months.

The three patients with clinical ascites at entry had complete resolution of the ascites after the first or second treatment. This effect lasted 3–7 months prior to the clinical relapse of the patients' disease. One of these patients achieved a PCR. All three patients had been treated with intravenous cisplatin in their primary chemotherapy; two were in clinical remission for 8 and 11 months prior to relapse, and one had progressive disease during primary treatment.

30.2.3 ALTERNATING INTRAPERITONEAL CISPLATIN AND INTERFERON-α

Nardi and co-workers[30] reported a trial of 14 evaluable patients who were treated with weekly doses that alternated with 50×10^6 iu of i.p. interferon-α and 90 mg/m² of cisplatin. In this trial, the surgically documented complete response rate was 50% (7 of 14 patients), and all of these responses were confined to those patients who started their treatment with microscopic or <5-mm residual disease. The toxicity was similar to that seen in the phase I–II trial outlined above, although the authors reported somewhat fewer symptoms of general malaise and gastrointestinal toxicity.

Therefore, the combination of i.p. cisplatin and interferon appeared to be tolerated in these patients, and had an appreciable response rate. The survival of those patients who had a complete response was generally longer than those patients who were non-responsive. However, since response rates are similar to those reported in other series of single-agent i.p. cisplatin, it is unclear that the addition of interferon to cisplatin had any additive affect on the response rate.

30.2.4 CO-OPERATIVE GROUP PHASE II TRIAL OF INTRAPERITONEAL INTERFERON-α AND CISPLATIN

Based on the experience in the combined trials as outlined above, the Gynecologic Oncology Group conducted a phase II trial of i.p. cisplatin plus rh interferon-α in patients with small volume persistent/recurrent ovarian cancer, based on the known single agent activity of the drugs administered by the i.p. route, and experimental evidence of cytotoxic synergy between the agents[34]. However, in 18 evaluable patients, only one partial response was observed (5.5% response rate).

Patients were scheduled to receive i.p. cisplatin (60 mg/m^2) and rh interferon-α (25 × 10^6 iu) every 3 weeks for a total of eight treatments unless there was evidence of disease progression or unacceptable toxicity developed. It was planned that all patients without evidence of progressive disease at the completion of therapy would undergo an exploratory laparotomy to assess response to the treatment programme. The treatment protocol was as follows: Patients were initially given 2 l of Dianeal (I.SS dextrose) intraperitoneally, which was followed by the rh interferon-α delivered separately over 1 hour in 100 ml of normal saline. Just prior to the rh interferon-α administration patients received oral acetaminophen (650 mg) which was continued every 4 hours for the first 24 hours following therapy. Patients were also started at this time on intravenous hydration with 1 litre of 5% dextrose half-normal saline (plus 20 mEq/l of potassium chloride) at 150 ml/hour.

Approximately 12–18 hours after the rh interferon-α administration, the cisplatin was instilled intraperitoneally in 500 ml of Dianeal. Immediately prior to the cisplatin delivery, patients received mannitol (12.5 g in 50 ml 5% dextrose) administered intravenously. Hydration continued for a minimum of 6 hours following cisplatin instillation. The choice of antiemetics was at the discretion of the treating physician, except that steroids were not permitted. All i.p. solutions were warmed to 37 °C prior to instillation. No effort was made to remove fluid from the peritoneal cavity following i.p. administration, unless essential for patient comfort.

Patients were considered evaluable for efficacy if they received two cycles of therapy, while patients administered even a single course of treatment were evaluable for toxicity. The side-effects of therapy were graded according to the Gynecologic Oncology Group toxicity scale[32]. To avoid haematological toxicity, treatments were not delivered until the WBC count was ≥3000/mm^3, granulocytes ≥1500/mm^3, and platelets ≥100 000/mm^3. If the blood counts did not meet these criteria, treatment was delayed 1 week. If treatment had to be delayed for >2 weeks due to inadequate blood counts, subsequent treatments were delivered at a reduced dose level (Table 30.7). Therapy was withheld if serum creatinine was >2.0 mg% or creatinine clearance <50 ml/min on the day of treatment. If the serum creatinine remained >2.0 mg% or creatinine clearance

Table 30.7 Dosage modifications for toxicity experienced during phase II trial of i.p. cisplatin plus rh interferon-α

	Interferon	Cisplatin
Starting dose level	25×10^6 U	60 mg/m^2
1st level reduction	10×10^6 U	45 mg/m^2
2nd level reduction	5×10^6 U	30 mg/m^2
3rd level reduction	1×10^6 U	15 mg/m^2

Reproduced, with permission, from reference 34.

<50 ml/min for longer than 4 weeks, the patient was removed from the study. If patients experienced grade 3–4 gastrointestinal toxicity with a previous course, subsequent treatments were delivered at the next lower dose level (Table 30.7). Therapy was reduced one dose level if any of the following blood tests increased more than 2.5 times normal: serum bilirubin, SGOT, alkaline phosphatase. Patients experiencing grade 2 neurotoxicity had therapy reduced by one dose level, and with grade 3–4 toxicity treatment was discontinued.

All responses were defined surgically. A partial response required at least a 50% regression of macroscopic disease, or the complete disappearance of visible lesions with the persistence of microscopic cancer.

A total of 48 patients were entered onto this phase II trial. Five patients were subsequently found to be ineligible for the protocol (largest tumour mass >1 cm, three patients; inadequate surgery prior to i.p. therapy, one patient; incorrect primary tumour, one patient). Table 30.8 summarizes the characteristics of the 43 eligible patients.

The toxicities observed with this i.p. treatment programme are presented in Table 30.9. The most commonly noted serious side-effects (grade 3–4) were neutropenia, anaemia, gastrointestinal toxicity (principally emesis), and neurotoxicity. For the 26 patients who experienced grade 1–4 leucopenia, the median WBC nadir was 2600 mm^3

Table 30.8 Characteristics of 43 patients treated on phase II trial of i.p., cisplatin plus rh interferon-α in persistent/recurrent ovarian cancer

Age Median, 56; range 30–78	
Cell type	
Serous	18 (42%)
Endometrioid	4 (9%)
Mixed epithelial	10 (23%)
Clear cell	6 (14%)
Grade	
1	6 (14%)
2	19 (44%)
3	18 (42%)
Performance status	
0	17 (40%)
1	18 (42%)
2	8 (19%)

Reproduced, with permission, from reference 34.

Table 30.9 Toxicity of i.p. cisplatin plus rh interferon-α in 43 evaluable patients

	Toxicity grade (% of patients affected)		
	2	3	4
Leukopenia	30	14	
Neutropenia	16	7	5
Thrombocytopenia	7	2	
Anaemia	26	9	
Gastrointestinal	63	14	2
Renal	16		5
Neurological	7	14	
Fever	30	5	
Fatigue	5		
Bacterial peritonitis	2		2
Pain	2		

Reproduced, with permission, from reference 34.

(range 1100–3700/mm^3). For the 15 patients developing grades 1–3 thrombocytopenia, the median platelet nadir was 116 000/mm^3 (range 38 000–146 000/mm^3). The neurotoxicity observed in the trial (grade 3 in 14% of patients) was predominantly a worsening of a pre-existing cisplatin-induced peripheral neuropathy. As this was not a randomized controlled trial, it is not possible to know for certain if the incidence of grade 3 neuropathy observed with the i.p. combination of cisplatin and rh interferon-α was greater than would have been anticipated with i.p. cisplatin administered alone or with the agent in combination with other cytotoxic drugs.

Of the original 43 eligible patients 25 (58%) were not evaluable for response. Seven patients (16%) refused third-look surgery. In an additional 4 patients (9%) the attending physician elected not to proceed with the assessment surgery at the completion of the treatment. Fourteen patients (33%) were removed for serious (grade 3–4) toxicity (Table 30.9). In these 25 patients there was no clinical evidence of disease (physical examination, symptoms, etc.) at either the completion of therapy or when the patient was removed from the treatment programme for unacceptable toxicity.

Of the 18 evaluable patients, 13 (72%) demonstrated persistent or progressive disease prior to the performance of the scheduled third-look laparotomy. One of the remaining five patients undergoing surgery demonstrated a partial response when the findings were compared to the bulk of disease remaining at the completion of the second-look laparotomy. Thus, the overall objective response rate in the 18 evaluable patients was 5.5%.

30.3 DISCUSSION

Patients have limited treatment options after they fail to achieve a surgically confirmed complete response after multi-agent chemotherapy that includes cisplatin. Intraperitoneal therapies, employed as a 'salvage' treatment in patients with persistent disease, offer an attractive opportunity to maximize the exposure of the agents to the residual carcinoma cells [4–11,20,21,30,33–35]. Cisplatin is one of the most well-tolerated agents that has been administered intraperitoneally, and it has a significant response rate in patients treated after primary systemic cisplatin chemotherapy [7–10,33]. Interferon-α likewise has been given intraperitoneally as a single agent in two separate phase I–II trials [20,21]. In these two studies, a significant rate of surgically documented responses was reported, although responses were confined to patients whose disease was minimal residual (<5 mm)[11,20,21]. The rationale for i.p. immunotherapy with a cytokine includes the stimulation of regional effector cells that can augment tumour rejection, as well as direct cytostatic and cytotoxic effects[11].

Because the majority of patients treated with second-line i.p. therapies ultimately relapse[1–3], and there appears to be significant *in vitro* synergy between cisplatin and other agents[22–27], a search for clinically tolerable and effective combinations is being undertaken. The demonstration that there is synergy *in vitro* between cisplatin and interferon-α provided the impetus for this phase I–II trial[24]. The clinical phase I study[29] demonstrated that combined i.p. therapy with cisplatin and interferon-α given within several hours of one another every cycle can be administered safely to patients with residual ovarian carcinoma after systemic chemotherapy. The MTD of the combination of interferon-α and cisplatin as determined by the phase I trial are 25×10^6 iu interferon-α and 60 mg/m^2 cisplatin. The therapy was best tolerated given as one cycle every 3 weeks, and of the 8 patients who were treated with this precise dose, the median number of treatment cycles was six courses. The complete responses were noted in two patients after five and six treatment cycles, respectively; and partial responses seen in patients treated with four and eight cycles, respectively.

Therefore, for phase II trials, we recommended that the MTD be given every 3 weeks for six cycles. Doses given in excess of the MTD doses, or those more frequently administered, were not well tolerated, principally because of persistent general malaise such that patients elected to receive no further therapy. Also, there were several catheter-related complications that may have been potentiated by either the particular agents used or the number of treatments administered, as most problems occurred in patients after seven or eight cycles of therapy. Presumably, progressive fibrosis around the catheters was induced as successive treatments were given.

It was encouraging that objective antitumour responses were seen in three of six evaluable patients with pretherapy residual disease measuring <5 mm, two of which were complete pathological responses. A phase II trial performed in patients with MRD <5 mm is warranted to further evaluate the efficacy of this combination. It was also noteworthy that this combination was capable of controlling refractory malignant ascites in all three of the studies within weeks of initiating therapy. The precise mechanism for these responses is unknown, although interferon induces some inflammatory responses that might produce a regional sclerosing effect similar to that produced by non-specific immunomodulators[6,11,20,34].

The experience of Nardi and colleagues[30] corroborates these data, as a 50% CPR rate was observed in their trial of alternating interferon-α and cisplatin. The toxicity in this series appeared to be somewhat less than that observed in the former trial, and that schedule may be more appropriate for future phase II trials.

However, when the two-institution trial combining i.p. interferon-α and cisplatin was applied to the co-operative group setting, a very low response rate was seen[34]. In view of the extremely low response rate observed in the Gynecologic Oncology Group trial, it is important to attempt to determine why this group of individuals did so poorly compared to previously reported phase I–II trials of cisplatin-based or interferon-α i.p. therapy in patients with persistent small-volume residual ovarian cancer where response rates of 20–40% have been noted[12,20,21].

In an effort to explain the disappointing findings, we retrospectively divided the patients into two prognostic categories, called **favourable** and **unfavourable**. Favourable patients were those who not only had small volume disease (no tumour mass >1 cm in diameter), but who also had previously demonstrated at least a partial response to intravenous cisplatin. In addition, to be considered in the favourable disease category, patients could not be described at the time of second-look laparotomy as having findings suggestive of diffuse carcinomatosis, even if each individual tumour lesion was stated to be ≤ 1cm in diameter. Thus, a patient was stated to fall into the unfavourable category if her best response to intravenous cisplatin was less than a partial response (minimal or no response, progressive or stable disease) or if the surgical findings suggested diffuse carcinomatosis. The favourable patient population would be predicted to have a far greater chance of responding to local therapy, but only three of the evaluable patients fell into this category. In the 15 unfavourable patients, only one partial response was observed (7% response rate).

Using these criteria, only 15 (35%) of the 43 evaluable patients entered into in this trial fell into the favourable category, while 28 (65%) patients were considered to be unfavourable. There was no significant difference between the two groups in terms of age, histological cell type, tumour grade or performance status. Unfortunately, only three (20%) of the 15 favourable patients were evaluable for response (all failed prior to third-look surgery). Of the remaining 12 patients in the favourable category, five did not undergo surgical re-evaluation and seven were removed for toxicity at which time they had no evidence of disease.

Of the 28 patients in the unfavourable category 15 (54%) were evaluable for response, with only one of the five patients undergoing surgical re-assessment demonstrating a surgically defined response (10 patients failed prior to the performance of third-look surgery). Thus, the overall response rate in this group of patients with unfavourable disease was 7%. Of the 13 non-evaluable patients in this category, six did not undergo surgical assessment despite completing the treatment regimen without evidence of disease, and seven were removed for excessive toxicity.

Intraperitoneal antineoplastic therapy has been demonstrated to result in surgically defined responses, including complete remissions, in patients with small-volume persistent or recurrent ovarian cancer following platinum-based intravenous chemotherapy [12,20,21]. Investigators have appropriately focused attention on the size of the largest tumour nodule(s) at the initiation of therapy as being a critical factor in determining the chances that an individual patient will experience a response to this treatment strategy, with several series reporting response rates of 40–50% when the largest tumour nodule measures <5 mm in diameter, i.e. MRD[12,20,21]. Response rates with larger tumour masses have generally been reported to be <10–15%.

While the criteria for entry onto this phase II trial was somewhat more liberal than that noted above (up to 1 cm tumour masses acceptable for protocol entry), they were otherwise quite reasonable based on the published data suggesting patient characteristics would predict a 20–30% response rate to a cisplatin-based i.p. chemotherapy programme[12]. Yet, only one of 18 evaluable patients responded. What is the explanation for these disappointing findings?

Two points should be made in answering this important question. First, as previously noted, the emphasis of much of the published literature on i.p. therapy has been on the size of the largest residual tumour nodule at i.p. treatment initiation. This is quite appropriate considering both the substantial preclinical[36,37], and recent clinical data[12], demonstrating the limited direct penetration of antineoplastic agents, including cisplatin[36], directly into tumour tissue by free surface diffusion.

However, it is clearly not reasonable to consider a patient with a relatively limited number of very small tumour nodules (plus documented or suspected diffuse microscopic disease) to have the same chances for therapeutic success following i.p. therapy as a patient with diffuse carcinomatosis. This would be the case even if all 'individual tumour nodules' measured ≤0.5 to 1 cm in diameter. The diffuse nodules will essentially coalesce into a large tumour mass making it difficult or impossible for the locally administered cytotoxic agent(s) to penetrate.

Second, a major justification for the use of i.p. therapy in the management of ovarian cancer is preclinical data demonstrating both concentration-dependent cytotoxicity for a number of antineoplastic agents against the tumour, and the ability of higher concentrations of certain drugs to overcome resistance which develops to the lower plasma levels following systemic drug delivery [14,37,38]. However, an important extension of this interpretation of the same experimental data is that the potential advantage for local drug delivery over systemic administration is relative rather than absolute.

Thus, if a patient experiences minimal or no response to a cisplatin-based programme, it can seriously be questioned whether that individual would be anticipated to experience any benefit from the modestly increased concentrations (10-fold) of the drug achievable following i.p. delivery. In contrast, the patient who has achieved an excellent partial response to treatment, or who has recurred following a surgically or clinically defined complete response, is probably more likely to have tumour which can respond to the concentrations of cisplatin achievable within the peritoneal cavity following local administration.

In our retrospective analysis, we have separated patients with small-volume residual ovarian cancer into two groups, based on these two factors; prior response to cisplatin-based therapy and the presence or absence of diffuse carcinomatosis. Unfortunately, only three patients fell into the favourable category (no responses observed), and it is not possible to make any statement about the response rate of this group of individuals to the i.p. combination regimen of cisplatin and rh interferon-α.

However, it is likely that investigators from single institutions experienced with i.p. therapy and reporting results of phase 2 trials of cisplatin-based i.p. programmes would appropriately select this category of patients for entry into their trials, based on the logical arguments presented above. Unfortunately, this reasonable 'selection bias' for patient inclusion into clinical trials of cisplatin-based i.p. has not been thoroughly discussed in the published literature on i.p. chemotherapy of ovarian cancer.

In contrast, in this Gynecologic Oncology Group trial, as in other co-operative group studies, selection bias to include the more favourable group of patients was less likely to be operative, as the participating institutions in the group were given no specific guidance in the eligibility criteria to focus their efforts at patient selection on the more favourable patient population. Thus, 15 (83%) of the 18 evaluable patients had unfavourable characteristics and only one partial response was observed in this group of individuals.

The 14% incidence of grade 3 neurotoxicity was unexpected. There are several possible explanations for this high incidence of serious treatment-related neurological dysfunction. First, the patient population selected for entry into this study may had been more susceptible to the neurotoxic effects of therapy, secondary to pre-existing mild cisplatin-induced neuropathies. Second, it is possible that the incidence of severe neurotoxicity may have been inflated due to the peculiar circumstances of the timing of the study and the particular trial design. At approximately the same time this study was initiated, carboplatin was approved by the FDA for standard clinical use in recurrent ovarian cancer. Carboplatin, which has an activity profile similar to cisplatin in ovarian cancer, is associated with a far lower incidence of neurotoxic side-effects[38]. However, in this phase II i.p. study design, even patients experiencing grade 2 cisplatin-induced neurotoxicity would be expected to continue this protocol, rather than switching to the alternative drug regimen. Thus, it is possible that individual physicians concerned about the potential morbidity associated with continuing cisplatin might have increased the grading of the severity of this subjective toxicity for the purpose of discontinuing cisplatin and initiating carboplatin. Future trials employing second-line cisplatin must seriously consider this issue in the study design.

Finally, it is possible that the high incidence of neurotoxicity was a direct effect of the combination or scheduling of cisplatin and rh interferon-α. A recent trial of i.p. cisplatin and rh interferon-α, employing higher single doses of each drug than used in this trial but with the agents separated by several weeks, reported no serious neurotoxicity[30].

This regimen was also associated with a high surgically documented complete response rate (7/14 evaluable patients) in individuals with very small volume persistent ovarian cancer (microscopic, tumour nodules <5 mm)[30]. Unfortunately, this study design, which separates cisplatin and rh interferon-α delivery to tumour and appears to result in a lower incidence of neurotoxicity than that reported in the current trial, fails to take advantage of the potential for cytotoxic synergy between the two antineoplastic agents[24,28]. Future studies will be required to determine if the synergy demonstrated in experimental systems can be translated into improved therapeutic effectiveness or only increased toxicity.

In conclusion, while this study has failed to confirm the clinical activity of either single-agent i.p. cisplatin and rh interferon-α[12,20,21], or the combination of the two agents[29,30], our explanation for this finding is not that the drugs are inactive when administered by the i.p. route, but rather that the patient population included in this trial was inappropriate to test the potential utility of a cisplatin-based i.p. regimen in ovarian cancer. The patient characteristics identified in this study which predict for the failure of cisplatin-based i.p. therapy in small-volume persistent ovarian cancer should be considered in future clinical trials of i.p. therapy and in the selection of second-line therapeutic programmes for patients being treated outside an investigative study setting.

ACKNOWLEDGEMENTS

Portions of this manuscript are reproduced and modified, with permission, from references 29 and 34.

REFERENCES

1. Berek, J.S. and Hacker N.F. (1990) Ovarian cancer, in *Cancer Therapy*, 3rd edn (ed. C.M. Haskell), W.B. Saunders, Philadelphia.
2. Berek, J.S. (1989) Epithelial ovarian cancer, in *Practical Gynecologic Oncology* (eds J.S. Berek and N.F. Hacker), Williams and Wilkins, Baltimore, pp.327–64.
3. Ozols, R.F. and Young, R.C. (1987) Ovarian cancer, in *Current Problems in Cancer*, Vol. 11 (ed. C.M. Haskell), Year Book Medical, Chicago, pp.74–102.
4. Myers, C. (1984) The use of intraperitoneal chemotherapy in the treatment of ovarian cancer. *Semin. Oncol.*, **11**, 275–84.
5. Dedrick, R., Myers, C., Bungay, P. and DeVita, V.T. (1978) Pharmacokinetic rationale for peritoneal drug administration in the treatment of ovarian cancer. *Cancer Treat. Rep.*, **62**, 1–11.
6. Berek, J.S., Martínez-Maza, O. and Montz, F. (1991) Immunology, immunotherapy and monoclonal antibodies, in *Gynecologic Oncology*, (eds M. Coppleson, M. Tattersall and C.P. Morrow), Churchill Livingstone, London.
7. Howell, S., Pfeifle, C., Wung, W. *et al.* (1982) Intraperitoneal cisplatin with systemic thiosulfate protection. *Ann. Intern. Med.*, **97**, 845–51.
8. Pretorius, G., Hacker, N.F., Berek, J.S. *et al.* (1983) The pharmacokinetics of IP cisplatin in refractory ovarian carcinoma. *Cancer Treat. Rep.*, **67**, 1085–92.
9. ten Bokkel Huinink, W.W., Dubbelman, R., Aarsten, E. *et al.* (1990) Experimental and clinical results with intraperitoneal cisplatin. *Semin. Oncol.*, **12**, (Suppl. 4), 43–6.
10. Hacker, N.F., Berek, J.S., Pretorius, G. *et al.* (1987) Intraperitoneal cisplatinum for salvage therapy in advanced ovarian carcinoma. *Obstet. Gynecol.*, **70**, 759–64.
11. Berek, J.S. (1990) Intraperitoneal adoptive immunotherapy for peritoneal cancer. *J. Clin. Oncol.*, **10**, 1610–12.
12. Markman, H., Hakes, T., Reichman, M. *et al.* (1989) Intraperitoneal therapy in the management of ovarian carcinoma. *Yale J. Biol. Med.*, **62**, 393–403.
13. Ozols, R.F., Corden, B.J., Jacob, J. *et al.* (1984) High-dose cisplatin in hypertonic saline. *Ann. Intern. Med.*, **100**, 19–24.
14. Alberts, D.S., Young, D.S., Mason, N. *et al.* (1985) *In vitro* evaluation of drugs against ovarian cancer at concentrations achievable by intraperitoneal administration. *Semin. Oncol.* **12**, (Suppl. 4), 38–42.
15. Levin, L. and Hryniuk, W.M. (1987) Dose

15. intensity analysis of chemotherapy regimens in ovarian carcinoma. *J. Clin. Oncol.*, **5**, 756–67.
16. Young, R.C., von Hoff, D.D., Gormley, P. *et al.* (1979) Cis-dichlorodiammineplatinum (II) for the treatment of advanced ovarian cancer. *Cancer Treat. Rep.*, **63**, 1539–44.
17. Epstein, L.B., Shen, J.T., Abele, J.S. and Reese, C.C. (1980) Sensitivity of human ovarian carcinoma cells to IFN and other antitumor agents as assessed by an *in vitro* semi-solid agar technique. *Ann. N.Y. Acad. Sci.*, **350**, 228–35.
18. Einhorn, N., Cantell, K., Einhorn, S. and Strander, H. (1982) Human leukocyte interferon therapy for advanced ovarian carcinoma. *Am. J. Clin. Oncol.*, **5**, 167–72.
19. Freedman, R.S., Gutterman, J.U., Wharton, J.T. and Rutledge, F.N. (1983) Leukocyte interferon (IFN alpha) in patients with epithelial ovarian carcinoma. *J. Biol. Response Mod.*, **2**, 133–8.
20. Berek, J.S., Hacker, N.F., Lichtenstein, A. *et al.* (1985) Intraperitoneal recombinant alpha-interferon for 'salvage' immunotherapy in stage III epithelial ovarian cancer: A Gynecologic Oncology Group study. *Cancer Res.*, **45**, 4447–53.
21. Willemse, P.H.B., de Vries, E.G.E., Mulder, N.H. *et al.* (1990) Intraperitoneal human recombinant interferon alpha-2b in minimal residual ovarian cancer. *Eur. J. Clin. Oncol.*, **26**, 353–8.
22. Balkwill, F.R. and Moodie, E.M. (1984) Positive interactions between human interferon and cyclophosphamide or Adriamycin in a human tumor system. *Cancer Res.*, **44**, 904–8.
23. Welander, C., Gaines, J., Homesley, H. and Rudnick, S. (1983) *In vitro* synergistic effects of recombinant human interferon alpha 2 (rIFN-a2) and doxorubicin on human tumor cell lines. *Proc. Am. Soc. Clin. Oncol.*, **2**, 42.
24. Inoue, M. and Tan, Y.H. (1983) Enhancement of actinomycin-D and cis-diamminedichloroplatinum (II) induced killing of human fibroblasts by human beta interferon. *Cancer Res.*, **43**, 5484–8.
25. Aapro, M.S., Salmon, S.E. and Alberts, D.C. (1981) Schedule dependent synergism of vinblastine and cloned leukocyte interferon A. *Stem Cells*, **1**, 303–4.
26. Le, J., Yip, Y.K. and Vilcek, J. (1984) Cytologic activity of interferon gamma and its synergism with 5-FU. *Int. J. Cancer*, **34**, 495–500.
27. Harabayshi, N., Nishiyama, M. and Yamaguchi, M. (1982) Assessment of the combined effects of mitomycin C with alpha interferon by the clonogenic assay technique. *Gan To Kagaku Ryoho*, **12**, 1056–62.
28. Carmichael, J., Fergusson, R.J., Wolf, C.R. *et al.* (1986) Augmentation of cytotoxicity of chemotherapy by human alpha interferons in human non-small-cell lung cancer xenografts. *Cancer Res.*, **46**, 4916–20.
29. Berek, J.S., Welander, C., Schink, J.C. *et al.* (1991) A phase I–II trial of intraperitoneal cisplatin and alpha-interferon in patients with residual epithelial ovarian cancer. *Gynecol. Oncol.*, **40**, 237–43.
30. Nardi, M., Cognetti, F., Pollera, C.F. *et al.* (1990) Intraperitoneal recombinant alpha 2-interferon alternating with cisplatin as salvage therapy for minimal residual disease ovarian cancer: a phase II study. *J. Clin. Oncol.*, **8**, 1036–41.
31. Lucas, W.E., Markman, M. and Howell, S.B. (1985) Intraperitoneal chemotherapy for advanced ovarian cancer. *Am. J. Obstet. Gynecol.*, **152**, 474–81.
32. Blessing, J.A. (1984) Design, analysis, and interpretation of chemotherapy trials in gynecologic cancer, in *Chemotherapy of Gynecologic Cancer* (ed. G. Deppe), Alan R. Liss, New York, pp.49–83.
33. Howell, S., Zimm, S., Markman, M. *et al.* (1987) Long-term survival of advanced refractory ovarian carcinoma patients with small-volume disease treated with intraperitoneal chemotherapy. *J. Clin. Oncol.*, **10**, 1607–12.
34. Markman, M., Berek, J.S., Blessing, J.A. *et al.* (1991) Characteristics of patients with small-volume residual ovarian cancer resistant to platinum-based intraperitoneal chemotherapy: lessons learned from a phase II Gynecology Oncology Group trial of intraperitoneal cisplatin and alpha-interferon. *Gynecol. Oncol.*
35. Berek, J.S., Knapp, R.C., Hacker, N.F. *et al.* (1985) Intraperitoneal immunotherapy of epithelial ovarian carcinoma with *Corynebacter-*

ium parvum. Am. J. Obstet. Gynecol., **152**, 1003–10.
36. Los, G., Mutsaers, P.H.A., van der Vijgh, W.J.F. *et al.* (1989) Direct diffusion of cis-diamminedichloroplatinum(II) in intraperitoneal rat tumors after intraperitoneal chemotherapy: a comparison with systemic chemotherapy. *Cancer Res.*, **49**, 3380–4.
37. Ozols, R.F., Locker, G.Y., Doroshow, J.H. *et al.* (1979) Pharmacokinetics of adriamycin and tissue penetration in murine ovarian cancer. *Cancer Res.*, **39**, 3209–14.
38. Andrews, P.A., Velury, S., Mann, S.C. *et al.* (1988) Cis-diamminedichloroplatinum(II) accumulation in sensitive and resistant human ovarian carcinoma cells. *Cancer Res.*, **48**, 68–73.
39. Muggia, F.M. (1989) Overview of carboplatin: replacing, complementing, and extending the therapeutic horizons of cisplatin. *Semin. Oncol.*, **16**, (Suppl. 5), 7–13.

Chapter 31

Overview of randomized chemotherapy trials for advanced ovarian cancer

ADVANCED OVARIAN CANCER TRIALISTS GROUP (AOCTG)

Members: K. Aabo, M. Adams, P. Adnitt, D.S. Alberts, V. Barley, D.R. Bell, M. Buyse, G. Bolis, H.S. Brodovsky, H. Bruckner, V. Chylak, C.J. Cohen, N. Colombo, P.F. Conte, J. Crowley, D. Crowther, J.H. Edmonson, E. Gilbey, D. Guthrie, T. Kato, S.B. Kaye, A.H. Laing, R.C. Leonard, C. Lewis, A. Liberatti, C. Mangioni, S. Marsoni, P. Mauro, H. Meerpohl, U. Muller, G.A. Omura, M.K.B. Parmar, J. Pater, S. Pecorelli, E. Reed, W. Sauerbrei, R.V. Smalley, H.J. Solomon, L.A. Stewart, J.F.G. Sturgeon, M.H.N. Tattersall, J. Taylor-Wharton, W.W. ten Bokkel Huinink, M. Tomirotti, C. Trope, M.M. Turbow, J.B. Vermorken, M.J. Webb, D.W. Wilbur, C.J. Williams, E. Wiltshaw, R. Wood, L. Yeap

31.1 INTRODUCTION

Ovarian carcinoma is the seventh commonest cancer in women and it is responsible for the greatest number of deaths from gynaecological malignancy in Europe and North America[1]. Despite more than 50 randomized clinical trials, testing the relative efficacy of different chemotherapy regimens in advanced disease (FIGO stages III and IV), individually none of these has had sufficient patient numbers to produce reliable results. However, many of these trials have had a major influence on clinical practice.

Ovarian cancer was one of the first solid malignant tumours to be regularly treated with chemotherapy and single alkylating agents were first used over 30 years ago and were considered optimal therapy. This situation changed following the publication in 1978 of a small randomized trial[2] that reported that Hexa-CAF (hexamethylmelamine, cyclophosphamide, methotrexate and 5-fluorouracil) had achieved higher response rates than the single alkylating agent melphalan and which showed an apparent improvement in survival. At this time phase II studies showing that cisplatin was a highly active agent were published[3]. These publications rapidly led to the standard use of cisplatin in combination with other cytotoxic drugs, often doxorubicin and cyclophosphamide with or without hexamethylmelamine[4,5]. The effectiveness of doxorubicin was, however, questioned by results both from randomized phase III trials[6,7] and from phase II studies in patients failing cisplatin[8], and many clinicians adopted cisplatin plus cyclophosphamide as standard therapy. The failure of other trials to demonstrate significant survival differences between single-agent cisplatin and cisplatin in combination with other drugs[9,10], led other centres to use single-agent platinum as routine first-line therapy. Despite this, none of these trials had sufficient patient numbers to answer the question asked with any sense of security.

If optimal therapy for advanced disease is unclear and treatment varies nationally and internationally, what is clear is that no clinical trial has been large enough to demonstrate survival differences of the magnitude that could be reasonably expected with available therapy. Because of this, the diversity of results reported in the literature could be consistent with moderate treatment benefits[11]. The UK Medical Research Council (MRC) Gynaecological Cancer Working Party decided to undertake a systematic overview of the data from these 50 or more trials in order both to consolidate existing information and to act as a precursor to designing a new generation of clinical trials in ovarian carcinoma. Because of the potential bias of a review of published results[11], they initiated a systematic overview which utilized formal quantitative methods of combining the results from all randomized trials addressing the role of platinum and of combination chemotherapy in the treatment of advanced ovarian cancer. The MRC overview secretariat contacted the investigators responsible for each trial, inviting their collaboration in this project and in so doing established the Advanced Ovarian Cancer Trialists Group (AOCTG), under whose name the overview was conducted.

31.2 METHODS AND DATA

The relative merits of single-agent and combination chemotherapy and the role of platinum in disease management were addressed so that a number of comparisons between different forms of chemotherapy could be examined.

I. Single non-platinum agent vs. non-platinum combination.
II. Single non-platinum agent vs. platinum combination.
III. Addition of platinum to a regimen.
IV. Single-agent platinum vs. platinum combination.
V. Carboplatin vs. cisplatin.

Trials were eligible for inclusion in the overview if they were unconfounded, believed to have been randomized in a manner that precluded prior knowledge of the next treatment assignment, examined first line therapy for advanced disease and made one or more of the comparisons listed above. In order to avoid publication bias[12] both published and unpublished studies were included and a number of methods were employed to identify as many relevant trials as possible. A bibliographic review, using MEDLINE, CancerLit and published texts, was carried out. This was supplemented by examining the trial registers produced by the NCI[13] and UKCCCR[14] as well as the proceedings of relevant clinical meetings. In addition, questionnaires were sent to the principal authors of published trials and members of international societies and colleges concerned with gynaecological cancer asking them to supplement a provisional list of trials. Pharmaceutical companies involved in ovarian cancer therapy were also approached in this way.

Methods and data

So far 53 eligible randomized trials (listed in Appendix A) have been identified. Three potentially eligible studies had to be excluded on the grounds that they did not appear to be appropriately blinded. Information was available from 45 studies accounting for 95% of all known randomized patients. At the time of data collection approximately 30% of these had not been published fully. Information was not available for the following reasons: destruction of data (one study), loss of data (one study), current unavailable data which is promised later (two studies) and the inability or unwillingness of authors to collaborate in the overview (four studies). Trials not included in the overview are listed in Appendix B. Information was sought for each individual patient randomized in these studies and the 45 available trials comprise data on 8139 patients with 6408 deaths originating from 11 different countries. Appendices A and B include all trials from all five comparisons though this report concentrates on comparisons though this report concentrates on comparisons IV and V (p. 316).

Where trials had multiple treatment arms and made more than one of the comparisons defined above, the patients from the relevant arms were included in each appropriate comparison, provided that there had been a direct randomization between the arms used. Thus a number of trials are included in more than one comparison. In trials where more than one arm was of the same type, the patients in these were grouped together for the purposes of analysis. This has led to an apparent imbalance in numbers of randomized patients.

Individual patient information for all studies, apart from two, was updated at the time of data collection. Thus the overview had available information from an extended period with a median follow-up varying from 3 to 10 years depending on the comparison.

Data that had been supplied was checked for any obvious flaws or inconsistencies such as missing values, dates out of sequence and apparent differences between the data set and published papers. Problems were rectified by correspondence with the principal investigator. Two trials stopped randomizing to certain arms early and a further two studies employed two different registrations whereby patients were randomized through two separate schemes. Each of these trials was subdivided into the appropriate number of data sets and for the purpose of analysis each were treated as an independent trial.

All analyses were carried out on the intention to treat; thus all patients were analysed according to the treatment they were allocated irrespective of whether or not they actually received that treatment. Virtually all the data were available and included in the overview analyses avoiding potential bias of post-randomization exclusion.

Survival analyses were stratified across all trials for each comparison to generate the logrank statistics that are reported. The individual observed and expected numbers of deaths as calculated in these actuarial survival analyses were pooled to provide an overall hazard ratio (HR) for the relevant comparison. The HR represents the relative risk of death associated with two treatments. For example, in a trial comparing treatment A with treatment B, an HR of 0.8 represents a 20% decrease in the relative risk of death when using treatment A rather than B, while an HR of 1.2 represents an increase in the relative risk of death of 20% when using treatment A. An HR of unity represents no difference in the relative risk of death.

Results are presented graphically by means of HR plots and survival curves. In the HR plots the hazard ratio for each study is plotted as an open square, whose size is directly proportional to the amount of information in that trial. Horizontal lines extend laterally from this to display the confidence intervals (CI), 99% CI for individual trials and 95% CI for the overall HR

Overview of randomized chemotherapy trials for advanced ovarian cancer

Table 31.1 Number of trials, deaths and patients per comparison

Comparison	Available trials	Unavailable trials	Number of deaths	Number of patients
I	19	6	2817	3146
II	11	2	1136	1329
III	8	0	1134	1408
IV	7	0	712	925
V	11	0	1771	2061
Total	45	8	6408	8139

The total number of deaths and patients are not a simple sum of those given for comparisons I to V since some trials were included in two comparisons but only counted once for the totals.

value. Wider confidence intervals are imposed on the individual trials because the problems of multiplicity increase the chance of observing a false positive result. The vertical line drawn through unity indicates the point where there is no difference between treatments. Trials indicating an advantage for treatment A lie to the left of this line and those showing advantage to treatment B lie to the right. Trials indicating a significant result lie wholly to one side of the line, such that their CI will not straddle it. The overall HR is shown as an occluded lozenge whose horizontal extremities denote the overall CI. All P-values presented in this report are two sided and the χ^2-values reported are on 1 degree of freedom.

31.3 RESULTS

This report concentrates on comparisons IV and V since these are of most current interest. A full report is currently in preparation. Summary information of the numbers of patients and deaths in each of the two comparisons are given in Table 31.1.

31.3.1 COMPARISON IV: SINGLE-AGENT PLATINUM VS. A PLATINUM COMBINATION

Data from all of the seven eligible trials were available to the overview: six compared cisplatin as a single agent and in combination and one trial[15] compared carboplatin as a single agent and in combination. There is a major imbalance in the numbers randomized – 360 to single-agent platinum and 565 to platinum combinations – because the large GICOG study[9] utilized two platinum combination arms. A total of 712 deaths were observed and the median follow-up for this comparison was 6.5 years. One trial[16] had a slightly different objective to the others in that it compared high-dose cisplatin (100 mg/m^2) given as a single agent with low-dose cisplatin (20 mg/m^2) given in combination. In all other trials in this comparison the dose of platinum was approximately equal in both arms and cisplatin did not exceed a dose of 60 mg/m^2. The analysis was therefore performed both including and excluding this trial.

All trials

The survival curves (Figure 31.1) appear to show a survival difference that appears after 2 years and is maintained until about year 8. The overall difference between the two survival curves, however, is not statistically significant ($\chi^2 = 2.53$, $P = 0.11$). Likewise, the overall HR of 0.89 has a confidence interval straddling unity (CI 0.76, 1.04).

Discussion

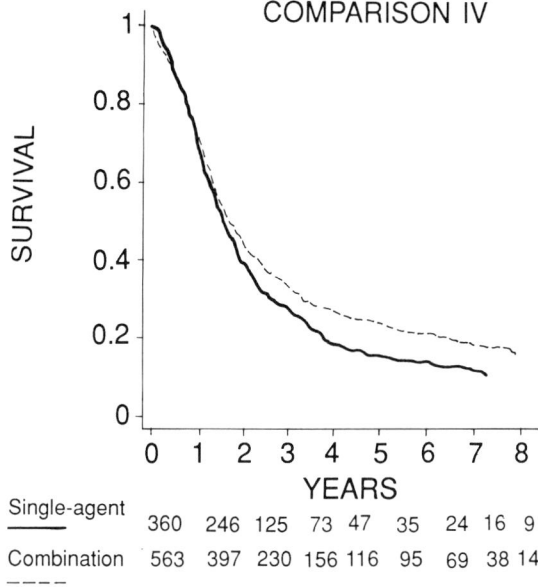

Figure 31.1 Comparison IV: single-agent platinum vs a platinum combination: overall survival.

Excluding high-dose/low-dose study

In this case the difference between the two curves achieves borderline statistical significance ($\chi^2 = 4.82$, $P = 0.03$) and the overall HR of 0.85 (95% CI 0.72, 1.0) suggests a 15% reduction in the relative risk in favour of the combination.

The HR plot (Figure 31.2) illustrates the difference between the two groupings. It should be noted that a large proportion of patients in this comparison (60%) are from the GICOG trial and that the results are therefore highly influenced by this study.

31.3.2 COMPARISON V: CARBOPLATIN VS. CISPLATIN

This comparison included trials comparing carboplatin with cisplatin either as single agents or in combination with other drugs. Data were available from all 11 eligible trials in which 1023 patients were randomized to cisplatin and 1038 to carboplatin. The total number of observed deaths was 1771 and in this case the median follow up is only 3 years reflecting the relatively recent nature of these trials. The doses of platinum used in these trials are given in Table 31.2. The survival curves (Figure 31.3) show no significant difference in overall survival between the two treatments ($\chi^2 = 0.91$, $P = 0.34$) and this is reflected in the overall hazard ratio of 0.98 (95% CI 0.89, 1.07) The HR plot (Figure 31.4) shows no obvious relationship between the estimate of treatment difference and whether single agents or combinations of cisplatin and carboplatin were compared. No study individually gives an indication of a significant benefit for either treatment and this is reflected in the overall results.

31.4 DISCUSSION

It is clear that since its introduction over 30 years ago, chemotherapy has not provided any major improvements in the survival for patients with advanced ovarian carcinoma. However, moderate improvements in survival may have been gained and while these may not be dramatic, given the relatively high incidence of the disease, such moderate improvements may be important in the public health context.

Because existing clinical trials were too small to detect reliably moderate survival differences, it was necessary to perform an overview to try to resolve outstanding

Table 31.2 Dose of platinum used in comparison V

Cisplatin dose (mg/m²)	Carboplatin dose (mg/m²)	Number of trials using dose
100	400	3
100	300	2
50	200	1
60	150	1
20	350	1
75	300	1
50	250	1
80	350	1

Overview of randomized chemotherapy trials for advanced ovarian cancer

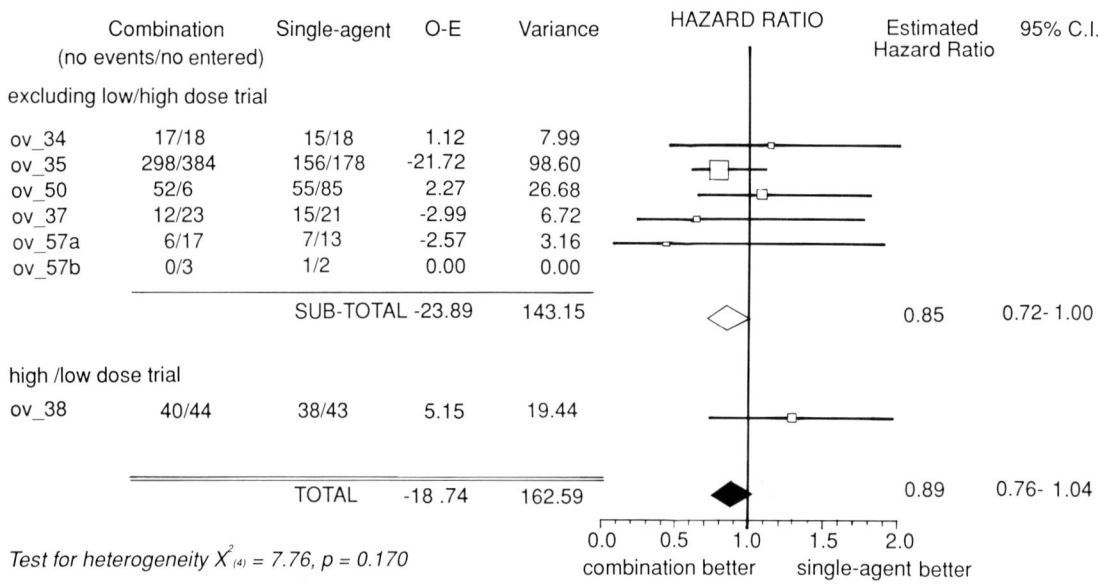

Figure 31.2 Comparison IV: single-agent platinum vs platinum combination: relative risk.

questions about chemotherapy; in particular to determine the role of platinum and combination chemotherapy in disease management. This overview represents the collation and analysis of updated individual information from more than 8000 patients from 45 trials carried out in 11 different countries. As such, it represents the most important compilation of data ever undertaken in advanced ovarian carcinoma. It is also one of the first overviews in oncology that has been able to collect individual patient data from every trial involved in the overview.

Before discussion of the results a few general points are necessary. The basic principle involved in the analysis of any overview is to make comparisons of treatments only within each individual trial and to combine the results of these comparisons to provide overall estimate of the average treatment effect of all the trials in the overview. It should be noted that since it is trial results

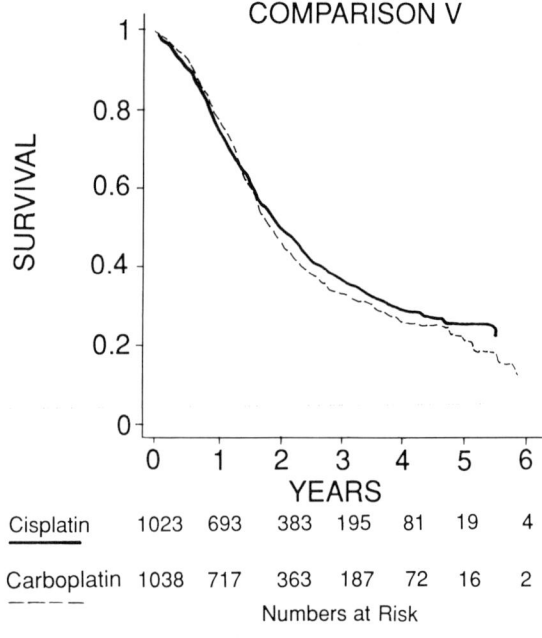

Figure 31.3 Comparison V: cisplatin vs carboplatin: overall survival.

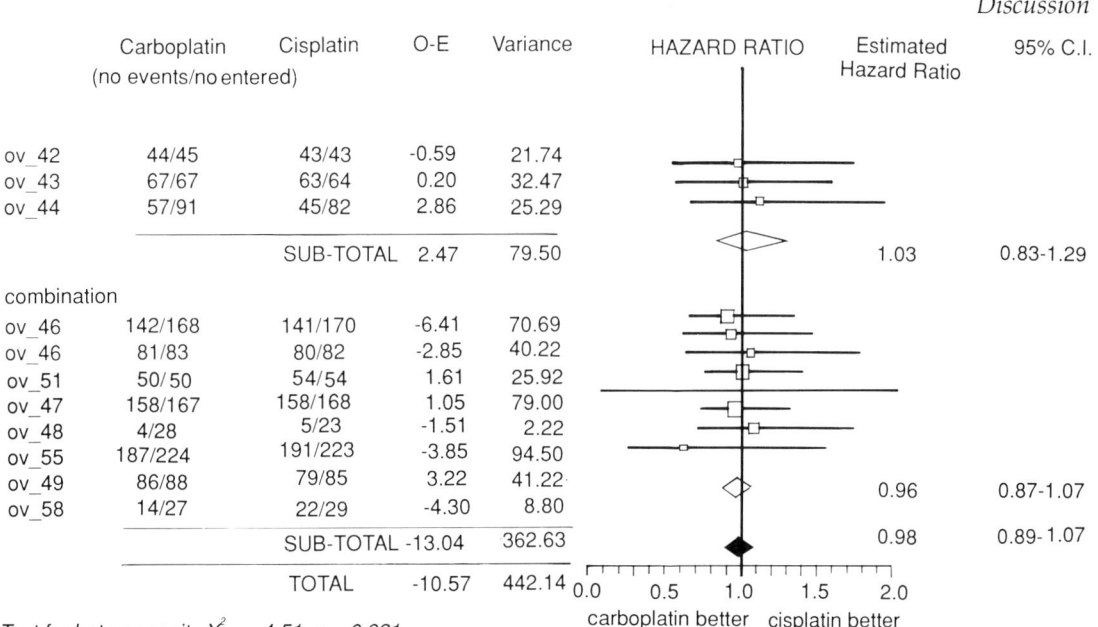

Figure 31.4 Comparison V: cisplatin vs carboplatin: relative risk.

that are being combined, the method does not implicitly assume that the patients in any one trial are directly comparable with those in another. Likewise it does not assume that any real treatment effects will be of the same magnitude, only that they will point in the same direction. It can be argued that utilizing a cross-section of patient types from a number of different trials is more likely to reflect the real world and therefore to estimate more accurately the type of treatment effects that are achievable outside of clinical trials. Even so, just as an individual trial cannot provide a prescription for treating any individual patient, neither can an overview. However, this overview provides the least biased and most accurate summary of existing information from clinical trials, from which the clinician can make his or her own decisions about disease management.

The survival curves for comparison IV show a trend favouring the combination arm that emerges only after 2 years, as opposed to the early treatment advantage shown in the two comparisons comparing treatment with or without platinum. This difference in pattern could be attributable to the fact that both patient groups received some form of immediate platinum treatment and therefore this comparison does not suffer from the same problems associated with second-line therapy. The results of the analyses imply that platinum in combination is more effective than single-agent platinum in the doses used. It is notable, however, that in most studies the dose of platinum used as a single agent was lower than is currently recommended. It will therefore be important to determine whether the apparent difference between treatments is due to the addition of other drugs *per se* or is in fact a question of dose intensity as implied by the Wiltshaw[17] study. If the latter were true, patients receiving drug combinations fared better simply because they received a higher dose of drug irrespective of type. Similar results and a similar problem of dose intensity were found in a separate overview[18] which examined the role of doxorubicin (Adriamycin) in ovarian carcinoma. This, the CAP/CP overview, found a significant survival advantage for patients treated with CAP (cyclophosphamide, doxorubicin and cisplatin) over those treated with CP (cyclophosphamide and cisplatin) but also concluded that the observed difference could simply be due to patients on the CAP arm receiving higher total doses of drugs.

It is highly desirable that this question of dose is addressed by comparing currently accepted doses of single-agent platinum with platinum combinations in a large prospective randomized clinical trial.

The results of comparison V offer no evidence that cisplatin is either superior or inferior to carboplatin in terms of survival, either as a single agent or in combination. However, some of the trials included in this comparison are still at a relatively early stage so that although the information presented on the first 3 years is reliable, further follow-up is required. It should also be pointed out that the trials included in this comparison used doses of carboplatin based on surface area rather than renal clearance, which is the method currently recommended[19].

It is clear from the confidence intervals associated with the overall HRs that even combining the results from many studies does not provide sufficient numbers of patients to make any firm statements about the results of the comparisons presented, except perhaps for comparison I (no difference in survival between the two arms). Consequently, although comparisons II–IV are suggestive of treatment benefit associated with first-line platinum-based combination chemotherapy, no conclusive evidence emerges. Trials of approximately double the size of these overviews would be required to establish reliably whether these trends do in fact represent actual treatment benefits. Comparison V does approach such numbers of patients and provides reliable information on relatively short-term survival, up to 3 years, although, as mentioned previously, longer follow-up is required. This uncertainty emphasizes the fact that in the past, clinical trials have been an order of magnitude too small to detect the sorts of differences which this overview suggests may be realistic. Thus, any future trials of chemotherapy in advanced ovarian cancer will have to accrue in the region of 2000 patients rather than, for example, the mean of 180 patients which the trials included in this overview achieved.

Despite not being able to provide any firm conclusions about the most efficacious forms of treatment for advanced ovarian cancer, this overview has helped to clarify the current state of knowledge of chemotherapy used in this disease and in doing so has generated questions that need to be answered. The results raise important hypotheses.

1. Platinum combinations are generally better than single-agent platinum when platinum is used at the same dose (comparison IV). This is supported by the results from the CAP/CP overview[18], since if CAP is superior to CP it is likely to be superior to cisplatin alone.
2. Cisplatin and carboplatin are equally efficacious (comparison V).

31.5 FUTURE STUDIES

This overview was initiated not only to assess the current status of chemotherapy in advanced ovarian cancer but also to consider future trials. There was a workshop consensus that trials have so far been wholly inadequate in size. Since current therapy has had only a modest impact on survival in this disease and because only modest gains for new treatments can be expected in the foreseeable future, large trials are needed. It was therefore suggested that an international trial should be set up to determine the current optimal therapy for this disease.

The proposed trial will compare CAP (cyclophosphamide/doxorubicin/cisplatin) with carboplatin alone. CAP was chosen as recommended standard therapy since a separate meta-analysis of the role of doxorubicin suggested that this regimen was more effective than cisplatin/cyclophosphamide[18]. This meta-analysis will be published separately, though representatives of the four studies analysed were present at the AOCTG overview meeting. Based on the randomized trials included in the current overview, it was felt that CAP and similar combinations had not been tested against adequate doses of a single platinum drug. To make the comparison most meaningful, carboplatin (given in full dose with adjustment for renal function) was chosen as the comparative arm because it is substantially less toxic than CAP or single-agent cisplatin. Even if the trial shows no overall survival difference between the two arms, an optimal therapy would nevertheless be identified since carboplatin would be better tolerated.

The design of this trial will be kept as simple as possible, with minimal data collection and disruption for participating clinicians. The trial will be run in parallel with another comparing no further therapy with platinum-based chemotherapy in patients with early disease for whom the responsible clinician is uncertain whether further therapy is required.

Details of these trials (ICON Studies) can be obtained from the MRC Cancer Trials Office in Cambridge, UK. These trials are intended to be a joint effort among clinical trial groups in the UK, Europe, North and South America and Australasia. Regardless of affiliation, any clinician who treats ovarian carcinoma will be encouraged to enter patients into these trials since it is likely that large single centres may prefer to concentrate on phase II studies of innovative approaches. The need for this type of large trial (2000 or more patients), designed to optimize current therapy, is quite clear. Even modest improvements in survival or reduction in toxicity are critical when the number of women dying of this disease is taken into account.

ACKNOWLEDGEMENTS

The advanced ovarian cancer overview was initiated by the British Medical Research Council who also funded the overview administration. We are indebted to the following institutions who funded the AOCTG meeting: Cancer Research Campaign, Imperial Cancer Research Fund, Medical Research Council, Sir Samuel Scott of Yews Trust, BUPA, Sterling Oncology, Bristol Myers Squibb Company, Epidemiology and Public Health Research Unit University of Surrey, Harold Smith Charitable Trust, Ciba-Geigy, Eli Lilley, Lederle, Schering-Plough Limited, Fitton Trust. We would also like to thank the staff of Bristol Myers Research and Development Division, Wallingford, USA for their help in identifying relevant studies and for providing data from trials that they had sponsored. We are indeed grateful to all the data centres around the world without whose efforts in providing individual patient data, this overview would not have been possible.

REFERENCES

1. Parkin, D.M., Laara, E. and Muir, C.S. (1988) Estimates of the worldwide frequency of sixteen major cancers in 1980. *Int. J. Cancer*, **41**, 184–97.
2. Young, R.C., Chabner, B.A., Hubbard, S.P. *et al.* (1978) Advanced ovarian carcinoma: A prospective clinical trial of melphalan (L-PAM) versus combination chemotherapy. *N. Engl. J. Med.*, **299**, 1261–6.
3. Wiltshaw, E. and Kroner, T. (1976) Phase II study of cisdichlorodiammineplatinum (II) (NSC 119875). *Cancer Treat. Rep.*, **60**, 55–60.
4. Williams, C.J., Stevensen, K., Buchanan, R. *et al.* (1979) A pilot study of cisdichlorodiammineplatinum (II) in combination with Adriamycin and cyclophosphamide in previously

untreated patients and as a single agent in previously untreated patients. *Cancer Treat. Rep.*, **63**, 1745–53.

5. Williams, C.J., Mead, G.M. and Arnold, A. (1982) Chemotherapy of advanced ovarian carcinoma: initial experiences of a platinum based combination. *Cancer Treat. Rep.*, **63**, 311–17.

6. Bolis, G., Bortolozzi, G., Carnelli, G et al. (1980) Low-dose cyclophosphamide versus Adriamycin plus cyclophosphamide in advanced ovarian cancer. *Cancer Chemother. Pharmacol.*, **4**, 129–32.

7. Bertelsen, K., Jakobsen, J., Andersen, J.E. et al. (1978) A randomised study of cyclophosphamide and cis-platinum with or without doxorubicin in advanced ovarian carcinoma. *Gynecol. Oncol.*, **62**, 1375–7.

8. Hubbard, S., Barker, P. and Young, R.C. (1978) Adriamycin therapy for advanced ovarian cancer recurrent after chemotherapy. *Cancer Treat. Rep.*, **62**, 1375–7.

9. Gruppo Interegionale Cooperativo Oncologico Ginecologia (1987) Randomised comparison of cisplatin with cyclophosphamide/cisplatin and with cyclophosphamide/doxorubicin/cisplatin in advanced ovarian cancer. *Lancet*, **ii**, 353–9.

10. Tomirotti, M., Perrone, S., Gie, P. et al. (1988) Cisplatin (P) versus cyclophosphamide adriamycin and cisplatin (CAP) for stage III–IV epithelial ovarian carcinoma: A prospective randomised trial. *Tumori*, **74**, 573–7.

11. MRC Gynaecological Cancer Working Party (1990) An overview in the treatment of advanced ovarian cancer. *Br.J. Cancer*, **61**, 495–6.

12. Simes, R.J. (1986) Publication bias: The case of an international registry of clinical trials. *J. Clin. Oncol.*, **4**, 1529–41.

13. USA National Cancer Institute. Physicians Database Query System clinical trials on-line register-CLINPROT.

14. UK Coordinating Committee on Cancer Research (1986) UK Cancer Trials Register.

15. Kaye, S. Randomised study of carboplatin and chlorambucil/carboplatin in advanced ovarian cancer (unpublished).

16. Wiltshaw, E., Evans, B. Rustin, G. et al. (1986) A prospective randomised trial comparing high-dose cisplatin with low-dose cisplatin and chlorambucil in advanced ovarian carcinoma. *J. Clin. Oncol.*, **4** 722–9.

17. Masding, J., Sarker, T., White, J.F. et al. Intravenous treosulphan plus cisplatinum in advanced ovarian carcinoma. (in preparation).

18. Omura, G.A., Buyse, M., Marsoni, S. et al. CP versus CAP chemotherapy of ovarian carcinoma: A meta-analysis (in press).

19. Calvert, A.H., Newell, D.R., Gumbrell, L. et al. (1989) Carboplatin usage: Prospective evaluation of a simple formula based on renal function. *J. Clin. Oncol.* **5**, 1748–56.

20. Aabo, K., Hald, I., Horbrow, S. et al. (1985) A randomised study of single agent vs combination chemotherapy in FIGO Stages IIB, III and IV ovarian adenocarcinoma. *Eur. J. Cancer Clin. Oncol.*, **21**, 475–81.

21. Bruckner, H., Pagano, M., Falkson, G. et al. (1979) Controlled prospective trial of combination chemotherapy with cyclophosphamide, Adriamycin and 5-fluorouracil for the treatment of advanced ovarian cancer: a preliminary report. *Cancer Treat. Rep.* **63**, 297–9.

22. Brodovsky, H.S., Bauer, M., Horton, J., and Elson, P.J. (1984) Comparison of melphalan with cyclophosphamide, methotrexate and 5-fluorouracil in patients with ovarian cancer. *Cancer*, **53**, 844–52.

23. Chylak, V., Ilijasm, L., Krusic, J. and Krusic, K. (1986) Controlled clinical trial of chemotherapy in advanced ovarian cancer – cyclophosphamide mono-chemotherapy versus a combination of Adriamycin, cyclophosphamide, 5-fluorouracil and methotrexate. *Lijek Vjsen*, **109**, 230–4.

24. Edmonson, J.H., Flemming, T.R., Decker, D.G. et al. (1979) Different chemotherapeutic sensitivities and host factors affecting prognosis in advanced ovarian carcinoma versus minimal residual disease. *Cancer Treat. Rep.*, **61**, 355–7.

25. Miller A., Klaassen, D.J., Boyes, D.A. et al. (1980) Combination v. sequential therapy with melphalan, 5-fluorouracil and methotrexate for advanced ovarian cancer. *Can. Med. Assoc. J.*, **123**, 363–71.

26. Medical Research Council's Working Party on Ovarian Cancer. (1981) Medical Research Council study on chemotherapy in advanced

ovarian cancer. *Br.J. Obstet. Gynaecol.*, **88**, 1174–85.
27. British Medical Research Council Gynaecological Cancer Working Party. Inadequacy of trials of chemotherapy in advanced ovarian carcinoma: a randomised trial of three regimens (in preparation).
28. Omura,G.A., Morrow, C.P., Blessing, J.A. *et al.* (1983) A randomised comparison of melphalan versus melphalan plus hexamethylmelamine versus Adriamycin plus cyclophosphamide in ovarian carcinoma. *Cancer*, **51**, 783–9.
29. Park, R.C., Blom, J., Disaia, P. *et al.* (1980) Treatment of women with disseminated or recurrent advanced ovarian cancer with melphalan alone in combination with 5-fluorouracil and dactinomycin or with the combination of Cytotoxan, 5-fluorouracil and dactinomycin. *Cancer*, **45**, 2529–42.
30. Sturgeon, J.F.C., Fine, S., Gospodarowicz, M.K. *et al.* (1982) A randomised trial of melphalan alone vs combination chemotherapy in advanced ovarian cancer. *Proc. Am. Soc. Clin. Oncol.*, **1**, 108.
31. Trope, C. (1987) Melphalan with and without doxorubicin in advanced ovarian cancer. *Obstet. Gynecol.*, **70**, 582–6.
32. Turbow, M.M., Jones, H., Yu, V.K. *et al.* (1980) Chemotherapy of ovarian carcinoma: a comparison of melphalan vs Adriamycin-cyclophosphamide. *Proc. Am. Assoc. Cancer Res. Am. Soc. Clin. Oncol.*, **21**, 196.
33. Wharton, J.T. Edwards, and Rutledge, F.N. (1984) Long-term survival after chemotherapy for advanced epithelial ovarian carcinoma. *Am. J. Obstet. Gynecol.*, **148**, 997–1005.
34. Bell, D.R., Woods, R.L., Levi, J.A. and Fox, R.M. (1982) Advanced ovarian cancer: a prospective randomised trial of chlorambucil versus combined cyclophosphamide and cis-diamminedichloro-platinum. *Aust. N.Z. J. Med.*, **12**, 245–9.
35. Crowther, D. (1986) A randomised trial of chemotherapy in advanced residual (stage IIB–IV) ovarian cancer. Manchester Ovarian Cancer Clinical Study Group protocol (unpublished).
36. Decker, D.G., Thomas, M.D., Fleming, R. *et al.* (1982) Cyclophosphamide plus cis-platinum in combination: treatment program for stage III or IV ovarian carcinoma. *Obstet. Gynecol.*, **60**, 481–7.
37. Gynaecological Group, Clinical Oncological Society of Australia and the Sydney Branch, Ludwig Institute for Cancer (1988) Chemotherapy of advanced ovarian adenocarcinoma: a randomised comparison of combination versus sequential therapy using chlorambucil and cisplatin. *Gynecol. Oncol.*, **6**, 282–90.
38. Leonard, R.C., Smart, G.E., Livingston, J.R.B. *et al.* (1989) Randomised trial comparing predinimustine with combination chemotherapy in advanced ovarian carcinoma. *Cancer Chemother. Pharmacol.*, **23**, 105–10.
39. Wilbur, D.W., Rentscler, R.E. Wagner, R.J. *et al.* (1987) Randomized trial of the addition of cis-platin (DDP) and/or BCG to cyclophosphamide (CTX) chemotherapy for ovarian carcinoma. *J. Surg. Oncol.*, **6**, 165–9.
40. Williams, C.J., Mead, G.M., Macbeth, F.R. *et al.* (1985) Cisplatin combination chemotherapy versus chlorambucil in advanced ovarian carcinoma: mature results of a randomized trial. *J. Clin. Oncol.*, **3**, 1455–62.
41. Vogl, D., Kaplan, B. and Pagano, M. (1982) Diamminedichloroplatinum-based combination chemotherapy is superior to melphalan for advanced ovarian cancer when age >50 and tumour diameter >2cm. *Proc. Am. Soc. Clin. Oncol.*, **1**, 119.
42. EORTC (1978) Randomised controlled clinical trial comparing cyclophosphamide plus Adriamycin plus cis-DDplatinum in the treatment of patients with stage III and IV epithelial carcinoma of the ovary. Protocol 55781 (unpublished).
43. Omura, G. Blessing, J.A., Ehrlich, C.E. *et al.* (1986) A randomized trial of cyclophosphamide and doxorubicin with or without cisplatin in advanced ovarian carcinoma. A Gynecologic Oncology Group study. *Cancer*, **57**, 1725–30.
44. Turbow, M.M. (1980) Chemotherapy of advanced ovarian cancer. Adriamycin-cyclophosphamide versus platinum-Adriamycin-cyclophosphamide. Northern California Oncology Group protocol 5091 (unpublished).
45. Cohen, C.J., Goldberg, J.D., Holland, J.F., *et*

al. (1983) Improved therapy with cisplatin regimens for patients with ovarian carcinoma (FIGO stages III and IV) as measured by surgical end-staging (second-look) operation. *Am. J. Obstet. Gynecol.*, **145**, 955–65.
46. Gilbey, E., Pollard, W., Barley, V. et al. (1986) Ovarian Cancer Trial. South West Oncology Study (unpublished).
47. Adams, M., Kerby, I.J., Rocker, I. et al. (1987) Cisplatin (CDDP) vs carboplatin (JM8) in advanced adenocarcinoma of the ovary. First Meeting of the International Gynecologic Cancer Society, 4–8 October, p. 48.
48. Anderson, H., Wagstaff, J., Crowther, D. et al. (1988) Comparative toxicity of cisplatin, carboplatin (CBDSA) and iproplatin (CHIP) in combination with cyclophosphamide in patients with advanced epithelial ovarian cancer. *Eur. J. Cancer Clin. Oncol.*, **24**, 1471–9.
49. Mangioni, C., Bolis, G., Pecorelli, S. et al. (1989) Randomized trial in advanced ovarian cancer comparing cisplatin and carboplatin. *J. Natl Cancer Inst.*, **81**, 1464–71.
50. Wiltshaw, E., Evans, B. and Harland, S. (1985) Phase III randomised trial cisplatin versus JM8 (carboplatin) in 112 ovarian cancer patients, stages III and IV. *Proc. Am. Soc. Clin. Oncol.*, **4**, 121.
51. Alberts D.D., Green, S.J., Hannigan, E.V. et al. (1990) Improved efficacy of carboplatin plus cyclophosphamide versus cisplatin plus cyclophosphamide: preliminary report by the Southwest Oncology Group of a phase II randomized trial in stages III and IV suboptimal ovarian cancer, in *Carboplatin: Current Perspectives and Future Directions* (eds P.A. Bunn Jr, R. Canetta, R.F. Ozols and M. Rozencweig), W.B. Saunders, Philadelphia, pp.163–4.
52. Chiara, S. Bruzzone, M., Calciari, C. et al. (1987) A randomized study comparing two combination chemotherapy regimens containing cisplatin or carboplatin (CAC or PAC) in advanced ovarian carcinoma. European School of Gynecol Oncol **6**, A463.
53. Edmonson, J.H., McCormack, G.M., Wieand, H.S. et al. (1989) Cyclophosphamide-cisplatin vs cyclophosphamide-carboplatin in stage III–IV ovarian carcinoma: a comparison of equally myelosuppressive regimens. *J. Natl Cancer Inst.*, **81**, 1500–4.
54. Kato, T., Nishimura, H., Yamage, T. et al. (1988) Phase III study of carboplatin for ovarian cancer. *Jpn. J. Cancer Chemother.*, **15**, 2297–304.
55. Meerphol, H.G., Kuhnie, H., Sauerbrei, W. et al. (1990) Cyclophosphamide/cisplatin (CTX/PT) vs CTX/carboplatin (carbopt) in advanced ovarian carcinoma: A randomised multicenter study. 15th International Cancer Congress, Hamburg, 1990.
56. Pater, J. (1990) Cyclophosphamide/cisplatin (CDDP) versus cyclophosphamide/carboplatin (CBDCA) in macroscopic residual ovarian cancer. Initial results of a National Cancer Institute of Canada Clinical Trials Group trial. *Proc. Am. Soc. Clin. Oncol.*, **4**, 121.
57. ten Bokkel Huinink, W.W., van der Burg, M.E., van Oosterom, A.T. et al. (1988) Carboplatin in combination chemotherapy for ovarian cancer. *Cancer Treat Rev.*, **15** (Suppl. B), 9–15.
58. Adams, M. Johansen, K.A., James, K.W. and Rocker, I. (1982) A controlled clinical trial in advanced ovarian cancer. *Clin. Radiol.*, **33**, 161–3.
59. Barlow, J.J., Lele, S.B., and Emrich, L.J. (1985) Long-term survival rates with various chemotherapeutic regimens in stage III and IV ovarian adenocarcinoma. *Am. J. Obstet. Gynecol.*, **152**, 310–14.
60. Carmo-Pereira, J., Oliveira Costa, F., Henriques, E. and Almeida Ricardo, J. (1981) Advanced ovarian carcinoma: a prospective and randomised clinical trial of cyclophosphamide vs combination cytotoxic chemotherapy (Hexa-CAF) *Cancer*, **48**, 1947–51.
61. Delgado, G., Smith, F.P., McLaughlin, E.K. and Tuholski, N. (1985) Single agent vs combination chemotherapy for ovarian cancer. *Am. J. Clin. Oncol.*, **8**, 33–7.
62. De Palo, G., De Lena, M. and Bonadonna, G. (1977) Adriamycin vs Adriamycin plus melphalan in advanced ovarian carcinoma. *Cancer Treat. Rep.*, **61**, 355–7.
63. Gronroos, M., Nieminen, U., Kauppila, A. et al. (1984) A prospective randomised national

trial for treatment of ovarian cancer: the role of chemotherapy and external radiation. *Eur. J. Obstet. Gynecol. Reprod. Biol.*, **17**, 33–42.

64. Senn, H.J., Lei, D., Castano-Almendral, A. *et al.* (1980) Chemo-(hormom)-therapie fortgeschrittener ovarialkarzinome der FIGO-stadien III und IV. Prospektive SAKK-studie 20/71. *Schweiz. Med. Wochenschr.*, **110**, 1202–8.

65. Carmo-Pereira, J., Oliveira Costa, F. and Henriques, E. (1983) Cis-platinum, Adriamycin and hexamethylmelamine vs. cyclophosphamide in advanced ovarian carcinoma. *Cancer Chemother. Pharmacol.*, **10**, 100–3.

66. Harvey, H.A., Lipton, A., Simmonds, M. *et al.* (1982) A randomised trial of alkeran versus cyclophosphamide, hexamethylmelamine, Adriamycin and cis-platinum combination chemotherapy in advanced ovarian carcinoma. *Clin. Res.*, **30**, 418.

67. ZFG (1989) Arbeitsgruppe tumorchemotherapie in der gynakologie der DDR die effektivitat von platin als mono und kominationstherapie bei prima resistentem oder rezidivierendem ovarialkarzinome. Ergebnisse der DDR prospectiven Multizenterstudie *Zentralbl. Gynakol.*, **111**, 938–46.

APPENDIX A: DETAILS OF RANDOMIZED CLINICAL TRIALS INCLUDED IN THE ADVANCED OVARIAN CANCER OVERVIEW

Reference	Author	Single drug	Combination	Patients in overview
Comparison I				
20	Aabo	CTX BU	ADR, CTX, FU	179
6	Bolis et al.	CTX	ADR, CTX	74
21	Bruckner et al.	L-PAM	MTX, TSPA ADR, CTX, FU	331
22	Brodovsky et al.	L-PAM	CTX, FU, MTX	409
23	Chylak et al.	CTX	ADR, CTX, FU, MTX	69
24	Edmonson et al.	CTX	ADR, CTX	111
25	Miller et al.	L-PAM	FU, L-PAM, MTX	254
26	MRC	CTX	CTX, HMM, MTX	344
27	MRC	CTX	ADR, CTX	116
28	Omura et al.	L-PAM	ADR, CTX HMM, L-PAM	339
29	Park et al.	L-PAM	FU, L-PAM ACT-D, FU, L-PAM ACT-D, CTX, FU	418
30	Sturgeon et al.	L-PAM	CTX, FU, HMM, MTX	83
31	Trope	L-PAM	ADR, L-PAM	168
32	Turbow	L-PAM	ADR, CTX	48
33	Wharton et al.	L-PAM	CTX, HMM ADR, CTX, HMM	75
33	Wharton et al.	L-PAM ADR HMM	CTX, HMM	128
Comparison II				
34	Bell	CLB	CTX, CACP	38
35	Crowther et al.	CTX	BLE, CTX, CACP	109
36	Decker et al.	CTX	CTX, CACP	42
37	COSA	CLB	CLB, CACP	370
38	Leonard et al.	PRED	FU, HMM, PRED, CACP	80
17	Masding et al.	TREO	TREO, CACP	157
27	MRC	CTX	CTX, CACP	100
30	Sturgeon et al.	L-PAM	ADR, CTX, CACP	83
39	Wilbur et al.	CTX	CTX, CACP	11
40	Williams et al.	CLB	ADR, CTX, CACP	89

Appendix A

Reference	Author	Single drug	Combination	Patients in overview
41	Vogl et al.	L-PAM	ADR, CTX, HMM, CACP	250
Comparison III				
37	COSA	CLB	CLB, CACP	370
36	Decker et al.	CTX	CTX, CACP	42
42	EORTC	ADR, CTX	ADR, CTX, CACP	149
17	Masding et al.	TREO	TREO, CACP	157
27	MRC	CTX	CTX, CACP	100
43	Omura et al.	ADR, CTX	ADR, CTX, CACP	495
44	Turbow et al.	ADR, CTX	ADR, CTX, CACP	84
39	Wilbur et al.	CTX	CTX, CACP	11
Comparison IV				
45	Cohen et al.	CACP	ADR, CACP	36
9	GICOG	CACP	ADR, CACP ADR, CTX, CACP	562
46	Gilbey et al.	CACP CBDSA	IFOS, CACP IFOS, CBDSA	30 5
15	Kaye et al.	CBDSA	CLB, CBDSA	161
10	Tomirotti et al.	CACP	ADR, CTX, CACP	44
16	Wiltshaw et al.	CACP	CLB, CACP	87
Comparison V				
47	Adams	CBDSA	CACP	88
48	Anderson et al.	CBDSA	CACP	56
49	Mangioni et al.	CBDSA	CACP	173
50	Wiltshaw	CBDSA	CACP	131
51	Alberts et al.	CTX, CBDSA	CTX, CACP	338
52	Chiara et al.	ADR, CTX, CBDSA	ADR, CTX, CACP	165
53	Edmonson et al.	CTX, CBDSA	CTX, CACP	104
54	Kato et al.	ADR, CTX, CBDSA	ADR, CTX, CACP	51
55	Meerpohl et al.	CTX, CBDSA	CTX, CACP	173
56	Pater et al.	CTX, CBDSA	CTX, CACP	447
57	ten Bokkel Huinink et al.	ADR, CTX, HMM, CBDSA	ADR, CTX, HMM, CACP	335

ACT-D, actinomycin D; ADR, doxorubicin; BLE, bleomycin; BU, busulphan; CACP, cisplatin; CBDSA, carboplatin; CLB, chlorambucil; CTX, cyclophosphamide; FU, 5-fluorouracil; HMM, hexamethylmelamine; IFOS, ifosfamide; L-PAM, melphalan; MTX, methotrexate; PRED, predinimustine; TSPA, triethylenethiophosphoramide; TREO, treosulphan.

APPENDIX B: TRIALS NOT INCLUDED IN THE ADVANCED OVARIAN CANCER OVERVIEW

Reference	Author	Single drug	Combination	Patients randomized
Comparison I				
58	Adams et al.	L-PAM	ADR, CTX, FU	40
59	Barlow et al.	L-PAM	ACT-D, CTX, FU	108
60	Carmo-Pereira et al.	CTX	CTX, FU, HMM, MTX	57
61	Delgado et al.*	L-PAM	CTX, FU, HMM	27
62	De Palo et al.	ADR	ADR, L-PAM	29
63	Gronroos et al.*	TREO	CTX, TREO ADR, FU, TREO	108
64	Senn et al.	CTX	CTX, FU	89
2	Young et al.	L-PAM	CTX, FU, HMM, MTX	80
Comparison II				
65	Carmo-Pereira et al.	CTX	CACP, ADR, HMM	59
66	Harvey et al.	L-PAM	CACP, ADR, CTX, HMM	40
Comparison III				
67	ZFG*	CACP	CACP, CTX CACP, ADR, CTX	133

For explanation of abbreviations, see Appendix A.
* Trials ineligible because they did not appear to be appropriately blinded. Other trials eligible but not available.

Chapter 32
Meta-analysis of treatment for advanced ovarian cancer

R.J. OSBORNE

32.1 INTRODUCTION

Ovarian cancer is now the leading cause of death from gynaecological malignancy in North America. Unlike cancer of the cervix, which can be identified by screening at a pre-invasive stage, or endometrial cancer which invariably results in vaginal bleeding early in the invasive process, ovarian cancer is typically diagnosed when the tumour has already spread from the ovary to contiguous and distant structures in the pelvis and abdomen. At this advanced stage, a patient's chance of surviving 2 years from the point of surgical diagnosis is approximately 15%, even with aggressive surgical intervention and intensive chemotherapy.

Historically, several post-surgical strategies have been available to physicians who treat patients with advanced ovarian cancer (FIGO stages III and IV). Until the middle 1970s, the treatment of choice for this disease was a single alkylating agent (AA), usually thiotepa, melphalan, chlorambucil or cyclophosphamide. Alkylating agents were taken orally with the exception of cyclophosphamide, which was also administered intravenously. The AAs were well tolerated with little cost to the patient in terms of diminished quality of life. Their only significant drawbacks were myelosuppression and a small risk of late-occurring haematological cancer which was dose and time dependent and was typically observed only after at least 12 months of treatment. Robert Young, at the National Cancer Institute, suggested in a 1978 paper that the use of several drugs might improve survival for these patients[1].

In 1979, Eve Wiltshaw at the Royal Marsden Hospital, London reported on a new drug, cisplatin, which appeared to be more active in ovarian cancer than any previous drug or drug combination (response rates in the order of 80% compared to 30% for AAs)[1]. Further evidence from both descriptive studies and randomized controlled trials quickly confirmed this finding. As a result, clinical investigation during the last 10 years has largely been directed towards the definition of how this particular drug can best be employed in the management of ovarian cancer. Recent trials have examined cisplatin alone, cisplatin in combination, cisplatin in different dose schedules, and more recently, new methods of drug delivery, i.e. intraperitoneal therapy. However, this drug is not without side-effects and complications. Even in moderate doses (50 mg/m^2) it may often produce significant nausea and vomiting and, less often, renal dysfunction and neurological toxicity or ototoxicity. At higher doses (>75 mg/m^2), there is universal and often significant dysfunction in some or all of these same organ systems.

Given the understandable enthusiasm to evaluate cisplatin, AAs have been ignored as primary treatment for these patients for more than a decade. Oncologists have neglected AAs in favour of cisplatin therapy because of the latter's higher response rate, which is often observed as early as the first few treatment cycles. Forgotten is the fact that the most important end-point for this disease is not response but survival, since even patients with stage III disease who do respond to cisplatin are unlikely to be cured.

Not only do most contemporary studies use an 'inferior' outcome measure (response) to assess the efficacy of chemotherapy in this disease, but the subjective 'costs' of cisplatin therapy for these patients have not yet been quantified. Epidemiological adjuncts called 'utilities' or 'worths' for ovarian cancer are currently being developed which could be used to impute an objective value for the 'cost' of treatment for patients, measured as quality adjusted life-years (quantity of life × quality of life)[2] With such information, the therapeutic worth of AAs vs. cisplatin or cisplatin-containing, combinations of drugs (CC) could be more meaningfully evaluated, since the subjective 'costs' of treatment (nausea, fatigue, neurotoxicity etc.) could all be reliably quantified for individual patients.

Should AAs be used in the primary management of some subsets of patients with advanced ovarian cancer, specifically in situations where bulky advanced disease or an infirm patient precludes aggressive 'toxic' initial treatment? Such a question would be difficult to pose prospectively since it would have to be formulated so that it addressed both the efficacy of the treatment and its toxicity. A trial which met both these objectives could not be mounted today, given the widespread belief that cisplatin could not ethically be excluded from the primary treatment of advanced ovarian cancer.

The question which this meta-analysis attempts to address is: How do several outcome measures (response rate, progression-free interval and survival) differ when patients with advanced ovarian (stage III/IV) are initially treated with AAs as opposed to cisplatin combination chemotherapy?

Simes, in an earlier review of the more general topic of AAs versus any parenteral combination chemotherapy (with or without cisplatin) came to the conclusion that the case for drug combinations as primary treatment for ovarian cancer was not proved[3]. He observed that the published data on ovarian cancer were significantly more likely to report a positive effect and a larger effect size than unpublished data, which presumably had been 'file-drawered' (not published) when not as great an effect, or an opposite effect to what the author(s) felt was clinically significant, was observed. This epidemiological phenomenon of 'publication bias' is addressed in the technique of meta-analysis which seeks to identify all the relevant clinical material that exists on a topic, including all published and unpublished prospective information.

Since cisplatin is presently considered optimal therapy for this disease, the follow-on question from Sime's article is not AAs versus any combination chemotherapy, but rather AAs versus cisplatin alone or cisplatin in combination. If it could be demonstrated that non-toxic primary single agent therapy such as AA was a real alternative in terms of 'cost benefit' or 'cost utility', perhaps for subsets of patients with bulky residuum, toxic primary drug therapies such as cisplatin could be withheld until patients developed symptomatic recurrence, unless it could be conclusively determined that the toxic treatment resulted in a demonstrable survival advantage. This approach would decrease the over-

Table 32.1 The studies identified by the information search are listed

Author	Year	Sample	Intervention	
			AA	CC
Bruckner	1974–80	53/0	TT + MTX	PL + A
Decker	1977–79	42/0	C	PL + C
Carmo	1978–81	53/6	C	PL + HA
Williams	1979–83	89/6	CHL	PL + CA
Tattersall	1979–85	369/85	CHL	PL + CHL
Bell	1978–79	37/1	CHL	PL + C
Vogl	1980–82	253/8	M	PL + CHA
Sturgeon	1982	115/0	M	PL + CA
Wilbur	1983	24/ns	C/NCG	PL + C/BCG

TT, thiotepa; MTX, methotrexate; C, cyclophosphamide; CHL, chlorambucil; M, melphalan; PL, platinum; H, hexamethylmelamine; A, Adriamycin.

all cost of chemotherapy (fewer courses of expensive parenteral drug combinations) and produce a great deal less toxicity for the patient, especially in the first 6–18 months after primary surgery when this disease is least likely to alter a patient's performance status.

32.2 METHODOLOGY

'Meta-analysis' is a relatively recent epidemiological technique for pooling data across the entire literature. By systematically combining all the clinical material available, this chapter attempts to draw inferences about the magnitude and direction of any treatment difference between the two primary treatment options (AA and CC) in the management of FIGO stage III and IV ovarian cancer. This meta-analysis includes all studies completed within a specific timeframe, beginning with the introduction of cisplatin in 1975 and concluding in 1988.

A systematic 'search strategy' was developed a priori to identify all the existing relevant literature. This literature search began with the bibliography of the Simes article since this was both a recent and an exhaustive report on the treatment of advanced ovarian cancer[3]. Those studies which did not contain a cisplatin arm were excluded leaving six articles and three abstracts for analysis (Table 32.1).

The next step was a hand search of the bibliographies of each of these six articles which yielded no further material, followed by a search of three computerized databases SCISEARCH, CLINPROT in PDQ of the National Cancer Institute and MEDLINE. No additional references were identified in any of these databases.

Next, the bibliographies of eight review articles on the more general topic of the treatment of advanced ovarian cancer were hand searched, but this yielded no new information. In an effort to locate unpublished and unregistered trials, the principal author of each of the six articles and three abstracts was contacted by mail and/or by telephone. This yielded no new information although two primary authors offered their data for secondary analysis (analysis of the actual data from each of the individual trials).

Three acknowledged experts in the treatment of advanced ovarian cancer were identified. They were similarly contacted and asked for additional information on the topic. These 'content experts' were Carmel Cohen, Mount Sinai Hospital New York, Maurie

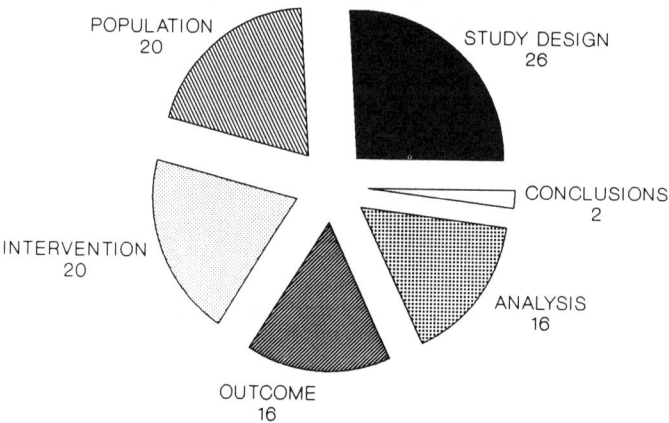

Figure 32.1 The validity criteria. The major categories and their proportional scores are shown.

Markman, Memorial Sloan Kettering Hospital New York and John Kavanaugh at the M. D. Anderson Hospital Houston. With the help of these three individuals, one of the initial studies (Vogl abstract) was found to have been published in the proceedings of a European meeting on ovarian cancer[4]. This report yielded significant additional information about the study.

In a further effort to identify relevant but unpublished data, Bristol Pharmaceuticals, the company which held the original Canadian patent on cisplatin was contacted through their Canadian medical director, but no new information was obtained.

Separate criteria for **relevance** (study inclusion) and **validity** (scientific merit) were developed before the formal search for information had been undertaken. Relevance criteria emphasize that, to be considered for inclusion, a study must first have been randomized and must therefore have been prospective. Secondly, it must have addressed the specific question of AA versus CC, and thirdly, it must have reported at least one relevant outcome measure (response, progression-free interval or survival). The need to restrict the search for clinical information to only randomized trials is particularly important in a retrospective study like a meta-analysis since this effectively limits systematic bias by assuring that all patients from all trials have been randomly allocated to treatment.

Each study was also assessed for scientific merit (validity). Despite the fact that all the studies were randomized controlled trials, the level of attention to methodological detail within the individual trials was not uniform. It is therefore appropriate to attempt to quantify the methodological quality of each study, and to attempt to weight the studies' outcome measures accordingly.

To assess the scientific merit of the individual studies, a list of validity criteria was developed using Chalmers' model of validity criteria for meta-analysis, modified for this specific tumour site[5]. This instrument was divided into six general categories: study design, population, intervention, outcome measures, analysis, and conclusions, and each of these was assigned a percentage score to reflect its relative importance in the assessment of scientific merit (Figure 32.1). Similarly, each major category was further broken down into between six to nine category-specific variables and each was allocated a proportion of the major category score.

This instrument and its weighting schema have not been validated but this methodological approach to the question of differential scientific merit in a cohort of trials has been previously reported.

To further improve the generalizability of this instrument, three individuals at the study institution were identified who were willing to examine the material for validity. It was thought advisable to involve oncologists with an epidemiological background but without a specific knowledge of the cited studies. These individuals were asked to develop their own independent validity criteria and score. Analysis of variance (ANOVA) was then used to develop the final composite scoring system (Figure 32.1). The score for individual items ranged from 'blinded randomization' (score = 8) to several smaller items such as 'study consent' (score = 1).

An arbitrary cut-off score of 35 was chosen in group discussion as the lowest level of clinical validity acceptable for inclusion in the analysis. Of the nine studies identified, two were excluded. The reasons for the rejection of these studies are enumerated in the next section.

32.3 ANALYSIS

Nine studies were retrieved. The study by Bruckner was began in 1974[6] and the Decker trial in 1977[7]. Temporally, the last trial in this analysis, the Tattersall study, was begun as recently as 1984[8]. All of the studies found were randomized, blinded trials of AA vs. CC, except for the study by Bruckner which was a three-arm study comparing AA to cisplatin alone and to CC. For the purposes of this meta-analysis, the latter two arms of the Bruckner study have been combined to give a composite cisplatin-containing arm.

The Tattersall study is the largest and most current. It examined AA (chlorambucil) vs. CC (cisplatin plus chlorambucil) in a trial of 284 subjects. The outcomes reported were response rate, disease-free interval and median survival. One hundred and thirty-four patients (68 arm A/66 arm B) had measurable disease. All three response outcomes could be evaluated and all favoured CC, but none achieved significance. Median survival also favoured CC (70 vs. 65 months).

The Williams study from the UK[9] ended a four-year accrual period at the end of 1983. In a sample of 85 patients, AA (chlorambucil) was compared to CC (cisplatin plus cyclophosphamide and Adriamycin), the standard CC in Europe and North America in the early 1980s. All outcomes of interest were reported, although the number of patients with a surgically documented response was not explicitly stated. There were not sufficient data to analyse this trial for response in patients with measured disease, the only truly valid assessment of response to treatment, but overall response (described in section 32.3.1) favoured CC and this difference was significant. There was a trend to CC for both median survival and disease-free interval but neither was significant.

The study from Portugal by Carmo and colleagues was well designed and evaluated 59 women with stage III/IV disease. The accrual period ran from 1978 to mid-1981. Four (0.07) were lost to follow-up and two (0.04) discontinued treatment for unstated reasons. This was the only study to report response which favoured AA but only overall response could be evaluated. Median survival and disease-free interval also favoured AA but these differences were not statistically different. This was the only study to use high dose parenteral AA (cyclophosphamide), possibly a more effective drug delivery regimen than oral administration (increased dose intensity). Patients in this trial also received hexamethylmelamine and Adriamycin.

The Decker study, begun in 1977, was

terminated prematurely 24 months later when a significant difference was observed which favoured CC. This study had been designed to test whether a difference existed between AA and CC (cisplatin and cyclophosphamide). Twenty-one patients were randomized to each treatment arm. The study did not have pre-established 'stopping rules' but was ended early when a significant difference in progression-free interval ($P<0.0013$) and a substantial difference in median survival (40 est. vs. 16.5 months), both favouring CC, were observed. Overall response also favoured CC but did not reach significance. The other response end-points could not be evaluated.

The study by Bruckner was undertaken in 1974 before cisplatin and was modified following the introduction of this drug in the late 1970s. The study was three armed and contrasted 17 patients who received AA (thiotepa and methotrexate, a combination rarely used in the last 15 years), with cisplatin alone (18 patients) and cisplatin plus Adriamycin (18 patients). For the purposes of this analysis, the two cisplatin arms have been combined to produce a single cisplatin (CC) group. The two groups were different with regard to prognostic factors. The cisplatin arm had significantly less residual tumour than the other arms (0.55 vs. 0.27 and 0.35) and a much higher percentage of well-differentiated tumours (0.44 vs. 0.11 and 0.29) which might have resulted in a better response in that group. However, when the two CC arms are combined, these prognostic advantages are no longer significant. There was a trend favouring CC for response, disease-free interval and median survival but only disease-free interval approached statistical significance.

The Vogl study compared cisplatin vs. cisplatin plus Adriamycin, cyclophosphamide and hexamethylmelamine. One hundred and twenty-three patients were randomized to each arm. Vogl reported all three outcomes of interest and demonstrated a difference favouring CC in each instance, the three response parameters and disease-free interval reaching significance.

The Bell study[11] compared AA (chlorambucil) and cisplatin plus cyclophosphamide in a trial which was a pilot study for the larger Tattersal co-operative trial. Nineteen women received AA and 17 were given CC. Response for measured disease significantly favoured CC. Progression-free interval and median survival also favoured CC, progression-free interval achieving marginal significance ($P<0.05$).

The abstracts by Wilbur[12] and Sturgeon[13] compared AA and CC (cisplatin, Adriamycin and cyclophosphamide). Both abstracts suffered from lack of information. Neither was apparently ever published in other than an American Society of Clinical Oncology meeting abstract. The Sturgeon abstract consisted of 38 patients in the AA arm and 40 in the CC group. Only complete response (non-measured) was reported, the trend favouring CC. Progression-free survival was statistically significant ($P<0.01$) but no survival advantage was observed. Wilbur compared 13 patients given cyclophosphamide plus a passive immunostimulant (BCG) to 11 women who were randomized to receive cisplatin and cyclophosphamide plus BCG. No response data were provided. No significant difference in median survival or progression-free interval was observed although both outcomes favoured AA.

32.3.1 RESPONSE

For patients with measured disease, complete response and complete plus partial response were evaluated. Overall response (complete plus partial response in patients with both measurable and non-measurable disease) was also examined. All of the studies employed standard response criteria with the exception that the period of observation used

Analysis

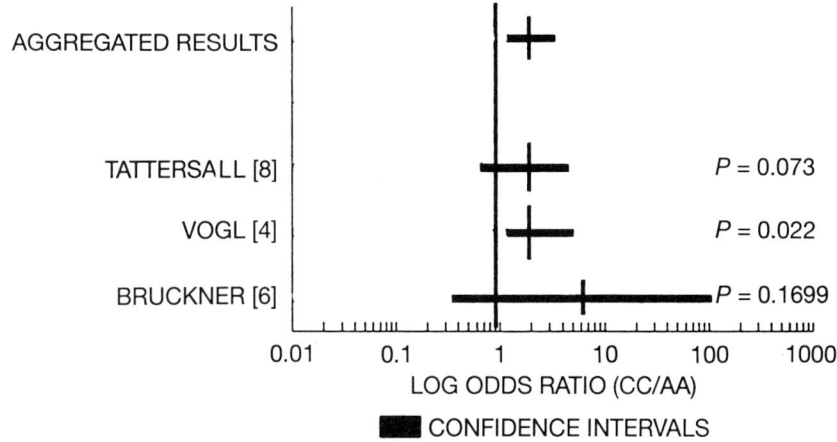

Figure 32.2 Data for complete response in patients with measurable disease (log odds ratios with the point estimates, confidence intervals and P values).

to define a response varied from 1 to 3 months. As a study endpoint, response in patients with measurable disease is preferred to response in patients with non-measurable disease since the latter includes (i) patients who may have been cured by surgery alone, and (ii) patients with small-volume microscopic disease who will likely develop progressive disease much later than patients who begin the observation period with clinically apparent disease.

Information for assessment of response was available from all of the studies except the abstracts by Wilbur and Sturgeon, which were not considered in the analysis of response because they lacked sufficient validity. In four of the remaining studies, information was provided about response in patients with measurable disease, although in some this was surgically staged data ('second look'), laparotomy and in others the method of measurement was not specified.

For the pooled analysis of response, odds ratio (OR) was used to compare the individual studies. The OR is a common statistical measure of association which can be used in both prospective and retrospective data analysis. It can also be used to estimate the magnitude of the difference between the two interventions. In studies where the event rate is small (low incidence), the OR closely approximates relative risk (RR) or risk reduction (1−RR). The null hypothesis of no difference of the pooled ORs (OR = 1) can be tested with a chi-square statistic. Rejection of the null hypothesis suggests that a patient's risk of developing the disease or outcome of interest is dependent to some extent on the group to which she has been allocated, once again emphasizing the need to limit the possibility of selection bias in trials by considering only blinded and randomized trials in the analysis.

The 'OR $2 \times 2 \times kA$' software program, written specifically for meta-analysis by the Department of Clinical Epidemiology and Biostatistics at McMaster University, Hamilton, was used to calculate the point estimators and test statistics for the 2×2 tables prepared from the seven studies. This software program calculates several statistics for each aspect of a meta-analysis. The author's preference for individual statistics in this analysis was determined *a priori*.

Of the several estimators of association calculated by the program, the corrected Mantel-Hansel statistic was preferred as the most accurate point estimator of the overall OR because it is consistent even with a large number of small studies and does not assume 'homogeneity' (consistent results across the trials) of the individual ORs. The logarithm of the ORs of the individual studies and their respective confidence intervals (two standard deviations) are demonstrated for each of the response criteria (Figures 32.2–32.4).

The ORs for the end-points of each study were tested for inter-study 'homogeneity'. A conservative level of significance ($P<0.01$) was chosen because tests of homogeneity generally are of low power, making it difficult to accurately detect study differences. These tests are based on the assumption of no interaction, i.e. the individual ORs of the studies are consistent. The Miettinen standardized estimate of the common OR was used to calculate the confidence intervals around the OR for the Carmo and Wilbur trials where homogeneity could not be assumed.

For complete response in patients with measured disease, three studies comprising a total of 583 randomized patients were available for analysis. The Vogl study alone claimed significance (OR did not include one). The aggregated OR of the three studies was 2.25 (1.32–3.70) which was also significant and therefore suggests that a significant response advantage does exist for this group of patients who were treated with cisplatin. The Rosenthal 'fail-safe n', a measure of the number of additional studies of similar methodological quality and an opposite conclusion which would have to exist to reverse the conclusion of the analysis, was five, suggesting that the result of this analysis is probably correct.

For complete plus partial response in patients with measurable disease, four studies were considered for analysis (620 patients). Each study favoured CC and the ORs for the Bell and Vogl trials were statistically significant. The aggregated result of the four studies was also significant (Figure 32.4). The 'fail-safe n' was 11.

Overall response was a compilation of all the response data presented in each trial. From a methodological perspective, the data across the studies are not uniform. All seven papers were considered to have sufficient

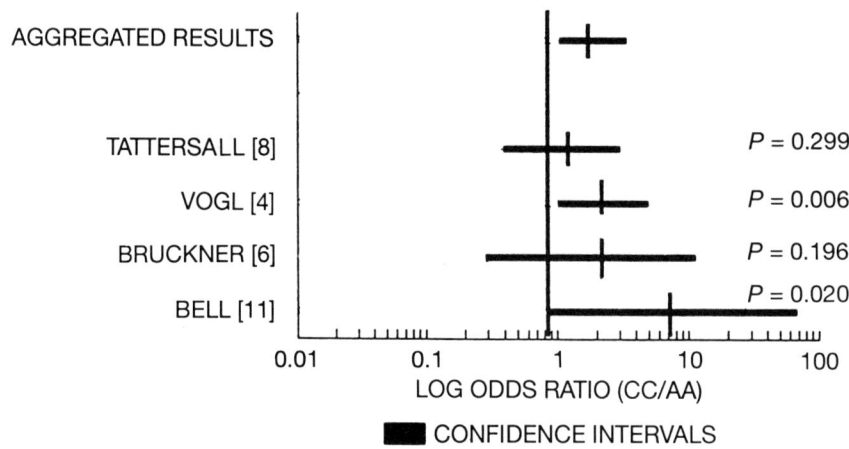

Figure 32.3 Data for response (complete + partial) in patients with measurable disease (log odds ratios).

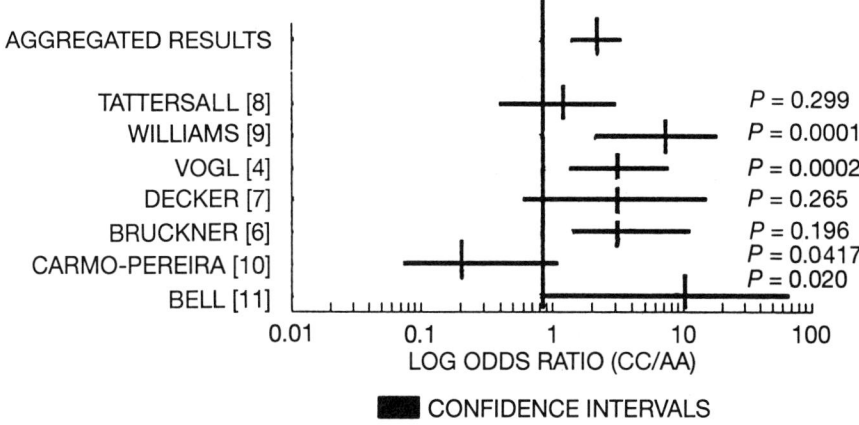

Figure 32.4 Data for overall response (measurable and non-measurable disease) (log odds ratios).

information to be included in the assessment of overall response. Only the Carmo paper suggested that the direction of the effect (response) favoured AA. The Bell, Bruckner, Vogl and Williams studies all reported a statistically significant difference favouring CC. The aggregated result also favoured CC and was statistically significant. The 'fail-safe n' was 29.

32.3.2 PROGRESSION-FREE INTERVAL

In some of the studies, progression-free interval was provided, and in others, disease-free interval was reported. These are not identical statistics. Disease-free interval is a measure of the time a patient is clinically free of tumour, whilst progression-free interval is a measure of absence of tumour growth in patients who may or may not have clinically apparent disease. Since these were randomized trials and systematic bias in the allocation of patients to treatment was therefore minimized, we have combined these two outcome parameters, even though they are not strictly identical, into a single 'progression-free interval' solely for the purposes of this analysis. All of the studies demonstrated a trend in 'progression-free interval' which favoured CC, except the Wilbur and Carmo studies which identified an opposite trend in favour of AA (Figure 32.5). A summary statistic for 'progression-free interval' was not calculated, given the methodological difficulty of combining these two similar but different outcome measures.

32.3.3 SURVIVAL

No information was available with regard to long-term survival, arguably the most useful and accurate outcome measure for this disease. Information about median survival was available from each of the seven studies and from the Wilbur abstract. The Wilbur and Carmo studies suggested a trend which favoured AA; however, all the remaining studies preferred CC, the Tattersall, Decker and Bruckner papers claiming that the difference was statistically significant (Figure 32.6). A summary statistic for median survival was calculated[3] and expressed as an OR (1.23) which did not achieve significance. This statistic strongly suggests that these two interventions (AA and CC) are not significantly different in terms of their ability

Meta-analysis of treatment for advanced ovarian cancer

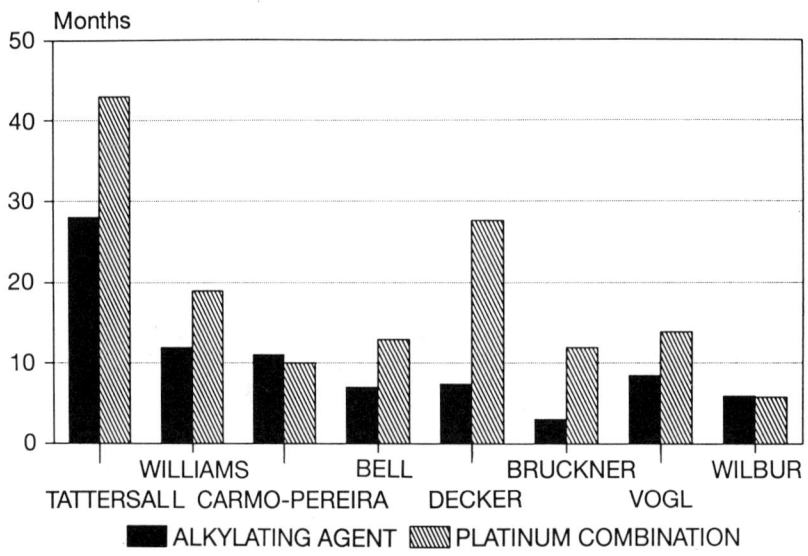

Figure 32.5 An inter-study and intra-study comprison of 'progression-free interval' by treatment.

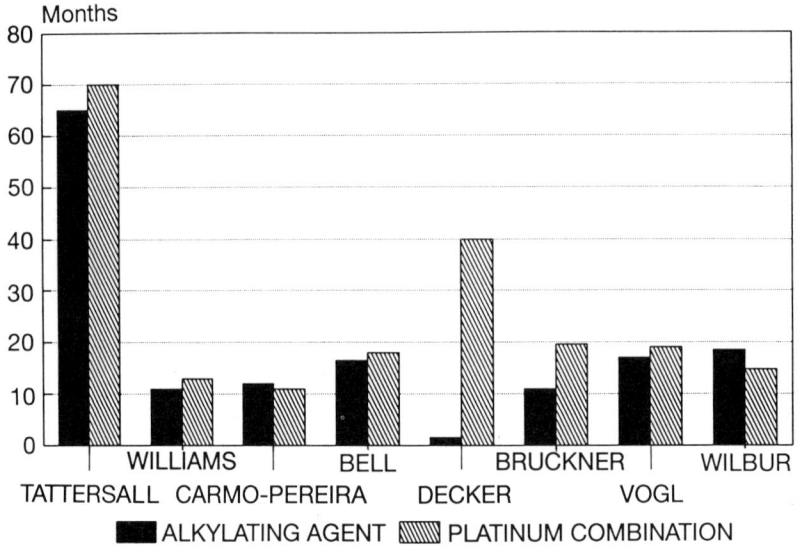

Figure 32.6 Inter-study and intra-study comparison (AA vs. CC) of median survival (all patients).

to improve short-term (2 years), and presumably long-term, survival for patients with advanced ovarian cancer. This conclusion is further strengthened when the 'co-intervention' which was commonplace for patients in the AA arm is also considered.

To test the rigour of the above conclusions, the survival data were next weighted by the sample size of each trial, on the assumption that larger trials should have a proportionately greater power to detect a treatment effect if one really did exist. The OR was 1.14

Analysis

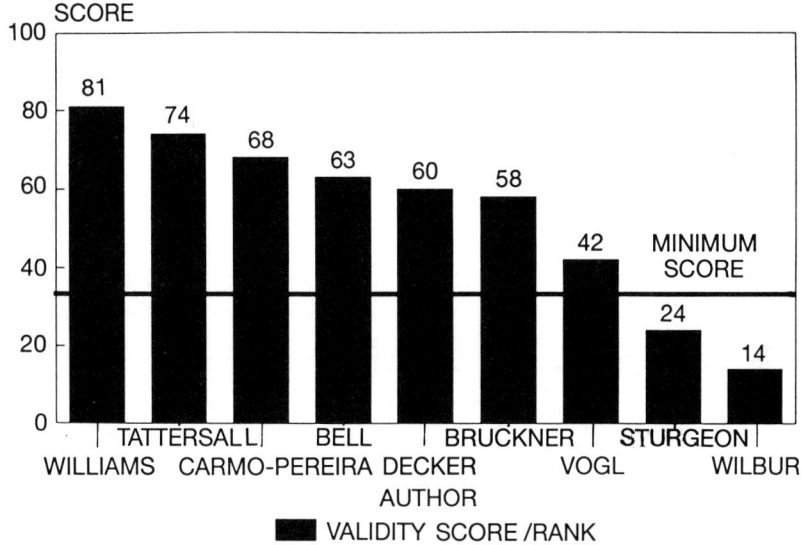

Figure 32.7 The results of the validity assessment by rank order with the minimum score (35).

but the confidence intervals included one implying non-significance[3].

The studies were then ranked according to their mean validity scores on the further assumption that this score represented a reasonable measure of the scientific merit of each trial. The Williams and Tattersall studies received the highest validity scores, a reflection of the overall quality of the studies' design and methodology. The Wilbur and Sturgeon abstracts did not receive sufficient marks for inclusion in the pooled analysis (Figure 32.7). These scores were then used to weight the analysis of median survival. Once again the OR favoured CC (1.28) but did not achieve significance (Figure 32.8).

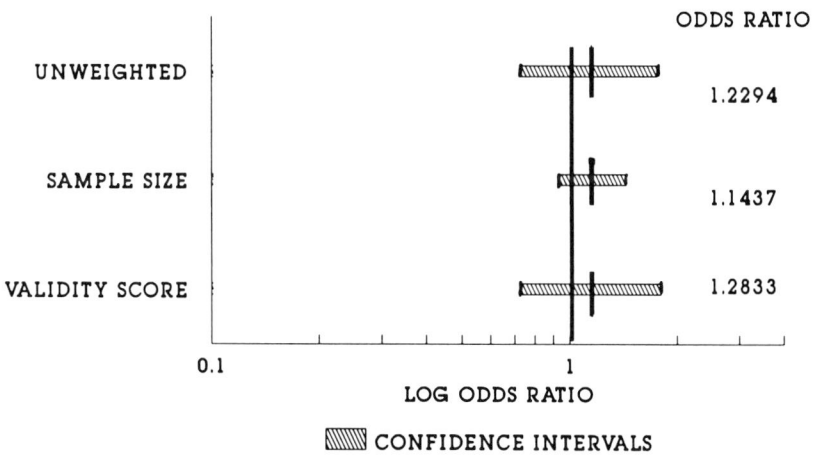

Figure 32.8 Median survival: log odds ratios of the unweighted and weighted comparison (AA vs. CC).

341

32.4 DISCUSSION

This review of the chemotherapeutic treatment of advanced ovarian cancer has attempted to address the question of whether initial therapy should consist of a single AA or cisplatin-based combination (CC) chemotherapy. The method of analysis, a meta-analysis, has been conducted by undertaking an exhaustive retrospective review of the published and unpublished literature on this topic.

Following a methodical search, seven papers and two abstracts were identified. A committee of local experts predefined a set of relevance and validity criteria which were used to both validate and rank the nine studies. The two abstracts failed to meet the minimum criteria for inclusion and were not considered in the pooled analysis of response. The remaining studies have demonstrated a statistically significant difference in complete and total response for patients with measurable disease, and a similar result for overall response (complete and partial) in patients with both measurable and non-measurable disease.

For progression-free interval and median survival, this analysis has demonstrated a trend favouring cisplatin-based combination chemotherapy but neither achieved statistical significance. Two trials (Wilbur and Carmo) suggested a contrary effect (AA preferred). When an attempt was made to partially correct the summary statistic for median survival based on methodological quality by weighting the OR on the sample size and the validity score, the analysis still favoured CC but still failed to achieve significance.

'Co-intervention' or the subsequent use of the alternative treatment for some patients enrolled in a clinical trial is something which must be considered when analysing data on the patients who received primary AA. In excess of 85% of the patients will eventually fail primary therapy and will invariably be treated with second or third line regimens containing cisplatin either immediately after AA, or at the time of symptomatic recurrence. The real question in this meta-analysis therefore is not AA vs. cisplatin (CC), but cisplatin as primary treatment vs. AA followed by cisplatin for recurrence.

In conclusion, the case for cisplatin-based combination chemotherapy as primary treatment for all women with advanced ovarian cancer remains unproven. The initial tumour response to cisplatin therapy is greater than that observed with AAs but the clinically important outcome measure, long-term survival, does not appear to be different, as inferred from the median survival data in this study. Almost all patients will receive cisplatin at some time in the course of their treatment for ovarian cancer.

For some patients with little chance of long-term remission, such as those with suboptimally debulked advanced disease or the elderly and medically infirm, primary AA or other relatively non-toxic interventions followed at the time of symptomatic recurrence with more aggressive and toxic treatments such as cisplatin may offer the most 'cost-effective' form of therapy. Given the newer serotonin antagonists, the significant nausea seen in patients on cisplatin has been very much diminished, and the newer anti-emetics to some extent modify this analysis' conclusions since moderate dose cisplatin can now be administered with much less significant toxicity. However, for high dose cisplatin, diminished quality of life will remain a very important issue. Based on this analysis, it is suggested that although a survival advantage may be demonstrated for these women, it is not in their interest to continue to use a treatment which carries such a high subjective 'cost'. We must not be seduced by rates of response in these patients even though response is much easier to measure and the information is available more rapidly than long-term survival data, which may take many years to mature but which are substantially more meaningful. Unless these patients have significant symptoms related to their tumour, response is of little value for these people, particularly given that to achieve this response, more drug and much more toxicity can usually be anticipated. We must learn to reliably measure the quality and the quantity of the life remaining for these patients and make informed recommendations to them based on both of these important measures of patient well-being and response to treatment.

REFERENCES

1. Wiltshaw, E., Subramarian, S., Alexopoulos, C. *et al.* (1979) A summary of experience with cis-dichlorodiammine platinum (II) at the Royal Marsden Hospital. *Cancer Treat. Rep.*, **63**, 1545–9.
2. Simes, R. (1985) Treatment selection for cancer patients: application of statistical decision theory to the treatment of advanced ovarian cancer. *J. Chronic Dis.*, **38**, 171–86.
3. Simes, R. (1987) Confronting publication bias: a cohort design for meta-analysis. *Stat. Med.*, **6**, 11–29.
4. Vogl, S., Pagano M., Davis T. *et al.* (1983) Platinum-based combination chemotherapy versus melphalan for advanced ovarian cancer. Proceedings Thirteenth International Congress Chemotherapy, Vienna.
5. Chalmers, T. Harvard School of Public Health (based on work supported by a National Library of Medicine grant no. LM3116) (personal communication).
6. Bruckner, H.W., Cohen, C.J., Goldberg, J.D. *et al.* (1981) Improved chemotherapy for ovarian cancer with cis-diamminedichloroplatinum and adriamycin. *Cancer*, **47**, 2288–94.
7. Decker, D.G., Fleming, T.R., Malkasian, G.B. and Webb, M.J. (1982) Cyclophosphamide plus cis-platinum in combination: treatment program for stage III or IV ovarian carcinoma. *Obstet. Gynecol.*, **60**, 481–7.
8. Gynaecological Oncology Group of Australia and the Sydney branch of the Ludwig Institute for Cancer Research (1986) Chemotherapy of advanced ovarian adenocarcinoma: a randomised comparison of combination versus sequential therapy using chlorambucil and cisplatin. *Gynecol. Oncol.*, **23**, 1–13.
9. Williams, C., Mead, G., Macbeth, F. *et al.*, Cisplatin combination chemotherapy versus chlorambucil in advanced ovarian carcinoma: mature results of a randomised trial, *J. Clin. Oncol.* **3**, 1455–62.
10. Carmo-Pereira, J., Costa, F.O. and Henriques, E. (1983) Cis-platinum, adriamycin and hexamethylmelamine versus cyclophosphamide in advanced ovarian carcinoma. *Cancer Chemother. Pharmacol.*, **10**, 100–3.
11. Bell, D., Woods, R., Levi, J. *et al.* (1982) Advanced ovarian cancer: a prospective randomised trial of chlorambucil versus combined cyclophosphamide and cis-diamminedichloroplatinum. *Aust. N. Z. J. Med.*, **12**, 245–9.
12. Wilbur, D., Rentschler, R., Wagner, R. *et al.* (1983) Randomised trial of the addition of cis-diamminodichloroplatinum (DDP) and/or BCG to cyclophosphamide (CTX) chemotherapy for ovarian carcinoma. *Proc. Am. Soc. Clin. Oncol.*, **2**, 147.
13. Sturgeon, J., Fine, S., Gospodarowicz, M. *et al.* (1982) A randomised trial of melphalan alone versus combination chemotherapy in advanced ovarian cancer. *Proc. Am. Soc. Clin. Oncol.*, **1**, 108.

Chapter 33
Impact of maximal cytoreductive surgery on survival in advanced ovarian cancer

W.P. SOUTTER, R.W. HUNTER
and N.D.E. ALEXANDER

33.1 A DEFINITION OF MAXIMAL CYTOREDUCTIVE SURGERY

Maximal cytoreductive surgery (MCS) may be defined as surgery which aims to remove all visible tumour or, if that is not feasible, to leave the smallest residuum possible. The maximum size of residuum that is acceptable varies between studies. In some patients with advanced ovarian cancer MCS may be achieved with reasonable ease but others may require small or large bowel resection, colostomy, partial cystectomy, splenectomy, removal of lymph node masses, or resection of involved diaphragm[1,2]. Routine, radical pelvic and para-aortic lymphadenectomy which aims to remove all lymph nodes in these regions has been advocated as a part of MCS[3].

33.2 THE RATIONALE BEHIND MCS

This approach appears to breach the principle of cancer surgery, which requires that the tumour be removed intact with an adequate margin of normal tissue. However, it is believed that by reducing the mass of residual disease the malignancy is made more sensitive to chemotherapy.

33.3 STUDIES WITHOUT CONTROLS

There is a wealth of data from uncontrolled studies which show that women with minimal residual disease at the end of primary surgery have a much better prognosis than those who do not[1–6]. All of these studies suffer from selection bias. The patients were all operated upon by surgeons with the same objective in mind. Indeed, in some studies the patients were all operated upon by the same surgeon[4]. The fact that some women ended up with more residual disease than others indicates that these women or their tumours were different. That difference implies a worse prognosis from the outset. Hence comparisons of the effect of any treatment on the outcomes of these different groups of patients are invalid. However, just because these data do not prove the value of surgery does not imply that the surgery is ineffective.

33.4 A NON-RANDOMIZED STUDY WITH CONTROLS

Only one retrospective study has attempted to include controls, using for this women who had minimal amounts of metastatic disease prior to surgery[1]. The survival of this group was compared with that of a cohort who had minimal residual disease as a result of surgical effort. The survival curves were identical. However, the surgical effort required to effect MCS in this group was probably not great as these patients were all treated before MCS was widely practiced. Others have found that women whose tumour was reduced to 2 cm or less in diameter by simple tumour removal had a similar median survival time (MST) to those who required extensive tumour reductive surgery to achieve the same end[7]. However, in that study the patients were treated with melphalan and only the grade of the tumour proved to have an independent, significant effect on survival. Tumour residuum did not.

33.5 EVIDENCE AGAINST SURVIVAL BENEFIT

Women with large metastatic tumours have a poor prognosis in spite of optimal cytoreduction[2]. When bowel resection is required to achieve MCS the prognosis is very poor[8]. Similarly, patients requiring resection of bladder or ureter do badly, having an MST of only 12 months[9]. This is reduced to 7 months if the pre-operative intravenous urogram shows obstruction. All of these studies illustrate that the higher the price which has to be paid in achieving MCS, the less is the apparent benefit to the patient. This suggests that the women who do well after MCS have less aggressive, more easily resected tumours and an intrinsically better prognosis.

33.6 A META-ANALYSIS OF THE EFFECT OF MCS ON MEDIAN SURVIVAL TIME

In the absence of adequate controlled trials, meta-analysis of the available data can provide useful information about difficult clinical problems and there are several examples of meta-analysis revealing important therapeutic information[11,12].

33.6.1 RATIONALE

This meta-analysis is unusual in that it does not use previously published controlled studies of MCS because no such studies exist. Instead, it depends upon the simple principle that, all other things being equal, the more often an effective therapy is employed in a population the better the survival should be in that population.

Rates of MCS vary widely in published studies. While this may represent patient selection, it is far more likely to result from different management policies. If patient selection plays an important part in this variation, any beneficial effects of MCS will be exaggerated because selection of patients with small volume or easily resected disease will increase the MCS rate.

If MCS does improve the survival prospects of women with advanced ovarian cancer, the overall survival rate should be higher in studies with a high rate of MCS (%MCS) than in those where MCS was less frequently achieved. However, other factors may affect survival and their impact must be calculated to determine the part played by MCS.

33.6.2 OTHER IMPORTANT VARIABLES

Chemotherapy is likely to have an effect on survival rates. The most potent single agents are the platinum drugs and women treated with platinum agents might be expected to have a better survival than those who are not. In a similar way, patients treated with more intensive regimens might have a better outlook. A formula for expressing the intensiveness of a regimen is the dose intensity (DI) and this has been shown to correlate with MST[13]. Other variables which might well have a bearing on the outcome are the percentage of FIGO stage IV patients (%IV) [6,7], and the year the results were published as a reflection of possible improvements in general management. By calculating the size of the effect of these variables on MST it becomes possible to determine any residual relationship between %MCS and MST.

33.6.3 METHOD

A literature search was performed on MEDLINE from 1976 to 1989, and in Index Medicus from 1967 to 1989. Unpublished data from the third North Thames Ovarian Cancer Trial and from a multicentre study run by the Cancer Research Campaign Trial Office in Glasgow were included.

Studies composed of patients with predominantly stage III and stage IV ovarian cancer were chosen if the overall MST and proportion of patients undergoing MCS were shown. MCS was defined in different ways and a note was made of the definition used. DI was calculated as previously described[13].

The effect on MST of the following six variables was investigated by multiple linear regression: percentage of women with stage III disease (%III); percentage of women with stage IV disease (%IV); percentage of women who underwent MCS (%MCS); DI; year of publication; and whether or not the patients received platinum-containing chemotherapy.

33.6.4 PRECAUTIONS TAKEN

The regression was weighted by the number in each cohort. Because the variability of MST was found to increase with MST itself, the data was transformed logarithmically. Anscombe's test[14] was used to confirm objectively that this problem had been resolved and the Shapiro-Francia W' test[15] confirmed that the data were now normally distributed. It is unlikely that publication bias has affected the conclusions of this analysis as most of the studies included were investigating chemotherapy rather than surgery. If publication bias does exist it would tend to exaggerate any beneficial effects of surgery.

33.6.5 RESULTS

The results are reported in detail elsewhere[10]. The search identified 58 studies with 97 cohorts and 6962 patients. DI could be calculated in 76 of the cohorts. There were no interactions between the variables and no non-linear effects. Leverage did not bias the results.

The single most important variable associated with increased MST was the use of platinum chemotherapy (Figure 33.1). This is similar to the results of another, recently published meta-analysis[12] but others have failed to detect this difference (Chapters 31 and 32). This is probably due to the option in controlled trials of crossing patients over to the platinum arm when they relapse on alkylating therapy. The clear advantage for platinum in this analysis may be because platinum was not available to be used as salvage therapy in many of these studies. Even when platinum is used as salvage therapy, there is an early advantage to the platinum group and the curves only converge at 6 years (Chapter 31).

Increasing DI was associated with a modest improvement in MST. However, this analysis covers a wide time span and a large range of DIs. It is possible that this finding may not be relevant to modern practice, However, it does show that this was an appropriate way

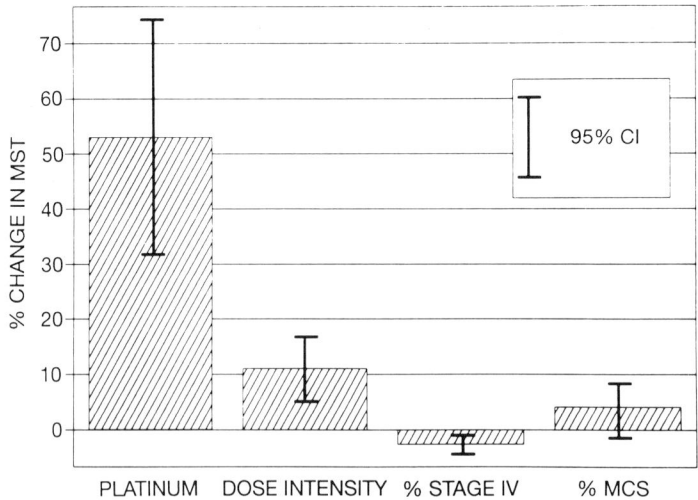

Figure 33.1 Results of multiple linear regression analysis showing the percentage change in median survival time (MST) and the 95% confidence intervals (CI) associated with the different variables. %MCS, percentage of patients undergoing maximum cytoreductive surgery.

of controlling for possible differences in the effects of different regimens.

Increasing the %MCS had only a very small beneficial effect on MST. This remained true regardless of the precise definition of MCS, even when MCS was described as 'no residual disease'. If the effect of DI was ignored and data from all 97 cohorts were used, the effect of increasing the %MCS became statistically significant ($P = 0.014$) but the benefit remained small (6.1% increase in MST for each 10 point increase in %MCS).

33.6.6 CONCLUSIONS

This analysis suggests that MCS is unlikely to have a major impact upon MST of women with advanced ovarian cancer. The use of platinum chemotherapy is of greater importance. This study does not address the effects of surgery upon quality of life nor on either short-term or long-term survival.

33.7 RELIEF OF SYMPTOMS BY MCS

Evidence suggesting that cytoreductive surgery provides a good quality of life may also be criticized for being non-randomized[16]. It seems likely that the short-term palliation achieved in that study could equally well be obtained in most patients by removing the main tumour masses without the extensive dissection, bowel surgery or lymphadenectomy often undertaken in an effort to remove all visible tumour in MCS.

33.8 COMMENT

In spite of the attractive hypothesis underlying the use of MCS, there is little substantive evidence to suggest that the survival of patients with advanced ovarian cancer is improved. The meta-analysis indicates that MCS has, at best, only a small effect on MST. It remains possible that longer or shorter term survival may be influenced by MCS.

While it is unlikely that MCS makes a major contribution to the survival of these unfortunate women, cytoreductive surgery does provide short-term relief of symptoms. To achieve this limited objective, it is neither necessary nor appropriate to undertake extensive surgery to remove all visible tumour. A more limited objective of excision of the main tumour masses is likely to be as effective but with less morbidity.

REFERENCES

1. Griffiths, C.T.H. and Fuller, A.F. (1978) Intensive surgical and chemotherapeutic management of advanced ovarian cancer. *Surg. Clin. N. Am.*, **58**, 131–42.
2. Hacker, N.F., Berek, J.S., Lagasse, L.D. *et al.* (1983) Primary cytoreductive surgery for epithelial ovarian cancer. *Obstet. Gynecol.*, **61**, 413–20.
3. Burghardt, E., Lahousen, M. and Stettner, H. (1989) The role of lymphadenectomy in ovarian cancer, in *Ovarian Cancer. Biological and Therapeutic Challenges* (eds F. Sharp, W.P. Mason and R.E. Leake), Chapman & Hall, London, pp.425–33.
4. Griffiths, C.T. (1987) Carcinoma of the ovary: surgical objectives, in *Ovarian Cancer – the Way Ahead* (eds F. Sharp and W.P. Soutter), Royal College of Obstetricians and Gynaecologists, London, pp.235–44.
5. Hainsworth, J.D., Grosh, W., Burnett, L.S. *et al.*, (1988) Advanced ovarian cancer: long-term results of treatment with intensive cisplatin-based chemotherapy of brief duration. *Ann. Intern. Med.*, **108**, 165–70.
6. Louie, K.G., Ozols, R.F., Myers, C.E. *et al.* (1986) Long-term results of a cisplatin-containing combination chemotherapy regimen for the treatment of advanced ovarian carcinoma. *J. Clin. Oncol.*, **4**, 1579–85.
7. Wharton, J.T. and Herson, J. (1981) Surgery for common epithelial tumors of the ovary. *Cancer*, **48**, 582–9.
8. Webb, M.J. (1989) Cytoreduction in ovarian cancer: achievability and results, in *Ballière's Clinical Obstetrics and Gynaecology; Operative Treatment of Ovarian Cancer* (eds. E. Burghardt and J.M. Monaghan), Ballière Tindall, pp.83–94.
9. Berek, J.S., Hacker, N.F., Lagasse, L.D. and Leuchter, R.S. (1982) Lower urinary tract resection as part of cytoreductive surgery for ovarian cancer. *Gynecol. Oncol.*, **13**, 87–92.
10. Hunter, R.W., Alexander, N.D.E. and Soutter, W.P. Meta-analysis of surgery in advanced ovarian carcinoma: is maximum cytoreductive surgery an independent determinant of prognosis? *Am. J. Obstet. & Gynecol.* (in press).
11. Chalmers, T.C., Levin, H., Sacks, H.S. *et al.* (1987) Meta-analysis of clinical trials as a scientific discipline. I: control of bias and comparison with large co-operative trials. *Stat. Med.*, **6**, 313–25.
12. Peto, J. and Easton, D. (1990) Randomised cancer trials – past failures, current progress and future prospects? *Cancer Surv.*, **8**, 511–33.
13. Levin, L. and Hryniuk, W.M. (1987) Dose intensity analysis of chemotherapy regimens in ovarian carcinoma. *J. Clin. Oncol.*, **5**, 756–67.
14. Wetherill, G.B. (1986) *Regression Analysis with Applications*, Chapman & Hall, London.
15. Royston, J.P. (1983) A simple method for evaluating the Shapiro-Francia W' test for non-normality. *The Statistician*, **32**, 297–300.
16. Blythe, J.G. and Wahl, T. (1982) Debulking surgery: does it increase the quality of survival? *Gynecol. Oncol.*, **14**, 396–406.

Chapter 34

Management and outcome of stage III epithelial ovarian cancer

N.F. HACKER, G.V. WAIN and J.P. TRIMBO

34.1 INTRODUCTION

The management of patients with advanced ovarian cancer is one of the major challenges confronting members of a gynaecological oncology department. The initial surgery is often difficult for the surgeon, cisplatin-based chemotherapy is usually distressing for the patient, the psychological sequelae of the diagnosis and treatment are onerous for all concerned, and the ultimate prognosis is poor. Not surprisingly, philosophies of management often differ significantly from one clinician to another.

At the Gynaecological Oncology Department of the Royal Hospital for Women, we have endeavoured to pursue an aggressive policy in all medically fit patients with advanced ovarian cancer. Primary cytoreductive surgery is performed at initial presentation, even laparotomy and biopsy have been performed elsewhere before referral. Chemotherapy with cisplatin and cyclophosphamide is offered as first-line chemotherapy, and the first cycle is usually given prior to discharge on about the tenth postoperative day. Patients who have no clinical or tumour-marker evidence of disease are offered second-look laparotomy after six cycles of chemotherapy, and a variety of second-line treatments have been offered to patients with persistent disease.

We have recently reviewed our experience with stage III ovarian cancer to determine the outcome of this approach, and to evaluate the need for modification of this approach in some patient subsets.

34.2 MATERIALS AND METHODS

Between November 1986 and August 1990, 122 patients with primary invasive epithelial ovarian cancer were referred to the Gynaecological Oncology Department at the Royal Hospital for Women in Sydney. Of these patients 59 had stage III disease and had their initial cytoreductive operation performed in our unit; these patients form the basis for this study. None of these patients had stage III disease on the basis of surgical staging.

All clinical, surgical and pathological data were prospectively entered into a computer data bank, and were retrieved for this review. Medical records were also reviewed on all

Table 34.1 Residual disease following primary cytoreductive surgery

Residual disease	Number	Percentage
No residual	7	11.9
≤ 10 mm	25	42.4
11–20 mm	19	32.2
≥ 20 mm	8	13.5

cases to verify the recorded data. Follow-up was available on all patients. Kaplan-Meier survival curves were constructed and significance determined by rank analysis.

The age of the patients ranged from 35 to 86 years with a median age of 63 years. In 47 patients, the primary laparotomy was performed at the Royal Hospital for Women. In five patients, the initial laparotomy and biopsy was carried out elsewhere and the patient referred prior to initiation of chemotherapy. In seven cases, the patient received up to three cycles of chemotherapy prior to referral and cytoreductive surgery.

In 14 patients (23.7%) the ovaries were not enlarged, but the surface was studded with small tumour nodules. These cases were considered to be primary peritoneal carcinomas. Histologically 48 patients had serous carcinomas, three mucinous, three endometrioid, three clear cell and there were two undifferentiated tumours. The histology was not reviewed specifically, but four patients were reported to have Grade 1 carcinomas, 13 Grade 2 and 42 Grade 3.

The extent of residual disease following cytoreductive surgery is shown in Table 34.1. Bowel resection was performed in 11 patients (18%). This included small bowel resection in six patients, and large bowel resection in seven.

Complications of the primary cytoreduction were recorded in 10 patients (17%). These included wound infection (4), necrotizing fasciitis (1), pulmonary embolism (2), pneumonia (1), intraperitoneal haemorrhage (1), and septicaemia (1). There was one postoperative death (from pulmonary embolism) for an operative mortality of 1.6%. One patient required re-laparotomy because of haemorrhage.

Throughout the study period, bulky retroperitoneal lymph nodes were resected. However, during the past two years, we have performed pelvic lymphadenectomy at least on the side of the tumour, more commonly, even if the lymph nodes were not obviously enlarged. Twenty-seven patients had at least a unilateral pelvic lymphadenectomy, while 19 patients had resection of enlarged nodes. Positive lymph nodes were identified in 22 patients (37.2%) with the number of positive nodes ranging from 1 to 32 (mean 6).

Following laparotomy, the first cycle of chemotherapy was given prior to discharge from hospital, usually on postoperative day 10–12. Planned first-line chemotherapy was cisplatin 75 mg/m^2 and cyclophosphamide 750 mg/m^2 and this was administered to 29 patients. In frail patients, carboplatin 400 mg/m^2 was often substituted for cisplatin (nine patients). Because many patients were referred from country areas where optimal facilities for administration of chemotherapy were not available, alternative regimens usually alkylating agent therapy, were used in 19 patients. Two patients received no postoperative therapy.

Second-look laparotomy was performed on 21 patients. All of these patients were clinically free of disease and had negative CA-125 titres. Eight patients (38.0%) had no histological evidence of disease. Of the eight patients with a negative second-look, four have recurred with a median follow-up of 24.3 months.

Patients with a positive second-look had as much tumour resected as possible, and then received a variety of second-line therapies, including carboplatin and etoposide, tamoxifen, intraperitoneal cisplatin, and high-dose chemotherapy (carboplatin and etoposide)

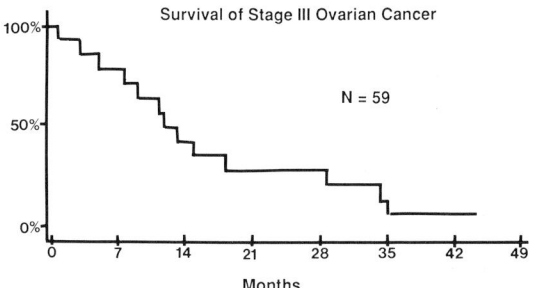

Figure 34.1 Overall survival for stage III epithelial ovarian cancer.

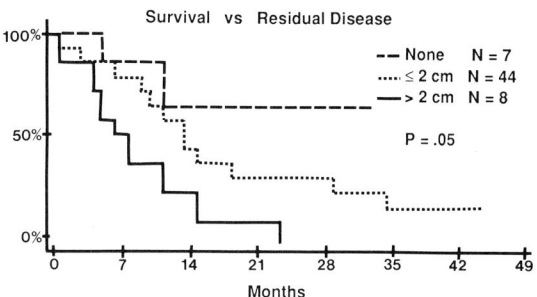

Figure 34.3 Survival vs. residual disease.

with autologous bone marrow transplantation.

34.3 RESULTS

At the time of writing, 39 of the 59 patients are dead of disease. The median time to death was 11 months, with a range of 5 days to 35 months. The median duration of follow-up for the 20 survivors is 21 months, with a range of 8–45 months.

The overall survival for the 59 patients is shown in Figure 34.1. The median survival was 13 months. Only one patient followed for at least 40 months was still alive. Survival for patients with primary peritoneal carcinoma was identical to that of patients with stage III ovarian cancer (Figure 34.2).

Survival by residual disease is shown in Figure 34.3. The median survival for the seven patients with no residual disease will exceed 36 months, while the median survival for patients with greater than 2-cm residual disease was 7 months ($P<0.02$). Thirty patients received both optimal cytoreduction and cycles of platinum combination chemotherapy. These patients were considered to have had 'optimal' therapy. Their median survival was 34 months, compared to 9 months for the other 29 patients ($P<0.001$) (Figure 34.4). Reasons for suboptimal therapy included progression on platinum-containing therapy (seven patients), non-platinum-containing therapy (19 patients), death from myelosuppression following the first cycle of cisplatin and cyclophosphamide (one patient) and early demise (two patients).

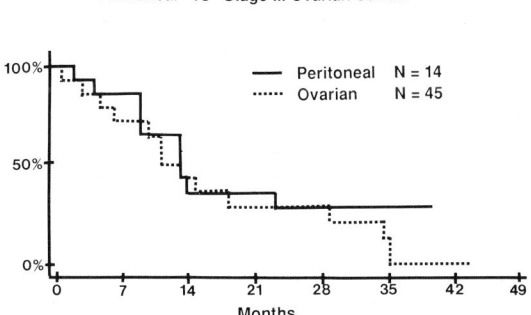

Figure 34.2 Survival for stage III ovarian carcinoma vs. primary peritoneal carcinoma.

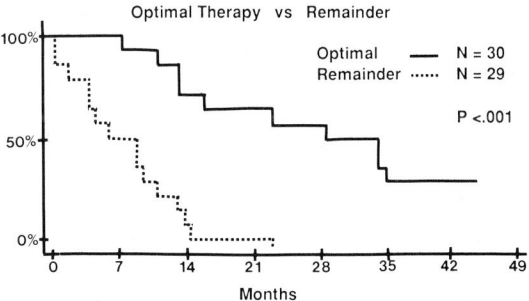

Figure 34.4 Survival for patients having optimal cytoreduction and six cycles of platinum combination chemotherapy ($n = 30$) vs. remainder of the group ($n = 29$).

For the entire group, the median survival for the 27 patients having a pelvic lymphadenectomy was 34 months, compared to 11 months for the 32 patients having nodal debulking or no lymph node resection ($P <0.03$). For the group of 30 patients considered to have received optimal therapy, 16 had a pelvic lymphadenectomy and their median survival was 35 months, compared to 10 months for the 14 patients who had nodal debulking or no nodal resection ($P = 0.07$).

34.4 DISCUSSION

The results of this study underline the grim survival rate for patients with advanced ovarian cancer. In spite of attempted aggressive initial surgery, the median survival was only 13 months, with 39 of the 59 patients already dead of disease. In fact, only one patient followed for at least 40 months is still alive. This patient, who had a negative second-look laparotomy, was followed with CA-125 titres, and at 40 months she developed rising titres (to 250 u/ml). She responded to carboplatin and etoposide as second-line therapy, and remains alive and clinically and tumour-marker free of disease at 46 months. Although the median survival is shorter than usually reported in chemotherapy trials, it probably more accurately reflects the outcome for this disease, because all patients presenting to the unit with stage III disease were included in the study, regardless of age, medical condition, or early demise.

The poor long-term survival is consistent with other reports which have suggested that although cisplatin-based chemotherapy increases response rates and survival times, there are very few patients still alive at 5 years[1,2]. Superiority over alkylating agent therapy has not been clearly demonstrated in terms of long-term survival. However, in a disease in which the treatment should be considered palliative, the median duration of good quality life is a more important index of the effectiveness of therapy than is the 5-year survival.

Much of the poor overall median survival was related to the inability to optimally debulk the disease (eight patients) and the tumour's inherent resistance to cisplatin and cyclophosphamide (eight patients). The inability to administer intensive chemotherapy postoperatively because of the patient's age, medical status or residence in a remote part of the state, and the early postoperative demise of three patients because of pulmonary embolism, rapid tumour progression, and death from myelosuppression, were also significant factors. For the subset of 30 patients who received both optimal cytoreduction and six cycles of postoperative chemotherapy, the median survival of 34 months was encouraging.

Adequate initial surgery remains the linchpin of treatment, and this study confirms earlier reports that in a specialized referral unit about 85% of patients can have their disease cytoreduced to individual nodules no greater than 2 cm in diameter, with acceptably low morbidity[3,4]. However, it is apparent that long-term survival (and possible cure) is likely only in patients in whom all macroscopic disease can be removed. This is feasible in a minority of patients only (12% in this series) and is a reflection of the biology of the tumour, rather than the technical skill of the surgeon. Although the ultimate prognosis for advanced ovarian cancer appears to be predetermined and related to the extent of metastatic disease[5], additional disease-free survival will accrue from aggressive surgical debulking. In addition, the quality of live will usually be improved by removal of the pelvic disease and omental metastases[6].

Throughout the study period, all bulky retroperitoneal lymph nodes were resected. However, for the last 2 years of the study, following the report of Burghardt and colleagues[7], we have usually performed a formal pelvic lymphadenectomy on patients with optimally cytoreduced disease. Our overall incidence of positive lymph nodes was 37%, which is consistent with that reported by Chen[8], but only about half the incidence reported by Burghardt's group. The higher incidence reported by the Graz group may reflect a more diligent search for micrometastases by multiple sectioning of all nodes in their gynaecological pathology laboratory.

The role of lymphadenectomy in the surgical management of patients with ovarian cancer remains to be clarified, but the results in the group of 30 patients having optimal therapy suggest that a randomized prospective study of lymphadenectomy in optimally cytoreduced patients would be justified. Although any benefit in patients with macroscopic residual disease is likely to be only in terms of progression-free interval and median survival, it is possible that some survival advantage may be conferred on the group of patients with no residual disease following primary cytoreduction.

This study confirms that in patients with Stage III disease, no more than 40% will have a negative second-look laparotomy, in spite of negative tumour markers[9] and at least 50% of these patients will eventually recur[10]. Hence, there is adequate justification for consolidation therapy in patients with a negative second-look laparotomy, and randomized prospective studies such as the current EORTC study of intraperitoneal cisplatin vs. no additional treatment should be encouraged.

The prognosis for the group having suboptimal initial surgery is so poor that less toxic chemotherapy should be considered. However, many of these patients also have gross ascites, and our feeling is that effusions are controlled more rapidly and more effectively with cisplatin-containing regimens, and hence the quality of life is usually improved. Quality of life issues for patients having either aggressive surgery or chemotherapy for advanced ovarian cancer have been inadequately studied up to the present time.

In view of these results, where should we be heading with advanced ovarian cancer? Clearly, earlier diagnosis is of fundamental importance, and the continued search for more sensitive tumour markers is essential. All patients should be centrally referred, and all should be entered onto study protocols so that new approaches can be tried. Surgically, the role of retroperitoneal lymphadenectomy should be adequately defined by a multi-institutional randomized study. Medically, with no new drugs on the horizon, innovative approaches with previously available drugs are clearly justified. High dose chemotherapy with autologous bone marrow transplantation has been mainly used in salvage situations previously, and it may be justifiable to introduce it earlier in the course of the disease. For example, if tumour markers return to normal within three cycle of chemotherapy, high dose chemotherapy and autologous marrow transplantation may be of value at that stage. Finally, quality of life issues have been inadequately addressed, and studies to evaluate the effect of aggressive surgery and chemotherapy on quality of life should be initiated.

REFERENCES

1. Demob, A.J. (1986) Controversy over combination chemotherapy in advanced ovarian cancer. What we learn from reports of matured data. *J. Clin. Oncol.*, **4**, 1573–6.
2. Louie, K.G., Ozols, R.F., Myers, C.E. *et al.* (1986) Long-term results of a cisplatin containing combination chemotherapy regimen for the treatment of advanced ovarian carcinoma. *J. Clin. Oncol.*, **4**, 1579–85.
3. Griffiths, C.T., Parker, L.M. and Fuller, A.F. Jr. (1979) Role of cytoreductive surgical treatment in the management of advanced ovarian cancer. *Cancer Treat. Rep.*, **63**, 235–40.
4. Heintz, A.P.M., Hacker, N.F., Berek, J.S. *et al.*

(1986) Cytoreductive surgery in ovarian carcinoma: feasibility and morbidity. *Obstet. Gynecol.*, **67**, 783–8.
5. Hacker, N.F., Berek, J.S., Lagasse, L.D. *et al.* (1983) Primary cytoreductive surgery for epithelial ovarian cancer. *Obstet. Gynecol.*, **61**, 413–20.
6. Blythe, J.G. and Wahl, T.P. (1982) Debulking surgery. Does it increase the quality of survival? *Gynecol. Oncol.*, **14**, 396–408.
7. Burghardt, E., Pickel, H., Lahousen, M. and Stettner, H. (1986) Pelvic lymphadenectomy in operative treatment of ovarian cancer. *Am. J. Obstet. Gynecol.*, **155**, 315–19.
8. Chen, S.S. and Lee, L. (1983) Incidence of paraaortic and pelvic lymph node metastases in epithelial carcinoma of the ovary. *Gynecol. Oncol.*, **16**, 95–100.
9. Berek, J.S., Hacker, N.F. and Lagasse, L.D. (1984) Second-look laparotomy in Stage III epithelial ovarian cancer: clinical variables associated with disease status. *Obstet. Gynecol.*, **64**, 207–19.
10. Cain, J.M., Saigo, P.E., Pierce, V.K. *et al.* (1986) A review of second-look laparotomy for ovarian cancer. *Gynecol. Oncol.*, **23**, 14–25.

Chapter 35

The role of surgery in epithelial ovarian cancer

D. LUESLEY, C. FINN and R. VARMA

35.1 INTRODUCTION

The role of surgery is a frequently debated issue in epithelial ovarian cancer. This, in part, testifies to the central role of surgery in the management of this disease but also reflects the underlying doubt concerning its use.

The thrust of most, if not all, discussion is the requirement for appropriate primary laparotomy, appropriate meaning extensive staging and radical excision. Most clinicians agree that optimal or total debulking at the primary laparotomy is of major survival benefit. This statement has been made so frequently and vehemently that challenging it might be considered heresy. It is important to remember that second-look laparotomy, usually performed in complete responders after completion of chemotherapy, was also a well-accepted form of management yet experience has taught us that secondary surgical intervention is unlikely, on its own, to improve survival. If effective therapy might be accurately directed because of second look then the outcome might be improved. It is this, and only this, that justifies the inclusion of second-look laparotomy in some clinical trials.

There are of course other surgical interventions in this disease (summarized in Table 35.1). These include third and subsequent laparatomies which might be regarded as an understandable extension of the thinking that lead to the second-look operation. It must be said that such thinking can hardly be described as innovative or lateral in any form.

Surgical restaging and debulking should be regarded as unnecessary as they reflect an

Table 35.1 The various roles of surgery in epithelial ovarian cancer

Achieving a histological diagnosis
Defining the extent of disease (surgical staging)
Primary cytoreduction
Restaging*
Interval cytoreduction†
Second-look laparotomy
Third (and subsequent) assessment laparotomies
Salvage laparotomy
Miscellaneous palliative procedures
   (defunctioning, etc.)

*Implies a laparotomy repeated because the first laparotomy was thought to be suboptimal in terms of definition of extent or resection.
† Implies a laparotomy repeated at a preplanned point during chemotherapy with the intention of achieving further cytoreduction in a patient who was not amenable to such cytoreduction at the outset.

inadequate primary operation. However, we do not live in an ideal world. As long as we believe that accurate staging and maximal debulking are prerequisites of the first operation we must continue to strive to reach these end-points. To this end we should endeavour to ensure that appropriately trained specialists manage patients with ovarian cancer.

Palliative and salvage procedures also have an important role in management and whilst long-term survival is unlikely to be influenced by such operations, quality of life is. In a disease where only 25–30% of patients will enjoy long-term survival, quality of life is a most valuable outcome measure and should be incorporated in all clinical trials.

What more can be added to the debate on surgery that has not already been said? Are there any further useful additions to the surgical arsenal, or have we reached a plateau where we must await a novel therapeutic upturn? Perhaps now is the time for reflection. It is an oft forgotten truism that progress is largely dependent upon a careful, critical appraisal of what has gone before. The major objective of the Helene Harris Memorial Trust is to guide further research and clinical practice through the promotion of discussion and therefore we would wish to evaluate as critically as possible our current level of knowledge regarding some of the surgical procedures employed in the management of epithelial ovarian cancer.

35.2 THE OBJECTIVES OF THE PRIMARY LAPAROTOMY

35.2.1 ASSESSING THE EXTENT OF DISEASE

The current consensus is that a full laparotomy is performed through a midline incision. If ascites is present, peritoneal lavage should be performed with up to 1 litre of warmed normal saline. This lavage or the ascites should be assessed cytologically for the presence of malignant cells. We have found that multiple brush smear specimens taken from all the peritoneal surfaces is a suitable alternative to fluid cytology. This technique has the advantage of fixing the specimen immediately and minimizes unsatisfactory samples[1]. Next, a thorough inspection of the peritoneal surfaces is necessary, supplemented by biopsies from any suspicious areas when indicated. In addition it is advisable to explore the retroperitoneum, particularly the para-aortic and pelvic lymph nodes. Some clinicians have recommended routine pelvic and para-aortic lymphadenectomies[2]. Omentectomy is also recommended as part of the procedure. Whether total omentectomy is always necessary in cases with no macroscopic disease is debatable; nevertheless, multiple biopsies should be taken from as many omental sites as possible. Infracolic omentectomy with the specimen divided into 8–10 samples is a useful compromise if the whole omentum is macroscopically normal. Griffiths[3] has summarized staging comprehensively elsewhere and for the sake of brevity further detail is omitted. Total abdominal hysterectomy and bilateral salpingo-oophorectomy should be performed besides any other procedures required to debulk tumour.

35.2.2 SUPPORT FOR THE STAGING STRATEGY

Many workers have shown that some cases can be upstaged by utilizing this type of staging approach. Young and colleagues[4] showed that 31% of stage I and II patients could be restaged, the majority as stage III by the application of this meticulous schedule. By inference the preceding laparotomies must have inaccurately documented disease. Whilst this type of exercise improves the survival for individual stages, the overall survival remains unchanged as a greater number of cases are allocated to higher stage categories. How can staging influence survival?

If understaging were to result in patients not receiving effective therapy then such an error might well result in a worse survival. The converse is equally alarming. Upstaging might result in more patients receiving ineffective therapy. The new variable is effective therapy and its introduction increases the philosophical complexity of the debate. If effective and safe therapy were available, would there be a need for elaborate staging procedures? As such therapies have become available in other disease situations (*Salmonella typhi*, Hodgkin's disease) the emphasis on staging has receded. This is not the case in ovarian cancer. Effective, non-toxic therapy is not yet available so the possibility of beneficial and harmful outcomes exists. In accepting an important role for staging one must also be aware of its limitations.

Many have championed the use of meticulous staging in ovarian cancer, particularly in early or localized disease[4–6]. They have presented data suggesting that survival improves if staging is rigorous although there are few other published data that corroborate these conclusions. The value of accurate documentation must not be measured solely in survival terms. All oncologists would agree that correct documentation is a prerequisite to comparison, standardization and stratification, these being essential components of controlled clinical evaluation. It is unfortunate that the level of 'documentation' activity far outstrips the level of clinical trial activity and thus diminishes much of the value of accurate documentation. One might hope that good epidemiological units could salvage something from this type of activity.

35.2.3 CRITICISM OF THE STAGING STRATEGY

Staging is probably of most clinical importance in early epithelial ovarian cancer as it is in this context that postoperative therapy may or may not be given. Wiltshaw and co-workers[7] have presented data suggesting that surgery alone is curative in cases where disease is confined to the ovary. Some consider that it is appropriate to give adjuvant therapy in these situation[8,9]. The justification for this is that up to 30% of stage I epithelial ovarian cancer will eventually relapse despite apparent surgical cure. There are two possible explanations for this observation: first, occult disease was present elsewhere in the peritoneal cavity at the time of primary laparotomy; secondly, malignant transformation occurred after the laparotomy. Both ideas are tenable although the latter seems less likely. The former implies understaging and it is not difficult to appreciate even a 10% sampling error with the most rigorous staging protocols. The surgeon can only minimize but never totally negate this risk.

The relationship between adequacy of staging and subsequent survival deserves further examination. We have examined the case records of all patients with ovarian cancer registered at the West Midlands Regional Cancer Registry between 1 January 1980 and 31 December 1984. There were 2395 patients with ovarian cancer during this period. In this group there were 457 stage I cases. After excluding non-epithelial tumours and epithelial tumours of borderline malignancy 373 cases remained for analysis (Table 35.2 lists the complete dataset). Adequate surgical staging was defined according to FIGO guidelines[10], with 30% of our sample satisfying these criteria.

Initial univariate analysis identified factors significantly associated with survival (Table 35.3). Neither the adequacy of surgical staging nor the use of adjuvant therapy significantly influenced survival. Cox regression analysis, using all considered variables, identified histological grade and type, adjuvant chemotherapy, age, peritoneal lavage

Table 35.2 Population-based survey of stage I epithelial ovarian cancer

Median age	58 years	
Median duration of symptoms	2 months	
Stage		
Ia	227	61%
Ib	31	8%
Ic	112	30%
Adequate staging*	113	30%
Histology		
Serous	167	45%
Mucinous	125	35%
Clear cell	34	9%
Endometrioid	28	7%
Not specified	16	4%
Differentiation		
Well differentiated	256	69%
Moderately differentiated	27	7%
Poorly differentiated	40	11%
Not specified	50	13%
Given adjuvant chemotherapy	132	35%
Given adjuvant radiotherapy	53	14%
Recurrence	105	28%
Deaths	126	34%
Non-cancer death	13	3%

* Adequate staging as defined by FIGO guidelines.

and surgical rupture as independent variables.

These data are retrospective and must obviously be treated with caution and we also recognize that platinum and its analogues were not in widespread use at the time of our review. However, the population-based nature of the data and large patient numbers must at least draw attention to these important issues. Given these findings we must question the relationship between 'adequate staging' and survival. In addition adjuvant chemotherapy in early ovarian cancer requires thorough and careful evaluation considering its potential toxicity[11].

Finally, we must question the role of peritoneal lavage in staging procedures. The increased risk seen with peritoneal lavage in our patients might reflect the spreading of tumour cells within the peritoneal cavity by the lavage fluid. Alternatively, or in addition, there may be an effect on host immunity. The identification of rupture at operation as an independent, adverse prognostic factor supports the former hypothesis[12].

The observations focus the finite limits of surgical staging. Many other factors 'within stage' contribute to prognosis. The biological behaviour of ovarian cancer is of increasing interest and deserves more attention. Most observers will readily acknowledge the different biological behaviour of borderline tumours yet all too often include these within the broad terminology of epithelial ovarian cancer. It is natural that such inclusions will introduce a bias into any analysis. As staging should allow a balanced comparison system, is there not scope to include further variables? As a crude yet simple example, can two population groups be compared unless the frequency with which aneuploidy occurs in each is similar? The known adverse effect of tumour aneuploidy has been well documented[13,14].

We believe there has been too much emphasis on technique of staging of ovarian cancer and little understanding of purpose. One is often left thinking that staging might be a substitute for therapy in situations where therapy is somewhat ineffective. Clinical staging should be a transitory phenomenon in any disease scenario and therefore must evolve to meet the needs of management; this is clinically most obvious in those ascribed stage I status. Stage I does not adequately describe those at risk of relapse. Any adjuvant strategy must accept that most patients will be overtreated and defining the risk of relapse requires more information. Without this information we cannot identify populations in which adjuvant therapy should be tested. Is it time to incorporate clinical staging into a more comprehensive risk assessment schedule?

Table 35.3 Factors significantly associated with survival

Variable	Group	No. dead	DF	χ^2	P-value	Survival improved in
Histological grade	Well Mod. Poor	66/249	2	39.4	0.00001	Well/Mod.
Adjuvant chemotherapy	Yes No	60/127 64/232	1	14.8	0.0001	No
Histological type	Serous Mucinous Clear	73/190 27/121 24/48	2	16.6	0.0002	Mucinous
Stage	Ia Ib Ic	61/223 13/30 50/106	2	15.2	0.0004	Ia
Washings	Yes No	46/100 78/259	1	11.09	0.0009	No
Intact capsule	Yes No	76/261 41/90	1	9.2	0.002	Yes
Ascites	Yes No	19/35 105/322	1	7.73	0.005	No
Surgical rupture	Yes No	22/44 102/315	1	6.3	0.012	No

35.3 CYTOREDUCTION

Histological diagnosis and staging require a surgical procedure, yet in themselves are not directly therapeutic. Surgical removal of tumour is therapeutic, at least in some situations. The challenge is to define the situations that are appropriate for cytoreduction. Few would argue that complete resection of localized disease results in a better outcome than cases with residual disease. Even in cases of extrapelvic disease, complete resection of all macroscopic tumour would appear to be useful in terms of survival (Figure 35.1). Difficulty arises when residual tumour remains after a surgical procedure. This type of situation would appear to contradict the basic surgical rules of oncology resection, i.e. to extirpate tumour with a surrounding zone of healthy tissue.

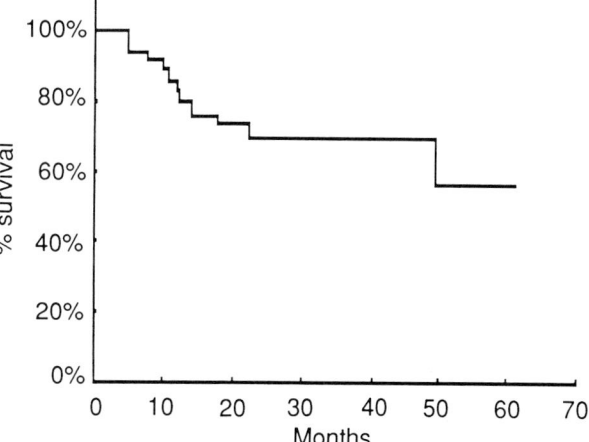

Figure 35.1 Actuarial survival in patients with epithelial ovarian cancer after total disease resection at primary laparotomy ($n = 44$: 37, stage II and III, 7 stage Ic).

Despite this, many gynaecologists regard ovarian cancer as an exception to this rule, this anomaly having arisen because of observations of improved survival associated with bulk disease resection, albeit incomplete [15–19].

35.3.1 THE CASE FOR RADICAL RESECTION OF DISEASE

The idea of maximal surgical endeavour in the management of advanced stage epithelial ovarian cancer largely evolved from the early work of Meigs and others. As early as 1935, Meigs[20] espoused optimal surgical debulking, as subsequent adjuvant radiotherapy was likely to be more effective in these circumstances. Much later Munnell reported an increased survival with adjuvant radiotherapy (from 26 to 40%) in patients who had their disease radically debulked[21]. Griffiths documented the prognostic significance of residual disease status after primary surgery, showing a clear relationship between maximum size of residual disease and survival[22]. Since then many others including ourselves have shown a strong correlation between postoperative residuum and survival. Not surprisingly, along with FIGO stage, these observations have achieved prominence in current management protocols for advanced ovarian cancer and form the basis for most therapeutic decisions.

35.3.2 WHAT IS OPTIMAL DEBULKING?

Realistically, the only optimal debulking is total debulking but as this only applies to a minority of patients with stage III and IV disease it is essential to try and quantify residual disease. In some situations there are discrete, measurable nodules and these should not pose any problems. In other cases tumour involvement is less readily quantified either because it is confluent or because an infiltrative process has occurred. This is not unusual in the involvement of structures like the pelvic sidewall or mesenteries. It is also worth considering the nature of disease spread as we have found this variable to influence survival significantly.

35.3.3 THE EFFECT OF SPREAD PATTERN

We have characterized advanced ovarian cancer into two spread patterns. First, those characterized by widespread intraperitoneal seedling formation and secondly those predominantly bulky with relative sparing of the peritoneal surfaces. We have analysed 166 patients recruited over 4 years into prospective trials: 91 had bulk disease and 75 had seedling disease with or without bulky disease. The median survival for the whole group was 20 months. There was a significant association between estimated residual volume and survival ($P < 0.0001$)(Figure 35.2). Patients

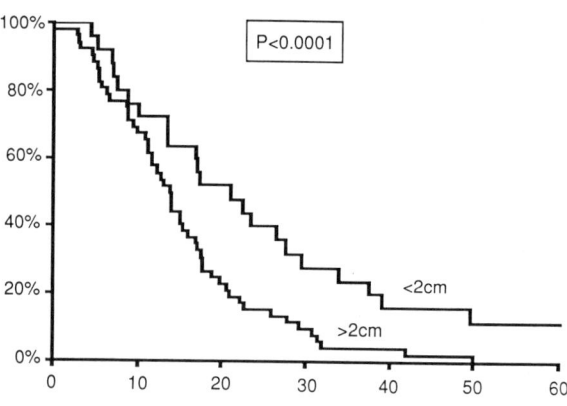

Figure 35.2 Actuarial survival in patients with epithelial ovarian cancer after optimal (<2 cm residual) ($n = 80$) and suboptimal (>2 cm residual) ($n = 96$) disease resection at primary laparotomy.

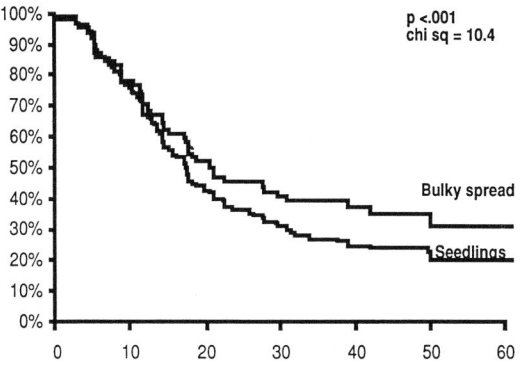

Figure 35.3 Actuarial survival in patients with epithelial ovarian cancer comparing patients with localized bulky disease (preoperative) ($n = 91$) and patients with widespread seedlings (with or without additional bulky disease preoperatively) ($n = 75$).

Table 35.4 Residual disease status in relation to disease topography

	Bulk disease group	Seedling disease group
Numbers	91	75
> 2-cm residual disease	50 (55%)	30 (40%)
< 2-cm residual disease	41 (45%)	45 (60%)

with bulk disease had a better survival than those with seedlings ($P < 0.001$)(Figure 35.3). Examining residual disease status, the proportion of patients with gross residual disease (>2 cm) was greater in the 'bulk' group than in the 'seedling group' (Table 35.4). Cox multiple regression, assessing topography with other factors, demonstrated that this characteristic had an independent effect on survival. This study has suggested that bulky disease in the absence of widespread seedlings has a more favourable prognosis. This may be because although optimally cytoreduced by present definition, patients with seedling disease still have large amounts of total tumour volume. Other unquantified biological factors such as tumour adherence molecules or host immune compromise might also explain the differences. Since, in this study, we always attempted maximal surgical excision it is unclear whether the survival difference is a result of surgery or tumour biology. We would suggest, however, that future studies should consider this variable.

35.3.4 RESPONSE TO CHEMOTHERAPY

Following Smith and Rutledge's observation 20 years ago[23] most observers still feel that patients will respond more favourably to chemotherapy following optimum debulking. Consideration of tumour kinetics, notably the Goldie–Coldman hypothesis[24] and the log kill effect, lend further credence to this strategy. The greatest chance of eradicating disease entirely by chemotherapy would appear to be in situations where only microscopic disease remains. This is a very difficult idea to prove in a clinical context.

35.3.5 QUALITY OF LIFE

Patients with ovarian cancer usually present because of symptoms. These are all too often vague and in part explain the delay in diagnosis that often is a feature of this disease. Relief of symptoms is an important and essential objective, although it might not be necessary to undertake radical cytoreduction to reach this goal. If radical surgery conferred further and more sustained improvements in quality of life, although still essentially palliative, these would be laudable goals in themselves. Blythe and Wahl[25] compared two groups of patients with regard to objective quality of life and survival. One group ($n = 19$) had extensive surgical debulking, removing all disease greater than 2 cm in maximum diameter (15 had colostomies); the other group had far less radical procedures (biopsy only in seven). Both survival and quality of life were better in the first group. The study was small and like most surgical studies did not address the question of disease biology. Was the former group debulked because of permissive biology and the latter not debulkable from the outset?

35.3.6 THE CASE AGAINST RADICAL SURGERY

This subheading requires clarification. It is not the intention to suggest avoiding tumour resection if all tumour can be removed but more to re-examine the therapeutic rationale where residual disease is inevitable. This would apply to most stage III and IV ovarian cancer patients. The previously mentioned tumour kinetic considerations, wealth of published material and commonsense feeling are strong 'pro-radical' forces. There are reasons to doubt this approach.

35.4 WEST MIDLANDS' STUDIES

Our first study, launched in the West Midlands 10 years ago, evaluated the survival benefit associated with a second-look operation. The study included patients with residual ovarian cancer who had not shown any evidence of progression on single agent cisplatin. The study entailed randomization to a second-look arm followed by consolidation therapy or consolidation therapy alone. Of 45 patients 29 (65%) had macroscopic residual disease at the time of second-look surgery and had optimal cytoreduction to a maximum residual diameter of less than 2 cm. Despite this, survival was similar to the non-second-look arm[26]. Berek and co-workers[27] have suggested that secondary debulking at second look is useful although there was no control group in this study. Raju and colleagues[28] did not show any survival advantage from secondary debulking but again there were no control patients.

One might argue that an evaluation of secondary cytoreduction does not provide a valid test of debulking. These patients have been exposed to large doses of chemotherapy and are therefore likely to have developed drug resistance. By bringing the cytoreductive surgery forward in the management schedule might this problem be avoided? This was the premise that prompted our next study. This randomized patients with known residual disease who might be further debulked pending a response to chemotherapy. Patients had an intervention debulking laparotomy or continued with chemotherapy. The surgically managed group continued with chemotherapy following their debulking operation. This study has just closed having recruited its initial target of 200 patients and although the survival data are preliminary they do not show a significant improvement associated with secondary cytoreduction (Figure 35.4). Comparing those who were optimally debulked with patients debulked at their first laparotomy (historical controls) shows that significant differences in survival persist in the long term. These data suggest that surgery only achieves a delay in the inevitable outcome.

Is optimal disease resection a function of permissive tumour biology? The observations of improved survival and enhanced response to chemotherapy might just as easily be interpreted in this way as by considering a direct surgical effect. An academic would argue that in a situation such as this, where significant doubt exists, one must consider a randomized clinical trial. This has not been done primarily because of practical considerations but equally because of the widespread conviction that surgery must be beneficial.

Palliative surgery and conclusion

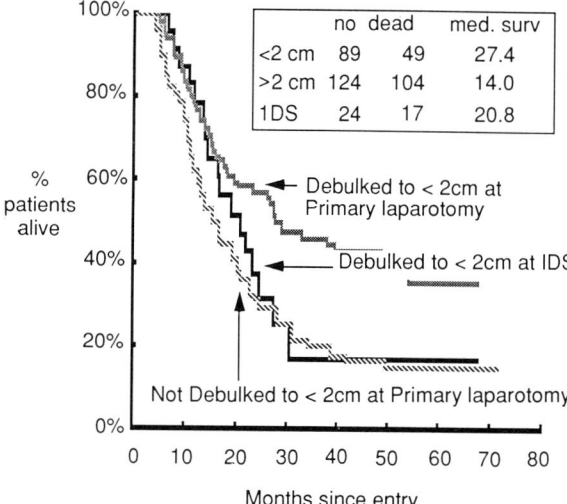

Figure 35.4 Actuarial survival comparing optimal debulking at primary laparotomy ($n = 89$), optimal debulking at intervention debulking surgery ($n = 24$), and suboptimal debulking at primary laparotomy ($n = 124$). IDS performed after two to three cycles of platinum containing combination chemotherapy.

35.5 SUPER-RADICAL SURGERY

There is a weight of circumstantial evidence in favour of radical debulking. Burghardt's data for instance could be regarded as compelling evidence for super-radical resection combined with pelvic and para-aortic lymphadenectomy[2]. Yet even this group, whose surgical expertise cannot be doubted, has fewer than 30% survivors at 5 years. Furthermore their actuarial data show a similar slope in the survival curve after 2 years. Ever increasing surgical endeavour is unlikely to provide the answer for all patients but undoubtedly benefits some.

Could super-radical surgery be harmful to some patients? Naturally the immediate morbidity from these procedures is going to be greater but there are also data to suggest a possible adverse influence on disease behaviour. There are data showing, albeit indirectly, an increased rate of tumour proliferation in some patients after laparotomy[29–32]. We have also reported, with others, that subgroups of totally debulked patients at second look seem to relapse and die in intervals much shorter than one would expect[33,34]. Finally, animal studies have shown major changes in intraperitoneal immune surveillance following laparotomy[35,36]. One can only conclude that if some patients do well because of radical surgery, some might do badly. This and this alone is a justification for a more critical approach to the primary laparotomy.

35.6 PALLIATIVE SURGERY AND CONCLUSION

Palliative surgery has a well-defined goal and was discussed in depth at a previous meeting of this group[37]. One of the most important messages conveyed was the need for selection. Properly selected patients, managed appropriately, will frequently achieve useful palliation. Cynics would suggest that the primary laparotomy is also palliative and indeed in some patients it is. Perhaps a more important message is to apply the same rigorous selection approach to primary laparotomies as are applied to palliative procedures. If one accepts this approach then one also must accept the need for further research to define those who will benefit from cytoreduction attempts at their first operation.

What then are the important messages regarding surgery in the management of epithelial ovarian carcinoma? Surgery remains the most reliable diagnostic and assessment procedure. The necessity for histological diagnosis still remains paramount and therefore some form of operative procedure whereby adequate histological material can be collected will remain central to the management strategy. Similarly the assessment of disease extent is crucial although unlikely to be of major import in survival terms. As long as controlled clinical evaluation is necessary then accurate documentation will remain a part of the initial assessment. As our quest for knowledge of disease biology expands it will become more obvious to incorporate such new variables in an expanded risk assessment, particularly in early disease situations.

There is a clear need for further evaluation of the role of cytoreduction. This task has been made more difficult by the efforts of gynaecologists and oncologists over the last 20 years who have concentrated their efforts on maximal cytoreduction for all patients. This strategy was certainly well intentioned but may have obscured vital information. Surgical resection of disease is vitally important for some patients, its role is limited in others. The task that gynaecological oncologists must address is firstly to properly evaluate the contribution of cytoreduction and then to work toward more critical case selection.

REFERENCES

1. Luesley, D.M., Williams, D.R., Ward, K. and Lawton, F.G. (1990) Peritoneal smears and lavage fluids in patients with epithelial ovarian cancer and benign gynaecological disease. *Acta Cytol.*, **34**, 539–42.
2. Burghardt, E., Pickel, H., Lahousen, M. and Stettner, H. (1986) Pelvic lymphadenectomy in operative treatment of ovarian cancer. *Am. J. Obstet. Gynecol.*, **155**, 315–19.
3. Griffiths, C.T. (1987) Carcinomas of the ovary: surgical objectives, in *Ovarian Cancer – The Way Ahead* (eds F. Sharp and W.P. Soutter), Royal College of Obstetricians and Gynaecologists, London, pp.235–44.
4. Young, R.C. (1987) Initial therapy for early ovarian cancer. *Cancer*, **60**, 2042–9.
5. Trimbos, J.B., Scheuler, J.A., van Lent, M. *et al.* (1990) Reasons for incomplete surgical staging in early ovarian carcinoma. *Gynecol. Oncol.*, **37**, 374–7.
6. Dembo, A.J., Davy, M.L.J., Stenwig, A.E. *et al.* (1990) Prognostic factors in patients with stage 1 epithelial ovarian cancer. *Obstet. Gynecol.*, **75**, 263–73.
7. Wiltshaw, E., Osborne, J. and Gallagher, C. (1987) Laparoscopic follow-up of early ovarian cancer, in *Ovarian Cancer – The Way Ahead* (eds F. Sharp and W.P. Soutter), Royal College of Obstetricians and Gynaecologists, London, pp.277–82.
8. Hreshchyshyn, M.M., Park, R.C., Blessing, J.A. *et al.* (1980) The role of adjuvant therapy in Stage 1 ovarian cancer. *Am. J. Obstet. Gynecol.*, **138**, 139–45.
9. Young, R.C., Walton, L.A., Ellenberg, S.S. *et al.* (1990) Adjuvant therapy in stage I and stage II epithelial ovarian cancer. Results of two prospective randomized trials. *N. Engl. J. Med.*, **322**, 1021–7.
10. Scully, R.E. (1975) World Health Organization classification and nomenclature of ovarian cancer. *Natl Cancer Inst. Monogr.*, **42**, 5.
11. Sell, A., Bertelsen, K., Andersen, J.E. *et al.* (1990) Randomized study of whole abdominal irradiation versus pelvic irradiation plus cyclophosphamide in treatment of early ovarian cancer. *Gynecol. Oncol.*, **37**, 367–73.
12. Petterson, F. (1988) *Annual Report on the Results of Treatment in Gynecological Cancer.* FIGO.
13. Iverson, O. and Skaarland, E. (1987) Ploidy assessment of benign and malignant ovarian tumours by flow cytometry. A clinicopathologic study. *Cancer*, **60**, 82–7.
14. Redman, C.W.E., Finn, C., Ward, K. *et al.* (1990) Tumour cell activity markers in epithelial ovarian cancer: Are biochemical and cytometric indices complementary? *Br. J. Cancer*, **61**, 755–8.
15. Delgado, G., O, K.H., Petrilli, E.S. (1984) Stage III epithelial ovarian cancer: the role of maximal surgical reduction. *Gynecol. Oncol.*, **18**, 293–8.
16. Gallion, H.H., Van Nagell, J.R., Donaldson, E.S. *et al.* (1987) Prognostic implications of large volume residual disease in patients with advanced stage epithelial ovarian cancer. *Gynecol. Oncol.*, **27**, 220–5.
17. Griffiths, C.T., Parker, L.M. and Fuller, A.F. (1979) Role of cytoreductive surgical treatment in the management of advanced ovarian cancer. *Cancer Treat. Rep.*, **63**, 235–40.
18. Hacker, N.F., Berek, J.S., Lagasse, L.D. *et al.*

(1983) Primary cytoreductive surgery for epithelial ovarian cancer. *Obstet. Gynecol.*, **61**, 413–20.
19. Wharton, J.T., Edward, C.L. and Rutledge, F.N. (1984) Long term survival after chemotherapy for advanced epithelial ovarian cancer. *Am. J. Obstet. Gynecol.*, **148**, 997–1004.
20. Griffiths, C.T. (1986) Surgery at the time of diagnosis in ovarian cancer, in *Management of Ovarian Cancer* (eds G. Blackledge and K.K. Chan), Butterworth, London, pp. 60–75.
21. Munnell, E.W. (1968) The changing prognosis and treatment of cancer of the ovary. A report of 235 patients with primary ovarian carcinoma, 1952–1961. *Am. J. Obstet. Gynecol.*, **100**, 790–805.
22. Griffiths, C.T. (1975) Surgical resection of tumour bulk in the primary treatment of ovarian carcinoma. *Natl Cancer Inst. Monogr.*, **42**, 101–4.
23. Smith, J.P. and Rutledge, F. (1970) Chemotherapy in the treatment of cancer of the ovary. *Am. J. Obstet. Gynecol.*, **107**, 691–703.
24. Goldie, J.H. and Coldman, A.J. (1979) A mathematic model correlating the drug sensitivity of tumours to their spontaneous mutation rate. *Cancer Treat. Rep.*, **63**, 1727–33.
25. Blythe, J.G. and Wahl, T.P. (1982) Debulking surgery: does it improve the quality of survival? *Gynecol. Oncol.*, **14**, 396–408.
26. Luesley, D.M., Lawton, F.G., Blackledge, G.R. *et al.* (1988) Failure of second look laparotomy to influence survival in epithelial ovarian cancer. *Lancet*, **ii**, 599–603.
27. Berek, J.S., Hacker, N.F., Lagasse, L.D. *et al.* (1983) Survival of patients following secondary cytoreductive surgery in ovarian cancer. *Obstet. Gynecol.*, **61**, 189–93.
28. Raju, K.F., McKinna, J.A., Barker, G.H. *et al.* (1982) Second look operations in the planned management of advanced ovarian carcinoma. *Am. J. Obstet. Gynecol.*, **144**, 650–4.
29. Luesley, D.M., Blackledge, G.R., Chan, K.K. and Newton, J.R. (1986) Random urinary cyclic guanosine monophosphate in ovarian cancer: relationship with other prognostic variables. *Br. J. Obstet. Gynaecol.*, **93**, 380–5.
30. Romsdhal, M.M. (1964) Influence of surgical procedures on the development of spontaneous lung metastases. *J. Surg. Res.*, **4**, 363–70.
31. Schiffer, L.M. and Braunschweiger, P.G. (1983) Proliferative recovery after surgical cytoreduction: mechanistic studies. *Cell Tissue Kinet.*, **16**, 603–17.
32. Simpson-Herren, L., Sandford, A.H. and Holmquist, J.P. (1976) Effects of surgery on the cell kinetics of residual tumour. *Cancer Treat. Rep.*, **60**, 1749–60.
33. Gershenson, D.M., Copeland, L.J., Wharton, J.T. *et al.* (1985) Prognosis of surgically determined complete responders in advanced ovarian cancer. *Cancer*, **55**, 1129–35.
34. Luesley, D.M., Chan, K.K., Lawton, F.G. *et al.* (1989) Survival after negative second look laparotomy. *Eur. J. Surg. Oncol.*, **15**, 205–10.
35. Lovett, E.J., Varani, J. and Lundy, J. (1983) Suppressor cells and increased primary tumour growth rate induced by thiopental. *J. Surg. Oncol.*, **22**, 26–32.
36. Whitney, R.B., Levy, J.G. and Smith, A.G. (1974) Influence of tumour size and surgical resection on cell mediated immunity in mice. *J. Natl Cancer Inst.*, **53**, 111–16.
37. Wiltshaw, E., Shepherd, J.H. and Crowther, M. (1987) Salvage surgery – results, in *Ovarian Cancer – The Way Ahead* (eds F. Sharp and W.P. Soutter), Royal College of Obstetricians and Gynaecologists, London, pp. 313–19.

Chapter 36

Debulking surgery at second-look laparotomy

W.J. HOSKINS

In 1991, 21 000 new cases of ovarian cancer were diagnosed in the USA[1]. About 70% of these patients presented with advanced disease at the time of initial diagnosis. The importance of cytoreductive (debulking) surgery as part of the initial therapy for ovarian cancer is well established. However, despite initial cytoreduction and platinum-based multi-drug chemotherapy, only one-half of these patients will be clinically free of disease and therefore candidates for second-look laparotomy. Of the patients who undergo second-look laparotomy, 40% will be free of disease (negative second look), and 60% will be found to have residual cancer (positive second look)[2]. The usual clinical course of patients with stages III and IV epithelial ovarian cancer is outlined in Figure 36.1.

Although the importance of initial cytoreductive surgery in epithelial ovarian cancer seems to have been defined, the importance of secondary cytoreductive surgery following a positive second look is less established. In a review of the literature, Williams and Hoskins[3] reported that 18.5% of patients with a positive second look would be found to have microscopic residual disease. Rubin and Lewis[2], in a similar review, found 17.3% with microscopic disease. Table 36.1 lists several reports which indicate that an average of 24% of patients have microscopic disease and 76% have macroscopic disease. In all of these reports, patients were apparently negative at surgical exploration but were found on pathological or cytological analysis to have microscopic evidence of residual malignant cells.

Copeland and colleagues[7] reported on 50 patients with microscopic disease at second-look laparotomy. Of these patients 45 were treated with chemotherapy, one received radiation therapy, and four were not treated. The 5-year survival rate was 71% in the entire series. Hoskins and co-workers[15] reported a 60% 5-year survival rate in a similar group of 17 patients. Table 36.2 reviews four reports which confirm the excellent 5-year survival rates of patients found to have microscopic disease at second-look laparotomy.

Although there are fewer data regarding secondary cytoreduction, the survival rate of patients reduced to microscopic disease after a positive second look appears to be similar to that of patients who were initially cytoreduced to microscopic disease. Hoskins and co-workers[15] reported a 50% 5-year survival rate and Podratz and colleagues[10]

Debulking surgery at second-look laparotomy

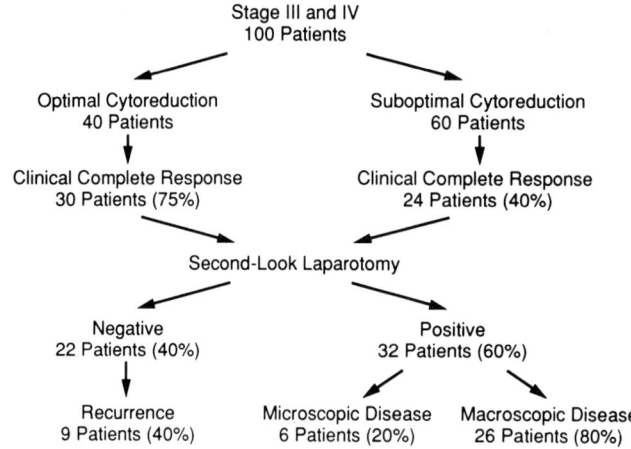

Figure 36.1 The usual clinical course of patients with stages III and IV epithelial ovarian cancer.

reported a 55% 4-year survival rate for patients reduced to microscopic disease by secondary debulking at second-look laparotomy. Lippman and co-workers[11] reported a significant improvement in survival for patients who could be secondarily debulked to less than 2 cm, with 40% of the patients surviving longer than 5 years with subsequent therapy. On the other hand, in the series by Lippman, no patient survived beyond 3 years if secondary reduction to less than 3 cm was not possible. Copeland and colleagues[7] and Berek and colleagues[17] reported similar results. Raju and co-workers[18] and Luesley and co-workers[13] reported either no benefit or only modest benefit from secondary cytoreduction.

Several authors have addressed the fre-

Table 36.1 Residual disease in patients with epithelial ovarian cancer at second-look laparotomy

Author	Year published	No. of patients with microscopic disease	No. of patients with macroscopic disease
Schwartz [4]	1980	26	57
Phibbs [5]	1983	3	22
Berek [6]	1984	8	30
Copeland [7]	1985	50	111
Smirz [8]	1985	14	35
Duplat [9]	1986	0	27
Podratz [10]	1988	15	43
Lippman [11]	1988	7	41
Chambers [12]	1988	7	22
Luesley [13]	1988	8	43
Fuks [14]	1988	2	17
Hoskins [15]	1989	17	50
Total/percentage		157 (24%)	498 (76%)

Table 36.2 Survival of ovarian cancer patients found to have microscopic disease at second look

Author	Year published	No. of patients	% 5-year survival
Schwartz [4]	1980	27	38
Copeland [7]	1985	50	71
Ho [16]	1987	6	42
Hoskins [15]	1989	16	60
Total/weighted mean		99	61

quency with which secondary cytoreduction is possible at second-look laparotomy[9–15]. Table 36.3 shows that approximately one-third of patients can be reduced to microscopic disease at second-look laparotomy. Lippman and colleagues[11] and Hoskins and co-workers[15] reported that 45% and 88% of patients, respectively, could be cytoreduced to less than 2 cm of residual disease at second-look laparotomy. The latter authors, however, did not report a significant survival advantage to secondary cytoreduction if any gross disease remained following surgery.

Table 36.3 Rate of 100% cytoreduction in patients found to have macroscopic disease at second-look laparotomy

Author	No. of patients with macroscopic disease at second look	No. of patients who had 100% secondary cytoreduction (%)
Podratz [10]	43	9 (21)
Lippman [11]	42	14 (33)
Duplat [9]	27	8 (30)
Chambers [12]	12	16 (73)
Luesley [13]	13	15 (33)
Fuks [14]	17	9 (53)
Hoskins [15]	50	16 (32)
Total	247	87 (35)

Reprinted, with permission, from reference [3].

It has become increasingly clear over the past few years that many of the salvage regimens for patients who fail to achieve a complete pathological response (negative second look) after primary therapy are effective only against minimal residual disease. Several authors[19–23] have reported negative third-look surgical reassessments following the use of intraperitoneal platinum-based therapy in patients who have failed primary therapy. In all of these reports, the primary therapy consisted of platinum-based multidrug chemotherapy. Also, patients who achieved a complete, surgically documented response began intraperitoneal therapy with minimal residual disease. Other authors [14,24–27] have reported a 13–40% 2-year survival rate following whole abdominal radiation therapy as salvage treatment. Again, this therapy was only beneficial in patients with minimal residual disease.

While it is apparent that many of the salvage regimens for patients with persistent ovarian cancer are only beneficial in patients who have minimal disease, it is equally clear that up to one-third of patients who undergo second-look laparotomy can be reduced to microscopic disease. An even larger percentage can be reduced to disease that is less than 2 cm. Secondary cytoreduction has the potential to become extremely important in providing prolonged survival after failure of primary treatment. Figure 36.2 illustrates the projected outcome for 100 patients who

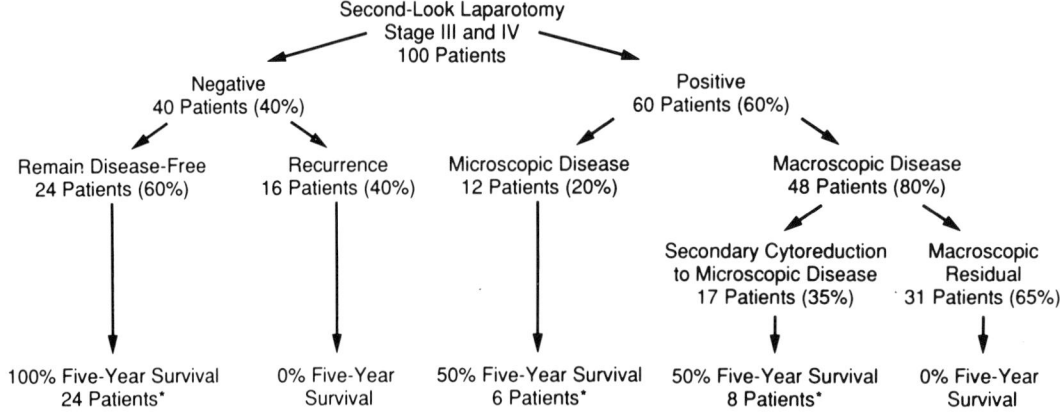

Figure 36.2 The projected outcome for 100 patients who complete primary therapy for epithelial ovarian cancer without clinical evidence of disease. * Total 5-year survival in 38 patients (38%).

complete primary therapy without clinical evidence of disease. Using percentages available in the literature, 38 patients can be expected to survive for 5 years. Of these 24 (63%) will have a negative second-look surgical reassessment. The remaining 14 (37%) will be salvaged after a positive second-look operation. Of these 14 patients, 6 (16%) will be found to have microscopic disease at second look and 8 (21%) will benefit from secondary cytoreduction to macroscopic disease. If salvage therapies improve over the next decade, it is clear that the patients who benefit the most from salvage therapy will be found to have either minimal disease or disease that can be reduced to a minimal amount by secondary cytoreductive surgery. It is likely that secondary cytoreductive surgery at the time of second-look operation will become increasingly important as better salvage therapies become available.

REFERENCES

1. Boring, C.C., Squires, T.S. and Tong, T. (1991) Cancer statistics, 1991. *CA*, **41**, 19–36.
2. Rubin, S.C. and Lewis, J.L. Jr (1988) Second-look surgery in ovarian carcinoma. *Crit. Rev. Oncol. Hematol.*, **8**, 75–91.
3. Williams, L. and Hoskins, W.J. (1990) The influence of secondary cytoreductive surgery at second-look laparotomy on survival in patients with epithelial ovarian cancer. *Contemp. Obstet. Gynecol.*, July, 13–24.
4. Schwartz, P.E. and Smith, J.P. (1980) Second-look operations in ovarian cancer. *Am. J. Obstet. Gynecol.*, **138**, 1124–30.
5. Phibbs, G.D., Smith, J.P. and Stanhope, C.R. (1983) Analysis of sites of persistent cancer at second-look laparotomy in patients with ovarian cancer. *Am. J. Obstet. Gynecol.*, **147**, 611–17.
6. Berek, J.S., Hacker, N.J., Lagasse, L.D. *et al.* (1984) Second-look laparotomy in stage III epithelial ovarian cancer: Clinical variables associated with disease status. *Obstet. Gynecol.*, **64**, 207–12.
7. Copeland, L.J., Gershenson, D.M., Wharton, J.T. *et al.* (1985) Microscopic disease at second-look laparotomy in advanced ovarian cancer. *Cancer*, **55**, 472–8.
8. Smirz, L.R., Stehman, F.B., Ulbright, T.M. *et al.* (1985) Second-look laparotomy after chemotherapy in the management of ovarian malignancy. *Am. J. Obstet. Gynecol.*, **152**, 661–8.
9. Duplat, J., Ferriere, J.P., Gorbinet, M. *et al.* (1986) Second-look laparotomy in managing epithelial ovarian carcinoma. *Cancer*, **57**, 1627–31.
10. Podratz, K.E., Schrag, M.F., Wieand, H.S. *et*

al. (1988) Evaluation of treatment and survival after positive second-look laparotomy. *Gynecol. Oncol.*, **318**, 9–21.

11. Lippman, S.M., Alberts, D.S., Slyman, D.J. *et al.* (1988) Second-look laparotomy in epithelial ovarian carcinoma: Prognostic factors associated with survival duration. *Cancer*, **61**, 2571–7.

12. Chambers, S.K., Chambers, J.T., Kohorn, E.I. *et al.* (1988) Evaluation of the role of second-look surgery in ovarian cancer. *Obstet. Gynecol.*, **72**, 404–8.

13. Luesley, D.M., Chan, K.K., Fielding, J.W.L. *et al.* (1984) Second-look laparotomy in the management of epithelial ovarian carcinoma: An evaluation of fifty cases. *Obstet. Gynecol.*, **64**, 421–6.

14. Fuks, Z., Rizel, S. and Biran, S. (1988) Chemotherapeutic and surgical induction of pathologic complete remission and whole abdominal radiation for consolidation does not enhance the cure of stage III ovarian carcinoma. *J. Clin. Oncol.*, **6**, 509–16.

15. Hoskins, W.J., Rubin, S.C., Dulaney, E. *et al.* (1989) Influence of secondary cytoreduction at the time of second-look laparotomy on the survival of patients with epithelial ovarian carcinoma. *Gynecol. Oncol.*, **34**, 365–71.

16. Ho, A.G., Beller, U., Speyer, J.L. *et al.* (1987) A reassessment of the role of second-look laparotomy in advanced ovarian cancer. *J. Clin. Oncol.*, **5**, 1316–21.

17. Berek, J.S., Hacker, N.F., Lagasse, L.D. *et al.* (1983) Survival of patients following secondary cytoreductive surgery in ovarian cancer. *Obstet. Gynecol.*, **61**, 189–93.

18. Raju, K.S., McKinna, J.A., Barker, G.N. *et al.* (1982) Second-look operations in the planned management of advanced ovarian carcinoma. *Am. J. Obstet. Gynecol.*, **144**, 650–4.

19. Cohen, C.J. (1985) Surgical considerations in ovarian cancer. *Semin. Oncol.*, **12**, (Suppl. 4), 53–6.

20. ten Bokkel Huinink, W.W., Dubbelman, R., Aartsen, E. *et al.* (1985) Experimental and clinical results with intraperitoneal cisplatin. *Semin. Oncol.*, **12**, (Suppl 4), 43–6.

21. Hacker, N.J., Berek, J.S., Pretorius, R.G. *et al.* (1987) Intraperitoneal cisplatinum as salvage therapy for refractory epithelial ovarian cancer. *Obstet. Gynecol.*, **70**, 759–64.

22. Reichman, B., Markman, M., Hakes, T. *et al.* (1989) Intraperitoneal cisplatin and etoposide in the treatment of refractory/recurrent ovarian carcinoma. *J. Clin. Oncol.*, **7**, 1327–32.

23. Markman, M., Hakes, T., Reichman, B. *et al.* (1991) Intraperitoneal cisplatin and cytarabine in the treatment of refractory or recurrent ovarian carcinoma. *J. Clin. Oncol.*, **9**, 204–10.

24. Peters, W.A., Blacks, J.C., Bagley, C.M. *et al.* (1986) Salvage therapy with whole abdominal irradiation in patients with advanced carcinoma of the ovary previously treated by combination chemotherapy. *Cancer*, **58**, 880–2.

25. Piver, M.S., Barlow, J.J., Lee, F.T. and Vongtana, V. (1975) Sequential therapy for advanced ovarian adenocarcinoma: Operation, chemotherapy, second-look laparotomy, and radiation therapy. *Am. J. Obstet. Gynecol.*, **127**, 355–7.

26. Hainsworth, J.D., Malcolm, A. and Johnson, D.H. (1983) Treatment of minimal residual advanced ovarian carcinoma. Abdominopelvic irradiation following incomplete response to combination chemotherapy. *Obstet. Gynecol.*, **61**, 619–23.

27. Hacker, N.F., Berek, J.S., Barrison, C.M. *et al.* (1985) Whole abdominal radiation salvage therapy for epithelial ovarian cancer. *Obstet. Gynecol.*, **65**, 619–23.

Chapter 37

Evaluation of debulking surgery at second-look laparotomy

W.T. CREASMAN

37.1 INTRODUCTION

In the USA ovarian cancer is the second most common malignancy found in the female pelvic genital tract. However, it kills more women than uterine and cervical cancer (first and third most frequent cancers) combined. With the advent of intense surgical staging, that was introduced over a decade ago, exact extent of disease has been better defined and subsequent therapy more appropriately applied. With the introduction of the cisplatin era, patients with advanced disease have appreciated a greater response to chemotherapy particularly those with measurable disease after primary surgery, yet this has not been translated into significant increase in survival of these patients. It is now appreciated that a considerable number of patients with complete clinical response will have persistent disease at the time of second-look laparotomy. Because so many of our patients have persistent disease and second-line chemotherapy has been so poor, the role of the second-look laparotomy in regard to patient benefit has been brought into question. It has been suggested that secondary debulking at the time of second-look laparotomy may improve the chances of a small group of patients by applying subsequent therapy which may benefit the individual patient, e.g. intraperitoneal ^{32}P, intraperitoneal chemotherapy or external irradiation. The applicability of secondary debulking has been debated in the literature with some authors stating categorically that secondary debulking had no impact on overall survival, while a small number of investigators have suggested that in small subsets secondary debulking has increased survival. Unfortunately most of the studies in the literature involve small numbers of patients and definite conclusions have not been reached. In an attempt to analyse the data available to us, the literature was reviewed in this regard.

37.2 MATERIAL AND METHODS

The contemporary literature from 1980 to the present time on second-look laparotomy was reviewed. A total of 18 articles were identified which addressed at least in part, if not

Evaluation of debulking surgery at second-look laparotomy

Table 37.1 Survival of patients with clinical disease before second-look laparotomy

	Residual post second-look laparotomy							
	'Optimal' debulk				'Suboptimal' debulk			
	NED	<1 cm	<1.5 cm	<2 cm	>1 cm	>1.5 cm	>2 cm	'residual'
Raju	2/9							0/29
Michel				5/32			5/45	
Neijt*		9/30			6/17			
Berek			0/6			0/5		
Luesley	1/2			1/6				0/2
	3/11 (27%)	9/30 (30%)	0/6 (0%)	6/38 (17%)	6/17 (35%)	0/15 (0%)	5/45 (11%)	0/31 (0%)
Total	18/85 (21%)				9/108 (8.3%)			
				$P = 0.191$				

* Interventional surgery (see text).
'Optimal', 'suboptimal': authors' designation.
NED, no evident disease.

whole, the subject of secondary debulking. Some studies addressed the role of secondary debulking in both the patient with no clinical evidence of disease at the time of the second-look laparotomy as well as those with persistent clinical disease with the primary aim of hoping to secondarily debulk the tumour. Some of the studies reported their experience with 'interventional' debulking in which the plan was to give limited courses of chemotherapy before planned secondary debulking. Because the reports in the literature and even in the same article contain a 'mixed bag' of patients, conclusions reached may be applicable only to one set of circumstances although inferences have been incorrectly drawn in some instances that addressed both of these situations. Most of the data in the literature suggest in their conclusions that secondary debulking does not appear to be of clinical benefit in regards to prolonged survival. There are at least three exceptions which are frequently referred to in the literature as having positive beneficial effects from the secondary debulking and these papers will be examined in detail.

37.3 RESULTS

37.3.1 CLINICAL RESIDUAL DISEASE

Five studies[1–5] have addressed the role of secondary debulking in patients with clinical residual disease at the completion of prescribed chemotherapy or 'interventional' surgery after only a limited number of planned chemotherapy courses. In many instances criteria of optimal and suboptimal resections were not uniform. Because of the small number of patients, groupings were necessary in order to attempt to draw valid conclusions. A total of 193 patients underwent the second-look surgery with the desired attempt to optimally debulk these patients (Table 37.1). Only 11 (5.6%) had their disease reduced to no clinical evidence of gross disease at the end of the surgical procedure. In addition, there were 74 patients who were said to have optimal debulking (<2 cm residual) at the completion of the second-look laparotomy; 15 of these patients (20%) were alive at the time of publication. This compared to three of 11 (27%) who were alive and had microscopic disease only at the completion of their surgery.

There were 108 patients who were said to have suboptimal debulking although 32 of these patients had residual disease between 1 and 2 cm, and six (18.7%) survived. There were 76 patients in the suboptimal category who had greater than 2-cm residual disease remaining after the surgery, some of which were said to be 'unresectable' and apparently only a biopsy obtained and only five (6.5%) survived.

37.3.2 NON-CLINICAL RESIDUA

There were a total of 1207 patients reported in 16 articles[3–18] who underwent a second laparotomy in whom no detectable clinical disease was present prior to the second-look laparotomy. Several of the articles did not note the number of patients with negative second looks as their topic was restricted to management of those patients with residual disease. Therefore the denominator for this group of patients is unknown. In those reported, 414 (34%) were pathologically complete responders and 294 (71%) survived. There were 193 (16%) who were found to have microscopic disease only and 91 (47%) survived. Macroscopic disease was noted in 600 (50%) of the patients. Only 110 (18%) of these patients survived (Table 37.2). Of the 600 patients, 118 (19.5%) were debulked to

Table 37.2 Survival based on findings at second-look laparotomy

	Survival	
No evident disease	294/414 (71%)	
		$P = 0.001$
Microscopic disease only	91/193 (47%)	
		$P = 0.001$
Macroscopic disease	110/600 (18%)	

Table 37.3 Survival after debulking at second-look laparotomy

Debulked to	Survival	
Microscopic	37/118 (31%)	
vs.		$P = 0.0001$
Macroscopic	74/482 (15%)	
<5 mm	14/62 (22%)	
vs.		$P = 0.133$
>5 mm	60/420 (14%)	
<2 cm	45/213 (21%)	
vs.		$P = 0.002$
>2 cm	22/221 (10%)	

microscopic disease only and 37 (31%) survived (Table 37.3). In patients where debulking could be carried out to <5 mm, 62 patients fell into this category (all from one institution) and 14 (22.5%) survived. Of the 420 patients with greater than 5-mm residual, 60 (14%) survived.

37.4 DISCUSSION

This retrospective review has the same drawbacks as most such analyses. The indications for surgery were different. Some was done in patients in whom no clinical evidence of disease was present after the prescribed number of chemotherapeutic courses. Others had surgery after the same number of chemotherapeutic courses but with the sole intention of attempting to remove residual disease that was clinically present. In some instances both groups of patients were evaluated in the same paper. Still others performed 'interventional' surgery after less than the optimal number of courses of chemotherapy in an attempt to 'debulk' persistent tumour. In some instances we did not know the number of patients that underwent a second-look laparotomy for whatever indication and therefore the denominator is not present. Authors used different residual criteria in their evaluation, e.g. <5 mm, >1.5 cm, etc. The definition of optimal and suboptimal varied and in some instances 'partial' removal and 'unresectable' were not infrequent terms. Although an attempt was made to include only those patients with advanced disease, nevertheless many of the articles did include some patients with stage I and II disease. Resectability both at initial surgery as well as second-look laparotomy certainly seems to be dependent upon stage. Some authors noted the ability to debulk was related to age. This was not a common factor evaluated. Some studies noted that those patients with more undifferentiated lesions had a lower chance of debulking and therefore a worse survival; again it was not evaluated in all papers. Most of the studies used cisplatin-containing regimens although several had some patients in which single agent alkylators were used. The study by Schwartz and Smith, although large in number, contained no patients treated with cisplatin. In studies in which non-cisplatin agents were compared to cisplatin-containing regimens, there appears to be a greater response rate with the cisplatin but no difference in pathological complete responders between the two therapies. Some studies reported median survival for different groups while others gave crude survival in which the time interval was different. Because median survival was not available in all papers crude survival was analysed.

The role of the second-look laparotomy continues to be hotly debated. While much of the rest of the world has given up the second-look laparotomy, in the USA many continue to consider it as part of standard management. Historically, at least in the USA, essentially all patients with ovarian cancer, if they were treated subsequent to surgery with chemotherapy, were suggested to have a second-look laparotomy. This included all stages. In 1987 Walton and colleagues published the result of a large Gynecologic Oncology Group (GOG) study of early ovarian cancer in which second-look laparotomy was performed in many[19]. In this study 112 patients with stage I and II cancers underwent a second-look laparotomy. Only 5 (5.2%) of the 95 clinically asymptomatic patients were found to have recurrent disease. Among 11 patients with symptoms suggestive of recurrent disease, eight had disease at the time of laparotomy. Six patients were re-explored for intestinal obstruction and only one was found to have carcinoma. In this series of 112 patients, 14 (13%) had persistent disease and all died despite a variety of salvage therapy options. A large number of patients were not re-explored for various reasons, some of which had not reached the appropriate time in the protocol. Of the 105 patients not re-explored, 11 recurrences were diagnosed in the 89 patients at risk. The crude recurrence rate was 12% in the unexplored group vs. 13% in those undergoing the second-look laparotomy. None of the patients in the unexplored group who developed a recurrence survived despite salvage therapy. This study, which is by far the largest in the literature concerning early stage disease, would suggest that there is a very limited, if any, role for second-look laparotomy in early stage disease. Therefore in these patients the question of secondary debulking is moot.

The applicability of the second-look laparotomy is probably somewhat limited, even in those patients with stage III and IV disease. In a randomized clinical trial by the GOG, conducted in women with bulky (suboptimal) stage III and IV ovarian cancer, a second-look laparotomy was to be done in complete responders[20]. Of 440 evaluable cases, only 152 (34.5%) came to second-look laparotomy. Only 34 of the 152 (22%) were found to have no macroscopic or microscopic disease. Of these 34 patients with negative second-look laparotomy, 18 (52.9%) have recurred. Therefore, only 16 patients (3.6%) of all of the patients entered into the protocol survived. Stage III patients with optimal (<3 cm residual) disease will have a greater chance of having a negative second-look laparotomy. In a prospective randomized study done by the GOG of patients with this limited stage III disease, 186 patients were entered in the protocol[17]. Of these, 84 patients were eligible and underwent a second-look laparotomy with 41 (49%) having a negative finding for persistent malignancy. There were 15 who were eligible for surgery but refused. Therefore in this study 24% of the patients had a negative second-look laparotomy compared to only 7.7% of those with more bulky advanced disease. The recurrence rate was much less in those patients with optimal stage III disease (26.8%) compared to those with more advanced disease (52.9%). This difference between subsets of stage III and IV cancer affects the number of negative second looks as well as the overall survival; therefore data on second-look laparotomies which include stage I and II disease would tend to make the results better.

The results of secondary debulking remain dismal in patients with clinical disease after the prescribed number of chemotherapeutic courses have been given. The survival of 193 patients with clinical residual disease before second-look laparotomy resulted in only 27 (13.9%) surviving. Those patients who were classified as having 'optimal' resection (<2 cm residual at the completion of the surgery) had a 21% survival compared with those with 'suboptimal' debulking where the survival was only 8.3%. Some authors included patients with 1–2-cm residual in both categories depending upon their definition. The difference between the results of these two groups of patients is not significant ($P=0.191$). Included in this analysis were patients who underwent 'interventional' surgery, i.e. attempted debulking before the prescribed number of chemotherapeutic courses had been given. In this study, survival was equal irrespective of whether the disease was debulked to <1 cm or remained >1 cm. Overall survival in both of these groups was 30% and 35%. If the interventional patients are removed from the analysis, survival for the optimally debulked patients was 18% vs. 5.5% for the suboptimal group.

In those patients who underwent a second-look laparotomy without any clinical residual disease present prior to surgery, the findings at the second look were highly predictive of survival. Those patients without evidence of disease had a 70.7% chance of long-term survival vs. only 47% in those patients with microscopic disease only ($P=0.001$). There is also a highly statistically significant different ($P=0.001$) in survival between those patients with microscopic disease only and those with macroscopic disease irrespective of the efforts of debulking. Unfortunately only about one-third of the patients undergoing a second-look laparotomy were pathologically without evident disease and only 16% had microscopic disease.

Of the 600 patients who had disease present at the time of the second-look laparotomy, there were 118 (19.6%) who at the completion of the debulking had only microscopic disease present. Of these, 37 (31%) survived. The difference between those who were debulked to microscopic disease only vs. those with any size of macroscopic disease had a difference in survival that was highly significant ($P=0.0001$). Of interest is the fact that when those with microscopic disease only were compared to those who had residual of <5 mm, the survival difference was not significant between 31% and 22% ($P=0.133$). There was also no difference statistically between the survival of those patients who were debulked to <5 mm vs. those >5 mm residual ($P=0.133$). When survival was compared between those with gross disease of 2 cm or less vs. >2 cm the difference in survival between 21% vs. 10%

was significant ($P=0.0027$). This difference may be more apparent than real. The movement of less than 10 patients from one comparison cell to another loses statistical significance. Because of the large numbers in the different subsets based upon a total population of 600 patients, it would appear that survival is dependent upon the degree of debulking carried out at the time of the second-look laparotomy, although even debulking to microscopic disease does not approach the survival of those individuals who are found with only microscopic disease at the time of the second-look laparotomy (47% vs. 31%, $P=0.0086$). The factor that is unknown is whether or not the tumour itself (biological activity, extent, etc.) predetermines who may or may not be debulked. Certainly those patients who had 'unresectable' disease at the time of the second look had a minuscule survival (4%) and the surgery obviously did not appear to be of benefit. Would survival have been the same for those debulked if they had not been debulked? Obviously that is the unanswered question although these patients did appear to have better prognostic factors, i.e. age, lower initial stage, small tumour volume at second look, etc.

Berek and associates are one of the first and few groups who have suggested that secondary debulking may improve survival. Although their article alludes to 73 of 106 patients undergoing a second-look laparotomy, the database is really 32 patients with stage III and IV disease. Even with this small number the clinical population is different: 11 had no clinical evidence of disease and isolated tumour masses were noted and resected, nine had palpable pelvic recurrence/persistence and 12 had surgery to correct partial or complete bowel obstruction. Of the 11 patients without clinical disease, six had 'optimal' (<1.5 cm) resection with median survival of 22 months compared with five patients who had non-operable resection and a median survival of 10 months. In the 21 patients who had either persistent disease or were operated on to correct bowel obstructions, six had optimal debulking for an 18-month median survival vs. 15 patients with suboptimal debulking and median survival of 5 months. There were several factors, however, that seemed to influence optimal debulking. More patients with previous optimal cytoreduction at the time of the primary surgery were able to have optimal secondary debulking compared to those patients who previously had non-optimal debulking. Of the 32 patients, 16 (50%) had inapparent ascites (<1000 ml) and only one had optimal secondary resection and she survived only 5 months. The median survival for the other 15 patients was 5 months. The optimal secondary debulking also appeared to be dependent upon the size of the lesion. Less than one-quarter of those with tumours 5 cm or greater were optimally debulked compared to 57% of those who had tumours less than 5 cm. Survival was also better if the time interval between primary and secondary surgery was greater than 12 months. The authors state that the median survival between optimal vs. suboptimal patients is significant; however, one must remember that this is a comparison of 12 vs. 20 patients with at least three different indications for surgery. In view of the fact that there appear to be so many factors of significance that determine optimal debulking, one questions whether or not statistical significance with regard to median survival is clinically significant when only one of the 32 patients showed no evident disease at the time of the report. One questions the validity of the authors' statement, 'These data indicate that a secondary attempt at bulk removal of tumor appears justified in patients who have had incomplete response to primary chemotherapy and who have no clinical evidence of ascites'. This statement is based on 16 or fewer patients!

Lippman and colleagues have also suggested that debulking and second-look laparotomy improve survival[10]. Of 70 patients undergoing second-look laparotomy, 60 had stage III or IV disease. Twenty-one patients were pathologically complete responders with 49 having positive second-look laparotomy. In multivariate analysis they found that stage I and II as well as optimal disease after the primary surgery were predictors of negative second-look laparotomy. In their analysis, survival in those patients with 2-cm disease found at the time of second-look laparotomy was not significantly different than those patients with >2-cm disease irrespective of secondary debulking results. In 27 patients who had disease >2 cm at the time of second-look laparotomy, 14 were debulked to <2 cm and they had a better survival than 13 not debulked to <2 cm (5/14 vs. 0/13, $P=0.001$, Breslow). There is some confusion in the authors' numbers as they state that in those patients with disease >2 cm noted at the second-look laparotomy, 12/27 were alive (44%), yet later they state that only 5/27 (18%) were alive. In multivariate analysis the authors state that the presence or lack of residual disease at second-look laparotomy was the only independent factor related to survival. Based on the survival of 5/27 patients, the statement that 'this strongly suggests that there is survival advantages associated with optimal resection of second look laparotomy' must be taken with some reservation.

Hoskins and associates more recently have suggested that secondary debulking to microscopic disease may result in an improved survival[14]. They evaluated 67 patients who had disease present at the time of the second-look laparotomy who were all clinically free of disease at the time of the surgical procedure. During the same time interval 194 patients with stage I–IV disease underwent a second-look laparotomy, thereby suggesting that 127 patients were negative for disease. Of the 67 patients, 80% were stage III and IV. Seventeen (25%) had microscopic disease only. Of the 50 patients with macroscopic disease, 16 (32%) were debulked to microscopic disease only. Eight of these patients survived for 5 years, which was not statistically different from those who had microscopic disease only found at second-look laparotomy. A patient with any amount of macroscopic disease remaining after the second-look laparotomy had <10% 5 year survival. Cox evaluation of factors affecting survival noted that age was an independent factor. The percentage of tumour debulked at the second-look laparotomy appeared important but was based upon only 43 patients (64%) of total. The percentage debulked was not quantified in regard to starting tumour size. Cytoreduction to microscopic disease was also an independent risk factor but this was only accomplished in 16 patients (24%). In the 194 patients that underwent a second-look laparotomy it would appear that a potential benefit of the second-look laparotomy would affect only the 17 (8.7%) patients with microscopic disease and the 16 (8.2%) who were able to be debulked to microscopic disease. Of these 33 patients, 18 (9.2% of total patient population) appeared to benefit from the second-look operation and one-half of these might have benefited from secondary debulking. Based on 16 patients the authors are probably correct in stating that secondary cytoreductive surgery at the time of second-look laparotomy may result in improved survival if they are reduced to microscopic disease.

In all patients with ovarian cancer, the role of secondary debulking may be extremely limited. If these three papers are combined in regard to survival, the comparisons are as shown in Table 37.4. Optimal and suboptimal designation varies from author to author. Berek defines this at 1.5 cm whereas Hoskins' division is microscopic vs. macroscopic. Of the 169 patients which make up the database, 42 (25%) were 'optimally' debulked but only

Table 37.4 Data which suggest the benefit of secondary debulking

	Survival	
	Optimal	Suboptimal
Berek (n=32)	1/12	0/20
Lippman (n=70)	5/14	0/13
Hoskins (n=67)	8/16	3/34
Total (169)	14/42	3/67
	P=0.0002	

14 survived. This survival represents 33% of those optimally debulked compared to 3/67 (4%) suboptimally debulked; however, the 14 represents only 8% of the patients undergoing second-look laparotomy. This number would be considerably smaller as both Berek and Hoskins removed from their denominator those patients with negative second-look laparotomy.

Is surgery justified in a large group of patients in which debulking may have an impact on only 27% of the patients (331/1207)? Even when debulking can be achieved so that residual disease is reduced to <2 cm, only 82 (6.7% of total explored) survived. If microscopic residual disease is the true division, then only 9.7% of patients explored could be 'optimally' debulked with only 3% survival (Table 37.5). This number would be even smaller if all investigators had included their negative second-look patients. The benefit-risk ratio would suggest that surgery for the possibility of debulking persistent disease does not appear justified.

37.5 CONCLUSIONS

Based on a retrospective review with all of its inherent problems it would appear the following statements can be made.

1. There appears to be little if any role for second-look laparotomy in patients with stage I and II disease. As a result the question of secondary debulking in this group of patients is moot.
2. In patients with persistent clinical disease after prescribed chemotherapy 'optimal' (<2 cm) debulking did not statistically improve survival in comparison to patients with >2-cm residual disease after secondary debulking (21% vs. 8% respectively, $P=0.191$).
3. Interventional debulking did not appear to be of benefit in the one study so reported. Survival was 30% if tumour was debulked to <1 cm vs. 35% if debulked to >1 cm.
4. In patients who were clinically without evidence of disease at the time of second-look laparotomy, secondary debulking appeared to be beneficial if the patient was reduced to two different residual loads: microscopic disease only, and >2-cm residual disease. This latter group was accomplished in 331 patients (27%) of those undergoing second-look laparotomy. Survival in this group was 25% of those debulked but only 6.7% of those explored.
5. Is a second-look laparotomy justified in order to do a possible debulking procedure when the number of patients that

Table 37.5 Effects of debulking at second-look laparotomy (patients without clinically evident disease)

	Debulked	
	Microscopic	< 2 cm
Debulked	118/1207 (10%)	331/1207 (27%)
Survived/debulked	37/118 (31%)	82/331 (25%)
Survived/total second-look laparotomy	37/1207 (3%)	82/1207 (6.7%)

might benefit are small? Can prognosis factors identify those patients who may benefit from second-look laparotomy?

REFERENCES

1. Raju, K.S., McKinna, J.A., Barker, G.H. et al. (1982) Second-look operations in the planned management of advanced ovarian carcinoma. *Am. J. Obstet. Gynecol.*, **144**, 650–4.
2. Michel, G., Zarca, D., Castaigne, D. and Prade, M. (1989) Secondary cytoreductive surgery in ovarian cancer. *Eur. J. Surg. Oncol.*, **15**, 201–4.
3. Neijt, J.P., Huinink, T.B., van der Burg, M.E.L. et al. (1987) Randomized trial comparing two combination chemotherapy regimens (CHAP-5 v CP) in advanced ovarian carcinoma. *J. Clin. Oncol.*, **5**, 1157–68.
4. Berek, J.S., Hacker, N.F., Lagasse, L.D. et al. (1983) Survival of patients following secondary cytoreductive surgery in ovarian cancer. *Obstet. Gynecol.*, **61**, 189–93.
5. Luesley, D.M., Chan, K.K., Fielding, J.W.L. et al. (1984) Second-look laparotomy in the management of epithelial ovarian carcinoma: an evaluation of fifty cases. *Obstet. Gynecol.*, **64**, 421–6.
6. Free, K.E. and Webb, M.J. (1987) Second-look laparotomy – clinical correlations. *Gynecol. Oncol.*, **26**, 290–7.
7. Mead, G.M., Williams, C.J., MacBeth, F.R. et al. (1984) Second-look laparotomy in the management of epithelial cell carcinoma of the ovary. *Br. J. Cancer*, **50**, 185–91.
8. Podratz, K.C., Schray, M.F., Wieand, H.S. et al. (1988) Evaluation of treatment and survival after positive second-look laparotomy. *Gynecol. Oncol.*, **31**, 9–21.
9. Podczaski, E., Manetta, A., Kaminski, P. et al. (1990) Survival of patients with ovarian epithelial carcinomas after second-look laparotomy. *Gynecol. Oncol.*, **36**, 43–7.
10. Lippman, S.M., Alberts, D.S., Slymen, D.J. et al. (1988) Second-look laparotomy in epithelial ovarian carcinoma. *Cancer*, **61**, 2571–7.
11. Bertelsen, K., Hansen, M.K., Pedersen, P.H. et al. (1988) The prognostic and therapeutic value of second-look laparotomy in advanced ovarian cancer. *Br. J. Obstet. Gynaecol.*, **95**, 1231–6.
12. Luesley, D., Blackledge, G., Kelly, K. et al. (1988) Failure of second-look laparotomy to influence survival in epithelial ovarian cancer. *Lancet*, 599–603.
13. Sevelda, P., Dittrich, C. and Salzer, H. (1989) The value of second-look operation in patients with advanced epithelial ovarian carcinoma. *Arch. Gynecol. Obstet.*, **244**, 79–86.
14. Hoskins, W.J., Rubin, S.C., Dulaney, E. et al. (1989) Influence of secondary cytoreduction at the time of second-look laparotomy on the survival of patients with epithelial ovarian carcinoma. *Gynecol. Oncol.*, **34**, 365–71.
15. Dauplat, J., Ferriere, J.P., Gorbinet, M. et al. (1986) Second-look laparotomy in managing epithelial ovarian carcinoma. *Cancer*, **57**, 1627–31.
16. Hainsworth, J.D., Grosh, W.W., Burnett, L.S. et al. (1988) Advanced ovarian cancer: long-term results of treatment with intensive cisplatin-based chemotherapy of brief duration. *Ann. Intern. Med.*, **108**, 165–70.
17. Creasman, W.T., Gall, S., Bundy, B.N. et al. (1989) Second-look laparotomy in the patient with minimal residual stage III ovarian cancer. *Gynecol. Oncol.*, **35**, 378–82.
18. Schwartz, P.E. and Smith, J.P. (1980) Second-look operations in ovarian cancer. *Am. J. Obstet. Gynecol.*, **138**, 1124–30.
19. Omura, G., Blessing, J.A., Ehrlich, C. et al. (1986) A randomized trial of cyclophosphamide and doxorubicin with or without cisplatin in advanced ovarian carcinoma. *Cancer*, **57**, 1725–30.
20. Walton, L., Ellenberg, S.S., Major, F. et al. (1987) Results of second-look laparotomy in patients with early-staged ovarian carcinoma. *Obstet. Gynecol.*, **70**, 770–3.

Chapter 38

Second-look laparotomy in the routine management of ovarian cancer

A.J. FERRIER and A.D. DE PETRILLO

38.1 INTRODUCTION

Carcinoma of the ovary continues to be the major cause of death from gynaecological malignancies in the western world[1] and defies reliable early diagnosis with currently available technology. Surgical extirpation remains the cornerstone for initial treatment followed by adjuvant chemotherapy for advanced disease.

Historically, second-look procedures have had both a therapeutic and diagnostic function. The procedure was first described in the early 1950s[2] for patients with stomach and colon cancer who were symptom-free at the time of surgical re-exploration. Approximately 50% of the patients demonstrated residual disease in the retroperitoneal lymph nodes and 9% of these patients were salvaged. The purpose of the second-look operation in these patients was to assess the results of initial surgery, but a survival advantae was seen in a small subset of patients. A similar approach was adopted following radiation therapy for patients with ovarian cancer without demonstrating a therapeutic benefit[3–5]. In 1976, Smith *et al.*[6] published the results of 103 patients with ovarian cancer who had undergone a second-look procedure following a complete clinical response to initial therapy. The results suggested that the procedure was safe and permitted the discontinuation of further therapy in those patients who had no residual disease. Nevertheless, the therapeutic value of a second-look laparotomy remains controversial due to contradictory data regarding the survival benefit of a pathologically documented complete response. Disease-free survival following a complete remission at second look varies markedly from 74%[7] to 38%[8].

This chapter attempts to support the view that second-look laparotomy has no place in the standard management of ovarian cancer.

38.2 INDICATIONS FOR SECOND-LOOK LAPAROTOMY

The current indications for surgical re-exploration are outlined in Table 38.1 The most widely held indication is the first one listed, as it identifies patients who have responded optimally to therapy and minim-

Table 38.1 Indications for second-look laparotomy

1. Assess treatment efficacy
   (a) Standard treatment – stop treatment
   (b) Experimental treatment – pathological response
2. Select potentially curable patients for second-line chemotherapy
3. Undertake secondary cytoreductive surgery

izes the toxicity of further adjuvant treatment modalities prescribed in the absence of demonstrable disease.

Without a sensitive non-invasive diagnostic test, clinical investigators have also utilized the findings at second-look laparotomy to test the efficacy of new treatment modalities. Furthermore, it has been hypothesized that those patients with only microscopic or small volume residual disease should respond to salvage therapies better than those patients with large volume residual disease[9]. This has spawned a wide range of investigational approaches, including intraperitoneal therapies (chemotherapy, immunotherapy, radionucleotides), whole abdomen and pelvic radiotherapy and novel intravenous chemotherapy. It has yet to be demonstrated that any salvage therapy offers a survival advantage or significantly alters the natural history of advanced ovarian cancer.

38.3 OUTCOME OF SECOND-LOOK LAPAROTOMIES

Until recently, there were approximately 130 reports in the literature detailing the experiences at various institutions with second-look procedures between 1970 and 1990. The bulk of these publications were retrospective and demonstrated considerable heterogeneity among the inception cohort in terms of extent of initial surgical extirpation, stage, tumour histology and adjuvant treatment regimens. Furthermore, a precise definition of the indication for surgical re-exploration was often not stated or the end-points of interest (macroscopic, microscopic residuum or pathological complete response) were not clearly defined. Table 38.2 summarizes the inclusion criteria selected for this overview.

Table 38.2 Inclusion criteria

1. Epithelial ovarian carcinoma
2. >60% FIGO stage III/IV
3. Adjuvant chemotherapy
4. Criterion for second-look laparotomy is clinical complete response
5. End-points of response clearly defined

38.4 SECOND-LOOK LAPAROTOMY AND SURVIVAL

The greatest benefit that a second-look laparotomy could confer on a patient with ovarian cancer is the possibility of long-term survival or perhaps cure. The initial enthusiasm for a second surgical procedure followed the publication of data from the M.D. Anderson Hospital[6] which suggested long-term survival for patients who demonstrated complete pathological remission. These data are subject to selection bias, as most of the second-look procedures were performed when the patients were in a complete clinical remission after 10 cycles of adjuvant chemotherapy, thus excluding many subjects who had already demonstrated disease progression.

Since the introduction of cisplatin as part of the adjuvant combination chemotherapy regimen for advanced ovarian cancer, the duration of therapy following initial debulking surgery has been reduced to 6–9 months in most centres. The timing of a second-look laparotomy has subsequently been advanced to take place within 9 months of diagnosis, and although the incidence of clinical complete remission has increased to 40–50% [10,11], the recurrence rate for patients with a negative second-look laparatomy has also increased[12]. Thus, it has not been possible to demonstrate a survival benefit for patients subjected to a second-look laparotomy.

Proponents of the second-look procedure claim that the data showing no survival benefit after a second laparotomy are biased due to retrospective analysis[13], or that there is an inappropriate comparison between patients who accept and those who refuse a second-look procedure[14]. The Birmingham (UK) group[15] published their data from a clinical trial in which patients were randomized to identical adjuvant cisplatin-based chemotherapy regimens with or without a second-look laparotomy, for those patients who demonstrated a complete clinical response. This study shows conclusively that the incorporation of a second-look procedure into standard management for ovarian cancer does not improve survival.

38.5 PATHOLOGICAL FINDINGS AND OUTCOME AFTER SECOND-LOOK LAPAROTOMY

Approximately 50% of patients with advanced ovarian carcinoma who undergo surgical debulking and cisplatin-based combination chemotherapy will demonstrate a complete clinical response[16–18]. Figure 38.1 outlines the clinical scenarios of patients with advanced ovarian cancer from the time of presentation. Figure 38.2 depicts the findings at second-look laparotomy[19–22].

Those patients with a complete pathological response have traditionally been those who have had treatment stopped and have been considered the group with the greatest chance of long-term survival. Many of the early reports, mostly retrospective, claimed low recurrence rates among this subgroup of

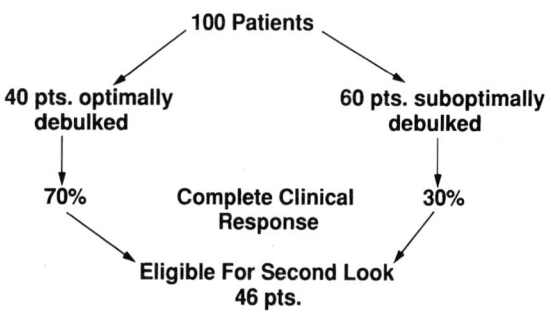

Figure 38.1 Clinical scenario of patients who present with advanced ovarian cancer and response to initial therapy.

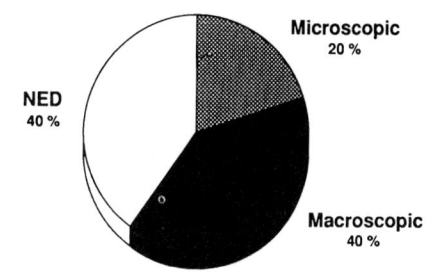

Figure 38.2 Findings at second-look laparotomy.

patients, ranging from 12–26%[23–25]. However, closer inspection reveals that some of these patients had early stage disease, tumours of low malignant potential, or were patients who remained in clinical remission after 12 months of adjuvant therapy (often single agent alkylating chemotherapy). These biases render an over-optimistic assessment of outcome for patients with advanced ovarian carcinoma.

More recent data, some of which were collected prospectively (Table 38.3), would suggest more cautious prognostication after a negative second-look operation. The recurrence rate following cisplatin-based adjuvant chemotherapy and a negative second look is approximately 50%. This disappointing observation can be attributed to two main causes.

Table 38.3 Outcome of negative second-look laparotomy

Reference	Study design	n	% Recurrence
Omura et al. [26]	Prospective	226	50
Luesley et al. [27]	Prospective	120	43
Omura et al. [28]	Prospective	62	59
Berlinson et al. [16]	Retrospective	103	45
Rubin et al. [12]	Retrospective	83	50
Ho et al. [13]	Retrospective	39	52
Podczaski et al. [21]	Retrospective	76	50
			49*

* Weighted for study design and size.

Table 38.4 Recurrence after negative second-look laparotomy

Author	n	Not assessable by second-look laparotomy (%)	Sites
Ho et al. [13]	9/17	56	Liver Retroperitoneum
Barnhill et al. [29]	6/48	50	Pleura Inguinal node Liver
Podratz et al. [22]	15/50	0	Concurrent disease in liver, chest
Rome and Fortune [30]	10/42	20	Not stated
Rubin et al. [12]	21/83	52	Liver, vagina pleura, supraclavical node

First the transcoelomic spread of ovarian cancer results in the microscopic involvement of an extensive peritoneal surface increasing the likelihood of sampling error. Secondly, intraperitoneal re-assessment will not adequately examine all the potential sites of recurrent disease, eg pleura, liver parenchyma (Table 38.4).

38.6 SECOND-LOOK LAPAROTOMY AS A DIAGNOSTIC TEST

The decision to perform a second-look procedure usually follows a thorough search for persistent disease including a careful physical examination, CA-125 measurement and an ultrasound or CT examination of the abdomen. Surgical re-exploration is therefore the final diagnostic test to assess the intraperitoneal status of the disease.

The information obtained from a second-look laparotomy may also be used for the following purposes: the data may indicate the response to therapy; patients who demonstrate a complete pathological response may be spared the morbidity and potential mortality of further treatment; the estimated volume of residual disease following first-line therapy may indicate the likely responsiveness to secondary treatment. Intraperitoneal chemotherapy has been shown to result in a prolonged progression-free interval when administered to patients with microscopic residual disease compared to residual disease with nodules greater than 2 cm[9]. Finally, the estimate of residual disease at the end of a second-look procedure is predictive of the subsequent clinical course and prognosis of the disease. Patients who have a large volume of unresectable disease at completion of the second-look procedure have a poor prognosis independent of salvage treatments with less than 10% survival at 5 years [19,31,32].

The overriding criterion to use when deciding which diagnostic data to seek should be their clinical usefulness and the patient benefit. The following guidelines are suggested to assess the usefulness of a diagnostic test.

1. Does the test incorporate a 'gold standard'?
2. What is the setting in which the test is used and what are the selection criteria for the subjects undergoing the procedure?
3. Does the description of the test procedure allow exact replication by other clinicians?
4. Has the utility of the test been determined?

The results of a second-look laparotomy are ultimately assessed by the histopathological description (gold standard) of the biopsy material collected by the surgeon. Hence, the false positive rate of the procedure is very low. The setting in which the test is performed is, however, more variable. A review of the literature reveals that the inception cohort of second-look procedures varies in many respects. First the FIGO stage may vary from stage I to stage IV. Clearly when these patients are grouped together for analysis, the patients with stage I disease (who have the higher probability of a negative second look but constitute only 15% of all patients with ovarian cancer) will bias the overall result and therefore overestimate the negative predictive value of the test. Secondly, the timing of surgical re-exploration is crucial as a delay of 12–18 months from diagnosis will self-select a more favourable group of patients who have a higher probability of long-term survival and therefore negative second-look procedures.

Although the literature is replete with detailed descriptions of second-look laparotomy, the completeness of the procedure varies amongst institutions. In our centre, a limited number of random biopsies are taken from the pelvic peritoneum, paracolic gutters, small bowel mesentery and diaphragms after the sampling of suspicious areas, while in other centres more extensive sampling is performed in an attempt to detect microscopic disease. The chance of detecting subclinical disease must increase with more extensive sampling, but this must be weighed against the likelihood of increased surgical morbidity with a more thorough surgical procedure.

The data obtained from a second-look procedure can be expressed as a dichotomous outcome, positive or negative for disease. Since recurrence mandates the presence of disease at second look, albeit subclinical, a table can be constructed to generate some indexes which assess the clinical usefulness

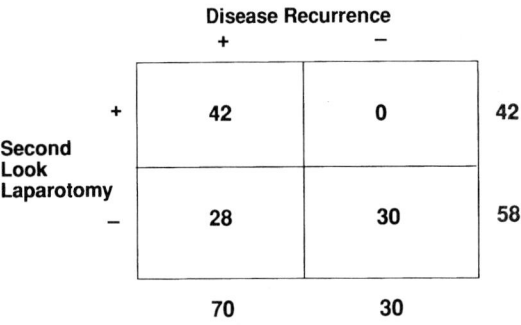

Figure 38.3 Disease recurrence vs. outcome of second-look laparotomy in 100 patients.

of second-look laparotomy as a diagnostic test (Figure 38.3).

In the theoretical model where 200 patients with advanced ovarian cancer undergo initial surgical debulking followed by cisplatin-based combination chemotherapy, 50% of subjects will have a complete clinical response and be eligible for a second-look procedure. Of these subjects, 30% will enjoy long-term disease-specific survival (This estimate assumes that only subjects who have a complete clinical response to initial therapy have the potential of long-term survival. Long-term disease-specific survival for the entire cohort (200 patients) would therefore be 15%.) The complement of disease-specific survival is therefore 70%, an estimate of disease recurrence or disease prevalence among the patients eligible for surgical re-exploration.

Second-look laparotomy will be positive in 42 or the 100 patients with a complete clinical response (60% of subjects will have either microscopic (20%) or macroscopic (40%) disease at second-look laparotomy: $70 \times 0.6 = 42$). Furthermore, approximately half of those patients with advanced ovarian cancer and a negative second-look procedure will experience disease recurrence. The sensitivity of a positive second-look procedure is therefore 60% and specificity is 100% (sensitivity = true positive/(true positive + false negative); specificity = true negative/(false positive + true negative)). Another index, predictive value, offers a more clinically relevant measure of the diagnostic test. In this example, the positive predictive value of a second-look laparotomy is 100% and the negative predictive value 52%. However, it is the negative predictive value that is the more important to clinicians.

The disadvantage of relying on the predictive value of a diagnostic test is its dependence on disease prevalence (Table 38.5). When the prevalence or pre-test probability of disease at the time of second look is high, i.e. 80–70% with advanced ovarian carcinoma, the negative predictive value of a second-look laparotomy is 52 and 38%, respectively.

The complement of the negative predictive value is the probability of disease despite the negative test, i.e. the post-test probability. For advanced disease, the range of post-test probabilities would be 48–62%. Therefore, a negative second-look procedure reduces the probability of disease recurrence in a setting of advanced disease by 18–22% (pre-test probability = 70%, post-test probability = 48%; the gain from undertaking a second-look procedure which results in a negative result is $70 - 48 = 22\%$). Nevertheless, a negative predictive value which may be 50% is no better than chance. Many clinicians would question the clinical usefulness of the data obtained from an invasive and expensive diagnostic procedure.

In contrast to advanced ovarian carcinoma,

Table 38.5 Effect of disease prevalence on negative predictive value

Disease prevalence (%)	90	80	70	60	50	40	30	20	10
Positive predictive value	100	100	100	100	100	100	100	100	100
Negative predictive value	22	38	52	63	71	79	85	91	96
Risk of disease with a negative test (%)	78	62	48	37	29	21	15	9	4
Gain from test in predicting disease over pre-test probability (%)	12	18	22	23	21	19	15	11	6

early stage disease has a pre-test probability of recurrence of approximately 20%. The negative predictive value of a second-look laparotomy increases to 91% and the post-test probability decreases to 9%. Negative findings at surgical re-exploration therefore offer an 11% reduction in the pre-test probability of detecting disease, and relatively little is learned about the status of the disease.

It is therefore apparent that only a modest amount of information is gained from a second-look procedure in patients with advanced ovarian carcinoma, and the diagnostic benefits obtained in patients with early stage disease are marginal. The diagnostic usefulness of a second-look laparotomy, expressed in terms of sensitivity, specificity and predictive value, remain limited and do not appear to justify even the modest surgical morbidity associated with procedure.

38.7 SECOND-LOOK LAPAROTOMY AS A THERAPEUTIC MANOEUVRE

A more recent indication for second-look operations is the purported survival benefit following secondary cytoreductive surgery (Table 38.6). There are several conflicting reports in the literature regarding the efficacy of such secondary debulking. Several retrospective reports compare the survival of patients whose disease could be secondarily debulked to small volume residuum with those with unresectable disease. The comparison fails to recognize the confounding effect of tumour biology which may render more aggressive tumours unsuitable for surgical extirpation. Other authors compare patients found to have microscopic disease at the beginning of the second-look operation with those resectable to microscopic residuum at the completion of the procedure, citing a comparable survival rate. Again this compar-

Table 38.6 Macroscopic disease at second-look laparotomy: effect of secondary cytoreductive surgery

Author	Design	Residuum at second-look laparotomy	Residuum after second-look laparotomy	% Survival (follow-up in years)
Bertelsen et al. [19]	Prospective	Micro	–	24 (4)
		>1 cm		17 (4)
			Micro	45 (4)
			<1 cm	13 (4)
			>1 cm	11
Luesley et al. [15]	Prospective	Micro		70 (3)
			Micro	20 (3)
			<2 cm	30 (3)
Hoskins et al. [31]	Retrospective	Micro		62 (5)
			Micro	51 (5)
			>2 cm	<10 (5)
Lippman et al. [32]	Retrospective	Micro		49 (5)
			Micro	40 (5)
			>2 cm	0 (5)

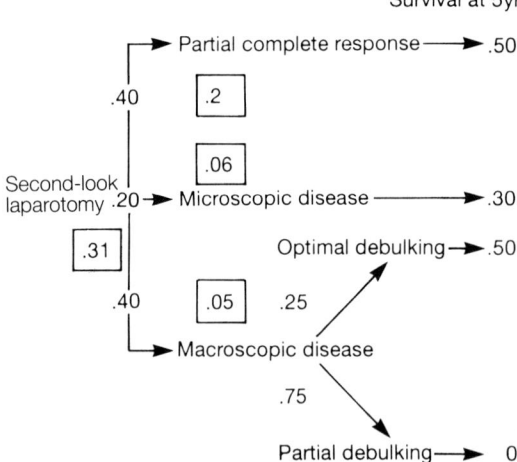

Figure 38.4 Outcome following second-look laparotomy.

ison ignores inherent tumour biology and assumes that the two groups are comparable in every other respect.

Several reports with data collected prospectively have analysed the effect of secondary cytoreductive surgery. While some have found no apparent benefit for such a manoeuvre[15], others have found striking results with marked survival benefit[19]. Closer analysis would suggest that the apparent improvement in survival is more likely due to a biologically less aggressive tumour than surgical prowess.

Figure 38.4 stratifies the findings at second-look laparotomy, the impact of secondary cytoreductive surgery and estimated survival at 5 years. It is assumed that no single non-surgical salvage therapy offers an unequivocal survival advantage in this setting. The majority of the long-term survivors have no pathological evidence of disease at second-look laparotomy. If it is assumed that the 50% 5-year survival estimate is accurate following secondary debulking, the apparent benefit on overall survival for patients undergoing a second-look laparotomy is small and the benefit for patients with advanced ovarian cancer is marginal. (Assuming only 50% of patients with advanced disease are eligible for a second-look procedure, then 2.5% of patients would benefit from secondary debulking.) Although the hypothesis supporting a survival advantage following secondary cytoreductive surgery can only be tested in the setting of a randomized controlled trial, the evidence available would not strongly support the implementation of such a study. It is clear, therefore, that the utility of a second-look procedure, a measure of patient benefit, has not been established to date.

38.8 CONCLUSION

Without a proven effective second-line therapy it would appear illogical to pursue second-look laparotomy as part of the standard management of ovarian cancer. Surgical re-exploration in these patients serves little more than to document the natural history of the disease process, while imposing both a physical/psychological and financial burden on the patient.

Alternative strategies aimed at investigating new treatments and improving current management of ovarian cancer include:

1. adhere to 'hard' end-points of disease-free progression or disease-specific survival;
2. recruit patients with measurable disease for new treatment protocols;
3. expand the number of phase III multi-centre trials;
4. pursue a philosophy that maximizes and measures the quality of life that many of these women forfeit as they desperately seek a cure from new, frequently toxic therapies.

REFERENCES

1. Silverberg, E. and Lubera, J.A. (1988) Cancer statistics, 1988. *Ca-Cancer J. Clin.*, **38**, 5.
2. Wangensteen, O.H., Lewis, F.J. and Tongen,

L.A. (1951) Second look in cancer surgery. *Lancet*, **71**, 303–7.
3. Parks, T.J. (1945) Carcinoma of the ovary treated preoperatively with deep X-ray: report of three cases. *Am. J. Obstet. Gynecol.*, **49**, 676–85.
4. Corsaden, J.A. (1962) in *Gynecologic Cancer*, 3rd edn., Williams & Wilkins, Baltimore, p. 493.
5. Kottmeier, H.L. (1961) Radiotherapy in the treatment of ovarian cancer. *Clin. Obstet. Gynecol.*, **4**, 865–74.
6. Smith, J.P., Delgado, G. and Rutledge, F. (1976) Second look operations in ovarian cancer. *Cancer*, **38** 1438–42.
7. Greco, F.A., Hande, K.A., Jones, H.W., et al. (1984) Advanced ovarian cancer: long-term follow up after brief intensive chemotherapy. *Proc. Am. Soc. Clin. Oncol.*, **3**, 166.
8. Sutton, G.P., Stehman, F.B., Ehrlich, C.E., et al. (1986) Seven-year follow up of patient receiving cisplatin, adriamycin and cyclophosphamide (PAC) chemotherapy for stage III and IV epithelial ovarian carcinoma. *Proc. Am. Soc. Clin. Oncol.*, **5**, 120.
9. Howell, S.B., Zimm, S., Markman, M., et al. (1987) Long-term survival of advanced refractory ovarian carcinoma patients with small volume disease treated with intraperitoneal chemotherapy. *J. Clin. Oncol.*, **5**, 1607–12.
10. Williams, C.J., Mead, G.M., Macbeth, F.R., et al. (1985) Cisplatin combination chemotherapy versus chlorambucil in advanced ovarian carcinoma: mature results of a randomised trial. *J. Clin. Oncol.*, **3**, 1455–62.
11. Vogl, S.E., Pagano, M., Davis T., et al. (1983) Platinum based combination chemotherapy versus melphalan for advanced ovarian cancer. *Proc. Int. Congr. Chemother.*, **207**, 9–13.
12. Rubin, S.C., Hoskins, W.J., Hakes, T.B., et al. (1988) Recurrence after negative second-look laparotomy for ovarian cancer: analysis of risk factors. *Am. J. Obstet. Gynecol.*, **159**, 1094–8.
13. Ho, A., Beller, U., Speyer, J.L., et al. (1987) A reassessment of the role of second look laparotomy in advanced ovarian cancer. *J. Clin. Oncol.*, **5**, 1316–21.
14. Cohen, C.J., Golgberg, J.D., Holland, J.F., et al. (1983) Improved therapy with cisplatin regimens for patients with ovarian carcinoma (FIGO stages III and IV) as measured by surgical end staging (second look). *Am. J. Obstet. Gynecol.*, **145**, 955.
15. Luesley, D., Lawton, F., Blackledge, G., et al. (1988) Failure of second look laparotomy to influence survival in epithelial ovarian cancer. *Lancet*, 599–602.
16. Berlinson, J.L., Lee, K.R., Jarrell, M.A., et al. (1990) Management of epithelial ovarian neoplasms using a platinum-based regimen: a 10-year experience. *Gynecol. Oncol.*, **37**, 66–73.
17. Creasman, W.T., Gall, S., Bundy, B.N., et al. (1989) Second look laparotomy in the patient with minimal residual stage III ovarian cancer (A Gynecologic Oncology Group Study). *Gynecol. Oncol.*, **35**, 378–82.
18. Schneider, J., Martin, M., Erasun, F., et al. (1990) Cisplatin containing versus cisplatin free adjuvant chemotherapy in ovarian cancer. *Oncology*, **47**, 109–11.
19. Bertelsen, K., Hansen, M.K., Pedersen, P.H., et al. (1988) The prognosis and therapeutic value of second look laparotomy in advanced ovarian cancer. *Br. J. Obstet. Gynaecol.*, **95**, 1231–6.
20. Gershensen, D.M., Copeland, L.J., Wharton, J.T., et al. (1985) Prognosis of surgically determined complete responders in advanced ovarian cancer. *Cancer*, **55**, 1129–35.
21. Podczaski, E., Manetta, A., Kaminski, P., et al. (1989) Survival of patients with ovarian carcinomas after second look laparotomy. *Gynecol. Oncol.*, **36**, 43–7.
22. Podratz, K.C., Malkasian, G.D., Wieand, H.S., et al. (1986) Recurrent disease after negative second look laparotomy in stages III and IV ovarian carcinoma. *Gynecol. Oncol.*, **29**, 274–82.
23. Berek, J.S., Hacker, N.F., Lagasse, L.D., et al. (1984) Second look laparotomy in stage III epithelial ovarian cancer: variables associated with disease status. *Obstet. Gynecol.*, **64**, 207.
24. Dauplat, J., Ferriere, J.-P., Gorginet, M., et al. (1986) Second look laparotomy in managing epithelial ovarian carcinoma. *Cancer*, **57**, 1627–31.
25. Schwartz, P.E. and Smith, J.P. (1980) Second look operations in ovarian cancer. *Am. J. Obstet. Gynecol.*, **138**, 1124.
26. Omura, G.A., Bundy, B.N., Berek, J.S. et al. (1989) Randomised trial of cyclophosphamide

plus cisplatin with or without doxorubicin in ovarian cancer: a gynaecologic oncology group study. *J. Clin. Oncol.*, **7**, 457–65.
27. Luesley, D.M., Chan, K.K., Lawton, F.G., et al. (1989) Survival after negative second look laparotomy. *Eur. J. Surg. Oncol.*, **15**, 205–10.
28. Omura, G.A., Blessing, J.A., Ehrlich, C.E., et al. (1986) A randomised trial of cyclophosphamide and doxorubicin with or without cisplatin in advanced ovarian cancer. *Cancer*, **57**, 1725–30.
29. Barnhill, D.R., Hoskins, W.J., Heller, P.B., et al. (1984) The second look surgical reassessment for epithelial ovarian carcinoma. *Gynecol. Oncol.*, **19**, 148–54.
30. Rome, R.M. and Fortune, D.W. (1988) The role of second look laparotomy in the management of patients with ovarian carcinoma. *Aust. NZ J. Obstet. Gynaecol.*, **28**, 318.
31. Hoskins, W.J., Rubin, S.C., Dulaney, E., et al. (1989) Influence of secondary cytoreduction at the time of second look laparotomy on the survival of patients with epithelial ovarian cancer. *Gynecol. Oncol.*, **34**, 365–71.
32. Lippman, S.M., Alberts, D.S., Slymen, D.J., et al. (1988) Second look laparotomy in epithelial carcinoma: prognostic factors associated with survival duration. *Cancer*, **61**, 2571–7.

Recommendations and conclusions

GENETIC ASPECTS

1. Women who have a first-degree relative with ovarian cancer fall into three groups:
   (a) those whose affected first-degree relative has herself another first- or second-degree relative affected by ovarian cancer;
   (b) those whose affected first-degree relative has first- or second-degree relatives with breast, colorectal or endometrial cancer (especially if before the age of 50 years);
   (c) those where there are no other first- or second-degree relatives with early onset or multiple cancers in the family.

2. Women in groups (a) and (b) above should have a careful family history taken, to evaluate in detail their likely risk of ovarian cancer. The risks of ovarian cancer in these families may be as high as 25–40% lifelong, or approximately 1% per year between the ages of 30 and 70 years.

   Members of some families may also be at similar or higher risk for other cancers, including colorectal and breast cancer.

   With respect to the risk of ovarian cancer, women in families where there is a strong indication of inherited predisposition should be offered the best currently available screening. As the efficacy of current methods of screening is unproven, prophylactic bilateral oophorectomy and hysterectomy is appropriate once childbearing has been completed.

3. Women in group (c) above may have a slightly elevated risk of ovarian cancer, but current epidemiological data suggest that this risk is still low, and therefore at present neither screening nor prophylactic oophorectomy is appropriate in routine practice.

4. Studies of families with two or more cases of ovarian cancer (or related cancers, such as breast, colorectal or endometrial cancer) in close family members, where inherited predisposition is likely to be present, have considerable implications for the study of ovarian cancers in general. Studies in these families should include the following.

   (a) Genetic linkage studies.
   (b) Pathology investigations of the peritoneum and ovarian epithelium from women at risk who have been treated by oophorectomy, as these may yield information about field change and/or pre-neoplastic changes.
   (c) Studies of screening methods. Such families are a high risk group in

Recommendations and conclusions

which screening methods can be evaluated, although the results may not necessarily be applicable to ovarian cancer in the general population.
(d) Evaluation of the efficacy of prophylactic oophorectomy.

5. A limiting factor in all the studies outlined in 4 above, will be the availability of and access to well-documented families. Overcoming this will require national and international collaborations, some of which are already in place.

   Clinicians are urged to take a short family history from every patient with ovarian cancer, to include relatives out to aunt and grandmother, and to contribute information about families with two or more cases of ovarian or associated cancers (see 1(b) above) in close relatives to the appropriate national registry. Dr. Bruce Ponder (Cambridge, UK) and Dr. Patrice Watson (Omaha, USA) maintain registers of such families, and are happy to provide information and advice, including contacts with national study groups, and are also willing to accept suitable families into their registers. Wherever possible, blood samples and tissue (including normal ovaries and peritoneal biopsies taken at prophylactic oophorectomy) should be collected and stored from family members.

6. Studies on somatic changes in ovarian tumour samples are well established and should continue. Clinicians and pathologists are urged to collaborate by providing matched frozen samples of tumour and blood from the same patient. Considerations should be given to the setting up in busy centres of archives of frozen and fixed tissues to support these studies.

7. So far as the genetic basis of ovarian cancer is concerned, major clues to the location of potential tumour suppressor genes have come from studying allele loss. Previously published data suggested allele loss for chromosomes 11p, 3p, 5p, 6q, 21q and both 17p and 17q. Losses on 6q (near the position of the oestrogen receptor gene), and on 17q (where there is a possible linkage with ovarian cancer familial susceptibility) have been confirmed. Furthermore, confirmation for losses on 11p and 17p has been provided.

8. It is likely that the allele losses on 17p are due to p53 gene mutations. Immunohistochemical studies using antibody to the p53 gene product, showed that up to 50% of ovarian tumours over-expressed p53. Sequencing studies of a small number of tumours have shown complete correlation between this over-expression and the occurrence of p53 mutations.

9. The antigen recognized by the ovarian carcinoma associated monoclonal antibody MOv18 has been cloned and shown to be the folate binding protein. Significant allele loss has been demonstrated for this gene on chromosome 11q.

GROWTH FACTORS AND CYTOKINES

1. Understanding the biology of growth regulation in ovarian cancer requires knowledge of the rate-limiting steps in growth factor and cytokine action.
2. *In vitro* and *in vivo* systems should be established and disseminated, allowing investigation of the interaction between ovarian cancer cells, stroma and infiltrating haematopoietic and lymphoid cells.
3. Techniques should be established to allow comparison of benign and malignant ovarian epithelial cells, and should include methods for purification and culture, as well as techniques to charac-

Immunology

terize freshly isolated cells, including polymerase chain reaction, immunohistochemistry and *in situ* hybridization.

4. It is important that future work in this area for ovarian cancer should attempt to establish how stroma will influence the effect of TGF-beta and other growth factors on the growth of epithelial cells.

5. Studies are required to elucidate why TGF-beta inhibits growth of normal ovarian epithelial cells but, at least in some cases, does not give corresponding inhibition of ovarian cancer epithelial cells.

6. The question whether clinical studies with folic acid blockers might throw any light on the growth advantage for tumour cells in having high levels of folic acid binding protein needs to be addressed.

7. The stimulation by TNF/IL of the coagulation cascade prompts further study of the mechanism by which anticoagulants can potentiate several different chemotherapy regimes, and whether this observation may be quantified properly.

8. The role of coagulation factors and angioneogenesis factors in the initiation and progression of ovarian cancer needs to be determined.

9. In cells expressing IL-6 protein but no detectable mRNA, the question is raised whether this is simply a matter of assay sensitivity, or whether there are other levels of control of cytokine production in ovarian cancer cells.

10. It is muted that some ovarian cancers may be due to repeated damage to the epithelial layer as a consequence of normal ovulation. This requires clarification, and the investigations should cover the role of growth factors released during ovulation in potential initiation and progression of ovarian epithelial tumours. The question whether hormonal control of ovulation may be used as a specific preventative measure in relation to subsequent development of ovarian cancer also needs further study.

11. Concerns are expressed about the observation of ovarian cancer developing in some patients who had undergone superovulation stimulation as a treatment for anovulatory infertility, and a controlled study on the true risk is required urgently.

12. Changes in M-CSF may be associated with response to progestins in ovarian cancer, and this area requires further attention.

13. Studies are required to determine whether distal mediators in growth factor signal transduction cascades, including TTK, MAP2 and RAF, are targets for therapy in ovarian cancer.

14. TGF-α levels, and those of other markers, may give better prognostic discrimination when assessed separately in hormone-sensitive and -insensitive tumours. Further investigations in this area are required. These observations may facilitate subgrouping of the heterogenous collection of lesions included under the term ovarian cancer.

IMMUNOLOGY

1. The identification of antigens associated with ovarian cancers has progressed with the advent of the monoclonal antibody technique.

2. Such antibodies have already been used for *in vivo* and *in vitro* diagnostic techniques, and are being investigated for use in specific targeting of toxic agents on cancers.

3. Where the antigen, or epitopes expressed on it, are relatively tumour-specific, it may be possible to use the antigen itself for active immunotherapy to generate a host response and tumour rejection.

4. Polymorphic epithelial mucin (PEM) is a tumour-associated antigen expressed

Recommendations and conclusions

and aberrantly glycosylated in more than 90% of ovarian cancers.
5. Studies of PEM have focused on molecular analysis of the antigen and the differences in glycosylation which result in the expression of novel tumour-associated epitopes.
6. Syngeneic and transgenic models have been developed for evaluating immune responses to PEM-based immunogens.
7. The proto-oncogene *erb*B2 is over-expressed in some ovarian cancers and toxin conjugates of antibodies to the extracellular domain are being considered as possible therapeutic agents.
8. Studies were presented analysing the response of the cell to binding of 11 antibodies, covering several epitopes on the *erb*B2 receptor, and examining blocking of ligand by the antibodies. Some antibodies appear to inhibit growth of the cells.
9. The antibodies were thought to hold some promise as candidates for targeting toxins, since the toxin conjugates did induce cell killing of cells expressing high levels of *erb*B2.

DRUG RESISTANCE

1. A number of strategies to overcome drug resistance during chemotherapy for ovarian cancer are being evaluated, based on principles of dose intensity, known mechanisms of drug resistance, and alternative biological approaches.
2. Although co-operative randomized studies have not yet documented an advantage for dose-intense vs. conventional therapy, individual groups have continued to explore more aggressive treatment regimes.
3. The availability of improved adjunctive measures, such as bone marrow, peripheral blood stem cell, cytokine, and blood product support, has allowed these studies to proceed with acceptable levels of toxicity. For example, current studies at Fox Chase Cancer Center, using a combination of cisplatin and carboplatin, delivering over 900 mg/m^2 of cumulative cisplatin equivalent, with a delivered dose intensity of 40–60 mg/m^2/week, have produced encouraging response rates, with a low incidence of neuropathy or other long-term toxicity.
4. The role of transmembrane signal transduction pathways in the process of drug resistance was reviewed, together with preliminary results of efforts to identify genes involved in cisplatin resistance, through subtractive hybridization and cloning. These studies emphasize the complex interaction of multiple cellular processes contributing to drug resistance, and suggest that interference with any single process may only achieve a small measure of reversal of drug resistance. These *in vitro* observations need extension to *in vivo* tumour models before eventually being applied to clinical trials.
5. Studies of the utilization of GM-CSF in combination with cyclophosphamide and cisplatin were reviewed. The possibility that the combined sequential use of GM-CSF before and after chemotherapy may ameliorate thrombocytopaenia, as well as neutropaenia, deserves further study.
6. The status of immunotherapy with IL-2, LAK cells, IL-4 and other cytokines in combination was reviewed. Initial enthusiasm for these approaches has not yet produced a high frequency of significant clinical responses. However, advances in understanding potential anti-tumour mechanisms, dose–response relationship, host toxicity, and cytokine interactions will promote the focused evaluation of these approaches in patients with ovarian cancer.
7. The development and current status of

monoclonal antibody–toxin conjugates was also reviewed. Collaborative clinical trials with two different immunotoxins have shown encouraging although minor anti-tumour responses at the expense of fatal unpredicted central neurological toxicity. These studies have illustrated the potential promise but also the substantial difficulties with antibody-guided therapy.

8. Recent efforts to develop anti-tumour immunotoxins have focused on the c-*erb*B2 antigen. Insufficient internalization of c-*erb*B2 may restrict immunotoxin activity to tumour cells with high-level antigen expression.
9. Application of recombinant DNA technology has been used to construct small single-chain chimeric proteins with both antigen-binding and toxin activity. Preclinical evaluations are in progress and will hopefully lead to new clinical trials.
10. Other strategies for overcoming drug resistance, not specifically addressed on this occasion, include ongoing studies with inhibitors of glutathione synthesis, the p-glycoprotein transmembrane drug efflux pump, and DNA repair. The results of these are awaited with interest.

EXPERIMENTAL THERAPEUTICS

1. One of the mechanisms of direct action of TNF is its activity on the mitochondrium. Mitrochondrial poisons may enhance this action. Using an *in vitro* cytotoxicity assay, the lipophilic, cationic antimicrobial agent dequalinium chloride (DECA) has been investigated, and displayed potent activity against a variety of ovarian cancer cell lines.
2. DECA was also shown to augment *in vitro* cytotoxic activity of TNF and cisplatin as well as reversal of resistance in cisplatin-resistant cell lines and 5-fluororacil-resistant cell lines.
3. These *in vivo* studies using DECA should be further explored and expanded and consideration should be given to phase I clinical trials.
4. Evidence was considered demonstrating that most gynaecological malignancy cell lines are resistant to lysis by TNF-α.
5. Such cell lines can become sensitive to lysis by TNF–α in the presence of protein synthesis inhibitors, such as actinomycin-D. TNF-α must bind to a higher affinity receptor which initiates the lytic mechanism. The exact reaction of the binding receptor is not known. Subsequent to this, phospholipase A_2 is activated, initiating the conversion of arachidonic acid to prostaglandin and leukotrienes. It is probable that this eventually results in the generation of free radicals and subsequently lysis.
6. Phase I clinical trials utilizing protein synthesis inhibitors, such as actinomycin-D and TNF should be considered.
7. The results of a study of 300 patients, previously treated with cytotoxic agents, and with minimal residual disease, treated with interleukin-2 (IL-2) intraperitoneally, were presented. Approximately 10% complete response rates were confirmed by laparotomy.
8. Similar studies using LAK cells in combination with IL-2 were also presented, with one complete responder and three partial responders out of 17 patients treated.
9. Rather disappointing results were experienced using treatment with tumour infiltrating lymphocytes.
10. Possibly more effective treatment regimes using IL-2 and blocking factors should be considered.
11. The ATP bioluminescence assay, which measures cellular ATP as a parameter of cell viability, was considered. This short-term assay may provide the ability to test a tumour specimen with a range of

Recommendations and conclusions

different drug concentrations and combinations *in vitro*. This assay should be considered for incorporation in clinical trials.
12. Using three different cancer cell lines inoculated intraperitoneally in mice the effect of combined cisplatin and recombinant IL-2 has been studied. The combination resulted in synergistic anti-tumour response. Cisplatin showed a dose-dependent effect on the host immune system. In low doses a stimulating effect on the immune system and increased macrophage toxicity were observed. These observations may lead to the development of less toxic therapy.

PATHOLOGY

1. Ovarian cancers not suspected by the operating surgeon or the pathologist on routine inspection of excised ovaries, seem to fall into two groups:
   (a) those apparently developing in the absence of other ovarian pathology;
   (b) those occuring in association with pre-existing ovarian pathology such as endometriosis, Brenner tumours, mucinous and serous cystadenomas.
2. There seems to be an association between endometriosis and endometrioid and clear cell carcinomas.
3. Where ovarian carcinoma developed in relation to pre-existing cystadenoma, the exact sequence was not clear, although there seemed to be a gradient in the mean age for development of benign tumours, borderline tumours and carcinoma.
4. Some borderline tumours have been noted, showing foci of microinvasion, and whose behaviour appears far less aggressive than is the case with stage I carcinomas.
5. Morphologically identifiable pre-neoplastic changes in the ovary still remain enigmatic and elusive.
6. Further work in this field is encouraged using a range of ancillary techniques such as immunocytochemistry, flow cytometry and ploidy measurement.

EARLY DETECTION

1. Population screening with ultrasound apparently has high sensitivity but poor specificity for early ovarian cancer due to benign and functional ovarian pathology.
2. If ultrasound screen is targeted on a high risk group, the specificity is unchanged, but, because of the higher prevalence of the condition in such a population, the positive predictive value is improved.
3. In the highest risk group. women with a family history of site-specific ovarian cancer, the chance of ovarian cancer being found at laparotomy in screen-positive individuals is one in three.
4. Targeted ultrasound screening in such defined high risk groups would have limited impact on the overall mortality from the disease in the general population.
5. To improve the specificity of ultrasound screening in the general population a morphological scoring system has been developed, and using transvaginal colour flow pulsed Doppler ultrasound, a pulsatility index has been defined.
6. The colour flow Doppler approach has been limited because of difficulties in distinguishing between the corpus luteum and early ovarian cancer. In this respect the morphological score may be a better discriminant, as may a combination of morphological scoring and blood flow alteration.
7. Using colour Doppler, the concept of angioneogenesis in solid ovarian tumours greater than 2–3 cm in diameter

is stressed and a resistance index described.
8. Measurement of serum CA-125 when used alone as a screening test for early ovarian cancer has inadequate specificity. Combined with ultrasound screening however a 99.9% specificity has been obtained by one group. The sensitivity of serum CA-125 measurement used alone or in combination with ultrasound is unlikely to exceed 50–60%.
9. The best estimates of lead time over clinical diagnosis associated with serum CA-125 screening are 1.9 and 3.3 years. With ultrasound the lead time is estimated to be 1.2 years.
10. Improved sensitivity may be possible by the use of complimentary markers, but it is unlikely that the use of such markers will enhance the sensitivity of serum CA-125 screening by more than 10%.
11. No ideal combination of screening markers exists at present.
12. The need for available bank serum from screened women to test new marker combinations rapidly in the future is considered to be of great importance.

NEW THERAPIES

1. In patients with microscopic or residual disease of less than 5 mm, and who have previously responded to cisplatin, intraperitoneal cisplatin can produce a 45% cure rate (surgically defined).
2. Patients who have not responded to previous intravenous cisplatin therapy should not receive intraperitoneal cisplatin, even if the residual disease is small volume.
3. Phase I and II trials with r-α-interferon showed a combined overall complete response rate of 32%, defined surgically. Where the disease was microscopic or less than 5 mm, the response was 50%. All complete responses were confined to the minimal residual disease group.
4. In two phase I and II trials of r-α-interferon plus cisplatin, the overall complete response rate defined surgically was 45%. Again all responses were confined to the minimal residual disease group. No complete responses were noted in another phase II trial, as a great majority of the evaluable patients had greater than minimal residual disease, were platinum refractory, or had extensive carcinomatosis. These treatments therefore should be confined to patients with favourable residual disease.
5. Radio-labelled antibodies are active therapeutic agents and significant responses have been seen in lymphoma, leukaemia and neuroblastoma. However, experience of this treatment approach in ovarian cancer is still very limited. Some groups have used the radio-labelled antibodies intraperitoneally. The use of ^{131}I-labelled B72.3 antibody in a phase I trial at a dose of 125 mCi was well tolerated.

PROGNOSTIC FACTORS

1. Measurement of tumour ploidy may be a useful way of providing an additional prognostic feature and should be considered as a possible routine where patients are being entered into formal clinical trials.
2. Diploid tumours are associated with longer survival than aneuploid tumours.
3. Aneuploid tumours are less often fully debulked than diploid tumours.
4. Flow cytometry studies seemed more sensitive than cytology in assessing malignancy of cells from peritoneal washings in patients with diploid cells.
5. In borderline tumours hyperdiploid status relates to high progression rates into frank malignancy.

Recommendations and conclusions

6. S-phase fractions greater than 18% are associated with a short time to relapse and low survival.
7. Ploidy analysis of 24-hour peritoneal washings following surgery suggests that patients with a high proliferative fraction may be more likely to respond to chemotherapy.

META-ANALYSIS OF TREATMENT

1. Meta-analysis provides a technique for reviewing data from clinical trials by pooling, to increase statistical power, resolve uncertainty, or reconcile differences in outcome between trials of similar design.
2. The results of the British Medical Research Council's meta-analysis of treatment of advanced ovarian cancer (Advanced Ovarian Cancer Trialists Group), were presented. Individual patient data were collected from 45 of 53 eligible trials, representing 95% of the patients randomized in the 53 studies.
3. Systematic meta-analysis including individual patient data from all published and unpublished randomized trials is a far more powerful tool than simple literature review or meta-analysis of published data.
4. There is an advantage for platinum combinations over platinum alone in the order of 7–10%, appearing after 2 years.
5. Carboplatin appears equivalent to cisplatin in the first 4 years of follow-up, but long-term follow-up is needed.
6. Clinical trials to date have been an order of magnitude too small to identify the modest effects of current chemotherapy.
7. The majority of ovarian cancer patients are still not entered in phase III clinical trials. While selected centres may provide needed innovative therapy, patient accession to true 'large-scale trials' should be encouraged. The International Collaborative Ovarian Cancer Group has started such trials.
8. A meta-analysis attempting to asess the role of cytoreductive surgery in patients with advanced ovarian cancer was also considered, reviewing 58 studies, 97 cohorts and covering almost 7000 patients. Initially there appeared to be a 16.3% increase in mean survival for each 10 point increase in cytoreduction. However, with corrections for close intensity, contribution of cisplatin and other variables, the impact of cytoreduction appeared to be reduced to statistical insignificance. The group raised questions about the design and methodology of the study.

SURGERY

1. Surgery still appears to have an important role in the management of patients with ovarian cancer.
2. Regarding aspiration of cystic ovarian masses for diagnostic purposes, it was generally felt that this practice was potentially dangerous, in view of the significant false-negative rate for cytology of the aspirate, and the potential for disseminating malignant cells into the peritoneal cavity.
3. The surgical management of advanced ovarian cancer remains controversial, because there has never been a prospective randomized study to adequately define the role of cytoreduction. Nevertheless, removal of the primary disease and bulky omental metastases will usually improve the quality of life, and the majority opinion is that every reasonable effort should be made to reduce the tumour to the smallest dimensions possible at the time of primary laparotomy.
4. There is probably a subset of patients with advanced disease who will have a

significant survival benefit from aggressive cytoreduction, but this group cannot be identified without prospective studies.
5. Similarly, the role of lymphadenectomy in this disease will not be identified without a randomized prospective study.
6. Surgery has an undeniable role in the palliation of patients with recurrent ovarian cancer, especially with associated bowel obstruction. The latter, especially with careful patient selection, may provide a short-term survival advantage and improved quality of life.

Selected arguments

FAMILIAL AND HEREDITARY OVARIAN CANCER

Bodmer: What is the population incidence of the Lynch syndrome? You have indicated that you think the site-specific ovarian cancer is relatively rare.

Watson: Just by looking at the number of families we have ascertained over the years, the number of site-specific ovarian cancer families is very small, compared to breast/ovary families for instance.

Taylor: Is it possible to know whether there is a higher incidence of benign ovarian neoplasms in these families and is that an indication that they are hereditary? Also, do you plan to look at hormone levels such as thyroid in the families with a higher incidence of ovarian/breast cancer?

Ponder: It is possible to discover whether there is a higher incidence of benign ovarian tumours or cysts in these families, but it would be quite a difficult study to set up because of the problems of bias in ascertaining this type of report. If you look at families which we have collected, there are quite a lot of women who appear to have ovarian cysts. But these families have been much more closely investigated than the general population.

Watson: We can look at ovarian morphology and benign ovarian abnormalities in women who are likely carriers and likely non-carriers. There have been some studies in the past looking at abnormalities in hormone levels associated with hereditary breast cancer and hereditary breast/ovarian cancer families.

Bookman: Is there any sign in these families that the type of tumours which develop are more of a higher grade phenotype with greater mortality in a shorter period of time?

Watson: There does seem to be evidence of change in terms of the age of onset of breast and ovarian cancers over time. There also seems to be some evidence of a shift with the generations in the predominance of ovarian and breast cancer. There seems to be some consistency to a pattern of a family, which in previous generations was heavily affected by ovarian cancer, and in current generations being more affected by breast cancer, and occurring at an earlier age. Some secular changes may have occurred in terms of exposures, and other environmental risk factors have caused breast cancer to occur at an earlier age with succeeding generations.

Bodmer: It seems to me that there could easily be an observational bias. You could get very important clues as to environmental factors involved in secular changes which could influence the expression of the genetic prevalence.

Selected arguments

Watson: Cancers which occur very early in life, or a cluster of such cancers in a family, are likely to bring people to a special clinic for family study. Whereas cancers occurring past the age of 55, or so on, are not seen as unusual by the family and therefore they are not as likely to be brought to our attention.

Scott: There has been a suggestion that chromosome-12 might be involved in a number of the benign ovarian neoplasms. It is also quite interesting to note that in hormonal terms and in terms of parity that group is really rather different, with higher parity and lower use of oral contraceptives.

Mills: Do you see the same genetic changes in the ovarian tumours that you see in the colon–rectal cancers, and do you see similar changes in ovarian cancers that you see in the sporadic diseases?

Watson: We have not done that.

Heintz: What do we advise when we find ovarian cancer families or when we diagnose hereditary syndrome?

Watson: We advise women in families like these to have ovarian ultrasound screens as long as they want to keep their ovaries and then to have prophylactic oophorectomy.

Ponder: Giving them information is terribly important. From a practical point of view we recommend screening by clinical examination and by ultrasound, and we consider prophylactic oophorectomy when the woman has completed her family.

Heintz: Do you advise them to use oral contraceptives at a young age?

Ponder: If I am asked about it I say that there is evidence that oral contraceptives protect against ovarian cancer incidence.

The question then is what if, as often is the case, the family also has breast cancer in it? What are the implications for the risk of breast cancer? I am sure that I know the answer to that question. Within the families pregnancies seem to have a protective effect.

Creasman: Do we have enough data right now that we know the actual risk of ovarian cancer in a family with one family member, two family members affected, etc.?

Ponder: The risk to a woman who has one first degree relative who has developed ovarian cancer is increased by a factor of about three-fold. It is rather more if the relative was young when she got the cancer and perhaps slightly less if she was older. That translates into about a 1:40 risk lifelong. If that woman has two first degree relatives who have developed ovarian cancer her risk life-long is something of the order of 30%. When you start to look at more complex situations where there is one relative with ovarian cancer and one with breast, or one with colon, then the data are not so firm at the moment.

Bodmer: It is very helpful to have a consultant geneticist involved in dealing with these families. We couple the oncologist with a genetic consultant who may spend up to a quarter or half of their time only on cancer family problems.

Ponder: You should never forget when you are talking to one of these women that the risk has an ageing element to it. Nobody has ovarian cancer when they are born. When they are 30 years [old] their period of risk is beginning – they may not develop their ovarian cancer until they are 70 years [old]. You also have to think about what is the risk in the next 10 years rather than confronting them with their life-long risk.

Soutter: To what extent is the age of the relatives who developed the cancer important?

Ponder: In general it is more significant if the relatives develop their cancer at an early age because one is always looking at the balance between this being a true genetic predisposition and it being a coincidence. Since ovarian cancer in the general population is commoner as people get older, you are more likely to have two elderly relatives with ovarian cancer by chance than you are two younger ones.

Sharp: This is clearly an important issue which is causing concern in the minds of the clinicians. In Graz in Austria we strongly recommend the setting-up of regional, and national, and international registries because it was from this source of cancer families that we felt the way forward would come.

GENETIC CHANGES

Bodmer: I wonder whether anyone has data on p53 mutations which really explain most of the events on 17p in ovarian cancer.

Berchuk: We have looked for p53 mutations and found them.

Bodmer: In about what percentage?

Berchuk: In about half of ovarian cancers.

Mills: People should also be aware of the concept of genetic imprinting. Many of these genes are irreversibly inactivated so that although they are there, constructurally normal, with their sequence completely normal, they cannot be expressed. So this is the other method in which a tumour suppressor gene could become evident in the system.

Bodmer: That's true, but I think the evidence of a widespread role for imprinting in that way is not yet clear. I wonder is there now some consistency in the allele loss patterns if you put the data together? Certainly for 17q and 6q, there seems to be clear consistency. What about the other chromosomes, do you see losses on 11p clearly?

Lowry: Yes.

Bodmer: What about 3p.

Lowry: We have not looked at 3p.

Gallion: When you see an advanced cancer which is established, you start to see lots of chromosome loss and the fact that we are able to pick up frequent chromosome losses of p6, q11, may just be a very late manifestation of the many chromosomal aberrations that occur.

Bodmer: That hasn't been the case in the colon where it has been looked at most carefully.

Gallion: Right.

Bodmer: And there you can actually define at which stage the allele loss is taking place.

Gallion: Certain statements have been made that tumour suppressor genes located in these chromosomes are probably involved in the genesis of ovarian cancer. I don't think one's anywhere near being able to make any conclusions like that.

Bodmer: Why do you say that?

Gallion: Because you see multiple chromosomal changes and particularly losses in advanced stage cancers.

Bodmer: But the losses are differential and

Selected arguments

the chromosome changes may not necessarily correlate with allele loss.

Gallion: Unfortunately, since no one really recognizes the premalignant stage, you can't really look at a progression from normal to malignant.

Soutter: As the number of allele losses proliferates I begin to be more uncertain about the true significance of this.

Bodmer: It is amazing how long it took for people to realize that chromosome changes, even like the Philadelphia chromosome, were significant for cancer. There is a great danger in assuming that this is just a background change that doesn't have specific significance.

Soutter: We should look at these data with a certain amount of caution.

Berchuk: I found Dr Lowry's data very interesting. For one marker he had about 29% of benign ovarian tumours having loss in a certain allele, I wonder if he would comment on the significance of that in terms of either the biology of those tumours or the technology involved in determining allele loss?

Lowry : I think I wouldn't go any further than saying that it is at present just a curiosity, because of very small numbers.

Berchuk: What marker was that?

Lowry: That was 17q. The loss of 17p seems to be a ubiquitous loss, possibly due to p53, which is found in colorectal cancer, breast cancer and other forms of malignant disease.

Bodmer: It is now very clearly established that p53 is probably the commonest gene to be mutated in all human cancer.

Ponder: Quite a high proportion of translocations on chromosome 19p have been reported in ovarian cancer.

Gallion: We rarely saw any. We did see some translations of chromosome 14 but they did not occur at the same break sites within those cases.

Ponder: The report mentioned 19p.

Gallion: We saw very few changes in 19, and would have involved either a p or a q. Chromosome 1 is a big chromosome. You could see it in most of the advanced, poorly differentiated tumours. You will have marker chromosomes that you cannot identify. We don't feel that chromosome 1 falls into that because chromosome 2 is almost of an equal length and fairly recognizable. We rarely see changes in chromosome 2. What we are doing now is loss of heterozygosity for similar markers that have been mentioned and comparing those to the karyotypic data on the individual patients. So on an individual patient we now have karyotypic of blood, tumour and the general screen of markers such as mentioned and then screening for loss of heterozygosity for chromosomal areas that have been implicated by that individual's patient's karyotypic finding.

Bodmer: Even if you have allele loss by non-dysjunction, you may have reduplication to bring up the chromosome number. I think that when it is simply chromosome number variation there is a problem in the correlation of that with allele loss. You said that you had 14 normal karyotypes. Do you have a way of knowing whether any of those were actually just in normal cells that had grown out rather than in a tumour cell?

Gallion: We have not solved that question at all. We observe our cultures as they are growing and look for the presence of what appear to be tumour cells growing. But there really is no way to tell, because in culture tumour cells can take on a fibroblastic appearance. Currently, and over the last several years, we have been obtaining flow cytometry on all of our samples and comparing ploidy to what we are seeing. In our initial series when we had a large number of advanced stage, we didn't really see much difference in ploidy. Now though, as we have got more benign tumours from more frequently seen normal karyotypes in the early stage and early grade tumours – in that case it may be one of two things. It could be that we are just simply growing the fibroblasts, like Dr Trowsdale showed.

Bodmer: There are many ways of getting rid of the fibroblasts.

Gallion: A normal karyotype may represent fibroblastic overgrowth and it also may represent a deletion, or an addition; or that it is just too small to be seen, because karyotypic analysis is a gross way of looking at chromosomes. Another thing that we tried is taking some cells from the flask, doing cytokeratin stains to try to distinguish the epithelial or stromal origin of the cells.

Gallion: You can't obviously use those markers on the metaphases that you are looking at to do the chromosome analysis because all you have are metaphase. It is just DNA. There is no cell membrane there, so to be able to do a marker for the individual cells that you are karyotyping as far as I know, is not really possible.

Bodmer: One thing you can do, is initially to isolate epithelial cells which you can do with available monoclonal antibodies. The other thing you can do is to specifically kill off fibroblasts, again with available monoclonal antibodies.

Bast: We have been able to get very clean preparations using immunomagnetic separations, and then digesting the RNA and DNA from this.

Bodmer: What monoclonal antibodies are you using?

Bast: There are five antibodies, three we have developed in our laboratory, and we would be glad to share those.

Bodmer: This is specific for the ovarian tumour cells?

Bast: We have screened those for reactivity against ovarian cancers and reactivity against mesothelial cells and cytokeratin. They are clean negatives on the mesothelial cells and macrophages, positives on ovarian cancer.

Cohen: I too was intrigued by the finding of allele loss in a small sampling of presumably benign contra-lateral ovaries and I wonder whether Dr Gallion, in collecting her samples, has studied the normal mesothelium far away from any of the tumour sites in the advanced ovarian cancers?

Gallion: No, we have not done that. I think it is an excellent idea and one that probably should be pursued.

Howell: To what extent do allele losses represent progression markers rather than markers of the individual original malignant transformation?

Gallion: We have not analysed such loss of heterozygosity data in this particular group of patients. But we did see that finding in

Selected arguments

ovarian tumours, that we look at cytogenetically. If you take the primary tumour then when that patient goes back to the operating room for a bowel obstruction or a tumour recurrence, you will see an accumulation of more cytogenetic abnormalities in that later specimen.

Howell: Is there any specificity?

Gallion: No.

Lowry: The problem there is that by that stage the patient has been interfered with to the extent that they are usually on chemotherapy and you really couldn't tell what the changes are due to.

Mills: How many of the karyotypes from the same patient will be abnormal. Do you have a lot of the cells having abnormalities?

Gallion: It is somewhat variable. We will usually karyotype 25 cells from an individual tumour.

Mills: There are several new culture medium conditions that can be set up to grow most of the ovarian cancer cells, very well and very efficiently.

Soutter: Could I ask Sidney Lowry a little bit more about the benign tissue which he was examining. He did say that it was benign material obtained from the contra-lateral ovary of women with ovarian cancer. What part of the benign ovary specifically was being examined? Was he attempting to look at the epithelium alone, or what?

Lowry: The results I showed referred to benign ovarian tumours which must be distinguished from normal ovaries. I can tell you also that we did look at some normal ovaries and found no changes in these normal ovaries, with the possible exception of one, but this had not been confirmed.

Bodmer: From the contra-lateral ovary?

Lowry: Yes.

Bodmer: Is there no possibility of transferring tumour cells from the affected to the contra-lateral ovary?

Lowry: That is a possibility.

Cohen: Is ovarian cancer really ovarian or a form of some other wider expression of an oncogene on a large mesothelial surface. It would be very fruitful to study the peritoneal surface away from the obvious ovary cancer.

Lowry: More interesting were the borderline tumours where the pathologist really couldn't say whether it was benign or malignant – and we found some loss in some of these cases. The other area of interest is the hormonal triangle, between the breast and the ovary and the endometrium, and we are not having a look at the endometrium because it seems to be part of the same spectrum of diseases.

Oram: How long does the cell replication for epithelial repair take?

Lowry: I don't know. I do know that the epithelial tears can be quite large because the follicular cysts, as any gynaecologist knows, can be up to 2 cm in diameter just before they burst, so presumably it might take some days to repair. This process is repeated each month for many years.

Kacinski: Is there really any evidence that benign cysts or borderline carcinomas ever actually progressed to invasive carcinomas, in the way that there is for intraductal carcinoma which progresses to invasive breast cancer?

Bodmer: That is an area that should be picked up specifically.

Luesley: Large numbers of women have been subjected to super-ovulation as a result of infertility treatment and one would expect therefore to see higher incidences within these groups. Does anyone have any data which can substantiate that?

Lowry: It is too early yet.

Scott: We have actually had two cases within the past six months of young women subjected to super-ovulation therapy who have subsequently presented with ovarian cancer. Now, of course, that is not suggesting for a minute that there is any direct cause and effect in that process.

Balkwill: During the epithelial repair of the ovary many different cytokines are going to be elaborated and may act in promoting invasive capacity, angiogenesis and things that would be necessary for the development of the tumour.

Bodmer: It is interesting to see that ovarian cancer joins the other cancers with p53 mutations. One wonders whether there will be any epithelial cancer that is not like that. You said that suppressor genes were always nuclear genes. That, of course, is not the case. There are examples of tumour suppressor genes, such as Fogelstein's DCC, which are clearly membrane acting. Also, what about p53?

Berchuk: You are absolutely right that not all tumour suppressor genes are nuclear genes. Regarding the p53 gene, apparently it is now known that its protein product may complex with one of the heatshock proteins and actually be translocated into the cytoplasm, and in some instances we actually have seen some cytoplasmic staining that may be indicative of this. With regard to other p53 mutations, or early or late events, I think Fogelstein has shown that in colon, at least where we can identify premalignant changes, it does not apear to be an early event. Even Fogelstein has come around to the thinking that it may not be so much whether a cancer causing gene is activated early or late, but that rather, to get a cancer it requires activation and multiple genes and that the sequence of whether one is an early or late event is actually less important than the accumulation of multiple activated genes.

Bodmer: There may be some genes which give you an advantage at a range of stages in carcinogenesis and others that only give you an advantage after some other events have first happened. All the evidence suggest that p53 comes into that latter category.

Berchuk: We have only sequenced about eight ovarian cancers. The ones that stained all had point mutations.

Bodmer: Was that all done with cDNA staining because now it is quite easy to do that on genomic DNA which is rather easier and of course can be done on rather small amounts of keratin block material.

Berchuck: No.

Bookman: Regarding c *erb*B2 or HER-2/*neu*, I wonder if Dr Karlan, might help reconcile some of the fundamental observations in this field. We think that this transmembrane molecule functions as a

Selected arguments

receptor because it has a tyrosine kinase domain, which in most human cancers is not activated constitutively but presumably is activated after ligand binding or some other events that occur at the cell surface. A number of monoclonal antibodies have been developed which either do not inhibit cell growth and are basically inert in terms of any *in vitro* or *in vivo* effects against the tumours, or they inhibit cell growth and in general the antibodies which inhibit cell growth, including the ligand, the proposed ligand GP30, increase the activity of the tyrosine kinase that is part of this molecule. So, here is a receptor on the cell surface that is highly expressed in tumours that grow aggressively and have an association with poor survival. But all the known ligands, or antibodies that might mimic ligand binary, that result in activation of the tyrosine kinase, inhibit the growth of the tumour. To my knowledge no one has described anything which stimulates or activates tumour growth through this receptor and that is a fundamental paradox. Data with transvected cells is at variance with Dr Karlan's data and some of the data from established cell lines that happen to express high levels. For example, the SKBR 3 line which is a breast cancer line which is among the highest known expressors of intact HER-2/*neu* on the surface, does not grow on any mouse. No one has been able to establish that, either in stimulo-deficient skid mice or in nude mice to my knowledge. You would think with such a high level of expression, if this is associated with increased growth, and increased soft agar growth, and increased tumour geneticity, that this might be the line that would grow well in mice.

Karlan: I think some of the concerns about GP 30 being the true ligand of HER-2/*neu* is just the fact that, if this is a growth factor receptor this ligand causes down regulation of growth. How can that be? The story I am sure is a little more complex than a simple growth factor receptor. We do see different antibodies. We have done electron microscopy on some of the tumour cells after binding with the antibodies and they are internalized at different degrees. Some of that is correlated with the inhibition of growth.

Bookman: Have you not seen any antibodies that activate growth?

Karlan: No.

Bookman: Most of the transmembrane receptor markers and T-cells can be quite well triggered or inhibited.

Bodmer: I don't think that's true for EGF.

Bookman: Correct.

Karlan: Is GP30 really the ligand for HER-2/*neu* that we are all looking for?

Taylor: One way of looking at these markers is as markers of a particular differentiation stage of the cell which may be a more primitive cell.

Kacinski: HER-2/*neu* is seen on normal surface epithelium. We have seen it in a reasonable number of borderline tumours, some as intense as an invasive carcinoma. There seems to be no prognostic significance, at least in the borderline carcinomas. We have also isolated cell lines, in particular from one patient whose tumour was heterogeneous to start, and with metastases. We had two clones, one of which is HER-2/*neu* amplified, the other which is HER-2/*neu* single copy number. The tumour geneticity in nude mice is identical.

Bodmer: Presumably when you see it in those normal tissues, you are looking at increased levels of expression that are not due to amplification.

Kacinski: Correct.

Mills: We have looked at ascitic fluid now from 32 independent patients for the presence of a ligand that would alter tyrosine phosphorylation of *neu*. It involved cells that expressed *neu* by transvection and in normal ovarian epithelial cells that have both the EGF receptor and the *neu* receptor that could crosstalk. In none of those cases did we find any evidence for a ligand that would alter *neu* activity. So there appear to be very low levels, of a putative *neu* ligand present in ovarian cancer ascitic fluid.

Bodmer: It could be an autocrine process.

Bast: We will be describing some of the work we have been doing with monoclonal antibodies with down regulation of breast and ovarian tumour cell growth by interacting with the extra-cellular domain of HER-2/*neu*. Mark Lipman's group have described a 30k ligand which down regulates the growth of SKVR3 and other cell lines that strongly over express HER-2/*neu*. They also state that with cell lines, where there is less strong expression of HER-2/*neu* with low levels of that ligand, it is possible to stimulate cell growth. That would be analogous to EGF and EGF receptors. We need to understand the substrates of the tyrosine kinase activity of both EGFR and *erb*B2 to really put this story together, as well as to understand the complex interactions of tyrosine auto-phosphorylation in several different sites in the same molecule. It is entirely possible that substrate affinities might change and that the kinase activities might change, depending upon the state of autophosphorylaton, and depending upon the cell type and setting in which the *erb*B2 finds itself and the local concentration of *erb*B2 receptors.

Bodmer: Regarding that family with amplification, it suggests that one should be looking for the possibility of inherited amplifications in other tumours. I wonder whether the usual techniques that we use, such as linkage, would have shown it? It could correlate it with the 7q21? Looking for abnormalities with southern blotting or sequencing might not pick up, say, a two-fold amplification, which might be enough to give a significant inherited susceptibility.

Karlan: With southern blotting we have picked up two-fold amplifications.

Bodmer: But then you probably do it rather more carefully than one normally would.

Karlan: We are doing these for quantitation.

Bodmer: When you know you are looking for it, you may find it. But don't you think that it is possible that it would have been missed?

Karlan: Clearly that could have been missed.

Creasman: How many of these genes overlap? I am looking at it from a clinical standpoint. We have said that a third have HER-2/*neu* and yet 85% of all far-advanced cancer patients die from their disease, so there has got to be something else. Are these overlappig or are they separate and have all significant impact on prognosis, or what?

Bodmer: There has to be a certain overlap. I don't know whether anyone has data where they have looked at the distribution on p53 mutations, or allele losses and of

Selected arguments

the *erb* and EGFR amplification, to look and see if there is any correlation between them.

Berchuk: We started to do that. Obviously you are getting into multivariate analysis, probably with an insignificant number of patients. What we found is that the things which are supposedly bad prognostic factors, when you add them one on top of another, do seem to add up to a worst prognosis. Vice versa, we did have a lot of favourable prognosis factors. They seem to do better. What we haven't had is the adequate number of patients really to prove that, because, as you know, most of these patients are dying – so if you are looking at something like a long-term survival analysis you don't have large numbers of patients who are living really to give you a good shot at statistical significance. So really to have adequate statistical power to do those types of studies you need larger numbers of patients than even we have been able to muster at this point.

Bodmer: It is inevitable that the more changes you have the worse the prognosis. That is just part of the underlying mechaism of carcinogenesis. You can look at the correlation of each change with some aspect of the clinical pattern, but you can also omit multivariate analysis and do a two-by-two analysis with pairs of markers and ask whether there is a tendency for them to occur together or not, or one before the other, and that may not need all that much data.

Kacinski: We have done that sort of analysis by *in situ* hybridization and can show strong correlations.

Bodmer: Strong correlations of what with what?

Kacinski: A variety of the genes that we discussed.

Bodmer: Are these the only ones that were amplified?

Kacinski: It was shown by over-expression.

Bodmer: Do you mean EGFR and *neu*?

Kacinski: Yes, and a few other genes as well. We looked at 17 different genes. Dr Karlan, you showed us an immunohistochemical slide and you say that there is gene amplification in the epithelium. How do you know that in that particular patient?

Karlan: In most of these patients, when we have frozen tumours, we do immunohistochemistry, southern and northern analysis in addition.

Kacinski: So you found that there was gene amplification in the normal tissue in this woman?

Karlan: Yes.

Bodmer: That was the woman who was normal in that family?

Karlan: That's correct.

Bodmer: And that was the evidence that the amplification may be inherited?

Karlan: Correct. She did have clear gene amplification in addition to the immunohistochemical data.

Mills: You state that the *erb*B2 gene in these patients is not abnormal in its sequence. Now the paper that was published showed transmembrane sequence only. Have you looked throughout the rest of the gene? Certainly with *erb*B itself, mutations can be in a lot of different places and have marked differences in their functions?

Karlan: We have sequenced multiple genes. There has not been any significant repeated mutation that we have seen in the human genes, which is different of course with the rat *neu* gene.

Mills: When you look at patients that have an over-expressed HER-2/*neu* in the breast, that protein is constitutively tyrosine phosphorylated as if it is active and functioning. That is not the case in ovary. With over expression we find the protein as far as we can tell in the cell is neutral. There is no evidence for increased tyrosine phosphorylation of that protein. So there are differences between the two tumour types.

GROWTH FACTORS AND CYTOKINES

Leake: How realistic is it to look at cells in primary culture compared with the way in which those same cells would behave in an intact tumour?

Bast: If there were no effect on cells in culture, that might discourage you from doing more difficult *in vivo* studies. Clearly we have to deal with tumours in the setting of a whole animal. Our ability to study the effects of these cytokines *in situ* in humans is always more difficult.

Taylor: Are you trying to immortalize normal cells or put oncogenes into the normal ovary?

Bast: We are certainly trying to. Two of the lines established from patients with apparently normal ovaries, have grown longterm, past 20 generations of culture.

Interestingly those seem to over-express p53. Preliminary data though suggest that the one cell line where the p53 is in sequence has wild type p53. Whether we selected a subpopulation of normal ovarian epithelium in culture that has over-expressed p53, or whether there has been a regulatory mutation in p53 during passage of culture, really remains to be sorted out. We are very much interested in looking at p53 in cells collected from patients, either by immunocytochemical analysis or removing sheets of the cells and staining them immediately.

Taylor: But these cells don't grow in agar or nude mice.

Bast: No.

Mills: If you look at the list of cytokines that are present in the peritoneal cavity you have a minimum of IL-1, IL-2, IL-3, IL-6, MCSF, interferon-γ, interferon-α, GMCSF, GCSF, interleukin-2, and TNFL for TGF-β. It is going to be very hard to figure out which ones are acting directly on the ovary and other cells and how they interact in a cascade. This is going to be an incredibly complex system to work out with the number of cytokines that appear to be there.

Balkwill: If I can just give a bit of information on *in situ* hybridization studies on the primary tumours. TNF was by far the most ubiquitous. We found IL-1 in tumour cell clumps where we find TNF. But IL-6 was within the limits of detection of our assay. In assays where we had, for instance, bone marrow with very high levels of expression of IL-6, we did not detect IL-6 mRNA in some 10 patients that we have looked at so far. Also, we must always be aware with cytokines that it is very easy to induce them by just a simple

Selected arguments

process of desegregating cells. The endotoxin is practically ubiquitous in a lot of tissue culture media, in supplements, and in cells attaching to substrate. It may stimulate the production of some cytokines, particularly IL-6 which does seem to be found in many different situations.

Leake: If you induce growth in breast cancer cells with oestrogen you can demonstrate that this is done through production of TGF-α. You can give the TGF-α and get the same dose response curve. But if you then block the EGF receptor and give oestrogen, the cells will grow just as well by a second mechanism. There is not going to be any single mechanism.

Bast: It has been shown that you can sub-divide a sub-set of cells where clearly you do inhibit growth at least *in vivo* in ascites cells taken directly from patients. There is also a sub-set of cells in patients where you can stimulate tumour growth with that primary material as well. So it is complex.

Mills: Although you might not see a direct effect on the ovarian cancer cells with GNG and CSF, these could very easily be altering lymphokine production by all of the stromal cells, the invading lymphocytes, and the macrophages. We really cannot consider the ovarian cancer cell in isolation, but must really consider it in its environment which is the stroma, the peritoneal cavity and all of these other cell types.

Bast: Some of these factors, like IL-6, can be contributed both by the tumour cell and also by mcrophages. So that may be true for IL-1 and TNF and several other factors. We may be seeing the end result of several different contributing sources.

Leake: It is certainly going to be useful when we have good systems for co-culturing stromal and epithelial cells.

Scott: I wonder whether Dr Balkwill has looked at the possibility of interfering effects from, for instance, anti-coagulation in terms of altering the metastatic potential of these tumours and the implantation potential?

Balkwell: We have done some experiments with anti-coagulants and at first we got some interesting results and then they were not reproducible.

Scott: I certainly have some *in vivo*, and I am sure several other people have anecdotal *in vivo* cases where anti-coagulation appears to have lead to a dramatic increase in responsiveness to particular therapeutic regimes – not necessarily sustained.

Mutch: TNF-α tends to select for anchorage. If your initial cell line was not anchorage independent then you may have just selected for anchorage independent and that would have expressed itself as a more malignant cell line.

Balkwill: With the TNF transvectors there was no difference in behaviour *in vitro* of the neo-transvectants or the TNF transvectors.

Leake: Have you actually done any experiments in terms of adding folate to cells to see if you do get proper stimulation of growth?

Trowsdale: No.

Kacinski: I think the interesting thing about your result would be if the expression of the folic acid receptor in any way correlated with relative resistance to methotrexate.

Trowsdale: According to some workers, 90% of these tumours over-express.

Leake: Amplification of the gene does not necessarily correlate with over-expression of the protein. We have certainly seen several cases where there is over-expression of a particular protein without amplification of the gene. What we are looking for may be that there are several of these cytokine or growth factors which can lead to a common point within individual target cells.

Carson: We have looked at cytokine stimulation assays for IL-1-α and TNF of these cells. Fibroinectin seems to be that inhibitor, that 2200 kDA protein. I think that is what seems to be inhibiting at least the growth, proliferation of the cells. We tried to demonstrate that levels of cytokines were physiologically active but with this inhibitor present it was very difficult.

Leake: You were obviously disappointed that your TNF-α/IL-1 measurements did not correlate with survival. Do you think this was due to tumour heterogeneity?

Carson: We did not assess tumour heterogeneity. It is hard to make any conclusions out of 11 individuals. We will need to collect some more information about that.

Leake: This question of heterogeneity is a very serious one.

Mills: Even at what we call stage 1 ovarian cancer, it is probably very late in the disease.

Bodmer: There is absolutely no doubt that if you can antagonize the functional effects of p53 you would have an effective treatment for tumours. So I don't think one should necessarily be negative about finding something which doesn't correlate with survival.

Leake: There are a number of markers coming through which do identify patients who do not respond to, say, platinum therapy. We have the potential to do a lot more with the cytokine therapies, or anti-cytokine therapies, and receptor blocking. It is a complex field and even once we have proper agents for identifying and blocking the various receptors, we may find we need to go further down the line, block tyrosine kinase, or even block at a later stage, in the various pathways.

Leake: This raises the possibility that at least some ovarian cancers are related to the extent of injury to the initial normal ovarian cells and that we may be looking at injury responses. We have seen cases in the familial ovarian cancers where having three or four children was protective. Of course, the more children you have the less opportunity you have of ovulation. The same thing might be said of the protective role of some of the contraceptive pills. Perhaps there is something for us to learn in terms of preventing injury to reduce the incidence of ovarian cancer.

Bodmer: I wonder if either Drs Kacinski or Mills would like to comment on whether they feel that what they are looking at involves some specific step that is switching on the carcinogenic process, or whether they are looking at a bi-product of the growth of cells?

Kacinski: What I think we are looking at is expression of a phenotype which allows the malignant cell both to be able to invade, metastasize and evoke immune responses.

Bodmer: Are immune responses relevant?

Selected arguments

What determinants are there and do you get down-regulation? I would suggest that it is unlikely that CSF has anything to do with the immune response to invasion. Most of these tumours will probably respond to γ interferon and express class II and be able to resent their own antigen.

Bast: It looks as if MCSF is constitutively expressed both by normal epithelium and by malignant epithelium, at least to the extent that it could be detected by immunohistochemical methods. It is not expressed in normal epithelium but is expressed in malignant. If the downstream effects of a fims/MCSF autocrine loop would be a more invasive phenotype through protease expression or whatever, then you have got a rationale for at least one step that would differ between normal and malignant epithelium.

Bodmer: I would accept that, although that is one of many explanations that have been put forward over many years for increased invasiveness. Then I would ask if you take a normal ovarian epithelium and see what happens to it when you put it in short term culture under appropriate conditions, do you then, for instance, have conditions under which you upregulate C fims. If you do, then again that would suggest that the upregulation in the carcinoma is not necessarily a distinctive feature of the carcinogenic process.

Kacinski: That is something we are actually doing.

Mills: The antisense data do suggest that it is at least one of the components that is involved in the proliferation, perhaps of many different cell types, including ovarian carcinoma cells. We have made a number of constructs of this gene which we think should be activating by analogy to some of the others that have been made. But our current data by no means suggest that this is an inductive phenomenon for ovarian carcinoma, rather one of the essential steps that happens in the proliferation of these cells.

Leake: But I think the exciting thing about your molecule is the fact that it seems to be further down the line than some of the standard tyrosine kinases.

Mills: We do know that the messenger-RNA levels and the protein levels are altered by activation through receptors that have tyrosine kinase activity, or activate tyrosine kinases. But we don't know what phosphorylation of this molecule does to its kinase activity yet. That is in progress.

Bookman: I would like Dr Mills to help me separate antisense from nonsense! That experiment is very interesting but it is a technology that has a lot of potential problems. In your case I am not familiar with the specific reagents you use but is there any possibility that some of the antisense compounds might have also effected the expression of other tyrosine kinase within the cell?

Kacinski: The particular sequence that we chose was to the potential ATG starred site which shows no homology with any other known molecule. Even the kinase domain is sufficiently far away from other known tyrosines or sero threonine kinases. Furthermore, we have taken the same antisense for all of the nucleotides and used them in mice where the start site is likely to be different and shown that they have no effect. So this is specific to the human gene in human cells.

Mills: There are now going to be a large number of members of this family of genes which will have the ability to phosphorylate threonine and tyrosine.

Leake: You mentioned a nuclear form of this molecule, is that functional at the nuclear level?

Mills: We have been able to show that when we isolate it from nuclei there is kinase activity, but we don't know anything more than that.

Leake: There is quite a lot of evidence with the EGF receptor that it too moves into the nucleus and may be functional at the nuclear level.

Mills: Microbiologists have now shown that you can make mutants that don't get internalized and they signal proliferation beautifully, so it is not an essential process for the EGF receptor.

Wells: I would like to make a point on the analogy between the invasion of extravillous trophoblast and ovarian cancer. The important thing of course about extravillous trophoblast is that the invasion stops at the end of the third of the myometrium around about the sixteenth to eighteenth week of pregnancy. The interesting speculation is why it stops and I wonder if you have any thoughts on that.

Kacinski: Obviously the possibility is, that some of the tumour suppressor genes are genes we have not even identified, genes which would suppress the invasive phenotype, involved both in that process and apparent expression in cancer. That is something we are looking at by subtraction hybridization.

IMMUNOLOGY

Bodmer: Aren't the blocking effects of the antibody likely to be due to some sort of steric interference with the ligand, which I understand is possibly produced internally by the cell and may even interact with the *neu* or *erb*B2, in some sort of intercellular phenomenon?

Bast: In collaborative studies, we have looked at which of our eleven antibodies block the 30 kd ligand and which don't. Only two, ID-5 and BD-5 blocked the binding of the ligand to the *erb*B2 extracellular domain. Seven of the 11 antibodies down-regulate growth and several of the antibodies that don't block the ligand binding are as effective as the ones that do.

Bodmer: But the ligand may not be the right ligand.

Bast: There could be more than one ligand.

Mills: It is possible that your FAb fragments are causing dimerization, not by cross-linking through the antibody but by the same mechanism that, say, the EGF does to the EGFR; or that *neu* presumably does with *neu*, if it works in the same way as the EGF receptor does. Also, p36 ligand, at reasonable concentrations, is growth inhibitory. This may not be a growth producing ligand under normal circumstances, it could be a growth inhibitor.

Bast: Possibly, although it has been reported that with low levels of receptor and with low levels of ligand you do see growth stimulation with that. I think the point that Dr Mills is making is a very important one. We have done cross-linking experiments and we need to do more cross-linking experiments with the FAbs to be sure that simply the shift of FAb binding isn't making it easier for the intra-membranous domains to associate and to dimerize.

Taylor: Did you say that you got different

epitope expression on cancer cells—versus what?

Bast: Yes, preliminary data with immunocytochemical analysis using the 11 different antibodies against *erb*B2 and a bank of normal tissues, suggest that there is some differential expression of the epitopes on normal tissues. How many of these epitopes are confirmational determinants, how many of them might be carbohydrate determinants, we are still trying to work out. One has to come up with a mechanism why *erb*B2 should be different in kidney and in bladder and in liver. Certainly differential glycosylation would be one possibility.

DRUG RESISTANCE

Taylor: Dr Howell, the protein that you cloned only had the carboxyl in. How can that function if there is only a small portion of it being expressed?

Howell: I don't think that it can. But I am not sure that it needs to in order to serve as a target for cisplatin in the cytoplasm itself. It is conceivable that it is the glycine-refining motive which is important. There is a fair amount of data by analogy. It is this motive which may be responsible for the molecular interactions.

Taylor: Can you transvect in the full length protein which you got from the database?

Howell: Those studies are underway now.

Mills: I know that you have cloned what is a shortened form of the HSP-60. In those cells though that are naturally resistant to cisplatinum, do they have the full length HSP-60 message.

Howell: Yes.

Mills: And probably are producing the full length form.

Howell: Yes.

Mills: This confers resistance when you transvect it, but in the normally resistant cell is it the full length?

Howell: There is no evidence of a shortened message. One sees only the full length message in the resistant cells.

Mills: It is fascinating to see that you can increase the accumulation of cisplatinum in the target cells by cyclic AMP treatment and other things. What is going to happen in places where cisplatinum is toxic, such as the kidney and other places which are even more exquisitely sensitive to the effect of cyclic AMP reagents? Should we be telling all of our patients to have coffee and tea before they have their drug courses?

Howell: You raise an interesting set of questions. We have no idea at this point what happens to the sensitivity *in vivo* when one uses cyclic AMP activator. What we do know though is that this phenomenon varies from cell type to cell type, and tissue to tissue. We already know, looking across a number of different cells lines, that there is heterogeneity in the ability of the KS to activate drug uptake in different cells. So one would not be surprised to discover that different tissues are differentially expressing the substrate which is presumably phosphate, or substrates that are phosphorylated by PKA, and may or may not show this kind of response to cisplatin. So I think there is reason to be, at least mildly, optimistic that it may be possible to modulate the selectivity of cisplatin. But that is not the major reason for being interested in this. This major reason is, that this simply opens up another biochemical pathway in which one then has new targets for either inhibitors or activators which may be able to increase selectivity.

Mills: You mentioned several times that this is uptake. Is it uptake or just accumulation?

Howell: That's a very complicated question. One of the problems in working with cisplatin is that, as soon as it goes through the membrane, it appears to undergo a number of different kinds of reactions that are not particularly well predicted by what is known of the equation chemistry of cisplatin. It has not yet been possible to fully define whether we are talking about membrane transport or whether we are talking about a combination of membrane transport followed by an immediate biochemical reaction once the drug gets in the cell.

Berek: Have you done any work with carboplatin?

Howell: Yes, carboplatin has virtually the same profile, that is we can sensitize synergistically to carboplatin using the PKA and EGF signal transduction patterns.

Berek: Is there any difference when you combine the two, or sequence the two?

Howell: We haven't done a lot of detailed studies.

Bast: Dr Markman made an important point, regarding the importance of biomarkers and their relationship to cisplatinum sensitivity or resistance. One of the biomarkers which was discussed was the expression of *neu*, and that brought to mind the observations that Dennis Sleighman made, where use of monoclonal antibodies against the extra-cellular domain of *neu*, apparently increased the sensitivity to platinum in breast and ovarian cancer cell lines. We presented some data that some of the antibodies against the extra cellular domain of *neu* can reduce, fairly dramatically, the diglycerol levels in cells and I wondered how important the PKC pathway was for platinum resistance? I think there is evidence, certainly for MDR, that PKC correlates with multi-drug resistance phenotype. But I was wondering what the evidence was for platinum resistance?

Howell: All the points that Dr Markman made are important. The issue of the PKC pathway is an interesting and confusing one. If you treat these cells with TPA you get a dramatic sensitization to platinum, exceeding three-to-four-fold. There is a classic time course which fits exactly with the time course for the expected activation and down regulation of PKC. Other people have reported that PKC activation will actually produce antagonism to cisplatinum and there is a well defined biochemistry in another kind of cell line in which you get exactly the opposite effect. I think the weight of evidence from our lab. is that in most cells activation of PKA pathway produces a sensitization to cisplatin. The mechanism of that is totally undefined at this point. It is not a change in drug uptake; it is not a change in DNA repair. So we don't really know how this is working at the present time, but it clearly can. There is every expectation that activation of *neu* may in fact produce a signal that is capable of sensitizing to cisplatinum and that work is ongoing.

Bodmer: You did one transvection with those selected cDNAs and selecting for

Selected arguments

cisplatinum resistance. Do you think that you have isolated the only clone that gave you that resistance, or could there be other clones?

Howell: Yes, my opinion is that there are probably other clones in there that are capable of doing this. I don't think that we have pulled out the one protein which is capable of mediating platinum resistance. We are in the middle of trying some additional strategies for pulling out other elements to see whether there are other proteins which could mediate the same change.

IMMUNOTOXINS

Howell: I would like to take issue with Dr McGuire's analysis of what's possible at this point. Clearly there is one setting in which it is possible to increase platinum dose intensity by a lot, and that is for free floating cell in the peritoneal cavity with IP drug administration. Now IP drug administration is currenty under attack across the board. For that component of the tumour which is successfully accessed when one puts drug in the environment of that tumour directly, it is very easy to achieve ten-fold increase in drug administration. I would submit that the challenge is in the arena of asking how can we increase local regional tumour drug delivery by an order of magnitude such as it is possible for, at least the free floating component of tumour cells in the peritoneal cavity? Now there is a major problem with drug penetration and that limits the usefulness of IP therapy to relatively small tumour volumes. But in that population of patients, it appears to be highly effective. For example, in a population of patients in which the second-look laparotomy was negative, following standard cisplatin cyclophosphamide therapy in San Diego, we have treated that population of patients with three cycles of IP high-dose therapy and we have only one relapse with now better than three years of median survival follow-up. In that population of patients the expected relapse is around 50 to 60%. So it is theoretically possible in a small subset of patients to achieve that goal and the challenge is more how to increase drug delivery for the rest of the tumour population.

McGuire: I don't disagree that one can achieve ten-fold to forty-fold platinum concentrations in the peritoneal cavity. I think you may be right that for a very, very small subset of patients it may be effective. But that is not most of the ovarian cancer patients we are dealing with. If you deal with the patient who has only microscopic disease at the time of second-look, the recurrence rate would be anticipated to be in the 50 to 70% range. But unless those patients are undergoing a third-look laparotomy, with biopsy, to prove that they don't have the disease, having those patients alive with 'no evidence of disease' two or three years out, is not verification that you really have got rid of the tumour, or may be not even evidence that you are going to change the natural history of the disease. I think the jury is still out on even the value of IP therapy in the patient with true minimal residual disease.

Luesley: We have preliminary data. I don't think that there is probably going to be any significant change in time to response. We also have a significant amount of marker data and, as has been published before, it is the patient who has a significant delta CA125 in the first one to two cycles, irrespective of whether they got the intense therapy or the non-intense therapy, that ended up with clinical and/or pathological complete response.

Howell: Let me come back to the issue of dose intensity and the clinical experience. Even with relatively drug resistant tumours in animals, one can find a dose response curve. With every cell line one can find a dose response curve. What is your thinking at this point on why, even when we are getting up to dose intensities in the close to two range or possibly even higher, we are not seeing results at the clinical level?

McGuire: Part of it is tumour cell heterogeneity. There may be cells that effectively have a significantly greater platinum resistance, either intrinsic or because of inability, when you are talking about bulky stage III disease, to deliver the intense therapy to the cell. If there was going to be an outcome effect, one should see it over the fairly narrow dose range of 0.5 to 1 in the sub-optimally debulked patient.

Bookman: Scheduling factors may be important and some of the efforts that have been utilized to try to deliver even higher dose intensity, such as support with autologous bone marrow or peripheral stem cell, often have been limited to just one cycle of high dose intense therapy combined with conventional cycles. It might be that, as we develop more skill in these techniques and can deliver more sustained and repeated cycles of high dose therapy, it might be possible to readdress this question in a more prospective fashion again in the future.

Markman: Regarding autologous marrow, what can we really do now in terms of dose intensification? Is it realistic to think about doubling or tripling the dose considering the toxicity and cost? Unless we have new drugs or a different approach, are we going to have a major impact?

McGuire: I don't think we will. I don't see that there is any way of doing it other than true high dose autologous rescue, or stem cell rescue. There are some data mounting in breast carcinoma which would suggest, at least, that the response rates are higher, that you can generate secondary response rates in patients who have 'drug refractory disease at standard doses'. It is not a totally unreasonable thing to do. But to do it in the patient with recurrent bulky stage III disease would seem, to me, to be the wrong patient population to try it in.

Bast: Clearly the jury is still out on autologous transplantation for ovarian cancer. Recent results in breast cancer are promising. Very high complete response rates have been obtained, if you combine adriamycin 5-fluororacil and methotrexate for three or four cycles before transplantation. Then give very high doses of cyclophosphamide, cisplatin and BCNU will return of the patient's own bone marrow. There is a 68% response rate in patients who have relapsed following adjuvant chemotherapy. There is currently a trial ongoing of similar therapy – CAF followed by CCB and autologous rescue in patients with 10 nodes positive. After three years there is about a 70% disease-free survival. The most impressive thing is how much safer those autologous transplants have become with increased experience and with the use of GM-CSF, G-SF, and primed peripheral blood stem cells, as well as priming of the bone marrow that is returned. With that, at Duke at least, the mortality of autologous transplantation in women generally less than 50 has decreased from 20% to 8%. Platelets are still a problem in autologous transplantation as they are, of course, in more conventional dose therapy. So given that at least half of ovarian cancer patients are less than 60, it may be possible to apply autologous transplantation more safely. So at the present time there are some encouraging results, both in terms of the efficacy and toxicity in breast cancer and some very preliminary data, not only from the US, but I believe from Australia and in Europe, on ovarian cancer, that need additional evaluation.

Hacker: We have treated about 12 patients now with high dose chemotherapy and autologous marrow transplant. These have been patients who have had a positive second-look, usually with small disease. They have usually had tumour maker negativity at the start of treatment. If their tumour markers have been elevated we have given them 3 cycles, usually, of carboplatinum and atopocide to try and get the markers down to zero. We have not yet had a mortality. With the use of growth factors, these treatments are surprisingly well tolerated. But the longterm results have been really quite disappointing. Most of the patients have relapsed and I think the longest survivor at the present time would be about 18 months, but most patients in fact have relapsed within 12 months.

Bookman: What was the high dose regimen that you used?

Hacker: Carboplatin and atopocide.

Bast: We may be able to use different kinds of alkylating agents in mycotoxic agents in these transplant protocols. In work with cell lines done in our lab., fairly impressive synergy by isobolic graphic analysis was obtained against ovarian cancer cell lines using thiotepa and ciplatin 4HC, thiotepa or cisplatin and also melphalan. This would have been more difficult to demonstrate at the lower concentrations of drug than one achieves with more conventional dose administration. Whether this will apply *in vivo* remains to be seen, but I think there is a lot of room for development in different combinations of drugs. At least in breast cancer where we have more experience, the very high response rates and more durable responses have only been seen when conventional dose therapy has been combined effectively with autologous bone marrow, when it has been possible to move back to patients with really minimal residual disease.

Kacinski: A question for Dr Bradley. Why do you think that biologicals, which only give you maybe less than one long kill, are going to be effective in an adjuvant setting if they are not effective for gross disease?

Bradley: It still might not work, but it will give you a better chance. It will solve some of the problems of access of activated lymphocytes to tumour. But it does not solve all of the problems.

Mills: Some very nice experiments have been reported looking at the ability of LAK cells, not to kill tumour lines, but rather to affect the colony forming units that are present in patients. Although they will kill a lot of tumour cells they have no activity whatsoever against the colony forming units. There are a lot of suppressors there. Ascitic fluid and even peritoneal fluid from patients with ovarian cancer with no apparent disease have high concentrations of TGF-β that completely prevent the ability to generate LAK cells or even TIL cells *in vivo*. When you put them back into the patient they are also inactivated, unable to kill the tumour, and in particular the colony forming units. So I think our whole model system has to be rethought in the context of what else is in the peritoneal cavity.

Bradley: A number of investigators did look for the peritoneal presence of NK and LAK cells and they were there. Then they looked at the ability to lyse both the standard targets and also autologous tumour cells and that also was seen. That doesn't say that there aren't suppressor factors or other factors in the peritoneal fluid.

Mills: The problem may be the target. It is not trying to kill all of the tumour cells, but rather those that can self-renew and they seem to be exquisitely resistant to the effect of LAK cells or NK cells, if those are the colony forming cells.

Bradley: Others have looked at colony formation using a variant of the salmon human clonogenic assay. There are many stem cells that are in fact sensitive to LAK, so it is probably very complex. Very high effector-to-target ratios may be required. Some stem cells appear to be resistant. We should not be surprised when we see, in many cases, patients not responding. We also shouldn't ignore the few patients who do respond. It is my guess that those are patients who do, in fact, have some tumour associated antigen that renders them more susceptible to a classical immunological mechanism. I think ten years from now we will look back at the percentage of patients who have responded to these kinds of therapy and we will be able to correlate them with surface markers that we are just not recognizing now.

Bodmer: Dr Bookman, how much of the toxicity in immunotoxin studies is due to inadequate conjugation of the toxin and release of it, and toxic effects of the toxin itself when it has been released?

Bookman: HO PE, if it was released in significant amounts in the circulation could cause some toxicity but the isolated severe neurologic toxicity in the absence of other systemic reactions at the very, very low measurable levels, would be hard to understand. It turns out that OVB3 does crossreact with some energin on cells in the cerebellum and other parts of the brain. Toxicity may have been due to specific targeting at very low level concentrations. The ricin a-chain by itself is a very non-glycosylated recombinant a-chain which can be present at high levels without much in the way of systemic effects. All the studies that have used ricin immunotoxins have shown some sort of non-specific toxicities, such as fluid retention, weight gain, some pulmonary capillary leak. Several studies have also shown transient aphasia and neurological problems, but none of them have been as carefully evaluated as the patient who died in the study. Transferrin receptor is expressed on capillary epithelium and the brain and it's possible that there was in fact specific targeting of the capillary endothelium by the anti-transferrin receptor immunotoxins. On the other hand the capillary endothelium which composes the blood–brain barrier basically takes up a lot of things from the blood and presents them to the glial foot processes which really compose the blood–brain barrier. It is possible that these cells are metabolically very active in terms of uptake and transport from the blood, and presenting things to the glial foot processes and the brain. They may be uniquely susceptible to damage by some of these reagents. It is hard to understand how it would have been binding to the transferrin receptor because there is free transferrin receptor in the serum of concentrations of up to 2 μg/ml. We had antibody and toxin concentrations of less than 10 ng/ml.

Selected arguments

Mills: You mentioned that all of your patients are developing neutralizing anti-immunotoxin antibodies. Are those antibodies against the toxin or against the common determinants on your antibodies?

Bookman: The antibodies develop against the different domains of PE which I think the domain 2 is the most immunogenic. Some of the antibodies are neutralizing and some are not and there are also anti-mouse antibodies that develop. We have not carefully evaluated the presence of anti-idiotypic antibodies in these patients. There are also studies that have been done with monoclonal antibodies using cyclosporin or other means of aggregating immune response transiently to increase the period of time during which you can deliver this therapy. It's not necessary that you be able to give therapy for two years without ever developing antibodies that would give clinicians a chance to look at efficacy in trials.

EXPERIMENTAL THERAPEUTICS

Bodmer: How did you purify your LAK cells?

Herberman: The most convenient way to purify them has been by adherence purification. When one exposes natural killer cells to IL-2, within a few hours they develop the property of plastic adherence and by pouring off the non-adherence cells one can get a population that is highly enriched in the NK phenotype. Those cells will then proliferate very well in response to IL-2 in the condition medium.

Bodmer: If you are right that it is mainly NK cells and there is no HLA restriction, then there is absolutely no case for using autologous cells. It should be possible to use allogenic cells or other approaches that can accumulate the relevant cell types. Any notion of going to tumour infiltrating lymphocytes would be really quite irrelevant and I wonder what your comments on that would be?

Herberman: I agree with you it is a waste of time and it is logistically much more complex to get the cells that way than from the blood. There is clearly the potential to use allogenic cells for the therapy as opposed to autologous cells. One would have to be very careful about that, because one has an immuno-compromised recipient. There would be the potential of having some T cells being transferred and having some problems with graft versus host disease.

Bodmer: If you are right about attack on the micro-vasculature it should be possible to verify it. It is not clear why there should be limitations on access to the tumour if that was the way things worked.

Herberman: We do have clear data, from various models that there is selective binding of these cells when they are transferred to the post-capillary endothelial cells, but only in some metastases and not in others. The basis for this selective interaction with the endothelial cells remains to be understood.

Bodmer: Isn't it a rather widespread phenomenon that these cells, granulocytes, no doubt including NK cells, can attach to activate endothelial cells which you might find in the tumour micro-vasculature. Is it specific for NK cells or has it anything to do with damage that they might inflict?

Herberman: In an animal model that we have, we can directly visualize the micro-vasculature by videomicroscopy. We have been able to demonstrate that with one cell attaching to the micro-vasculature, and only one cell in that particular area, within 24 hours there is a cessation of blood flow followed by haemorrhagic necrosis.

Howell: Could you state again the population of patients in which these two studies were being done? These were minimal residual disease patients, after failing first-line chemotherapy?

Herberman: That's correct.

Howell: Did you see in your studies the same overwhelming peritoneal fibrosis that was the basis on which the NCI studies were stopped?

Herberman: We did see a substantial amount of peritoneal fibrosis, mainly at the highest dose tier and although I think the NCI group, because they gave cells plus IL-2, thought that this might have been contributed by the cells, I think it is more likely, when we got up into that dose range, the IL-2 itself, did that in a fair proportion of the patients. We did not see that as a major problem down at the low dose range where we saw the clinical responses.

Howell: How do you determine that the patients had a complete clinical response or partial response, if they started out with minimal residual disease? Did you operate on these patients again?

Herberman: This was based on a second-look or a third-look laparotomy.

Berek: In our own work, using fresh ovarian cells, just in chromium release assays, as targets against combinations of different cytokines with platinum, pre-treatment with very low doses of TNF, for example, or IL-2, produced the maximum killing, with the lowest possible dose that you could titrate in pico-moles. I think it is an important observation and it may turn out that, as you pointed out, less is better and it may make the whole approach much more relevant and less toxic. Do you have other experimental data in your group to support the clinical observation?

Herberman: If one looks at generation or maintenance of cytotoxicity *in vitro* by these cells, it would not conform to that sort of a reverse dose response curve. What we would see with IL-2 is that one does generate more cytotoxic activity at doses above 100 units/ml than below that. There is a plateau effect at about 500 units/ml.

Hacker: We have treated about 14 patients with low dose interferon, 10 to 30 million units intraperitoneally. Two of these patients had microscopic disease at second-look and both patients recurred, and of the patients who had no disease at second-look we had the same sort of incidence of recurrence as patients not so treated.

Balkwell: With the IP IL-2 in the ovarian cancer patients, do you think that it was due to the generation of LAK cells or do you think it was due to the activation of a whole cytokine network within the peritoneum, resulting in a cocktail of cytokines that was directly toxic to the tumour cells?

Herberman: I wouldn't make that distinction very strongly. If one exposes a population of cells to IL-2 the major cell population that we have seen, that have made this cocktail of cytokines, in fact, are the NK cells or the LAK cells. I would agree that this is part of the hypothesis that we have now. It is quite possible that the mechanism by which one is getting the effects if from the cytokine production as

Selected arguments

opposed to the generation of cytotoxicity. We know in fact that in these patients we are generating LAK activity, but we are also generating detectable amounts of both γ-interferon and tumour necrosis factor.

Mills: Dr Mutch, where you were looking at PLA-2 and the inhibitors, quinacrine and BPB – those concentrations should totally block PLA-2, but in those cases, although there was a decrease in the activity of TNF, there was actually a marked residual activity of TNF that was not blocked. Is there clearly more than one mechanism of action or was the BPB used in the quinacrine non-effective on PLA-2 in these experiments?

Mutch: When we used higher doses of quinacrine and BPB they were very toxic to the cells. Whether or not there is a transport problem, whether the intracellular concentration is really high enough to totally inhibit phospholipase A-2, is unknown also, I think. It is the concentration in the media. We did show a dose dependent response. But in the higher doses, when we went up too far, in the presence of protein synthesis inhibitors, it was too toxic for the cells so we couldn't determine anything.

Mills: So you didn't test the effect on the TNF-induced PLA-2 activity? Which would suggest it is blocked. It should be blocked at about 10 to 100 times lower concentrations than you are using. I wonder if you are seeing non-specific toxic effects here?

Mutch: We controlled for that. We zeroed the toxicity.

Mills: You need to measure the effect on PLA-directly. Also I think there could be many mechanisms by which TNF-α works independent of PLA-2.

EARLY DETECTION

Bodmer: Using colour flow Doppler to distinguish the luteal cysts from what might be malignant growths, I wonder whether you had more of the mature cysts in the family cases and whether that phenomenon wouldn't actually disappear with age?

Campbell: There was a bigger bulge, in the 45 to 50 age group. We have three cancers who were above .41 in resistance index.

Hacker: One of the diagnostic techniques that seems to have crept into this country, and also Australia, is laparoscopic aspiration of these cystic structures that are picked up on ultrasound and we have certainly had one case where there seems fairly clear evidence that this disseminated the tumour, with the cytology on the cyst fluid negative. Within a month the patient presented with disseminated disease.

Campbell: I wasn't really worried about disseminating cancer, what I was more worried about was the false negatives. But I am a bit against needling.

Oram: But we still have to do the operation.

Campbell: If it is a simple cyst, with no solid areas and no colour Doppler change, you can say this is a simple cyst and there is no need to do anything. If you want to get rid of it then minimal invasive surgery would be an appropriate way of dealing with it.

Oram: With what level of confidence can you predict? 98–99%?

Kurjak: Yes.

Oram: How reproducible is all this?

Kurjak: Our own inter group reproducibility was ± 12%.

Sharp: You indicated that in your study of high risk families there had been one interval cancer observed in a scan-negative woman, eight months later. Do you have similar information for the population study which you completed in 1987 since now four years have passed? Have you any evidence of any interval cancers appearing in scan-negative women in the original study? In other words, what is the protective value of your negative-scan in terms of time?

Collins: Yes. We did undertake a follow-up study form one to two years after the last scan on the general population and there was one interval cancer which occurred 21 months after the last scan. We are now scheduled to undertake a second follow-up, because we are particularly keen to know if the removal of all of those benign masses in that particular group of women has reduced the incidence of cancer or not.

Leusley: How can we explain the relative lack of more advanced disease in this screened population? One would expect to have seen some patients with more advanced tumours.

Campbell: I would have expected, to have found some advanced stage II, stage III cancers, and we didn't. But they are self-referred. Women who are anxious about cancer are less likely to present with advanced disease. Exactly the same thing has happened in the family group – four stage Ia cancers, no advanced cancer.

Oram: Was the staging done accurately?

Campbell: Yes.

Cohen: If you are talking about the same population base, there was a 2000 patient waiting list and the patient with advanced cancer will already have been treated.

Taylor: Shouldn't you be doing something about the so-called benign lesions, particularly in the women who are being screened from the high risk group? Is the reason immunoscintigraphy is not being considered, mainly expense or time?

Collins: The false-positives that were found at surgery after the application of a morphological score and colour Doppler were in fact two cases of teratoma, one endometriosis and one cystadenoma. In general those were best removed. I believe that the gene which may indicate a predisposition to a cancer in some members of this family, induces a predisposition to corpus luteum cyst formation in another.

Campbell: Immunoscintigraphy hasn't been proven to be effective in stage 1 disease as a diagnostic modality. It involves an overnight stay.

Bookman: Please clarify the patient with the interval peritoneal carcinomatosis with normal ovaries. Was she from a true high risk family syndrome?

Collins: She had a pedigree classified as the multiple site cancer syndrome.

Luesley: Why were there no advanced cases when people were invited for an ultrasound screening and yet when you invited people you saw an increase in the prevalence of your advanced cases?

Jacobs: I can't explain that; the findings in the London Hospital study are what you would expect in a prevalence screen. Finding just stage I disease in King's College studies is very difficult to explain.

Selected arguments

Luesley: Unless you are finding a different type of disease.

Jacobs: Well that should reveal itself on follow-up. It hasn't done as yet.

Williams: One of the things we need to know is whether the stage I tumours are going to stay stage I even though they get bigger. Can you give us an idea of the size of stage Is you were seeing?

Jacobs: The stage I tumours in the London Hospital study were all between 5 and 7 cm in diameter on ultrasound scan.

Williams: Were the King's [College] tumours big?

Collins: Of our four cancers, two were detected in normal size ovaries and two were something like like ×3, ×5 increase in volume. Could I just make a comment about the two populations. We do exclude some people if we believe they are not asymptomatic, and certainly in the last study we excluded them if they do not have a family history.

Bast: Had there been serum levels of CA-125 determined in the patients who were false-positives and true-positives in the King's study?

Collins: We analysed the serum from all of the patients who were classified as having a positive ultrasound result for six tumour markers. CA-125 was only raised above 35 units/ml in one of the four cancers. However, we have found a raised level of at least one of the markers in each of the four cancers, in particular, the borderline cystadenocarcinoma, where we didn't see colour, had a raise HMFG1 and HMFG2, even though it was a very small tumour in a small ovary.

Bast: And the false positives?

Collins: We analysed them in the false-positives as well. With the same tumour markers, we found have odds of about one to five of finding cancer at surgery but it would be different patients. What I am looking at, at the moment, is whether we are more likely to have false-positives that are premalignant with one second stage test than another second stage test.

Bast: At least two of these were advanced stage within a year of diagnosis, all in premenopausal patients. The tissues have not been analysed for CA-125. There are a number of examples, though, where in borderline cancers, for example, CA-125 is strongly expressed in the tumour but doesn't find its way into the circulation.

Taylor: In searching for other markers there may be particular tumours which are not releasing into the serum.

Bast: Possibly.

Bookman: Is there any evidence that the molecular weight of the energin might vary in patients who have benign conditions versus those who have malignant conditions, or any other features that might be used to aid in increasing the specificity?

Bast: An excellent idea. Ian Jacobs actually looked for different isoforms of CA-125 in benign and malignant disease and at least in a limited study, a couple of years ago, hadn't been able to find that.

Campbell: If the combination of CA-125 and ultrasound seems to be an effective one, albeit low sensitivity, would it not be better to lower the threshold for diagnosis of CA-125 and do more ultrasound scans?

Jacobs: When we started our study we

used the lowest level of CA-125 that was compatible with reasonable assay performance. The coefficient of variation intra and inter assays is quite high when you start to use levels below 25 units/ml. From the data we have, it looks as if we were to think of using a cutoff of say 15 or 20 units/ml, we could substantially improve the sensitivity. But that is using an assay which doesn't really have the performance to look at levels in that area.

Bast: The radioimmunoassay is probably a bit better in its precision than the enzyme linked assay in that range.

Soutter: The purpose of your test is not to detect ovarian cancer, but rather to detect early ovarian cancer or potentially curable ovarian cancer. I wonder whether it might not be better to present the results of your sensitivity in those terms. In the women who subsequently developed ovarian cancer, there was a substantial number who appeared to be detected at early stage disease. How were these patients detected?

Jacobs: In a prevalence screen you cannot influence the stage of the disease in the population on the day that you screen them, so the value of CA-125 in detecting early stage disease will only become apparent on follow-up screens. We are currently following-up half of the 22 000 women, randomly selecting 50% of the 22 000 women initially screened by subsequent screens with CA-125 and ultrasound. That will give us more information on the value of the screening protocol. The false-negatives detected during the two year follow-up all presented clinically with symptomtic disease, either pain or abdominal distension. None of them were found incidentally at the time of cervical smear or routine pelvic examination.

Soutter: It is astonishing that such a high proportion of that group of women should have presented with early stage disease.

Bodmer: There may be mechanisms by which you could increase the secretion of products that could be detected; γ-interferon, for example, is one that is known to increase the secretion of various products. Possibly it could be justified in familial cases. One would do this in the cell cultures to start with.

Bast: That is an excellent idea. So far there are relatively few things that effect CA-125 levels in culture, although steroids are among them.

PROGNOSTIC FACTORS

Wells: I would like to raise a note of caution that aneuploidy *per se* is not invariably associated with a malignant phenotype.

McGuire: I would like to ask about the data from the changes in proliferative fraction. Maybe that supports Rocheski's data for biological approaches to the administration of chemotherapy or maybe even biological approaches to surgery. Maybe surgery should be scheduled for midnight instead of 8 a.m.

Braly: Rocheski's data are based more on trying to minimize toxicity than anything to do with tumour efficacy. In his ovarian study he saw a difference in longterm survival depending on the sequencing of adriamycin and cisplatinum, but it was based more on almost a dose intensity type of approach. I don't think anybody knows how free tumour cell studies correlate to solid tumour and cell cycle changes in the solid tumour. Just because a cell is thought to be in s-phase by flow cytometry, doesn't mean that the cell is actively synthesizing

Selected arguments

DNA. There maybe many cells that are being held up in s-phase.

Oram: If a borderline tumour is described as being tetraploid, does that necessarily put it in the poor prognosis group?

Braly: Everybody is agreed on what tetraploid means, but which tracing do you assign to being tetraploid? There needs to be standardization of flow cytometry terms. Some studies look at diploid versus non-diploid and then would put tetraploid in with non-diploid. There are others that say euploid which is diploid plus tetraploid and then everything else is aneuploid. I don't know in the specific instance of borderline whether a tetraploid tumour behaves poorly.

Soutter: You must have a considerable admixture of non-epithelial cells which might significantly influence your results.

Braly: Quality control, especially if you are looking at paraffin embedded tumours is very important. You need to look at a thin section from each block and pick out the area which has the highest concentration of actual tumour cells and then cut the paraffin block to just take out that part for the flow cytometry.

Soutter: You can hardly do that for borderline tumours.

Braly: That's what is actually being done in the GOG trial.

Bookman: With regard to tetraploid, there is a little more information in breast cancer than ovarian cancer. One of the more useful techniques has been to do Feulgen staining on thick sections of tissue and then do quantitative nuclear morphometry to count the amount of DNA in each cell individually rather than relying on the averaging that takes place through the flow cytometer. The prognostic subgroups seen most often indicate that tetraploid is not a bad prognostic feature. In people with intraductual tumours you can have very bizarre results with significantly aneuploid tumours with the very high s-phase fractions. With invasive disease you would describe these as a poor prognostic subgroup. Are there any prognostic factors at this point, that might contribute to clinical decision making?

Braly: Until we have completed clinical trials we won't have a definitive answer on that.

De Petrillo: What about the standardization of the surgery among the centres co-operating in the trial?

Torri: Over the period the time of second surgery was changed. The first surgery was always the same.

Braly: I think we do have enough information to use ploidy as a stratification in entry into clinical trials. Within five years or so we should be able to make treatment decisions on the basis of that information.

Heintz: Did you also consider the use of morphometry?

Braly: I have seen very few studies comparing the results of morphometry to flow cytometry. The advantage of flow cytometry is that it is theoretically very fast. It really depends almost on what you have available.

Heintz: True. There are clear advantages to morphometry in that you have the cell. Probably for diploid tumours the information you get from morphometry is easier to interpret and clearer than from flow cytometry.

Wells: Regarding sampling of ovarian tumours, the problems of flow cytometry are analogous to those of oestrogen receptors and biochemistry versus immunocytochemical assay. There are a number of other approaches which have not yet been mentioned. One is the antibody χ 67, which is a proliferation marker, and does appear to correlate with S and G2 phase. We have also published work in the endometrium on bromodioxiuridine labelling and its correlation with S and G2 phase. There is no doubt that the AgNor count also correlates with S and G2 phase as measured flow cytometrically.

META-ANALYSIS OF TREATMENT

Markman: did the two analyses come to a different conclusion regarding the influence of cisplatin?

Williams: These are very different analyses. The analysis that I presented was an analysis of all the individual patients in the studies which were looking at a specific question. The analysis of surgery I don't think really can be called a meta-analysis.

Soutter: The purpose of this study was to examine the effect of surgery. Our interest in the chemotherapy and the dose intensity and the use, or otherwise, of platinum, was simply as a means of labelling the fields. It is quite striking that the effect of platinum should have been so obvious. I am encouraged that another group led by two very distinguished statisticians should have come to very similar results. To call our approach meta-analysis is perhaps not strictly speaking accurate.

Williams: People don't like the idea of looking at the literature in this way, and have suspicions. Doug Easton's was a review of the published literature which excluded some 40% of the studies that we presented. There is clear publication bias in that study. There is no exclusion of those patients on intention to treat.

Cohen: Griffiths showed clearly a variability in survival from 0.5 cm up to 2 cm residual disease after the first surgery. He stated boldly that if the operating surgeon did not anticipate being able to reduce the disease to 1.5 cm he should not waste the patient's time. It may well be that platinum gives us a certain security when we have not been able to cytoreduce.

Lowry: I wonder whether cisplatinum became popular in ovarian cancer because of the dramatic results seen in testicular cancer? That meta-analysis is required at all, means of course that the differences are small. I wonder if the statisticians can tell us whether we really can detect 10% differences in survival? Large numbers are required and this introduces not just large numbers of patients but large numbers of physicians, and then the quality control begins to deteriorate.

Williams: I take your point that it is dilution of the good results. There are studies in the literature, in reinfarction studies with fibronlytics in heart disease where people have accrued 17 to 20 000 patients and showed modest differences in survival which are highly significant. The same should apply with medical oncology. The treatment may not be done quite as well, the surgery might not be quite as good, but if you can still show a 10% advantage then that is the minimum that you can expect to achieve.

Markman: We know that the majority of women with ovarian cancer cannot be cured with chemotherapy. The woman who has the best chance of long-term survival is that one with very small disease when we start the therapy. Most of the trials focused on those patients with bulk disease and looked for response rates as well as survival. Not very much focus has been on the smaller residual disease population. Have you been able to look at that smaller subset, to say whether the drugs are in fact equivalent in effectiveness?

Williams: We need long-term follow-up for these studies. We didn't examine in detail the bulk of disease in these patients. This first meeting of the ovarian overview group was really designed to look entirely at survival. A second meeting is planned in about two years' time and we will be able to answer that question. The key thing is to actually have 5+ years medium follow-up for those studies before we can be 100% certain of the equivalents between carboplatin and cisplatin.

Heintz: If I look through our own data at my own hospital then we saw that patients who were what we called 'natural optimal' from the beginning, with small initial advanced disease, they survived well. If you looked at the ones who need cytoreduction, they survived about 35 months medium survival. The ones who were not optimally cytoreduced had a medium survival of 23 months. You could operate on the last group better and bring them into the second group, with, theoretically, a medium survival improvement of about 12 months. Those 12 months I think are worthwhile in a human's life. I think that is the way in which we probably have to study cytoreduction in the coming years, not in terms of cure but in terms of quality of life.

Soutter: I think that what you do not know is what would have happened to these patients if you had not cytoreduced them. It seems almost that the women who we can't cytoreduce are a different group and have a different biological tumour from the ones whom we can cytoreduce.

Heintz: I have my doubts in how far just looking at optimal data and the literature really reflects the influence of cytoreduction.

Soutter: The range of maximum cytoreductive surgery in our study was very large. One study that claimed 90% maximum cytoreductive surgery. These patients were also treated extremely aggressively with chemotherapy.

Taylor: What Mr Soutter is trying to do is very important to assess whether the cytoreductive surgery should be used. If it is thought that he hasn't addressed that problem, how can it be addressed?

Williams: I think the only way that it could be scientifically addressed is if surgeons would agree to a randomized clinical trial. If you cannot do that study, I find it very difficult to know how you can really scientifically and rigorously show that surgery is important.

Cohen: I thin that one has to be very careful about this notion of the biology determining whether a tumour can be reduced. All of us in referral centres know that there are patients referred from the outside who were declared 'inoperable', who have been referred and have been optimally cytoreduced. So surely that was not a biologic determinant in the patient.

Scott: Maximal cytoreduction is obviously very much a surgeon determined endpoint.

Hacker: We did, if you recall, set-up a study through the Gynaecological Oncology

Group which was to take those patients referred in, said to be inoperable, and randomized those patients between reoperation with an attempt at aggressive cytoreduction versus chemotherapy. That protocol failed from lack of accrual.

Jacobs: I think we are in danger of underestimating the importance of Pat Soutter's study. The approach used is the closest that we can come to without doing a radomized controlled study in assessing the value of cytoreduction.

Williams: If you extract and show [that] platinum in fact doesn't make a huge difference as the other analyses have shown, then you end up with showing that there is quite a strong correlation between the amount of cytoreductive surgery and median survival. What worries me is that you have got some fudge factors in here which there are question marks over. How can you make a conclusion?

Soutter: We chose to use dose intensity and the effect of platinum, because we thought a priori that these might influence the median survival. The analysis that we did, confirmed that with the data set that we were looking at, that was true. We were tyring, by selecting these variables, to choose those variables which would muddy the waters from the point of view of our looking at that aspect of treatment which we wished to examine, i.e. surgery. Hence our analysis showed the platinum and dose intensity in this group of nearly 7000 patients did affect the median survival time. We used the figures that came from that to remove the apparent effect of the chemotherapy with these parameters. If you have a surgeon working in a district general hospital or a chemotherapy trial all of these fundamentally were studies of chemotherapy. There may be one or two other surgical studies, but they weren't controlled studies. The majority were chemotherapy studies, so the operations were done by a variety of different individuals. Those studies where the surgeons were working hard and getting great surgical results, these patients should have done better. The sad fact is that they weren't. The rationale for using cytoreductive surgery is emotional; it is uncontrolled. We are using criteria that we would not accept for any other form of therapy.

Heintz: The value of the work that Pat did is that it shows clearly one thing that cytoreductive surgery in this meta-analysis did not improve survival. But again I would like to underline the importance of the quality of life in the short-term.

Cohen: If a single institution has a large volume of patients all of whom have been treated identically with standard protocols, and the dose intensity is inspectable, and it is apparent that the protocols have delivered a large percent of the planned dose, the only variable appears to be the volume of residual disease at the end of the first surgical procedure. There is a difference in either progression-free interval, disease-free interval or survival. How does one critize that as being an uncontrolled study? There is almost certainly a selection bias in there not by the surgeon, but by the tumour. By sub-categorizing and sub-classifying patients as we did by improving the results of the sub-groups and sub-categories, without in any way influencing the survival of the overall group. It is my view that this is precisely what we are doing in this sub-set analysis of women who have minimal residual disease versus those who don't.

Selected arguments

SURGERY

De Petrillo: What was the survival in regards to lymph node status?

Hacker: It did not show any survival advantage for having negative nodes. We did just a pelvic lymphadenectomy, while others are doing the extensive para-aortic dissection as well. I think that we need to find out in some prospective fashion whether or not this has a role to play or not. Eric Burghardt is so convinced of it he won't enter into any prospective study himself. Most people are not quite so convinced.

Heintz: Dr Luesley had several factors listed that are independently significant for survival. What does he recommend as treatment now, having done this analysis?

Luesley: One of the reasons for doing this was if, hypothetically, we were going to do a trial in the future, we would not wish to include patients at a very low risk of relapse or expose them to chemotherapy. We wanted to define a high risk group. We'd see more events in that group and then probably would not need as large a study as well. At the moment our current treatment is no treatment at all. We don't routinely treat stage I, well-differentiated ovarian cancer. Stage Ic we would treat in the same protocol as patients who have more advanced disease but total cytoreduction. At the moment these are either randomized to no treatment at all or cisplatinum and cyclophosphamide. Should we now formalize this for our retrospective population study, and use it to construct a trial? The ICON are already running a trial.

Heintz: With grade 2 and 3 tumours?

Luesley: I would consider those at risk of relapse and therefore they are treated in the current protocol which is no treatment versus CP.

Edmonson: In your stage I prognostic factors study did I understand that adjuvant chemotherapy was a negative prognostic factor in the multi-variant analysis?

Luesley: Yes.

Edmonson: Do you think you can make this kind of judgement in a group of people in which the decision to give or not to give was a deliberate medical judgement type decision.

Luesley: That is one of the reasons why it is a multi-variate analysis, to see if it was inter-dependent from other adverse factors.

Edmonson: Even so, there is strong bias the physician introduces when he or she decides to give or not to give treatment to an individual patient.

Luesley: We would expect that strong bias in patients who have high grade, Ics, etc., that should have come out in the multi-variate analysis; don't forget the time interval of this study was 1980 to 1984. A proportion of these patients had not have received cisplatinum as their chemotherapy, although towards the end, virtually all of them would have done.

Hacker: Just on that point of the stage I, I noticed that your clear-cell carcinomas did worse and that seems to be a controversial point in the literature. I wonder if Dr Scully would comment on that?

Scully: I don't have any information myself on clear-cell carcinoma. I know there are at least three studies showing that chemotherapy of stage III clear-cell is not as good as, for example, serous carcinoma, but then a study from the M.D. Anderson Hospital, showed no difference in chemosensitivity, clear-cell or other forms.

Williams: Whenever anyone has proposed an early stage study, there has never been a unanimous agreement as to what criteria should be developed. That was the reason for introducing the uncertainty principle, plus the fact that it has been very successful in large scale studies elsewhere. It does give you a broader spectrum of patients.

Luesley: It is going to be quite important that your documentation is fairly good. The concern that I would have, is that there are going to be a proportion of patients in this study exposed to chemotherapy and they are not at risk of relapse, because only three out of ten stage I cases are going to relapse. You are going to be treating these patients. Why not identify a high risk group? It can be done.

Williams: You know what the high risk group is? When we get the clinicians together from the MRC – and the EORTC did the same thing in an early stage study – none of them could agree what they felt were the appropriate criteria.

Luesley: You therefore would accept there are going to be well differentiated stage I patients in this study?

Williams: If the clinician really feels there is some reason why they should go in, then they would go in the study. I personally feel that it would be inappropriate and I think very few would go in.

Hacker: There will always be some reason; you are uncertain that you couldn't do the appropriate operation, or whatever.

Sevin: We are going to be faced with more and more cases where gynaecologists do laparoscopic surgery, rupture the cyst and then find either benign or malignant process. Your data indicate that those who have ruptured cysts have the worst prognosis – is that correct?

Luesley: Yes.

Sevin: Are those different than those who have positive cytology? What is the risk of these patients who are converted into a positive cytology by our surgical procedure? Are the surgeons who do laparoscopic surgery increasing the risk of failure in these patients or not?

Luesley: Staging was done according to FIGO guidelines, in other words, lavage or ascites was aspirated before surgical manipulation of the tumour. I can't answer the next question.

Hacker: We recently looked at the literature as we had one definite case and another case which I think probably also fitted this category of rupture of a cyst and then subsequent disseminated disease. It is unconvincing that it impairs the prognosis. It may well be that that's because you are able to lavage the peritoneal cavity and get rid of those malignant cells. On the other hand, when you rupture the cyst at laparoscopy, and that fluid slowly leaks out over a period of days or weeks, I think that it is very likely that those cells have an excellent opportunity to implant on the peritoneal cavity. There is probably a difference between intra-operative rupture and rupture at laparoscopy or for some other reason whether there is delayed intervention. That's admittedly a personal bias.

Sevin: So what you are saying is: if the

surgery that follows is an appropriate cancer type operation, you don't think there is a difference. On the other side if it is not recognized, and the patient persists to leak and then has time to disseminate, that would be the problem?

Hacker: Yes, I mean I think if someone does a laparoscopy, gets a frozen section and intervenes straight away, there is probably no real problem. But that doesn't usually happen; there is usually a delay of several days or in the report presented at the SGO there was frequently several weeks delay, between the laparoscopic diagnosis and the subsequent surgery. I would think it is highly likely that that would impair the prognosis and lead to dissemination of the disease.

Luesley: I think this might be one of the obvious applications for Professor Kurjak's colour Doppler. If there is any minimal risk of this being a possible carcinoma, then I would strongly recommend that it isn't aspirated, but removed. Because whilst I would agree it isn't conclusive data, it is not conclusive the other way either. You might just be adversely effecting prognosis when it would be quite easy to avoid doing it. The only reason you are doing it really is convenience. If there is any indication beforehand that it is not benign, then these sorts of techniques ought to be employed.

Hacker: On the other hand, I think if they are functional they'll disappear and if they are neoplasms they'll recur, so I really see no argument in favour of laparoscopic aspiration of these tumours.

Soutter: Your treatment results in your own series were really very disappointing; that was in part because you were unable to administer the chemotherapy which you would have preferred. Do you have any idea what the medium survival was in the patients who got your preferred chemotherapy?

Hacker: About 30 months.

Soutter: I meant your total group of patients, whether or not they had optimal cytoreduction, who received the full chemotherapy.

Hacker: I didn't analyse those.

Soutter: I think if you do that, you may well find your overall survival, inspite of all this tremendous surgical expertise that you have applied to your patients, is really quite disappointing and not very different from what can be achieved just with modest surgery and chemotherapy.

Hacker: That's certainly possible. That's why we and you have been advocating prospective studies.

Hacker: I think the problem with advanced ovarian cancer, is that there are a lot of patients to whom you can do no real good, however hard you try. There may well be a group of patients who benefit very significantly from cytoreduction, but we haven't specifically identified that group, I think it would be wrong to dismiss the concept just because it is difficult in retrospective studies to identify clear benefits.

Oram: If we were going to consider doing the trial of the place of lymphadenectomy, do you think we should stop at the pelvic brim or should we do a para-aortic as well as the pelvic.

Hacker: It's clearly a fairly non-morbid thing to do a pelvic lymphadenectomy. Once you start doing the para-aortics as well, there is unquestionably an increased morbidity. I don't think I would personally advocate a full para-aortic lymphadenectomy, but probably one should go up to the inferior mesenteric artery.

Sevin: If we do lymphadenectomy and there is no tumour in the nodes, there is no theoretical or practical benefit of a lymphadenectomy. If we selected the debulk tumours that are palpable, like you did, but then put them into the group of not lymphadenectomy, then you are skewing your data. Is the sensitivity of the tumour in the lymph nodes different to the tumour that is in the peritoneal cavity or somewhere else?

Heintz: After 15 years of the introduction of cytoreductive surgery in gynaecological oncological practice, we still do not have the answer that the procedure is right or wrong. When we want to change surgical practice we have to walk along the same roads as is used in radiotherapy and in chemotherapy. Perform a good investigation before you advocate it to the whole medical community.

Edmonson: What is the incidence of peri-ureteral fibrosis in people who have had stripping of the posterior peritoneum for a para-aortic node dissection.

Hacker: You are talking about tumour fee stricture?

Edmonson: Yes.

Sevin: There are large data on radical hysterectomy involving node dissection in cervic cancer. The complication rate in those patients who are not radiated, is not a significant problem. It is those groups of patients that are then treated either with pelvic or pelvic and para-aortic radiation that seemingly have an increased risk and there it is a big problem.

Edmonson: Perhaps it's the radiation factor.

Sevin: Probably a combination. In your group where you had initial chemotherapy and then interval debulking, both of you had patients in that group that do worse. Were these responders or were these non-responders to chemotherapy?

Hacker: My patients were responders. For some time now I have adopted the policy that anybody who doesn't respond to cisplatinum is incurable, purely a palliative patient, and I would not operate on them. In fact, I have modified our approach, for example, in patients who come in with large effusions, large ascites; that group have a high post-operative morbidity and so we give those patients cisplatinum and cyclophosphamide upfront and only operate on the responders.

Luesley: All the ones in our study were responders. They didn't do wose than those who weren't debulked.

Sevin: They didn't do better, they had initial lag and then they all did the same as the poor group?

Luesley: That's right.

Sevin: One could interpret that we are potentially doing the patient a disfavour by inducing or selecting resistance. They are not doing better even being completely debulked after three months.

Luesley: That was the reason why I showed the data. There are certainly some patients who do benefit from being

Selected arguments

properly cytoreduced – I don't know how big that group is and I don't know how to identify them. If we are going to have to go away and do something from this meeting, it is to make attempts to try and find out which patients at primary laparotomy may well benefit from this type of operation. I don't believe at the moment any of the information says that there should be no cytoreduction. I think that would be a retrograde step, you'd probably put the cause of gynaecological oncology back 20 years if we said that.

Manetta: There are a couple of preliminary reports showing for instance a very high incidence of metastatic disease including the lymph node. I am not quite sure where you draw the line as far as the lymph node metastases is concerned. That is where immunoscintigraphy will perhaps be very useful. Are we really talking, as we have thought for many years, of a disease which is contained within the peritoneum, or actually have we been fooling ourselves all this time and are we talking about a systemic disease perhaps much earlier than we think?

Creasman: Just a point of order – let's not call those early stage patients that we treat with surgery in the category of no treatment, because obviously the treatment has been the surgery.

Hacker: I would like to get a sense of how many people here feel that the initial approach to patients with stage III disease should still be aggressive cytoreduction.

Cohen: I would like to ask whether any investigator in this room knows of a single tumour model, either *in vitro*, or in an animal, or in a human, that does better with large volume disease than with small volume disease?

De Petrillo: There is no study.

Hacker: Well, Steve Howells showed yesterday that one treatment was sufficient to induce sufficient acquired resistance to platinum to influence clinical outcome, so I think there is a difference.

Soutter: The breast surgeons are now doing lumpectomy rather than radical mastectomy; the gastrointestinal surgeons, long ago gave up doing cytoreductive surgery in extensive g.i. tract malignancy.

Hacker: That's a very poor analogy. You have a tumour that's not chemosensitive, in the case of bowel. In the case of breast you are talking about resecting the disease, the analogy there is vulva cancer. It is certainly not advanced ovarian cancer.

Cohen: I would support a trial only if it included a laparotomy at which time the surgeon made a decision about whether the tumour was reducible or not, and if it were reducible, then the patient were to be randomized, with the belly open.

Luesley: I must have seen many talks that put less than or greater than 2 cm up. I wonder how people are measuring this? A lot of this is dense infiltration of mesenteries or peritoneum that is not amenable to measurement and I suspect there is a huge error in these measurements. We are not really looking at homogeneous groups at all.

Scully: I would like to ask David Luesley, in his series of cases was there any difference in the histology or type of the tumours that could not be debulked versus those that could be debulked successfully?

Luesley: In terms of histological sub-type, no significant differences; there was a higher proportion of high grade tumours than the ones that couldn't be totally debulked.

Creasman: I want to reiterate what David has said about size of tumour. One of the studies that the GOG did was in stage III optimal disease. After debulking they went on chemotherapy; one of the criteria that we evaluated were the number of nodules remaining behind after they fell into that optimal group. There was no question that the number of nodules were directly related to how well that patient did, i.e. negative second-look, and even long-term survival. I think we have got hung up on size instead of tumour volume. Although we use size as tumour volume that's not the only criterion. We need to add up all of those nodules because many of those patients we are calling optimal are really sub-optimal when we look at tumour volume.

Index

Numbers in bold refer to figures, numbers in italic refer to tables.

Acetaminophen 305
Actinomycin D,
  phase I clinical trials 399
  pre-treatment 171, 175, 179, 298
Acute myeloblastic leukaemia, chromosome translocation 17
Adenocarcinoma, stroma proportion **27**
Adriamycin, see Doxorubicin
Advanced Ovarian Cancer Trialists Group (AOCTG) 315–23, 402
Age distribution and expected incidence study *269*
Alkylating agents (AA) 331–42, 378
  complete remission 387
  evaluation 332–3, *333*
  survival 339–40, **340**
  validity assessment **341**
  vs CC 335
  vs CDDP and doxorubicin 336
Allele loss 24–30
  on chromosome 6 25–6, **29**
  for chromosomes 396
  data compilation 26–30
  materials and methods 25
  ovarian cancer 35–7, 79
  p53 mutation 58
Amenorrhoea 268
American Type Culture Collection 105
Analysis of variance (ANOVA) 335
Aneuploid cells 214

Aneuploid tumours 401
Aneuploidy 360
Angioneogenesis 400
  factors 397
Anscombe's test 347
Anti-oncogenes, see Tumour suppressor genes
Antibodies
  anti-keratin 7 208
  host formation 158–9
  radio-labelled 401
Antibody binding, to extracellular domain 68
Antigens, tumour-associated 73
Antihistamines 165
Arachidonic acid 181, **183**
Ascitic fluid 93
  CDDP control 355
  growth factors 129–30
  intraperitoneal therapy control 308
Aspartate transaminase 300
ATP bioluminescence assay 399

Basal cell naevus syndrome 9
BCG 336
Bell study 336, 338–9, **339**
BEWO cells 142
Birmingham (UK) group 387
Bone marrow transplantation 355
Breast screening, integration with ovarian 281, 286
Bromodeoxyuridine (BUdR) 214–15, 219

Bromophenacylbromide 181–2, **182**
Bruckner study 335–6
  response data 337–8, **338**, **339**
Burkitt's lymphoma, chromosome translocation 17

CA-125 101
  annual assays 281
  antigen 265–6
  histological associations 110
  lead time over clinical presentation 274
  measurement in ovarian mass differentiation 251
  in ovarian cancer screening 265–74
  screening
    lead time 271–3, 401
    performance summary *271*
    studies 282–3
  sensitivity 274
  serum distribution in volunteers **269**
  serum elevation prior to diagnosis *283*
  serum estimation 283
  serum levels 265–6, **266**
  tissue barrier disruption 266–7
  serum measurement 401
  specificity 274
  test performance data summary *274*

443

Index

and ultrasound screening 268
Calmodulin 167
  inhibition by DECA 171
Cancer cell lines, CSF-1R expression 118–19, **119**, *120*
Cancer Family Study Group 31
Cancer Research Campaign Trial Office 347
Cancer and steriod hormone (CASH) study 3
Carboplatin (CBDCA) 161–6
  better toleration 323
  dosage 165
  FDA approval 310
  GM-CSF regimens *163*, 163–5
  intraperitoneal penetration 292–3
  intraperitoneal therapy *291*
  meta-analysis 402
  stage III EOC 352, 354
  vs CADP 323
Carcinoembryonic antigen (CEA) 208
Carcinogenesis, genetic alterations 17
Carmo study **339**, 342
Catheter-related complications 308
CC, *see* Combination chemotherapy
cDNA, *see* DNA, complementary
Cell division rate 179–80, **180**
Cell interaction in cancer 93–4
Cell surface receptor 53
Chemotherapy
  GM-CSF in *164*
  randomized trials 315–23
    data comparison *318*
    details 328–30
    methods and data 316–18
    survival curves 319–20, **320**
Chinese hamster ovary (CHO) cells, and TNF 87
Chlorambucil 331

vs CDDP plus chlorambucil 335
vs CDDP plus cyclophosphamide 336
vs CDDP plus cyclophosphamide and doxorubicin 335
Chou–Fasman secondary structure plot **45**
Chromosomal abnormalities in ovarian cancer **19**, **21**
  frequency **19**
Chromosome 6, allele loss 29
Chromosome 11 centromere probe, hybridization **81**
Chromosome 17, allele loss 29
Chromosome 17 linkage 4–5
Chromosome 17 marker, in HBOC 12
Chromosome 17q, allele loss 36–8
Chromosome p53
  gene mutations 396
  ovarian tumour over-expression 396
Chromosomes
  allele loss 396
  localization 78
  losses 407
  trisomy **20**
  and tumour function control 21
Chronic myelogenous leukaemia, chromosome translocation 17
Cisplatin (CDDP) 149, 168, 322–3
  alternating with interferon-α 304–5
  antitumour effect **191**, **193**
    synergistic *191*
  in combination chemotherapy 386–7, 390
  comparison with platinum 318
  as cost-effective therapy 342
  and cyclophosphamide study 336
  dose escalations 189

effects on
  cell division **180**
  later rIL-2 therapy **193**
  rIL-2 in mouse model 192–3
evaluation 331–42
immune system potentiation **191**, 191–2, 194
immune-modulating effects 190
intraperitoneal 297
  combined with interferon-α 299–304
  penetration 292–5
  therapy *291*
MTD 303
NC-mediated lysis effects 177–8, **178**
in randomized chemotherapy trials 315–23
  regimens 378
  response lack 401
  with rIL-2 400
stage III EOC 352–5
synergy with DECA 169, 179
toxicity 161, 331–2
vs carboplatin 319
vs CC 336
Cisplatin (CDDP), *see also* Platinum
Clear cell carcinoma 270
CLINPROT 333
Coagulation cascade, TNF/IL stimulation 397
Coagulation factors 397
Coefficient of variation (CV) 214
Coelomic endothelium 5, 7
Colonoscopy in HNPCC 14
Colorectal cancer, age-incidence curve 37
Colour velocity imaging (CVI) 246
Combination chemotherapy (CC) 332
  and drug resistance 398
  evaluation 333–5
  favourable studies 339
  progression-free interval 339
  survival 339–40, **340**, **341**
  validity assessment **341**

Index

vs AA, response data 337–8, **338**
Complete clinical response (CCR) 303
Complete pathological response (CPR) 304
Confidence intervals (CI) 317–18
Cox's model 210
CT scans 284
Cycloheximide (CHX) 179–80, **180**
Cyclophosphamide (CYCLO) 161, 315–16, 322–3, 331, 335–6
  stage III EOC 352–4
Cyclosporin A 159
Cyst, complex, colour flow imaging **242**
Cystadenocarcinoma 270
Cystadenoma, and ovarian cancer 400
Cystadenoma serosum **256**
Cytokeratins 208
Cytokines
  action 396–7
  discussion 415–19
  and ovarian cancer 87–92
  in tumour cells 96
Cytolysin/perforins 175–6
Cytoplasmic protein 53
Cytoreduction 361–4
  secondary 371

DECA, see Dequalinium chloride
Decker trial 335–6
Deoxyspergalin 159
Dequalinium chloride (DECA) 167
  combination therapy 169–70
  evaluation 399
  in vitro assay 168–9, **169**
  intraperitoneal toxicity 170
  as single agent **169**
  synergy with CDDP 169, 179
  toxicity studies 168–9, **169**, *170*, 170–1
Diacylglycerol 70
Diamminedichloroplatinum (DDP) 149–50, **150**, 151–2
Diarrhoea 303

DNA
  for allele loss 25
  complementary
    clone isolation 83–4
    clone 'panning' protocol **75**
    cloning by 'panning' 74–6
    cloning strategy 74
    libraries 74–5, **75**
    sequence analysis 76
  content measurement 213–14
  hybridization screening 74
  marker inheritance 4–5
  polymorphism 76–8
  probe hybridization **79**
  recombinant technology 399
  repair 399
  S phase content 209, 214
  Southern blot **80**
    autoradiograph **28**
DNA index (DI) 214, 216
DNA-specific fluorescent dyes 214
Doppler
  colour
    detection rate 241
    false positive rate 241
    intra-ovarian blood flow values *241*
    principles 237–8
    screening 285
    technological developments 245–6
    transvaginal, as second stage test 245
    use in postmenopausal or premenopausal women 239
Doppler blood flow measurement 6
Doppler colour flow approach, limitations 400
Doppler flow signals 238
Doppler frequencies, end-points 238
Doppler probes 229
Dose intensity (DI) 347–8, 398
Doxorubicin 168, 316, 322–3, 335–6

Drugs
  distribution in intraperitoneal therapy 291–2
  and MST 346–7
  penetration in intraperitoneal therapy 292–3
  resistance 398–9
    discussion 420–2

Electroporation 74
Emetine (EM) 179–80, **180**, **183**
Endometrial cancer, Doppler transvaginal ultrasonography detection 244–5, **255**
  in HNBCC families 9
Endometriosis 5, 227
  ovarian malignancies 201
  and ovarian tumours 400
  in stroma **201**
Endosalpingiosis 5
EOC, see Epithelial ovarian cancer
EORTC study 355
Eosin, assessment 207
Epidermal growth factor (EGF) 61, 133–7
  in ascitic fluid 131
  autocrine/paracrine growth regulation **62**
  structure 61–2
Epidermal growth factor (EGF) receptor 53, 62, 133–8
  abnormal expression 54
  expression 94
  sensitivity modulation 149
  signal transduction pathway 151
Episialin 41
Epithelial inclusion cysts 204
Epithelial membrane antigen (EMA) 41
Epithelial ovarian cancer (EOC) 128
  allele loss data compilation *30*
  cell culture 128
  chemotherapy response 363
  growth regulation 61–6
  histological subtypes *26*
  IL-6 levels *108*
  inherited 31

445

Index

microscopic disease at second-look laparotomy, survival *371*
optimal debulking 362
p53 expression 55–6, 58
persistent, chemotherapy 297–311
primary therapy outcome **372**
quality of life 363–4
radical resection 362
raised serum CA-125 266, 270
residual disease
  at second-look laparotomy 369–70, *370*
  at second-look laparotomy MCS *371*
  in relation to topography 363
screening trials justification 277–8
spread pattern effect 362–3
stage I, population-based survey 360
stage III
  management and outcome 351–5
  survival **353**
    after MCS and platinum CC **353**
    vs peritoneal carcinoma **353**
    vs residual disease **353**
stages III and IV, clinical course **370**
staging strategy criticism 359–61
staging strategy support 358–9
surgery 357–66
  roles 357
  and survival **365**
survival
  after MCS **361**
  in bulky vs seedlings disease 363
Epithelial ovarian cancer, *see* Ovarian cancer
Erythrocyte movement in Doppler ultrasound 263

Etoposide 352, 354
  intraperitoneal 294
Experimental therapeutics 399–400
  discussion 426–8

Fab fragments 69
Familial breast cancer, allele loss 26, 30
Familial ovarian cancer 3–7
  aggregation 11
  colour Doppler
    early screening 239–41
    false results *242*
  discussion 405–7
  registers 7
  relative risks *234*
  screening detection *233*
  transvaginal ultrasonography outcomes **231**
  ultrasound based screening procedure **231**
Familial ovarian cancer, *see* Ovarian cancer
Family Cancer Study Group 279
Family cancer syndrome 5
Fathalla hypothesis 36, 38
FBP, *see* Folate binding protein
Fibroblast growth factor (FGF) 61, 133
Flow cytometry 209, 213–19, 400
  effusion evaluation 217–18
  technique 214–15
  in treatment monitoring and modification 218–19
Flow velocity waveforms (FVWs) 238
Fluorouracil 315
Fogelstein's DCC 411
Folate binding protein (FBP) 76–84
  gene copy number 78–9
  genomic organization 84
  nucleotide sequence 76–7, **77**
  RNA splicing **78**
  role in ovarian cancer 82
  Southern analysis 76
Forskolin 149–50, **150**, 151
Fox Chase Cancer Center 398
Future years of life lost (FYLL) concept *278*

Gastrointestinal toxicity 303, 306
Gene
  dominance and recessivity 24
  p53 55–8
    expression and survival 56–7
    mutation in cancers 55
    mutation in EOC 65
    sequencing **57**
  p185
    extracellular domain epitope map **68**
    as growth factor receptor 67
    kinase activity 69
Genetic changes, discussion 407–15
Genetic counselling 14
Genetic heterogeneity 5
Genetic imprinting 407
Genetic linkage 4
  analysis **5**
  in HBOC 12
  studies 395
Genetic mapping 4
Germ cell tumour 269
Germline mutations 31
GICOG study 318–19
Glucocorticoids, CSF-1R expression control in cell lines 121
Glutathione synthesis inhibitors 399
Glycoproteins 53
  antibody recognition 73
Glycosyl–phosphatylinositol (GPI) linkage 81
Glycosylation, core protein variations 43–4
Goldie–Coldman hypothesis 363
Granulocyte colony stimulating factor (G-CSF) 131
Granulocyte macrophage colony stimulating factor (GM-CSF) 162–3, *163*, 164, 164–6
Granulocyte macrophage colony stimulating factor (GM-CSF), in combination chemotherapy 398
Granulomyelopoiesis 94

Growth factor(s), see Growth substances
Growth inhibitors 140–1
Growth substances
 and cytokines 396–7
 discussion 415–19
 expression 94, 97
 in female cancers 115–16
 local action 129
 ovarian epithelium 128
 pathways, aberrant activation 129
 receptor 208
Gynecologic Oncology Group 298, 300–1, 305, 308, 310, 378

Haematological toxicity versus dose level *302*
Haematoxylin, assessment 207
Hazard ratio (HR) 317–18
Hazard ratio (HR) plot
 in comparison trials 319, **321**
 in platinum trials 319–20, **320**
HBOC, see Hereditary breast and ovarian cancer
HELA cells 142
Heparin binding secretory transforming factor 83
HER-2/*neu*
 expression and survival **55**
 in ovarian cancer 94
 overexpression 101
 and tumour cell inhibition 67–71
Hereditary breast and ovarian cancer syndrome (HBOC) 11–14
 cumulative evidence *12*
 family pedigrees **13**
 surveillance 14
Hereditary non-polyposis colorectal cancer (HNPCC) 9–11
 family pedigrees with ovarian cancer **10**
Hereditary ovarian cancer (HOC) 9–15
 clinical implications 14
 discussion 405–7

Hereditary ovarian cancer, see Ovarian cancer
Hexa-CAF 315
Hexamethylmelamine 315–16, 335
Histopathological description of biopsy laparotomy material 389
HNPCC, see Hereditary non-polyposis colorectal cancer
HOC, see Hereditary ovarian cancer
Homofolate 82–3
Hybridoma/plasmacytoma growth factor, see Interleukin 6
Hydrosalpinx 252
Hyperamylasaemia 208
Hyperplasia, and tumours 201

Ibuprofen 165
Icon studies 323
ICRF Medical Oncology Unit 35
Ifosfamide 218
IgG 69
 rabbit anti-mouse 74
Immune system, CDDP potentiation **191**, 191–2
Immunocytochemistry 400
Immunohistochemistry 207–9
Immunology 397–8
 discussion 419–20
Immunomodulators 308
Immunomodulatory processes 116
Immunoscintigraphy 284
Immunotoxins
 anti-tumour 399
 discussion 422–6
 internalization 154
 intraperitoneal, clinical trials 156–9
 new targets 159
 peritoneal 155
 selection 153–5
 therapy 153–60
 regional versus systemic 155–6
Inclusion bodies 129

Insulin-like growth factor (IGF)-1 133
Inter-receptor 'cross-talk' 121–2
Interferon, MTD 303
Interferon-α
 alternating with CDDP 304–5
 and CDDP combination, MTD 307
 intraperitoneal 297–8, *298*, 299
 combined with CDDP 299–304
 toxicity 299
 recombinant 295, 401
 recombinant human (rh) 298
Interferon-γ 122
 recombinant 295
 therapeutic potential 90–1
Interleukin 1 (IL-1) 64, 130–1
 CSF-1 synthesization 122
 in tumour cells 96
 tumour production 89
 tumour promotion 87
Interleukin 2 (IL-2)
 in ascites 141
 evaluation 398
 'high-dose' 190
 intraperitoneal 399
 'low-dose' 190–2
Interleukin 2 (rIL-2), recombinant 189–90, 295
 CDDP effects in syngeneic mouse model 192–3
 mouse immunity **192**
 synergistic effect with CDDP *191*
Interleukin 4 (IL-4), evaluation 398
Interleukin 6 (IL-6) 130–1
 angiogenesis 102, 111
 in ascitic fluid 107, 109
 biological activities 102
 CSF-1 synthesization 122
 high levels in ovarian cancer *108*, 108–9
 in ovarian cancer 101–13
 production 107
 secretion modulation *107*, 110
 by cytokines 110

447

Index

Interleukin 6 (IL-6) receptor 102, 106
International Collaborative Ovarian Cancer Group 402
International Workshop on Carcinoma-associated Mucins, San Francisco 1990 40
Intraperitoneal 454A12-rRA 157-8, *158*
Intraperitoneal dose schedule 301
Intraperitoneal IL-2 399
Intraperitoneal immunotoxins, clinical trials 156–9
Intraperitoneal interferon-α 298, *298–9*
Intraperitoneal OBV3-PE clinical trial *156*, 156–7
toxicity 156–7, *157*
Intraperitoneal therapy 155–6, *156*, 291–5
alternating CDDP and interferon-α 304–5
CDDP 352
discontinuation reasons 302
discussion 422
in microscopic residual disease 389
patient characteristics *306*
for persistent ovarian cancer 297–311
platinum based, for EOC stage III 371
residual tumour 301
response evaluation *304*
rh interferon-α with CDDP 305–7
toxicity *306*
surgically defined responses 309
use justification 309
Isoamylase 208
Isobutyl-1-methylxanthine (IMBX) **150**

Jameson–Wolf algorithm **45**
JANUS serum bank, retrospective analysis 267, 273
Jurkat T cell leukaemia line 133

Kaplan–Meier survival curves 352
Kaposi's sarcoma, IL-6 autocrine growth factor 102
King's College Study 273

LAK cells 398–9
Laparotomy
critical approach 365
primary objectives 358–61
second-look
cytoreductive surgery effect 391
debulking
benefits 382
effects 382
evaluation 375–83
as diagnostic test 388–91
disease recurrence vs outcome **389**
findings 387, **388**
GOG randomized clinical trial 378–82
inclusion criteria 386
indications 385–6, *386*
MCS 369–72
negative
outcome 388
recurrence sites 388
outcome 386, **392**
predictive value 390
in routine management 385–92
survival 377
after debulking 377
as therapeutic manoeuvre 391–2
timing 389
unproved patient benefit 392
Lead time bias 279
Length bias 279
Li–Fraumeni familial cancer syndrome 58
Lipopolysaccharide (LPS) 102
Log kill effect 363
London Hospital Ovarian Cancer Screening Study 267–71
interim findings summary **270**

Lymphadenectomy 352, 354–5, 403
Lynch syndrome 9–11

M/V index 209–10
Macrophage colony-stimulating factor (M-CSF) 63–5, 130–1, *131*
autocrine/paracrine growth regulation 63–4, **64**
in cancer cell lines 94–6
changes 397
effects on cancer cells 121
effects on macrophage function 122
elevated plasma levels in cancer 117–18, **118**
expression by cancer cell lines 120
in female cancers 115–25
in HOC 89–90
immunoreactive 95
in ovarian cancer 139
in ovarian cancer diagnosis 95
properties 96
tumour expression *in vivo* 117
Macrophage colony-stimulating factor (M-CSF) receptor 101, 116
cancer cell line expression control 121
genetic rearrangement lack 120–1
interaction with other GFRs 121–2
Macrophage differentiation control 117
Macrophages, in tumour cell growth 93–4
Mannitol 305
Mantel–Hansel statistic 338
MAP2 397
Maximal cytoreductive surgery (MCS) 345–9
at second-look laparotomy 369–72, *391*
evaluation 366
evidence against survival benefit 346
meta-analysis 402
on MST 346–8
residual disease 352

448

Index

studies without controls 345–6
symptom relief 348
Maximum tolerable dose (MTD) 299, 303
MCS, *see* Maximal cytoreductive surgery
Median survival time (MST) 346–9
MEDLINE 333, 347
Melphalan 315, 331
Meta-analysis
 analysis 335–41
 methodology 333–5
 relevance criteria 334
 treatment 402
 validity criteria **334**
Meta-analysis of treatment, discussion 433–5
Methotrexate 315
Microcirculation of tumours 250
Miettinen standardized estimate 338
Minimal residual disease (MRD) 298–9
Mitoxantrone, intraperitoneal *291*, 291–2, 295
Monoclonal antibodies,
 454A12-rRA 154–5, 157–8, *158*
 against p185 68
 anti-CSF-1 receptor 117
 antigen coding 73–8
 cytotoxic potency increase 70
 Ki-67 208
 MOV 18 207
 OC125 207, 265
 OV-TL23 207
 OV-TL3 207
 OVB3 154, *156*, 156–7
 techniques 397
 tissue penetration evaluation 155
Monoclonal antibody-toxin conjugates 399
Monocyte chemotactic protein (MCP-1) 90
Mononuclear cells, growth inhibition 140
Morphometry 209–10
MOT, *see* Murine ovarian teratocarcinoma

Mouse models for antigen testing 47–8
MRC Cancer Trials Office 323
MRC Gynaecological Cancer Working Party 316
MRC Human Genetics Unit, Edinburgh 35
MRI 284
MST, *see* Median survival time
MUC1
 aberrant glycosylation 44–5
 animal models for antigen testing 47–8
 mouse homologue 42–3, **43**
 oligosaccharide side chains 44
MUC1 core protein 41–2, **42**
MUC1 gene 41–3
Mucin
 analysis 40
 general features 40–1
 possible functions 43
 as tumour-associated antigen 39
Mucin genes, tandem repeat units 41
Mucinous tumour, stroma proportion **27**
Multiple site cancer syndrome (MSCS) 232
Murine ovarian teratocarcinoma (MOT), CDDP and rIL-2 therapy 193–4, **194**
Mutation screening 4

Natural cytotoxic cell lysis (NCs) 176
Natural cytotoxic cell resistance (NCr) 176–7
Natural cytotoxic (NC) cells 175–8, **178**
 anti-tumour activity **176**
NC, *see* Natural cytotoxic cells
NCA (CEX) 208
Neo-angiogenesis **245**, 245–6
neu
 overexpression 138
neu receptor **137**
Neurofibromatosis type I 4
Neuropathy,
 cisplatin-induced 307, 310–11

No clear inheritance pattern (NCIP) 232
North Thames Ovarian Cancer Trial 347
Nuclear factor-IL6 (NF-IL6) 102
Nuclear proteins 53
Nuclear organizer regions (NOR) 210

Odds ratio (OR) 337–8, 340
Oestrogen receptor (ER) 29–30, 208
Office of Population Censuses and Surveys (OPCS) 3, 270, 273, **278**
Omentectomy 358
Oncogenes
 activation 30–1
 expression 208
 in female cancers 115–16
 functions 24
 pair correlations 116
Oophorectomy, prophylactic 6
Oral contraception, and ovarian cancer risk 36, 280
Ovarian adenocarcinoma 127
Ovarian cancer
 advanced
  response to initial therapy **387**
  surgical management 402–3
 age-incidence curve **37**
 borderline epithelial, clinical and cytogenetic findings 20
 borderline mucinous cystic 202–3, **203**
 chromosomal abnormalities **19**, **21**
 colour Doppler, test analysis 239
 CSF-1 expression *in vivo* 117
 cytokine production 88–90
 disease-specific survival complement 390
 early 199–204
  definitions 199
 early detection 400–1
 discussion 428–33

449

epidemiological data   36–7
epithelial cells and IL-6
    103–4, **104**, 105–7
familial, *see* Familial ovarian
    cancer
family history   199
gene mapping   25
genetic alterations   17–22
genetic aspects   395–6
growth factor secretion   97
hereditary predisposition
    395
hereditary site-specific   11,
    400
    family pedigrees   **12**
high IL-6 levels   108
histology   **202**
in HNPCC   **10**
imaging by SM-3 antibody
    45–6, *46*
inherited predisposition
    evidence   3–4
    implications   4–5
    recognition   5–6
intraperitoneal therapy
    291–5
MCS, benefit-risk
    unjustified   382
metastatic, combination
    chemotherapy   189–94
molecular genetics   23–33
molecular pathogenesis
    35–8
mortality increase   277–8
mouse survival   **89–90**
mucinous   **203–4**
new therapies   401
p53 mutations   56–7
pathological assessment
    207–10
ploidy vs survival   *215*
preclinical, elevated CA-125
    levels   267–8
premature death cause
    278
prevalance effect on
    predictive value   *390*
prognostic factors   54–8,
    401–2
progress model   271–2, **272**
randomized screening
    trial   284
recurrent versus
    persistent   293
registrations   279

residual disease   376–7
    survival   *376*
screening   6, 15
second-look laparotomy
    debulking evaluation
    375–83
    survival   *377*
    survival, after
    debulking   *377*
    stage III, secondary
    debulking results
    379–82
susceptibility gene   280
Ovarian cancer, *see* Epithelial
    ovarian cancer; Familial
    ovarian cancer;
    Hereditary ovarian
    cancer
Ovarian cancer cell lines
    62–3, 74, 76, **132**, *134*
    CaMOv 18 expression   83
    CDDP action   177–8, **178**,
    179–80, **180**
    CSF-1 expression   120
    DECA and CDDP effects
    170
    desqualinium chloride
    (DECA)   168
    FBP expression   78
    HEY   **130**, 136–7, **137**,
    141–2
    IL-6   102, *105, 106*
    Northern hybridization
    analysis   **82**
    OCC1   136–7, **137**, 138
    TNFα action   177–84
Ovarian cancer growth factor
    (OCGF)   131–3
Ovarian cancer-associated
    marker, *see* CA-125
Ovarian epithelium, growth
    regulation   *64*
Ovarian function, normal,
    colour flow imaging
    239, *240*
Ovarian masses
    clinical differentiation
    methods   251–2
    colour Doppler diagnosis
    238–9
    in ultrasound screening
    programmes   244
Ovarian pathophysiology   36
Ovarian tumour markers   74
Ovary, surface crevice   **200**

Ovulation
    incessant   36, 38
    and ovarian cancer
    128–30, 397

p-glycoprotein transmembrane
    drug efflux pump   399
Panning, for cDNA   74–6
Paracrine loop between
    macrophages and cancer
    cells   93
Partial clinical response
    (PCR)   303
Partial pathological response
    (PPR)   304
Peanut-lectin binding urinary
    mucin (PUM)   41
Pedigree analysis   232–3,
    234, 234–5
Pelvic inflammatory disease
    255
PEM, *see* Polymorphic epithelial
    mucin
Peritoneal carcinomatosis
    199, 201
Peritoneal fibrosis   427
Peritoneal lavage risk   360
Peutz–Jegher's syndrome   9
Pharmacokinetic advantage
    with intraperitoneal drug
    administration   291
Philadelphia chromosome
    17, 408
Phospholipase A2 (PLA2)
    activity   181
Phospholipase C (PLC)   70,
    132–3
Placental implantation   116
    regulation   117
Platelet-derived growth
    factor   53, 61, 130, 133
    expression   94
    tumour stroma   88
Platinol   190
Platinum   316
    combinations   322
    comparison dose   *319*
    comparison with platinum
    combination   318–19
    see also Cisplatin (CDDP)
Platinum chemotherapy, and
    MST   346–8, **348**
Platinum combinations   402

Platinum-based combination chemotherapy   189–94
Platinum-based intraperitoneal therapy   293–4
Ploidy analysis   402
Ploidy measurement   209, 213, 400–1
Ploidy studies   *215*, *216*, *217*
Ploidy vs residual disease   *217*
Polymorphic epithelial mucin (PEM)   397–8
  antibodies recognized by epitopes   *40*
  MAb reaction   *74*
  molecular structure and clinical applications   39–50
  T cell epitopes   46–7
Post-test probability   391
Predisposing genes, search for   4–5
Predisposition, inherited   31
Progestins   397
Proleukin   190
Prophylactic oophorectomy   395
  evaluation   396
Protein kinase A (PKA), pathway   149–51
Protein kinase C (PKC)   133
  inhibition by DECA   171
Protein kinase C (PKC) pathway   421
Protein phosphatase   83
Protein synthesis inhibitors, effect   **181**, **183**
Proto-oncogenes   53
  activation   94
  *erb*B2   398–9
  evaluation   63
  over-expression   93
Pseudomonas exotoxin (PE)   153, 159
Pulsatility index (PI)   238, 252, 400

Quality adjusted life years (QALY)   284
  benefit detection   285
  CC   332
Quinacrine   181–2, **182**

Radiation and ovarian cancer risk   280

Radiation therapy, survival rates   371
Radioimmunoassay, CA-125   268
Radionucleotides for imaging and therapy   47
RAF   397
Randomized trials   279
Receptor transphosphorylation cascades   121
Recessive oncogenes   53
Recombinant interleukin 2, *see* Interleukin 2 (IL-2), recombinant
Regional immunotoxin therapy   155
Relative risk (RR)   337
Resistance index (RI)   238, 252, 261
Restriction fragment length polymorphism   **80**
Restriction fragment length polymorphism, analysis   18
Retinoblastoma   4, 17–18
Retinoblastoma, carcinogenesis   23–4
Retroviral oncogenes, *see* Proto-oncogenes
Rhodamine 123   167
Ricin, composition   153
Ricin A chain conjugates   70
RNA, alternative processing   82
Rosenthal 'fail-safe *n*'   338
Royal Hospital for Women, Sydney, Gynaecological Oncology Department   351–2

S phase fraction (SPF)   216–17, 402
SCISEARCH   333
Screening
  CA-125   *271*, 271–3, 401
  CA-125 studies   282–3
  ultrasound   268
  cost-benefit considerations   280
  ovarian cancer families   285
  randomized controlled trials advantages and disadvantages   277–86

  purpose   285
  size determination   279
  target population   279–81
  timing   285
  targeted ultrasound   400
Screening, *see* Doppler; Transabdominal ultrasonography; Transvaginal ultrasonography
Secreted growth factor   53
Serum marker estimation   282–5
Shapiro–Francia W' test   347
Site-specific ovarian cancer (SSOC)   232
Site-specific risks   4
Stem cell growth factor (SCF)   131
Superovulation techniques and ovarian cancer risk   280, 397
Superoxide dismutase   183
Surgery
  discussion   436–41
  maximal cytoreductive, *see* Maximal cytoreductive surgery
  palliative   365–6
  super-radical   365
Surgical investigation results   *233*
Survival
  analyses   317
  associated factors   *361*
Synergy between interferons and cytotoxic agents   298

T activation antigen, *see* Interleukin 2 (IL-2)
T cell, epitopes in PEM   46–7
TAC, *see* Interleukin 2 (IL-2)
TAH-BSO, *see* Total abdominal hysterectomy and bilateral salpingo-oophorectomy
Tamoxifen   352
Tattersall study   335–7, **337**
  response data   **337–8**, **339**
  validity score   **341**
Taxol, intraperitoneal   295
Thiotepa   331
Tissue characterization by

451

Index

pulsed and colour Doppler 250–1
Tissue penetration depth 155
Total abdominal hysterectomy and bilateral salpingo-oophorectomy (TAH-BSO) 11
Toxicity
  in clinical trials 154
  dosage modifications 306
  non-haematological, versus dose level 303
Transabdominal ultrasonography screening 226–9
  morphological characterization 228–9
  outcomes 226–7
  retrospective analysis 227, 227–8
  screening studies 282
Transferrin receptor 154, 158, 425
Transforming growth factor-α (TGF) 61–4, 128, 131, 133–7
  levels in prognosis 397
Transforming growth factor-β (TGF), 61–4, 131
  action 397
  in ascitic fluid 140
Transforming growth factor-β (TGF) autocrine growth inhibition 63
Transmembrane signal transduction pathways 398
Transvaginal screening 229–32
  cancer families study 230–1, 232
  morphological scores 230
  population-based study 229–30
Transvaginal sonagrams 228
Transvaginal ultrasonography 237
  colour Doppler
    abnormal blood vessel identification 263
    adnexal tumour malignancy detection 261
    analysis 241

detection of early ovarian cancer 242
normal ovarian blood flow 257
ovarian cyst 258
ovarian mass differentiation 249–63
ovarian tumour blood flow 257, 259–60
ovarian tumour classification 262
screening specificity 400
tubo-ovarian abscess 258
general population screening by colour Doppler 241–3
TTK 397
  cloning and characterization 141–3
  as growth factor receptor 141
  structure 142
Tubo-ovarian abscess 258
Tumour aggression and outcome 391–2
Tumour cells, proliferative pool after surgery 219
Tumour growth inhibition, possible mechanisms 69
Tumour mass reduction 345–9
Tumour necrosis factor (TNF)
  experimental intraperitoneal tumours 87–8
  intraperitoneal 295
  mRNA 88, 88–9
  ovarian cancer cell lines 168
  procoagulant factors 88
  resistance mechanism 171
Tumour necrosis factor-α (TNFα)
  and cachexia 96
  combination therapy use 179–80, 180
  effects on arachidonic acid release 183
  lysis in ovarian cancer 175–84
  lytic pathway 183–4, 184
  NC cell use 177
Tumour necrosis factor-α (TNFα), protein

synthesis inhibitor 399
Tumour neovascularization 249–50
Tumour suppressor genes 17, 23–4, 53, 129
  allele loss 396
  functions 24
Tumours, histological characteristics 18–19, 19
Tyrosine kinases 131–2
  cloning and characterization 141–3
Tyrosine phosphorylation 131–2, 134–5, 151

Ultrasonography
  benign cystic mass 256
  cystic mass 254
  early ovarian cancer screening advances 244
  in early ovarian mass differentiation 251–2
  for early screening 225–35
  enlarged ovaries 253–4
  findings in volunteers 270
  first stage advances 234, 234–5
  ovarian cancer 255–6
  pelvic inflammatory disease 255
  for positive CA-125 283
  positive predictive values 243–4
  repeat timing 234
  technique 225–6
United Kingdom Committee for the Coordination of Cancer Research (UKCCCR) 279, 283
USA
  National Cancer Institute, screening studies 281
  PLCO trial 285

Variable number tandem repeat (VNTR) unit 42
Velocity flow scanners 246
Vimentin 208

Vogl study 336–7, **337**, **338–9**
Volume-corrected mitotic index (M/V index) 209–10

West Midlands Regional Cancer Registry 359, 364–5
West Midlands' Studies 364–5
Wilbur study 342
Williams study 335
response data **339**
validity score **341**
Wilms' tumour 4, 17

Years of cancer detected (YCD) 283